Quantum Physics and Linguistics

QUANTUM PHYSICS AND LINGUISTICS

A Compositional, Diagrammatic Discourse

Edited by
CHRIS HEUNEN
MEHRNOOSH SADRZADEH
and
EDWARD GREFENSTETTE
University of Oxford, Department of Computer Science

UNIVERSITY PRESS

Great Clarendon Street, Oxford, OX2 6DP,
United Kingdom

Oxford University Press is a department of the University of Oxford.
It furthers the University's objective of excellence in research, scholarship,
and education by publishing worldwide. Oxford is a registered trade mark of
Oxford University Press in the UK and in certain other countries

© Oxford University Press 2013

Chapter 3 © Ross Duncan

The moral rights of the authors have been asserted

First Edition published in 2013

Impression: 1

All rights reserved. No part of this publication may be reproduced, stored in
a retrieval system, or transmitted, in any form or by any means, without the
prior permission in writing of Oxford University Press, or as expressly permitted
by law, by licence or under terms agreed with the appropriate reprographics
rights organization. Enquiries concerning reproduction outside the scope of the
above should be sent to the Rights Department, Oxford University Press, at the
address above

You must not circulate this work in any other form
and you must impose this same condition on any acquirer

British Library Cataloguing in Publication Data

Data available

ISBN 978-0-19-964629-6

Printed and bound by
CPI Group (UK) Ltd, Croydon, CR0 4YY

Links to third party websites are provided by Oxford in good faith and
for information only. Oxford disclaims any responsibility for the materials
contained in any third party website referenced in this work.

PREFACE

Quantum mechanics and linguistics appear to be quite unrelated at first sight. Yet significant parts of both concern compositional reasoning about the way information flows among subsystems and the manner in which this flow gives rise to the properties of a system as a whole. This book is about the mathematics underlying this notion of compositionality, how it gives rise to intuitive diagrammatic calculi, and how these compositional methods are applied to reason about phenomena of both disciplines.

Scientific advances often consist in an expansion of the conceptual language used to describe the object of study, to increase expressivity and descriptive power, but also to add to the simplicity of their reasoning techniques. Over the past few decades, such a move has started taking place in the language and formalisms of theoretical physics and quantum information theory, by introducing new concepts and tools from the mathematics of category theory. Categorical methods allow for a more conceptual and insightful expression of elementary events such as measurements, teleportation, and entanglement operations, which are easily obscured in formalisms that are more directly based on vector spaces. They also provide a succinct way of putting these elementary events together and forming more complex protocols. The categories under consideration come equipped with a diagrammatic calculus that simplifies the corresponding reasoning to a great extent. This book contains many examples; a good one is the correctness proof of the so-called teleportation protocol. Originally it was pages long, scattered with words and matrix computations. Categorically, it simplifies to two simple diagrammatic equalities. Recent work in natural language semantics has begun to use similar categorical models and methods to relate grammatical analysis and semantic representations in a unified framework for analysing language meaning, formalizing meanings of sentences beyond true and false, and hence for the first time allowing us to reason about the notion of sentence synonymy. The diagrammatic tools allow a depiction of the flow of information that occurs among the words within the sentence and demonstrate how the grammatical structure and the meanings of words compose to form the meaning of a sentence.

Thus, in both physics and linguistics applications, compositionality and diagrammatic reasoning play key roles. They form the overarching theme of this volume.

A general expository and exemplary description of the categorical methods and diagrammatic calculi and how they apply to quantum theory and linguistics is presented in the rest of this preface, which ends with an overview of the layout of the overall book and a brief description of each of the chapters.

Diagrammatic algebra

The mathematical methods mentioned find their most natural expression in category theory (Mac Lane, 1971). This abstract branch of algebra makes relationships between objects the primary focus of study instead of the structure of individual objects themselves. Already this philosophy resonates well with intuition about a system consisting of constituent objects and information flowing between them. More precisely, what is needed are monoidal categories, i.e. categories that come equipped with a way to group objects together in compound systems. Technically, we have:

- *objects* $X, Y, \ldots$ related by *morphisms* $f\colon X \to Y$ between them, in particular *identity* morphisms $X \to X$ that 'do nothing';
- a way to *sequentially compose* morphisms $f\colon X \to Y$ and $g\colon Y \to Z$ into $g \circ f\colon X \to Z$;
- a way to group objects X and Y into a *compound* object $X \otimes Y$;
- a special object I to signify the 'empty' system;
- a way to compose morphisms *in parallel*, making $f_1\colon X_1 \to Y_1$ and $f_2\colon X_2 \to Y_2$ into $f_1 \otimes f_2\colon X_1 \otimes X_2 \to Y_1 \otimes Y_2$.

Of course, these data are required to cohere in the appropriate way.

A very neat feature of such categories is that manipulations of them can not only be performed algebraically, but also graphically, without losing any mathematical rigour (Selinger, 2011). Various chapters explain this graphical calculus in more detail, so that a cursory introduction can suffice here. A morphism $f\colon X \to Y$ is depicted as $\begin{smallmatrix}X\\ \boxed{f}\\ Y\end{smallmatrix}$, the identity on X simply becomes $\begin{smallmatrix}X\\ |\\ X\end{smallmatrix}$, and sequential and parallel composition look as follows.

To signify a lack of input, morphisms $f: I \to X$ are often depicted as $\overset{\triangle f}{\underset{X}{|}}$. Let us consider *compact* categories for a moment, in which every object X additionally has a partner object X^* and two special morphisms $\eta: I \to X^* \otimes X$ and $\varepsilon: X \otimes X^* \to I$ witnessing their duality, in the sense that the following equation and its dual hold (Kelly and Laplaza, 1980; Day, 1977).

$$\begin{array}{c} \end{array}$$

If we suggestively draw the special morphism ε as $\overset{X \ X^*}{\smile}$ and η as $\underset{X^* \ X}{\frown}$, we get a first glimpse of a sense of 'flow', as the equations become the following.

$$\begin{array}{c} \end{array} \tag{0.1}$$

It is hard to escape the reading that X and X^* are the 'input' and 'output' of some canonical procedure that links them together. Indeed the mathematics described so far is sufficiently abstract to allow many interpretations. We now go a bit deeper into the two branches of interest here: quantum information theory and linguistics.

Quantum theory

One of the counterintuitive aspects of quantum mechanics is that it is *non-local*. That is, particles can influence each other's behaviour at distances from which this should not be possible according to common sense (Einstein, Podolsky and Rosen, 1935). Such *entanglement* can be seen as a channel for passing information from one particle to another. Indeed, quantum computation and quantum information theory are based on this premise. A basic unit of information is in fact defined there, dubbed a *qubit*, and the change in its information content under various operations is studied, even going so far as to quantify how much information a channel can transmit (Nielsen and Chuang, 2000).

For example, one of the basic building blocks is the marvellous quantum *teleportation* protocol, in which the unknown state of a qubit is transferred across a channel (Bennett et al., 1993). This can be modelled categorically by letting objects signify physical systems—or rather the *vector space* that their states form—and letting morphisms be processes that systems can undergo. Compact structure then sets up channels: the morphism η denotes the preparation of an entangled state, physically distributed between the two systems X and X^*, and the morphism ε stands for measuring the pair and thereby destroying their entanglement. Looking back, we now see that eqn (0.1) precisely corresponds to the teleportation protocol being practicable: setting up an entangled pair and measuring one of them jointly with an unknown qubit results in the other assuming the state of the unknown qubit (Abramsky and Coecke, 2009). In effect, the state of an unknown qubit can be teleported across the channel.

A striking property of qubits is that they cannot be copied or deleted (Wootters and Zurek, 1982). It is physically impossible to build a machine that inputs unknown qubits and outputs two perfect copies in the same state as the original. In this sense, we can actually think of a qubit as a carrier of information flowing through diagrams such as eqn (0.1)! But then something about measurement is off: if qubits cannot be deleted, where does the information they contain go upon measurement? Indeed, the above explanation of quantum teleportation contains an omission: the cost of teleporting one qubit is two classical bits. This is still quite an achievement, since classical bits can only be 0 or 1, and hence a single qubit carries infinitely more information than two classical bits.

Luckily, this classical information, as opposed to the quantum channels, can also be modelled categorically. We simply pose a copying morphism $A \to A \otimes A$ and a deleting morphism $A \to I$, which are depicted in the graphical calculus as $\curlyvee$ and $?$. By a general principle of quantum physics, we also have upside-down versions $\curlywedge$ acting as conjunction of classical information and $\downarrow$, the latter picking out the two states 0 and 1. As it turns out, the crucial law that these morphisms must obey to truly model classical information is the following Frobenius equation (Coecke and Pavlovic, 2007; Coecke, Pavlovic and Vicary, 2011; Abramsky and Heunen, 2012).

Objects that embody classical, as opposed to quantum, information in such a way are called *Frobenius algebras*. They might look innocent, but their theory has implications reaching as far as quantum field theory (Kock, 2003).

Similar structures form the basis of quantum algebra (Kassel, 1995; Street, 2007). A *braiding* $\times$ is an additional structural ingredient here typically. Often the techniques used to study quantum groups are disguised algebraically, but can be recognized as dealing with patterns of flow in the diagrammatic perspective. For example, a multiplication $\curlywedge$ and a comultiplication $\curlyvee$ can interact in other ways than shown above. The following *bialgebra* law is, in a sense, opposite to the Frobenius equation.

The basic object in quantum algebra is a *Hopf algebra*. This is a bialgebra that additionally comes with a 'twist' morphism $A \to A$, drawn as ⌐, satisfying the following equation.

Hopf algebras are the quantum analogue of the venerable notion of group, used to mathematically study symmetries. With diagrammatic hindsight, connections between quantum algebra and knot theory are perhaps less surprising.

Linguistics

There are two major approaches to formal analysis of natural language: one of a *logical* nature (Dowty, Wall and Peters, 1981), and one of a *distributional* kind relying on *vector spaces* as models of meaning (Schütze, 1998). These two schemes have complementary features. The logical model is compositional—the meaning of a sentence is a function of the meanings of its words—but says nothing about the lexical meaning, i.e. the meanings of individual words. The vector space model constructs meanings of individual words by counting co-occurance with words used often in a given corpus of text, but does not address meanings of strings of words.

We already saw that vector spaces form a compact category, a structure shared with the recently introduced logic of *pregroups* (Lambek, 2010). This common categorical structure has been the basis for the development of the first compositional vector space model of meaning (Coecke, Sadrzadeh and Clark, 2010). In a bit more detail, a pregroup is a partially ordered monoid in which each element p has both a left adjoint p^l and a right adjoint p^r:

$$pp^r \leq 1 \leq p^r p, \qquad p^l p \leq 1 \leq pp^l.$$

Reading the partial ordering as morphisms, i.e. postulating a morphism $p \to q$ if and only if $p \leq q$, and the monoid multiplication as tensor product, the above equation comes down to eqn (0.1). Hence a pregroup is a compact closed category, but for the fact that the non-symmetric tensor product forces one to distinguish left and right duals. This gives four (co)units for the adjunctions instead of two:

$$\epsilon^r : pp^r \leq 1, \qquad \epsilon^l : p^l p \leq p,$$
$$\eta^r : 1 \leq p^r p, \qquad \eta^l : 1 \leq pp^l.$$

In linguistic applications, each element of the pregroup represents a grammatical type, and multiplication means concatenation (Lambek, 2008). For instance, the noun phrases 'John' and 'Mary' are of a basic type n. If the transitive verb 'likes' also had a basic type v, the type of the sentence would become nvn. This would lead to a static view of sentencehood, whereby a sentence was just a string of non-interacting words, and in which extra rules were needed to stipulate that strings of types nvn reduce to the type of a grammatical sentence s. Instead, we make the information flow between the types of words and automatically result in a sentence type through the use of the adjoint types. Typing 'likes' as $n^r s n^l$ makes the types of the words *interact* with each other as $n(n^r s n^l)n \leq 1s1 = s$. This computation means that 'likes' is a relator, inputting 'John' as its subject and 'Mary' as its object to produce the sentence 'John likes Mary' as output. Diagrammatically, the interaction becomes:

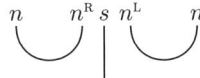

To analyse a negative transitive sentence such as 'John does not like Mary', type 'does' as $(n^r s j^l \sigma)$ and 'not' as $(\sigma^r j j^l \sigma)$ to get the following interaction.

$$n\,(n^r s j^l \sigma)\,(\sigma^r j j^l \sigma)\,(\sigma^r j n^l)\,n \to s$$

depicted as

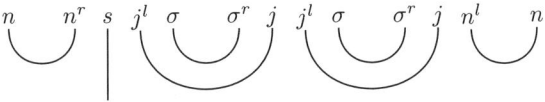

This method of analysis has been applied to a wide range of natural languages, from English, German, and French to Polish, Persian, and Arabic, and many more (Casadio and Lambek, 2002). The interaction diagrams provide a simple way to compare the structure of sentences in different languages. For example, here are the diagrams of a transitive sentence with indirect object in English and French (left), and in Persia and Hindi (right) (Sadrzadeh, 2006):

Without prior knowledge of the languages or their corresponding pregroup types, some information about their general structure can be gained just by looking at the diagrams. For example, from the location of the straight line that marks the position of the type *s* in the verb, one can infer that English and French are verb-in-the-middle (SVO) and Persian and Hindi are verb-at-the-end (SOV).

Interaction diagrams can also be used to study more sophisticated sentence types. Below, for example, is the diagram for two verses of a rubbayiat, a Persian poem, by Omar Khayyam. In Fitzgerald's English translation, the first verse reads 'How sweet is mortal Sovranty!—think some' (Sadrzadeh, 2006).

Like all logical approaches, the pregroup analysis provides a structured view of meaning but does not give us much information about the meaning of each word. This is overcome in the vector space model, best described by a frequently cited quotation of Firth (1968): 'You shall know a word by the company it keeps.' In this approach, a high-dimensional vector space is built from basis words, which are those most frequently used in a language (according to standard language corpora such as the British National Corpus). The meaning of a word is a vector, built by counting how often the word occurs within a prespecified distance of each basis word. This raw frequency is normalized in a certain way to minimize noise, resulting in a meaning vector $\vec{w}$ for each word w, giving the weight of the word with respect to all basic words.

As an example, consider 'dog' and 'cat'. Both are pets, both may be furry, both run and eat, and both may be bought. These facts are reflected in the texts in which the words 'dog' and 'cat' appear. They both occur with high frequencies close to 'pet', 'furry', 'run' and to a lesser extent to the words 'sea' and 'ship'. Hence their vectors have high weights for basis words 'pet', 'furry', and 'run' and low weights for 'sea' and 'ship'. Once the vectors are built, a degree of *synonymity* of words can be defined from the angle between their vectors. This degree has been successfully used to automatically derive synonymous words from corpora (Clark and Curran, 2007). For our example, 'cat' and 'dog' will be closer to each other than to 'sailor'.

However, as well as the vector space approach works for meanings of single works, it has problems with strings of words. The major problem is that the vector of 'cats sleep', for example, cannot be computed compositionally from the vectors of 'cats' and 'sleep'. Moreover, the vectors of logical words such as 'not' and 'does' are void of meaning, hence cannot contribute to computing the meaning of, for instance, 'cats do not sleep'.

The essence of compositionality can be carried over from the logical model to the vector space model by recognizing that the vectors of certain words, such as verbs, conjunctives, and adjectives, should not be at the same level as those of nouns. The argument is similar to the one above explaining why a verb does not have the simple type v. Instead, the compound type $n^r s n^l$ must be interpreted as a relation or matrix acting on the vectors of other words, such as nouns. In general, a prescription to decide which types are compound and act on others is desirable. This information is precisely what is encoded in the interaction diagrams.

The compositional vector space model of meaning (Coecke, Sadrzadeh and Clark, 2010) stipulates that the meaning vector of a sentence is the application of the morphism represented by its pregroup diagram to the tensor product of the meaning vectors of its constituent words. Hence, diagrammatically, the meaning of 'John likes Mary' is the following vector.

The types of the three wires above are N, S, and N, respectively; these are the vector spaces in which the meaning vectors of 'John' and 'Mary' live. The general properties of such vectors can be recovered, through morphisms $I \to N$, from the tensor unit. These morphisms are sometimes depicted as triangles instead of boxes to reflect the lack of input. To compute the meaning of a negative transitive sentence such as 'John does not like Mary', logical words are interpreted through the special η morphisms given by compact structure, 'does' is interpreted as an identity and therefore does not affect the meaning, and 'not' is interpreted as the matrix that swaps the meaning. Diagrammatically:

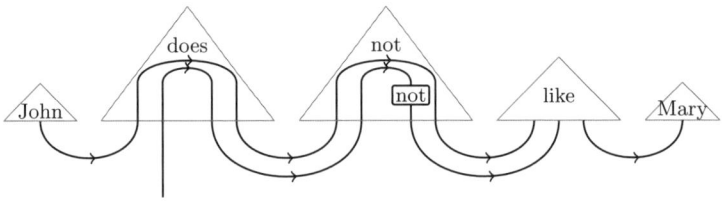

where the morphisms drawn as grey triangles signify the words of the sentence. According to eqn (0.1), this simplifies to:

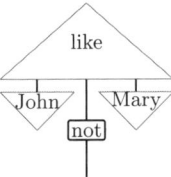

Notice how this is the negation of the meaning of the positive version of the sentence. A nice analogy with quantum information protocols such as teleportation arises. To obtain the meaning of the negative sentence, the verb first has to interact with its subject and object. But in this case the subject does not immediately precede the verb, as it did in the positive case. Instead, they are separated by 'does' and 'not'. The compact structure morphisms reconnect the two via a *non-local correlation* in such a way that the verb can still act on its subject from a distance, and the application of 'not' is delayed.

Once vectors for the meanings of sentences are built, one can use the cosine similarity measure to automatically derive synonymous sentences, a task which has applications in automating language-related tasks such as translation and paraphrasing.

Overview

The above notions of compositionality and diagrammatic reasoning set a solid basis on which formal models of the fields of quantum physics and computational linguistics rest. It was this solid base which, with funding by the British Council and the Dutch Platform Bèta Techniek, inspired the first two editors to organize a workshop in Oxford in October 2010 to bring these two seemingly separate research communities together. The idea for this volume originated there. Although the intuitions behind compositionality are far from novel in either area, the realization that categorical and diagrammatic methods can make this precise is rather recent, and there does seem to be a need for a somewhat self-contained text highlighting the connection. The majority of the chapters started as presentations at this workshop. Most chapters are a balanced mixture of theory and practice, presenting some abstract algebraic or categorical notions and motivating or hinting at applications in physics or linguistics. At the same time, some chapters, such as Chapters 3 and 12, are more application-oriented, while others are more on the theoretical side, such as Chapters 5, 6, 7, and 10. The chapters are roughly organized in the order of the title of the book: earlier chapters are more about quantum theory, whereas later chapters centre more on linguistics.

The book starts with an introductory chapter by Bob Coecke. This motivates the use of monoidal categories in physics and computer science in general, and in reasoning about information in particular, by starting with the basic mathematical structures of order, with or without composition. He highlights the role of compositionality by listing several natural incarnations, showing that such elementary structures can be very powerful when taken as primitive because of diagrammatic methods.

We proceed in Chapter 2, by Bertfried Fauser, with a survey of specific types of algebraic structures that have found applications in reasoning about quantum theory. He focuses on Frobenius algebras and Hopf algebras, the relationship between the two, and the diagrammatic reasoning developed for each.

Ross Duncan focuses more on the application side in Chapter 3, and shows how these algebras model observables and phase groups and can successfully reason about measurement-based quantum computing in terms of quantum circuits. He then presents a measurement calculus to replace the quantum circuit computations, develops its semantic underpinnings, and discusses some of its properties such as non-determinism and flow.

In Chapter 4, Shahn Majid presents an introduction to, and overview of, braided algebra, which is the natural setting to study Hopf algebras. This is the starting point for the definition of a quantum group, which is discussed in some detail. He considers several applications of braided groups, including braided Lie algebras.

Chapter 5, by Joost Vercruysse, continues this theme by studying reconstruction theorems for Hopf algebras. Using a categorical setting, it gives an overview of the many generalizations of characterizing these algebras in terms of their category of representations.

Then, in Chapter 6, Michael Müger gives a very clear survey of modular categories. These kinds of braided categories embody the theme of this book—compositionality—very well. Moreover, they also have several important applications in topological quantum computation and topological quantum field theory.

Roughly ending the quantum half of the book, in Chapter 7 Dion Coumans and Bart Jacobs observe that the categorical notion of a monoid of scalars corresponds in a triangle to monads and to Lawvere theories. They explain how additional structure on the scalars, such as that of a semiring, relates to the other two corners of the triangle.

The categorical trend is followed in Peter Hines's Chapter 8, where he shows how type-logical approaches to linguistics, formally introduced in the seminal work of Lambek in the 1950s, can be formalized in a categorical setting and how these more general categorical formalisms can also accommodate semantics and hence provide meaning for sentences, based on the meanings and the types of words within a sentence. He goes on to discuss how logical connectives can be formalized in this setting and how sentence similarity can be stated in terms of functors and Frobenius algebras.

Next, in Chapter 9, Anne Preller builds on the categorical setting by focusing on concept spaces and thesaurus-based models of meaning, and develops a compact categorical setting for a truth-based conceptual semantics; she shows how Lambek's newer pregroup grammars and their corresponding diagrammatic calculus can be combined with this truth-theoretic semantics and how this combination gives rise to a compositional model of conceptual semantics for sentences, one that offers a natural interpretation for some of the logical connectives too.

Michael Moortgat and Richard Moot's Chapter 10 is on the same type-logical trajectory as Hines and Preller, yet they take a complementary algebraic turn and focus on another kind of diagrammatic calculus, namely proof nets. They present the state of the art on the more expressive type logics of Lambek–Grishin algebras, develop a new proof-net semantics for this algebra, and show how meanings of sentences can be depicted using their proof nets and how inference can be formalized in this purely logical setting.

Daoud Clarke, in Chapter 11, also focuses on algebras, but this time algebras for meanings rather than grammatical structure; he shows how distributional models of meaning can be formalized in the setting of vector spaces, how these can be generalized from word meaning to sentence meaning by taking into account a broader notion of context using special infinitary algebras over vector spaces, and how in this generalized context algebras even give rise to a notion of semantic inference between sentences.

In Chapter 12, Stephen Pulman provides a more historical and philosophical take on the distributional models of meaning and traces back the origins of these models to the work of Harris and Schütze; he shows how these models were applied to language tasks such as learning grammar and word-sense disambiguation, and discusses various notions of compositionality that are appropriate for this specific setting and in particular in the light of their historical roots. He then moves on to present a brief review of recent approaches that attempt to bring in compositionality to distributional models and discusses their shortcomings. This chapter ends by suggesting features and experiments that any compositional distributional model of meaning must have and try to evaluate themselves with.

The book ends with Chapter 13, in which Stephen Clark provides an overview of a specific compositional distributional model of meaning, namely the categorical compositional distributional model of meaning developed by himself, Coecke, and the second editor; he reviews the abstract categorical setting in an easy-to-follow less technical language, summarizes its concrete initiations and experimental evaluations, and suggests a novel concrete model based on plausibility spaces that suggest degrees of truth for encoding meanings of sentences.

Oxford, May 2012

CONTENTS

1 An Alternative Gospel of Structure: Order, Composition, Processes   1
   BOB COECKE
  1.1 Order   4
  1.2 Orders and composition   9
  1.3 Processes witnessing existence   15
  1.4 Conclusion   21

2 Some Graphical Aspects of Frobenius Algebras   22
   BERTFRIED FAUSER
  2.1 Introduction   22
  2.2 Graphical calculus   24
  2.3 Frobenius and Hopf algebras   31
  2.4 A few pointers to further literature   46

3 A Graphical Approach to Measurement-Based Quantum Computing   50
   ROSS DUNCAN
  3.1 Introduction   50
  3.2 The rudiments of quantum computing   54
  3.3 Observables and strong complementarity   57
  3.4 The zx-calculus   65
  3.5 The measurement calculus   77
  3.6 Determinism and flow   81
  3.7 Conclusions   88

4 Quantum Groups and Braided Algebra   90
   SHAHN MAJID
  4.1 Introduction   90
  4.2 Braided categories   92

	4.3 Braided groups	96
	4.4 Applications of braided groups	104
	4.5 Braided-Lie algebras	109
	4.6 Diagrammatic geometry	112
5	**Hopf Algebras—Variant Notions and Reconstruction Theorems**	115
	JOOST VERCRUYSSE	
	5.1 Introduction	115
	5.2 Preliminaries	117
	5.3 Bialgebras and Hopf algebras in monoidal categories	119
	5.4 Reconstruction theorems	127
	5.5 Variations on the notion of Hopf algebra	132
6	**Modular Categories**	146
	MICHAEL MÜGER	
	6.1 Introduction	146
	6.2 Categories	147
	6.3 Tensor categories	148
	6.4 Braided tensor categories	155
	6.5 Modular categories	166
	6.6 Modularization of pre-modular categories: generalizations	175
	6.7 The braided centre of a fusion category	177
	6.8 The Witt group of modular categories	180
7	**Scalars, Monads, and Categories**	184
	DION COUMANS AND BART JACOBS	
	7.1 Introduction	184
	7.2 Preliminaries	186
	7.3 Monoids	189
	7.4 Additive monads	196
	7.5 Semirings and monads	205
	7.6 Semirings and Lawvere theories	210
	7.7 Semirings with involutions	213
	7.8 Conclusions	216
8	**Types and Forgetfulness in Categorical Linguistics and Quantum Mechanics**	217
	PETER HINES	
	8.1 Introduction	217
	8.2 Introducing typing to models of meaning	218
	8.3 What is a type?	218

8.4	Comparing words versus comparing sentences	229
8.5	Inner products, evaluation, and inverses	229
8.6	Does evaluation preserve inner products?	230
8.7	How to type connectives?	233
8.8	Self-similarity, categorically	238
8.9	Self-similarity and lax Frobenius algebras	244
8.10	Classical structures	245
8.11	A self-similar structure familiar in logic (and linguistics)	248
Appendix 8.A	Order-preserving bijections $\mathbb{N} \cong \mathbb{N} \uplus \mathbb{N}$ as interior points of the Cantor set	249
Appendix 8.B	Isbell's argument in a general setting	251

9 From Sentence to Concept — 252
ANNE PRELLER

9.1	Introduction	252
9.2	Notations, basic properties	254
9.3	Two-sorted first-order logic in compact closed categories	261
9.4	Semantics via pregroup grammars	267
9.5	Compositional semantics in concept spaces	274
9.6	Conclusion	281

10 Proof Nets for the Lambek–Grishin Calculus — 283
MICHAEL MOORTGAT AND RICHARD MOOT

10.1	Background, motivation	283
10.2	Display sequent calculus and proof nets	288
10.3	Proof nets and focused display calculus	301
10.4	Conclusions	319

11 Algebras over a Field and Semantics for Context-Based Reasoning — 321
DAOUD CLARKE

11.1	Introduction	321
11.2	Theory of meaning	322
11.3	From logical forms to algebra	325
11.4	Conclusion	332

12 Distributional Semantic Models — 333
STEPHEN PULMAN

12.1	Introduction	333
12.2	Compositional syntax and semantics	334
12.3	Harris's notion of distribution	336

12.4	Firth's notion of collocation	338
12.5	Distributional semantics	339
12.6	Compositional distributional semantics	343
12.7	A philosophical digression	344
12.8	Experiments in compositional interpretation	347
12.9	An empirical test of distributional compositionality	353
12.10	Compositionality vs disambiguation	355
12.11	Conclusion: a new program for compositional distributional semantics	357

13 Type-Driven Syntax and Semantics for Composing Meaning Vectors 359
STEPHEN CLARK

13.1	Introduction	359
13.2	Syntactic types and pregroup grammars	362
13.3	Semantic types and tensor products	365
13.4	Composition for an example sentence space	368
13.5	A real sentence space	372
13.6	Conclusion and further work	376

References 378
Index 407

CHAPTER 1

An Alternative Gospel of Structure: Order, Composition, Processes

BOB COECKE
(University of Oxford)

What are the fundamental mathematical structures? Evidently, in order to address this question one first needs to address another question, namely what a mathematical structure actually is. There are several options available.

Are mathematical structures the objects of mathematical theories? And is it then the mathematician who decides which ones are truly fundamental? It is indeed often said by mathematicians that good mathematics is what good mathematicians do. It has also been strongly argued by mathematicians, e.g. by Bourbaki, that good mathematics should be a discipline that exists in isolation from the (other) natural sciences and that the use of example applications and illustrations outside of straight mathematics is to be discouraged. We find these views somewhat circular and solipsistic, and even disrespectful to the other disciplines, in the light of the origins and history of mathematics.

From an alternative more reductionist perspective, one may think that the fundamental mathematical structures are the simple things from which one can build more complex things. Proposed candidates of fundamental structures include the Platonic solids from which the *classical elements* (Earth, Water, Air, and Fire) were constructed, and more recently, *sets*, which within the realm of set-theory are supposed to underpin all of mathematics. In our digital age, zeros and ones underpin all of our data. So do zeros and ones constitute the appropriate language of communicating our opinions about the movies that are encoded in terms of zeros and ones on a computer's hard drive? Of course not. The fact that one can 'code' everything by means of one particular structure, just as most modern mathematics can be encoded in sets, does not mean that it is the most convenient language for discourse.

Alternatively, are fundamental mathematical structures the objects which underpin the world in which we live, i.e. our reality? This raises the question, what do we mean by reality? Is this restricted to material reality or does this also include the stuff going on in our minds? In Aristotelianism numbers exist as concepts, which means that they do exist, but only in our mind. So where were those numbers before there were any minds that they could inhabit. Today, number-based calculus is still a pillar of modern mathematics, and many other fields of mathematics have some sort of numbers (be it either integer, real or complex) as their starting point. At school, first we learn natural numbers, then negatives, then rationals, reals, and eventually complex numbers. We can organize these as matrices yielding vector spaces, which is probably the most prominent mathematical structure at the moment. This observation points at these number systems maybe being the fundamental mathematical structures.

In this paper we will preach an alternative gospel, which qualifies structures in terms of their high-levelness in use. As a starting point, we try to analyse several forms of practice, starting in fact with 'counting' as underpinning trade, as well as basic mathematical practice and its relation to speech and reasoning, as well as nature. This is by no means a thorough treatment but just a few stone throws towards trying to develop a story of this kind.

Our preliminary treatment involves three stages:

(1) Firstly, we consider ordering of things.
(2) Secondly, we assume that these things can be composed.
  (2′) These ordered composable things also provide a stepping stone to the next stage. We can consider processes on a single thing that can be composed in time, and that moreover happen to be ordered. Dropping this ordering yields a stage (2′) which is complementary to (2). Together, stages (2) and (2′) combine into the remaining stage, that is:
(3) We consider processes, composable in time, for composable things.

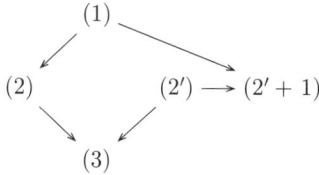

We will make a case for why we consider these structures as fundamental.

For each of these stages we provide a number of example applications in different areas, which indicate the general discipline-transcending nature of these structures. We consider this as additional evidence of the foundational nature of these structures.

But as a matter of convenience, we will still rely on set theory to provide us with a notion of equality in the formal definitions, as this will appeal to what the reader is used to. And of course, most existing example applications that we want to present have been formulated in set-theoretic terms, so we don't really have much choice here. However, it should be clear that these examples have a 'meaning in reality' that does not require this particular underpinning. We will illustrate this for one specific example. Consider the case of describing the possible states of three things in set theory versus in reality.

In set theory, each of these things will be some structured set; that is, a set X together with additional operations, which may encode topology (a collection of open subsets), geometry (lines etc.), or algebraic structure (e.g. a binary operation). The operation 'putting things together' would be encoded in most cases as a cartesian product. Focussing from now on only on the supporting sets, we can build a triple of things in three manners:

- first we combine X and Y into $X \times Y := \{(x,y) \mid x \in X, y \in Y\}$ and then we combine $X \times Y$ and Z into $(X \times Y) \times Z := \{((x,y),z) \mid x \in X, y \in Y, z \in Z\}$
- first we combine Y and Z into $Y \times Z := \{(y,z) \mid y \in Y, z \in Z\}$ and then we combine X and $Y \times Z$ into $X \times (Y \times Z) := \{(x,(y,z)) \mid x \in X, y \in Y, z \in Z\}$
- we do everything at once by considering triples $\{(x,y,z) \mid x \in X, y \in Y, z \in Z\}$.

Now, the three results are not equal! Instead, they are isomorphic in a very 'natural' manner, since we can pass from $\{((x,y),z) \mid x \in X, y \in Y, z \in Z\}$ to $\{(x,(y,z)) \mid x \in X, y \in Y, z \in Z\}$ by 're-bracketting', and to $\{(x,y,z) \mid x \in X, y \in Y, z \in Z\}$ by 'dumping brackets'.

Clearly 're-bracketting' and 'dumping brackets' are operations that have many nice properties, which boil down to saying that each of these three ways of combining three things are essentially the same for all practical purposes. But to properly state this in mathematical terms requires some pretty heavy machinery. For example, simply saying that these are isomorphic is not enough, since along such an isomorphism we may lose the fact that we truly are considering three things, e.g.

$$\{0,1\} \times \{0,1\} \times \{0,1\} \simeq \{0,1,2,3,4,5,6,7\}.$$

To capture the 'natural' manner in which constructions like

$$\{((x,y),z) \mid x \in X, y \in Y, z \in Z\} \text{ and } \{(x,(y,z)) \mid x \in X, y \in Y, z \in Z\}$$

are equivalent, Eilenberg and Mac Lane (1945) introduced category theory, which first required definition of categories, then functors, and finally natural transformations, a highly non-trivial concept. For something as simple as saying that there are three things, these concepts feel a bit like overkill. Indeed, in reality, given three objects, it simply doesn't matter that we first consider two of them together, and then the third one—or the other way around—in order to describe the end result. In other words, the bracketing in the above description has no counterpart in reality, it is merely imposed on us by set theory!

But make no mistake, it is not category theory that we blame in any manner here, but set theory. Certain branches of category theory have no problem in describing three things on an equal footing, in terms of so-called strict symmetric monoidal categories. True, in its usual presentations (e.g. Mac Lane 1971) it is built upon set theory, but not to do so is exactly what we are arguing for here. An elaborate discussion of this point can be found in Coecke and Paquette 2011. Also, category theory is a huge area and by no means all of it will be relevant here.

1.1 Order

Above we mentioned that number-based calculus is a pillar of modern mathematics. However, probably more primal than counting is the simple fact of whether one number is *larger than* another. If numbers characterize the cost of something, say a piece of meat, being able to afford that piece of meat requires one to have *more* money available than the cost of that piece of meat. The meaningfulness of the question 'Can I afford this piece of meat?' results from the fact that numbers are *ordered*: if x is the price and y is one's budget, then 'Yes I can!' means $x \leq y$ while 'No I cannot!' means $x \not\leq y$.

Definition 1.1 A *total ordering* on a set X is a relation $\leq$ on X, i.e. a collection of pairs $(x, y) \subseteq X \times X$ that is *anti-symmetric*, *transitive*, and *total*; that is, respectively,

- $\forall x, y \in X : x \leq y, y \leq x \Rightarrow x = y$
- $\forall x, y, z \in X : x \leq y, y \leq z \Rightarrow x \leq z$
- $\forall x, y, z \in X : x \leq y \text{ or } y \leq x$

Of course, one cannot compare apples and lemons if we don't consider monetary value. Three apples are less than four apples, but what about three apples versus three lemons?

Definition 1.2 A *preordering* on X is a transitive relation $\lesssim$, which is also *reflexive*; that is,

- $\forall x \in X : x \lesssim x$

A *partial ordering* is an anti-symmetric preordering, and denoted $\leq$.

By definition, each partial ordering is a preordering and also each total ordering is a partial ordering since totality implies reflexivity (apply totality when $x = y$).

Moreover, each preordering yields an *equivalence relation*, which is defined as a preordering $\simeq$ that is also *symmetric*; that is,

- $\forall x, y \in X : x \simeq y \Leftrightarrow y \simeq x$,

simply by setting $x \simeq y$ to be $x \lesssim y$ and $y \lesssim x$. Then, the corresponding set of equivalence classes $\{C_x \mid x \in X\}$, where $C_x = \{y \in X \mid y \simeq x\}$, forms a partial ordering when setting $C_x \leq C_y$ whenever $x \lesssim y$. For example, if we order things in terms of their cost we obtain a preordering, since there may be many things that have the same cost. This gives rise to a partial ordering, which is in fact a total ordering, of the occurring costs themselves.

Sometimes one encounters the notion of a *strict partial order* $<$, which is *irreflexive* ($x \not< x$), *asymmetric* ($a < b \Rightarrow b \not< a$), and transitive, but it is easily seen that each partial ordering as in defn 1.2 induces a strict partial ordering by setting $x < y$ when $x \leq y$ and $x \neq y$; conversely, $x \leq y$ when $x < y$ or $x = y$ turns a strict partial ordering into a partial ordering as in defn 1.2. So both notions are equivalent in their uses.

Preorderings and partial orderings arise in many everyday situations, as well as in important areas of science. Arguably, they are the most primitive mathematical concept imaginable, capturing the very basic notion of 'comparing'.

1.1.1 Reachability, causality, and relativity

Consider a collection of events, a certain rock concert in New Orleans on a particular day, the marriage of some old friend, Carnival in Rio this year, etc. Then there exists a partial ordering where the relation $\leq$ stands for: 'if I attend event a, can I also attend event b?' Otherwise put, the ordering $a \leq b$ captures whether one can reach event b from event a. The overall data in this partial ordering is imposed by the transport network of the world.

In physics, due to the velocity bound imposed by the speed of light, a similar partial order exists that encodes whether light can travel from one point x in space-time to another point y in space-time. In the case of yes we can write $x \leq y$. And since nothing travels faster than light, this partial order encodes which event in space-time can causally affect which other event in space-time. In fact, there exist results that enable one to reconstruct the entire geometric space-time manifold from this partial ordering (e.g. Martin and Panangaden, 2006). There are moreover several research programmes that take partial ordering not only as a framework for discussing special and general relativity, but also as the basis for crafting a theory of quantum gravity (Bombelli et al., 1987; Sorkin, 2003), the Holy Grail of modern physics.

Also in computer science, identical ideas on causality as partial orderings exist (Lamport, 1978), and through a pair of 'partial order'-glasses, both the areas of relativity theory and the organization of events in a distributed computational system look remarkably similar. In other words, in high-level terms they essentially coincide.

1.1.2 Information content, propositional knowledge, and domains

One may be interested in the information content of some pieces of data, for example in terms of entropy. Typical measures of information content will assign some real number, so that, since real numbers are totally ordered, we can decide which one of two of pieces of data is the most informative one. This then induces a total preordering on these pieces of data. In many cases the actual values don't really matter; merely what is more informative than what. So in fact, rather than taking a valuation into real numbers, one can just consider a total preordering on the pieces of data.

Now of course, if two pieces of data each constitute one bit, while their information content is the same, these may be incomparable in terms of their propositional content. For example, while 'x is an apple' implies 'x is fruit', it is incomparable with 'x is a lemon'. As propositions, these pieces of data form a partial order, where $a \leq b$ stands for the fact that a implies b, e.g. 'being an apple' implies 'being fruit'. This idea is the cornerstone of *algebraic logic* (Davey and Priestley, 1990), which is discussed in the next section.

For the purpose of combining informative and propositional content, domain theory was crafted (Scott, 1970) as a new mathematical foundation for computation. Here, domains are partial orders in which certain subsets have least upper bounds.

Definition 1.3 Given a partial ordering $(X, \leq)$, a subset $Y \subseteq X$ has a *least upper bound* (or *join*) if there is an element $x \in X$ that is such that:

- $\forall y \in Y : y \leq x$
- $\forall x' \in X : (\forall y \in Y : y \leq x') \Rightarrow x \leq x'$.

In this case we denote this element x as $\bigvee Y$. A *greatest lower bound* (or *meet*) $\bigwedge Y$ of $Y \subseteq X$ is defined similarly simply by replacing $\alpha \leq \beta$ by $\beta \leq \alpha$ in the above.

While we will not discuss the nature of those subsets that have least upper bounds in a domain, a particular example of a least upper in algebraic logic is *disjunction* $a \vee b$, where Y consists of two elements. *Conjunction* $a \wedge b$ is the corresponding greatest lower bound.

Interestingly, the partial orders underpinning space-time are in fact also domains (Martin and Panangaden, 2006). Hence at this order-theoretic level two seemingly disjoint subjects again become the same when taking a sufficiently high-level perspective.

Also in this context, while it fails to be a domain and only captures propositional content up to symmetries, the *majorization preordering* (Muirhead, 1903) on probabilities has a range of applications, ranging from economy (Marshall, Olkin and

Arnold, 2010) to quantum information theory (Nielsen, 1999), where it captures degrees of entanglement. What is ordered here are descending discrete probability distributions—that is, n-tuples $(x_1, \ldots, x_n)$ with $\sum_i x_i = 1$, and which are such that $x_i \geq x_{i+1}$. We say that an ordered n-tuple $(x_1, \ldots, x_n)$ is *majorized* by $(y_1, \ldots, y_n)$ if

$$\forall k \in \{1, \ldots, n-1\} : \sum_{i=1}^{i=k} x_i \leq \sum_{i=1}^{i=k} y_i.$$

Intuitively, this means that $(y_1, \ldots, y_n)$ is a 'narrower' or 'sharper' distribution than $(y_1, \ldots, y_n)$.

Unfortunately, majorization does not extend to a partial order on all probabilities, so it fails to capture propositional content. A genuine partial ordering on probabilities in which propositional structure naturally embeds is in Coecke and Martin (2011). It is 'desperately looking for more applications', so please let us know if you know about one!

We refer the reader to Abramsky and Jung (1994) for the role of domain theory in computer science, where it originated, and in particular to the tutorial by Martin (2011) for a much broader range of applications in a variety of disciplines.

1.1.3 Logic and the theory of mathematical proofs

In algebraic logic, which traces back to Leibniz, one typically would like to treat 'a implies b' itself as a proposition, something which is realized by an *implication connective*; that is, an operation $\Rightarrow : X \times X \to X$ on the partial ordering $(X, \leq)$. Typically, one also of course assumes conjunction and disjunction, and in its weakest form implication can be defined by the following order-theoretic stipulation:

$$(a \wedge b) \leq c \text{ if and only if } a \leq (b \Rightarrow c) \qquad (1.1)$$

One refers to partial orderings with such an implication as *Heyting algebras*. From eqn (1.1) it immediately follows that the *distributive law* holds; that is,

$$a \wedge (b \vee c) = (a \wedge b) \vee (a \wedge c),$$

or in terms of words,

$$a \text{ AND } (b \text{ OR } c) = (a \text{ AND } b) \text{ OR } (a \text{ AND } c).$$

A special case of implication is a *Boolean implication*, which is defined in terms of negation as $a \Rightarrow b := \neg a \vee b$. We will not go into the details of defining the negation operation $\neg : X \to X$, but one property that it has is that it reverses the ordering; that is:

- $a \leq b \Leftrightarrow \neg b \leq \neg a$.

In the case of a Boolean implication we speak of a *Boolean algebra*. The smallest Boolean algebra has two elements, *true* and *false*, with *false* $\leq$ *true*.

Closely related to algebraic logic is *algebraic proof theory*, a meta-theory of mathematical practice. One may indeed be interested from which assumptions one can deduce which conclusions, and this gives rise to a preordering, where $a \lesssim b$ stands for 'from a we can derive b'. Since we can possibly both derive b from a and a from b we are dealing with a proper preorder rather than a partial order. In this context, eqn (1.1) corresponds to the so-called *deduction theorem* (Kleene, 1967). It states that if one can derive c from the assumptions a and b then this is equivalent to deriving that b implies c given a.

We stress here that $\lesssim$ now represents the *existence* of a proof, but not a proof itself. We will start to care about the proofs themselves in Section 1.3.

1.1.4 General 'things' and processes

The idea of the existence of a mathematical proof that transforms certain assumptions into certain conclusions extends to general situations where processes transform certain things into other things (see Table 1.1). For example, if I have a raw carrot and a raw potato I can transform these into carrot–potato mash. On the other hand, one cannot transform an apple into a lemon, nor carrot–potato mash into an egg. So in general, when one considers a collection of things (or systems), and processes that transform things into other things, one obtains a preordering that expresses what can be transformed into what. More precisely, $a \leq b$ means that *there exists a process that transforms a into b*. The technical term for things in computer science would be *data-types*, and for processes *programs*; in physics things are *physical systems* and examples of processes are evolution and measurement; in cooking the thing are the *ingredients* while example processes are boiling, spicing, mashing, etc.

TABLE 1.1 Example 'things' and processes

	Thing/system	Process
Math. practice	Propositions	Proofs (e.g. lemma, theorem, etc.)
Physics	Physical system	Evolution, measurement etc.
Programming	Data type	Program
Chemistry	Chemical	Chemical reaction
Cooking	Ingredient	Boiling, spicing, mashing, etc.
Finance	Currencies	Money transactions
Engineering	Building materials	Construction work

1.2 Orders and composition

We mentioned that one cannot compare apples to lemons. However three apples and two lemons is clearly less that five apples and four lemons. To formalize this, we need to have a way of saying that we are 'adding apples and lemons'. We will refer to this 'adding' as composition. Clearly, this composition needs to interact in a particular way with the ordering such that either increasing the apples or the lemons increases the order of the composite. Note that adding apples to apples, or money to money, is just a special case of this, where the ordering is total, i.e. everything compares to everything, rather than being a proper partial ordering in which some things don't compare.

Definition 1.4 A *monoid* is a set X together with a binary operation $\cdot : X \times X \to X$ that is both associative and admits a two-sided unit $1 \in X$; that is, respectively:

- $\forall x, y, z \in X : x \cdot (y \cdot z) = (x \cdot y) \cdot z$
- $\forall x \in X : 1 \cdot x = x \cdot 1 = x$

A *totally/partially/preordered monoid* is a set X that is both a monoid and a total/partial/pre-order, and which, moreover, satisfies *monotonicity of the monoid multiplication*:

- $\forall x, y, x', y' \in X : x \lesssim y, x' \lesssim y' \Rightarrow x \cdot x' \lesssim y \cdot y'$

Many of the applications mentioned above extend to ordered monoids. For example, in algebraic logic, either the conjunction or the disjunction operation yields a partially ordered monoid. When thinking about things and processes, we can obtain an ordered monoid as in the case of apples and lemons. We can compose different things, and we can compose the processes acting thereon in parallel. In the case of mathematical proofs this would simply mean that we prove b given a as well as proving d given c.

While in the case of apples and lemons composition is *commutative*; that is:

- $\forall x, y \in X : x \cdot y = y \cdot x$,

composition of processes *in time* is typically non-commutative.

And we can indeed also think of the elements of an ordered monoid as processes themselves, the monoid composition then standing for process b *after* process a. The ordering is then an ordering on processes. For example, given a certain proof, the proof of either a stronger claim from the same assumptions or the same claim from weaker assumptions would be strictly below the given proof.

Summing the above up we obtain two distinct ordered monoids for the example of mathematical proofs (as well as for the other examples).

- *Composition of things* or *parallel composition*, where the ordered propositions can be composed; that is, we can consider collections of assumptions rather than individual ones, and comparing these may happen

component-wise, but does not have to (cf. conjunction and disjunction as composition).
- *Composition of processes* or *sequential composition*, where the elements of the monoid are the proofs themselves rather than propositions. The composition is now 'chaining' proofs; that is, a proof of b given a and a proof of c given b results in a proof of c given a. The ordering is then in a sense the quality of the proof, in that a better proof is one that achieves stronger conclusions from weaker assumptions.

More generally, non-commutative ordered monoids yield some interesting new applications, for example in *reasoning about knowledge* and in *natural language*.

1.2.1 Galois adjoints as assigning causes and consequences

Definition 1.5 For two order-preserving maps $f : A \to B$ and $g : B \to A$ between partially ordered sets $(A, \leq)$ and $(B, \leq)$, we say that f is *left adjoint* to g (or equivalently, that g is *right adjoint* to f), denoted $f \dashv g$, if we have that:

- $\forall a \in A, b \in B : f(a) \leq b \Leftrightarrow a \leq g(b)$,

or what is easily seen to be equivalent (Erné et al., 1993), if we have that:

- $\forall b \in B : f(g(b)) \leq b$ and $\forall a \in A : a \leq g(f(a))$.

This at first somewhat convoluted-looking definition has a very clear interpretation if we think of f as a process that transform propositions $a \in A$ of system A into propositions $b \in B$ of B. This goes as described next.

Assume we know that process f will take place (e.g. running it on a computer as a computer program) but we would want to make sure that after running it $b \in B$ holds, and the means we have to impose this is to make sure that before running it some $a \in A$ holds. Then, the Galois adjoint to f gives us the answer, namely that the necessary and sufficient condition is to take $a \leq g(b)$. In computer science one refers to $g(b)$ as the *weakest precondition* to realize b by means of f (Dijkstra, 1975; Hoare et al., 1987). More generally, one can think of g as assigning *causes*, while f assigns *consequences* for any kind of process (Coecke, Moore and Stubbe, 2001).

Note in particular that eqn (1.1) is also of the form of a Galois adjunction. Indeed, explicitly putting quantifiers and re-ordering symbols, eqn (1.1) can be rewritten as:

- $\forall c \in X, \big(\forall a \in X, b \in X : (c \wedge a) \leq b \Leftrightarrow a \leq (c \Rightarrow b)\big)$.

The expression within the large brackets is a Galois adjunction for $A = B = C := X$ and:

$$f := (c \wedge -) : X \to X \quad \text{and} \quad g := (c \Rightarrow -) : X \to X.$$

In eqn (1.1) we ask this Galois adjunction to be true for all $c \in X$, and all these conditions together then define an implication connective $(- \Rightarrow -) : X \times X \to X$.

More generally, adjointness provides a comprehensive foundation for logic (Lawvere, 1969; Lambek and Scott, 1986).

While the author does not subscribe (anymore) to this approach, so-called *quantum logic* (Birkhoff and von Neumann, 1936; Piron, 1976) is also an order-theoretic enterprise, and the corresponding 'weak' implication, the Sasaki hook (Sasaki, 1954), can best be understood in terms of Galois adjoints.

Birkhoff and von Neumann noted that observationally verifiable propositions of a quantum system still form a partial ordering, which admits least upper bounds and greatest lower bounds, but that the distributive law fails to hold, and as a consequence, that there is no connective $(- \Rightarrow -)$ as in eqn (1.1) since that would imply distributivity (see Section 1.1.3). However, there is an operation $(c \rightsquigarrow -) : X \times X \to X$ for all $c \in X$ that is such that:

- $\forall a \in X, b \in X : P_c(a) \leq b \Leftrightarrow a \leq (c \rightsquigarrow b)$

where $P_c : X \to X$ stands for the orthogonal projection on C or, in physical terminology, the collapse onto c. The collapse is an actual physical process that happens when measuring a quantum system, so in the light of the above discussion on causes and consequences, which, as discussed in Coecke, Moore and Stubbe 2001 extends to these quantum logics, the operation $(c \rightsquigarrow -) : X \to X$ should be understood as assigning the weakest precondition $(c \rightsquigarrow b) \in X$ that has to hold before the collapse in order for $b \in X$ to hold after the collapse.

While, as already mentioned, the author does not subscribe to quantum logic any longer, Constantin Piron's operational take on the subject (Piron, 1976; Moore, 1999) has greatly influenced the author's thinking. Unfortunately, Piron died during the final stages of writing of this chapter.

1.2.2 Dynamic (and) epistemic logic

In Section 1.1.3 we briefly discussed propositional logic; that is, describing the set of propositions about a system as a partially ordered set. So what about the change of these propositions? *Actions*, which change propositions, can be described as maps acting on these propositions, and form themselves an ordered monoid. It is easily seen that these maps should preserve disjunctions or, in order-theoretic terms, greatest lower bounds. Indeed, if a OR b holds then, after the action f, clearly $f(a)$ OR $f(b)$ should hold, so $f(a \vee b) = f(a) \vee f(b)$. This guarantees that these actions have left Galois adjoints, assigning causes.

Now, while propositions may be as they are, an agent may perceive them differently. This may, for example, be due to *lying actions* by some agent who is supposed to communicate these changes of propositions. These situations were considered in Baltag, Moss and Solecki (1998), using what is referred to as *dynamic epistemic logic*, and it was shown in Baltag, Coecke and Sadrzadeh (2006) that all of this is most naturally cast in order-theoretic terms.

This setting encompasses and stretches well beyond the fields of epistemic logic (Meyer and van der Hoek, 2004) and dynamic logic (Harel, Kozen and Tiuryn, 2000), both conceptual variations on so-called *modal logic* (Kripke, 1963), and all of which are part of modern algebraic logic. They have a wide range of applications in computer science, including soft- and hardware verification (Stirling, 1991). On the other hand, these logics also underpin Carnap's philosophy on semantics and ontology (Carnap, 1988).

1.2.3 Linguistic types

We can also compose words in order to build sentences. However, not all strings of words make up meaningful sentences, since meaningfulness imposes constraints on the grammatical types of the words in the sentence.

By having a partial order relation besides a composition operation we can encode how the overall grammatical type of a string of words evolves when composing it with other types and, ultimately, make up a sentence. The ordering $a_1 \cdot \ldots \cdot a_n \leq b$ then encodes the fact that the string of words of respective grammatical types $a_1, \ldots, a_n$ has as its overall type b. For example, if n denotes the type of a noun, tv the type of a transitive verb and s the type of a (well-formed) sentence, then $n \cdot tv \cdot n \leq s$ expresses the fact that a noun (= object), a transitive verb, and another noun (= subject), make up a well-formed sentence.

The remaining question is then how to establish a statement like $a_1 \cdot \ldots \cdot a_n \leq b$. The key idea is that some grammatical types are taken to be atomic, i.e. indecomposable, while others are compound, and one considers additional operations, which may either be unary or binary, subject to some laws that allow one to reduce type expressions. For example, assuming that one has left- and right-'pre-inverses', respectively denoted $^{-1}(-)$ and $(-)^{-1}$, and subject to $a \cdot {^{-1}a} \leq 1$ and $a^{-1} \cdot a \leq 1$, then for the compound transitive verb type:

$$tv = {^{-1}n} \cdot s \cdot n^{-1}$$

we have:

$$n \cdot tv \cdot n = n \cdot {^{-1}n} \cdot s \cdot n^{-1} \cdot n \leq 1 \cdot s \cdot 1 \leq s,$$

so we can indeed conclude that $n \cdot tv \cdot n$ forms a sentence.

As the area of type grammars has not yet reached a conclusion on the question of which partially ordered monoids (or *pomonoids* for short) best captures 'universal grammatical structure', we give a relatively comprehensive historical overview of the structures that have been proposed. Moreover, several of these will provide a stepping stone to the categorical structures in the next section. Historically, the idea of universal grammar of course traces back to Chomsky's work in the 1950s (Chomsky, 1956). The mathematical development was mainly driven by Lambek (Lambek, 1958; Lambek, 1999; Lambek, 2008; Moortgat, 1988), in many stages and spanning some 60 years of work.

Definition 1.6 A *protogroup* (Lambek, 1999) is a pomonoid

$$(X, \leq, {}^*(-), (-)^*)$$

where ${}^*(-) : X \to X$ and $(-)^* : X \to X$ are such that:

- $\forall a, b \in X : a \cdot {}^*a \leq 1$ and $b^* \cdot b \leq 1$.

Definition 1.7 An *Ajdukiewicz–Bar-Hillel pomonoid* (Ajdukiewicz, 1937; Bar-Hillel, 1953) is a pomonoid

$$(X, \leq, (-\multimap-), (-\olessthan-))$$

where $(-\multimap-) : X \times X \to X$ and $(-\olessthan-) : X \times X \to X$ are such that:

- $\forall a, b, c \in X : a \cdot (a \multimap c) \leq c$ and $(c \olessthan b) \cdot b \leq c$.

For 1 the unit of the monoid and setting ${}^*a := a \multimap 1$ and $b^* := 1 \olessthan b$, it then follows that each Ajdukiewicz–Bar-Hillel pomonoid is a protogroup (Lambek, 1999).

Definition 1.8 A *residuated pomonoid* (Lambek, 1958) is a pomonoid

$$(X, \leq, (-\multimap-), (-\olessthan-))$$

such that for all $a, b \in X$ we have two Galois adjunctions:

$$(a \cdot -) \dashv (a \multimap -) \quad \text{and} \quad (- \cdot b) \dashv (- \olessthan b);$$

that is, explicitly:

- $\forall a, b, c \in X : b \leq a \multimap c \Leftrightarrow a \cdot b \leq c \Leftrightarrow a \leq c \olessthan b,$

or, equivalently, using the alternative characterization of the adjunctions:

- $\forall a, b, c \in X : a \cdot (a \multimap c) \leq c, \ c \leq a \multimap (a \cdot c), \ (c \olessthan b) \cdot b \leq c, \ c \leq (c \cdot b) \olessthan b.$

From the second formulation in terms of four conditions it immediately follows that each residuated pomonoid is a Ajdukiewicz–Bar-Hillel pomonoid. However, note in particular also that what we have here is a *non-commutative* generalization of eqn (1.1), which defined an implication connective, conjunction being replaced by the (evidently) non-commutative composition of words, and the implication $(-\Rightarrow-)$ now having a right-directed and left-directed counterpart, respectively $(-\multimap-)$ and $(-\olessthan-)$.

Definition 1.9 A *Grishin pomonoid* (Grishin, 1983) is a residuated pomonoid

$$(X, \leq, (-\multimap-), (-\olessthan-), 0)$$

with a special element $0 \in X$ that is such that:

- $0 \multimap (a \multimap 0) = a = (0 \multimap a) \multimap 0$.

With some work one can show that every Grishin pomonoid is a residuated pomonoid too, for example see Lambek (1999). Now, anticipating the following definition we can set:

$$^*a := a \multimap 0, \quad a^* := 0 \multimap a \quad \text{and} \quad a + b := {}^*(b^* \cdot a^*) = (^*b \cdot {}^*a)^*,$$

(where the last equality is quite easy to prove) and then we have:

- $a \cdot {}^*a \leq 0, \ 1 \leq {}^*a + a, \ b^* \cdot b \leq 0, \ 1 \leq b + b^*$.

We also have that $a \multimap c = {}^*a + c$ and $c \multimap b = c + b^*$. So now the non-commutative implication resembles the Boolean implications that we discussed in Section 1.1.3, the *-operations playing the role of negation and the +-operation corresponding to the disjunction.

Definition 1.10 A *pregroup* (Lambek, 1999) is a pomonoid

$$(X, \leq, (-)^*, {}^*(-))$$

where $(-)^* : X \to X$ and $^*(-) : X \to X$ are such that:

- $a \cdot {}^*a \leq 1 \leq {}^*a \cdot a, \ b^* \cdot b \leq 1 \leq b \cdot b^*$.

Hence, each pregroup is a Grishin pomonoid with $\cdot = +$ and $0 = 1$, and each pregroup is a protogroup that satisfies two additional conditions. In the case that $(-)^* = {}^*(-)$ then we obtain a group with $a^* = a^{-1}$, hence the name 'pre'-group.

Right, that was a bit of a zoo! Still, there is a clear structural hierarchy. Below the arrows represent the increase in equational content:

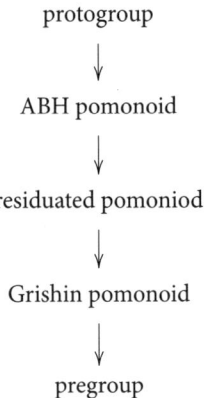

There are four kinds of inequalities that play a key role, which either reduce or introduce types, and do this either in terms of (left/right) unary or a (left/right)

TABLE 1.2 Four kinds of inequality

	Unary connective		Binary connective	
Type reduction	$a \cdot {}^*a \leq e$	$b^* \cdot b \leq e$	$a \cdot (a \multimap c) \leq c$	$(c \multimap b) \cdot b \leq c$
Type introduction	$1 \leq a \circ a^*$	$1 \leq {}^*b \circ b$	$c \leq a \multimap (a \circ c)$	$c \leq (c \circ b) \multimap b$

binary connective. Those in terms of a unary connective imply the corresponding ones involving a binary connectives. Table 1.2 depicts these rules, with $e \in \{0, 1\}$ and $\circ \in \{\cdot, +\}$.

1.3 Processes witnessing existence

In Section 1.1.4 we observed that, for the very general setting of things/systems and processes thereon, orders witness the existence of a process between two things/systems. In Section 1.2 we saw how composition of things interacts with ordering. On the other hand, in Section 1.2 we also saw that monoid structures naturally arise when we consider process composition in the sense of one process happening *after* another process.

Here we will make the passage from order witnessing existence of processes to explicitly describing these processes. Since processes themselves also come naturally with sequential composition, and if the systems on which these act also compose, we will obtain a structure with two interacting modes of composition. A dual perspective is that starting from a process structure on a fixed system, we allow for variation of the system:

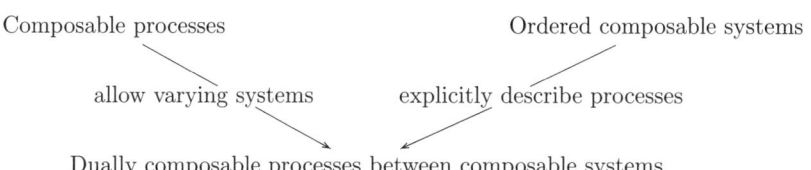

Historically, this structure traces back to Benabou (1963) and Mac Lane (1963). Symbolically, this is what one is dealing with:

Definition 1.11 A *strict symmetric monoidal category* $\mathcal{S}$ consists of:

- a collection (typically a 'class') of things/systems $|\mathcal{S}|$,
- with a monoid structure $(\mathcal{S}, \otimes, 1)$ thereon,

and for each pair $S, S' \in |\mathcal{S}|$

- a collection (typically a 'set') of processes $\mathcal{S}(S, S')$,

with two unital associative composition structures:

- $\forall S, S', S'' \in |\mathcal{S}|, (- \circ -) : \mathcal{S}(S, S') \times \mathcal{S}(S', S'') \to \mathcal{S}(S, S'')$,
- $\forall S, S', S'', S''' \in |\mathcal{S}|, (- \circ -) : \mathcal{S}(S, S') \times \mathcal{S}(S'', S''') \to \mathcal{S}(S \otimes S'', S' \otimes S''')$.

We denote $\mathcal{S}(S, S')$ also as $f : S \to S'$. Explicitly, associativity and unitality are:

- $\forall f : S \to S', g : S' \to S'', h : S'' \to S'''$ we have $(h \circ g) \circ f = h \circ (g \circ f)$,
- $\forall f : S \to S', g : S'' \to S''', h : S'''' \to S'''''$ we have $(f \otimes g) \otimes h = f \otimes (g \otimes h)$,
- $\forall S \in |\mathcal{S}|$ there exists an *identity process* $1_S : S \to S$ that is such that for all $f : S' \to S, g : S \to S'$ we have that $1_S \circ f = f$ and $g \circ 1_S = g$,
- there exists an *identity system* $I \in |\mathcal{S}|$ that is such that for all $S \in |\mathcal{S}|$ we have that $I \otimes S = S \otimes I = S$.

These composition structures moreover interact *bifunctorially*, that is:

- $\forall f : S \to S', f' : S' \to S'', g : S''' \to S'''', g' : S'''' \to S'''''$ we have that:

$$(f' \circ f) \otimes (g' \circ g) = (f' \otimes g') \circ (f \otimes g). \quad (1.2)$$

- $\forall S, S' \in |\mathcal{S}|$ we have that:

$$1_S \otimes 1_{S'} = 1_{S \otimes S'}. \quad (1.3)$$

Finally, we assume *symmetry*; that is,

- $\forall S, S' \in |\mathcal{S}|$ there exists a *symmetry process* $\sigma_{S,S'} : S \otimes S' \to S' \otimes S$, and these are such that for all $f : S \to S', g : S'' \to S'''$ we have that:

$$\sigma_{S',S'''} \circ (f \otimes g) = (g \otimes f) \circ \sigma_{S,S''}.$$

What a mess, or better, what a syntactic mess! The problem here is indeed of a syntactic nature. The concept behind a strict symmetric monoidal category is intuitively obvious but one would not get that intuition easily when reading the above definition. In fact, by presenting *strict* symmetric monoidal categories rather than general symmetric monoidal categories we have already enormously simplified the presentation. In the strict case we assume associativity and unitality of the ⊗-connective on the nose, while, as discussed in the introduction, set-theory-based mathematical models would typically be non-strict. However, the physical reality itself is strict, which points at an inadequacy of its typical mathematical models. Let us recall this physical conception of process theories.

There is a notion of system, to which we from now on will refer as *type*, and for each pair of types of systems there are processes that take a system of the first type to the system of the second type. These processes can be composed in two manners.

- *sequentially*—that is one process taking place *after* another process, the second process having the output type of the first process as its input type.
- *in parallel*—that is, one process takes place *while* the other one takes place, without any constraints on the input and output types.

Examples of particular systems and processes respectively are 'nothing' (cf. I in defn 1.11 above), and 'doing nothing' (cf. 1_S in defn 1.11 above).

But there is more, ... Even in the strict definition above much of the structure is about 'undoing' unavoidable syntactic features. For example, symmetry simply means that there is no significance to the list-ordering of systems when writing $S \otimes S'$; that is, $S \otimes S'$ and $S' \otimes S$ describe one and the same thing and we can use $\sigma_{S,S'}$ to pass from one description to the other.

Of course, sometimes the order does matter, as in the case of words making up a sentence. Swapping words evidently changes the meaning of a sentence, and in most cases would even make it meaningless, as swapping words of different grammatical types would typically destroy the grammatical structure. So here, rather than a strict symmetric monoidal category, we would consider a *strict monoidal category*, which boils down to defn 1.11 without the $\sigma_{S,S'}$-processes.

Turning our attention again on the 'syntactic mess', even more striking than the role of symmetry in undoing the ordering of one-dimensional linear syntax is the role played by eqn (1.2). To expose its 'undoing'-nature we will need to change language, from one-dimensional linear syntax to two-dimensional pictures. This will also bring us much closer to our desire to base our conception of foundational mathematical structure on the idea of high-levelness in use, given that the pictorial presentation gets rid of the artifacts of set-theoretical representation, as illustrated in the introduction in the example of 'three things'. The diagrammatic language indeed precisely captures the idea that strict symmetric monoidal categories aim to capture, but still within a syntactic realm. The study of these diagrammatic languages is becoming more and more prominent in a variety of areas of mathematics, including modern algebra and topology.

In the two-dimensional pictures processes will be represented by boxes and the input and output systems by wires:

$$\text{Box} := \begin{array}{c} \big| \leftarrow \text{output wire(s)} \\ \boxed{f} \\ \big| \leftarrow \text{input wire(s)} \end{array}$$

We can then immediately vary systems by varying the number of wires:

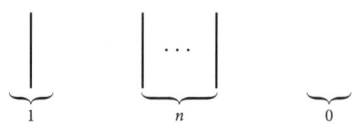

The two compositions boil down to either connecting the output wire of one process to the input wire of the other, or by simply putting the processes side-by-side:

Doing nothing is represented by a wire, and nothing, evidently, by nothing.

The rules of the game are: 'only topology matters'; that is, if two pictures are topologically equivalent then they represent the same situation, e.g.:

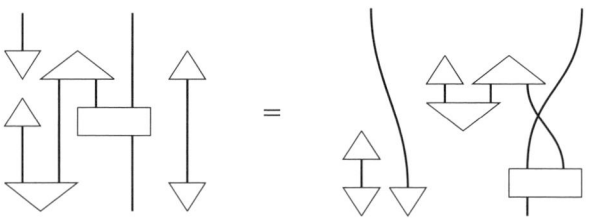

Now, the non-trivial symbolic equation $(f' \circ f) \otimes (g' \circ g) = (f' \otimes g') \circ (f \otimes g)$ becomes:

i.e. a tautology! In other words, in this diagrammatic language, which more closely captures the mathematical structure of processes than its symbolic counterpart, essential symbolic requirements become vacuous.

The reason is simple: there are two modes of composition that are in a sense 'orthogonal', but one tries to encode them in a single dimension. As a result, one needs to use brackets to keep the formulas well-formed, but these brackets obviously have no counterpart in reality. This is where they would be in the pictorial language:

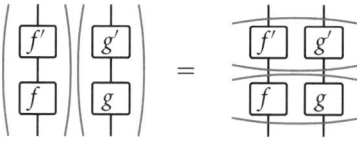

A more detailed discussion of the upshot of graphical languages is in Coecke and Paquette (2011). The development of graphical languages for a variety of structures is an active area of research. For a survey of the state of the art we refer to Selinger (2011). Also useful in this context are Kock (2003), Street (2007), and Baez and Stay (2011).

1.3.1 From-word-meaning-to-sentence-meaning processes

We can blow up the pomonoids of Section 1.2.3 to full-blown proper (typically non-symmetric) monoidal categories by replacing each relationship $a \leq b$ by a collection $\mathcal{S}(a, b)$ of processes. A residuated pomonoid then becomes a so-called *bi-closed monoidal category*, a Grishin pomonoid becomes a category of which the symmetric counterpart is called a *$*$-autonomous category* (Barr, 1979), and a pregroup then becomes a category of which the symmetric counterpart is called a *compact (closed) category* (Kelly, 2005; Kelly and Laplaza, 1980).

The non-symmetric case of compact (closed) categories has been referred to as planar autonomous categories (Joyal and Street, 1993; Selinger, 2011), and the non-symmetric case of $*$-autonomous categories as linearly distributive categories with negation (Barr, 1995; Cockett, Koslowski and Seely, 2000).

These symmetric categories have themselves been studied in great detail, since closed symmetric monoidal categories capture *multiplicative intuitionistic linear logic*, while $*$-autonomous categories capture classical linear logic (Girard, 1987; Seely, 1989). Compact closed categories model a degenerate logic that has found applications in quantum information processing (Abramsky and Coecke, 2004; Duncan, 2006). We will discuss this a bit more in the next section.

Table 1.3 summarizes the blow-up and symmetric restriction of some of the pomonoids that arises when describing grammatical structure:

TABLE 1.3 Blow-up and symmetric restriction of pomonoids

Grammatical types	Monoidal category	Symmetric case
Residuated pomonoid	Biclosed monoidal	Closed symmetric monoidal
Grishin pomonoid	Linearly distributive +negation	$*$-autonomous
Pregroup	Planar autonomous	Compact closed

While, as explained in Section 1.2.3, the ordered structures capture how grammatical structure evolves when composing words, the categorical structures capture how the meaning of words transforms into the meaning of sentences (Coecke, Sadrzadeh and Clark, 2010; Grefenstette and Sadrzadeh, 2011). The diagrammatic representation of these categories then explicitly shows how meaning 'flows' within sentences. Here is an example of such a 'meaning flow' taken from Coecke, Sadrzadeh and Clark 2010:

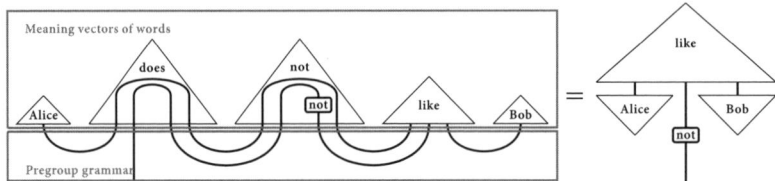

The verb 'like' receives the subject 'Alice' via the flow of meaning through 'does' and 'not', and also receives the subject 'Bob'. Then it produces the meaning of the sentence 'Alice does like Bob', which then is negated by the not-box.

The reader may verify that this particular explicit representation of meaning flow requires, at the grammatical level, the full structure of a Grishin pomonoid.

1.3.2 Discipline-transcending process structures

In fact, very similar pictures emerge when modeling information flows in quantum protocols, in work that inspired the above understanding of language meaning (Abramsky and Coecke, 2004; Coecke, 2010; Coecke, 2012):

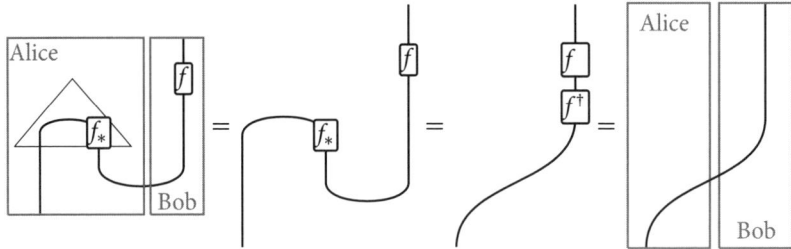

Here Alice and Bob share a Bell state (the 'cup'-shaped wire), Alice then performs an operation that depends on a discrete variable f, and also Bob does an operation that depends on f. The end-result is a perfect channel between Alice and Bob. This protocol is known as quantum teleportation (Bennett et al., 1993).

Another example is probabilistic Bayesian inference (Coecke and Spekkens, 2011):

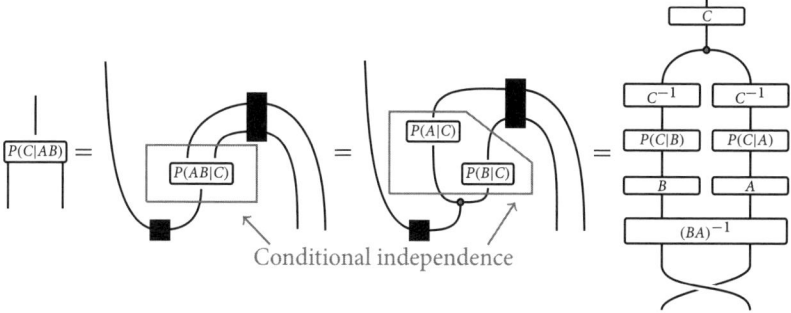
Conditional independence

Rather than just wires we now also have 'dots' connecting several wires. These structures are also highly relevant in quantum mechanical applications (e.g. see Coecke and Duncan 2011), and since recently, also in the linguistic applications, where they play the role of bases. They are also key to quantum algebra (Majid, 2002; Street, 2007) and topological quantum field theory (Kock, 2003; Turaev, 1994). The precise connections between these uses have yet to been fully explored.

1.4 Conclusion

We started by discussing ordering on things, to which we then adjoined composition of things, and by passing from existence of processes that take certain things to other things, to explicitly representing them, we ended with a structure that is most naturally represented, not by a syntactic, but by a diagrammatic language.

We gave examples of applications of these structures in a wide range of disciplines. The message that we have tried to pass to the reader is that these structures are very basic, in that they appear in a huge range of situations when taking a high-level perspective.

It would therefore be natural, before doing anything fancy, to give these structures a privileged status. Then, as a first next step, other well-understood structures could come in like, for example, so-called categorical enrichment (Kelly, 2005; Borceux and Stubbe, 2000). Evidently, it would be nice to take this even further, and see how far one could ultimately get by setting up this style of structural hierarchy, driven by high-levelness of actual phenomena.

Acknowledgements

The author enjoys support from the British Engineering and Physical Sciences Research Council, from the Foundational Questions Institute, and from the John Templeton Foundation. We thank Robin Cockett and Robert Seely for filling some holes in our background knowledge.

CHAPTER 2

Some Graphical Aspects of Frobenius Algebras

BERTFRIED FAUSER
(University of Birmingham)

2.1 Introduction

2.1.1 Scope of this chapter

Frobenius algebras surface at many places in mathematics and physics. Quite recently, using a convenient graphical notation, Frobenius algebras have been used to investigate foundational issues of quantum theory—references will be given below. Also, as shown elsewhere in this book, Frobenius algebras emerge in the semantic analysis of natural languages. The aim of this chapter is to present the basic results about Frobenius algebras, with a special emphasis on their relation to finite-dimensional Hopf algebras. Our major tool will be the consistent use of using graphical notation. Frobenius structures, which are related to ring theory, need to be considered too. Frobenius structures encode such notions as semisimplicity and separability of rings. We need occasionally to extend the graphical calculus to encode properties of underlying rings, but will not venture properly into 2-categorical notions. Moreover, no new results can be found in this chapter, and far from everything that is known about the subject is covered here. However, in passing we will give some pointers to the literature, which unfortunately is far too large to be considered completely. Note that references do not indicate an attribution, but unless otherwise stated we merely give the source we use.

2.1.2 Frobenius' problem

In the late 19th century Ferdinand August Frobenius (1849–1917) and his student Issai Schur studied the representations and characters of the symmetric groups S_n. Together with work by the English mathematicians, notably Alfred Young, this led to a breakthrough in finite group theory. In the early literature (Brauer and Nesbitt, 1937; Nesbitt, 1938; Littlewood, 1940), the group algebra $\mathbb{C}[G]$ of a finite group G was eponymously termed 'Frobenius algebra'. A finite group has finite order $|G|$ and one can form the free $\mathbb{k}$-vector space M over (the set underlying) G. The group structure then induces a left and right G action on the bimodule M from the algebra structure of $\mathbb{k}[G]$.

Definition 2.1 (**regular representations**) *Let G be a finite group, $x_i \in G$, with multiplication $x_i x_j = \sum_k f_{ij}^k x_k$, with $f_{ij}^k \in \{0,1\} \subset \mathbb{k}$, and multiplication table $[f_{ij}^k]$. We associate the following left $\mathbf{l}$/right $\mathbf{r}$ representations on ${}_A A / A_A$ and parastrophic matrix $\mathbf{p}_{(a)}$ to $A = \mathbb{k}[G]$.*

$\mathbf{l} : A \to \operatorname{End}_{\mathbb{k}}({}_A A) \quad \mathbf{l}_{x_i} \cong [f_i]_j^k = [f_{ij}^k]$ 'group matrix'
$\mathbf{r} : A \to \operatorname{End}_{\mathbb{k}}(A_A) \quad \mathbf{r}_{x_j} \cong [f_j]_i^k = [f_{ij}^k]$ 'antistrophic matrix'
$\mathbf{p}_{(a)} : A \otimes A \to \mathbb{k} \quad \sum_k [f_{ij}^k] a_k = [(P_{(a)})_{ij}]$ 'parastrophic matrix' $(a_k \in \mathbb{k})$

It is easy to see that $\mathbf{l}, \mathbf{r}$ are, in general reducible, representations induced by left and right multiplication. The parastrophic matrix is not a representation but is related to a linear form on A. It contains the following important information, solving Frobenius' problem of determining when $\mathbf{l}$ and $\mathbf{r}$ are equivalent:

Theorem 2.2 (Frobenius, 1903) *If there exist $a_k \in \mathbb{k}$ such that the parastrophic matrix $[P_{(a)}]$ is invertible, then the left and right regular representations are isomorphic ${}_A A \cong A_A$.*

Extending to algebras, we have:

Definition 2.3 *An algebra A is called Frobenius iff left and right regular representations are isomorphic.*

Example 2.4 The reader may check that the commutative polynomial rings $\mathbb{k}[X,Y]/\langle X^2, Y^2 \rangle$ and $\mathbb{k}[X]/\langle X^2+1 \rangle$ are Frobenius, while $\mathbb{k}[X,Y]/\langle X^2, XY^2, Y^3 \rangle$ is not Frobenius. Further examples for Frobenius algebras are the matrix algebras $A = M_n(\mathbb{k})$, where $\mathbb{k}$ is a division ring. In particular for G a finite group $A = \mathbb{C}[G]$ is Frobenius.

2.1.3 Finite-dimensional Hopf algebras

Studying the topology of group manifolds, Heinz Hopf (1894–1971) introduced in 1941 the concept of an *Umkehrabbildung*; that is, a comultiplication. The history of Hopf algebras is sketched in Cartier (2007). We only note that in the old days the term 'Hopf algebra' was what is now called 'bialgebra'. Coalgebra is the categorical dual notion to algebra; that is, we have a vector space C and two structure maps $\Delta : C \to C \otimes C$, an associative comultiplication, and $\epsilon : C \to I$ a counit, fulfilling the axioms obtained by 'reversing arrows' in the respective diagrams for an algebra (2.1). It is convenient to introduce the Heyneman–Sweedler (Heyneman and Sweedler, 1969; Heyneman and Sweedler, 1970) index notation for comultiplications. On an element $c \in C$ one sets $\Delta(c) = c_{(1)} \otimes c_{(2)} := \sum_i c_{1i} \otimes c_{2i}$.

Definition 2.5 *A finite-dimensional Hopf algebra H is the sextuple $(H, m, \eta, \Delta, \epsilon, S)$, where H is a finite-dimensional vector space, m, η are algebra multiplication and unit, Δ, ϵ are comultiplication and counit, and S is the antipode, defined as convolutional inverse of the identity $m(S \otimes \mathrm{Id})\Delta = \eta\epsilon = m(\mathrm{Id} \otimes S)\Delta$, fulfilling the compatibility condition: $\Delta(ab) = \Delta(a)\Delta(b)$, see diagram (2.51).*

The compatibility relation can be read as 'the comultiplication is an algebra homomorphism' (and vice versa). A bialgebra is the above structure without the antipode map. Any graded connected bialgebra is actually a Hopf algebra. We will see below that a Frobenius algebra can be described in a similar way using a comultiplication. The Frobenius compatibility law is different, saying that 'the comultiplication respects the module structure' (and vice versa). All this will be more obvious when we have the graphical notation available.

2.2 Graphical calculus

2.2.1 History and informal introduction

We work in a (strict) symmetric monoidal category $\mathcal{C}$, with a tensor as monoidal structure (Majid, 1995a; Balakov and Kirilov Jr., 2001; Street, 2007). Mainly we are interested in the case of **finVect**$_{\Bbbk}$, finite-dimensional $\Bbbk$-vector spaces, categories of finite-dimensional representations $\mathcal{R}$, or categories of projective finitely generated left(/right) modules $_R\mathcal{M}$ over a (not necessarily commutative unital) ring R. Category theory (Mac Lane, 1971) comes with the diagrammar of commutative diagrams (CDs), where objects are represented as vertices and morphisms as arrows (directed edges) between them. This is one way to define categories, see Lambek and Scott (1986). Graphical calculus was informally used for a long time, e.g. Brauer (1937). Usually its origin is attributed to Roger Penrose's seminal paper of 1971. The formal statement that graphical calculus, also called string diagrams, is a sound transformation of category theory is given in papers by Joyal and Street (1988, 1991a). A main thrust for developing graphical techniques came from low-dimensional topology; that is, knot theory (Kauffman, 1991; Turaev, 1994; Kassel, 1995; Ohtsuki, 2002) and topological quantum field theory (TQFT), e.g. Atiyah (1989); Kock (2003). A survey and further literature is in Selinger (2011). Graphical calculus is in some sense a (Poincaré) dual picture to commutative diagrams, where morphisms are depicted as labelled vertices (depicted also by boxes called *coupons*) and objects label the edges connecting them. Such a diagram is called a *tangle*. It is a representative of an isotopy class of equivalent such diagrams. Every (unoriented) cycle in a commutative diagram gives rise to a tangle equation, which establishes a rewriting rule also called a *move*. For example, the unit law for an algebra A in the monoidal category $\mathcal{C}$ has a CD with two triangles, and an equivalent description by two tangle equations

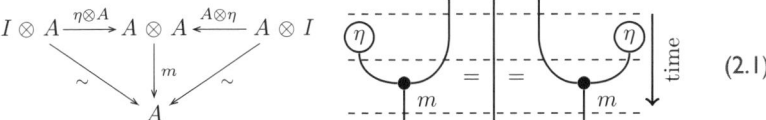
(2.1)

Here we have used A in the CD to denote the identity map 1_A, and we have dropped the edge label A in the tangles. The tangle is sliced by horizontal dotted lines so that in each slice only one non-identity operation is performed (Morse decomposition). At this time we also need to make clear that we read tangles downward using the *pessimistic arrow of time* (Oziewicz, talk at ICCA5, Ixtapa, 1999). Also if (non-trivial) crossings occur, we use the *left-handed* crossings, see (2.38). The reader needs to exercise caution when comparing tangle diagrams as some people are right-handed optimists. In that case (2.1) read upwards would describe the counit (relabel: $m \mapsto \Delta$, $\eta \mapsto \epsilon$). We find that reversing arrows in a CD and relabelling them is equivalent to changing the reading order of the tangle.

Graphical calculus can be interpreted as a 'language' built out of basic *letters*, which form *words* by horizontally (tensoring) or vertically (composition of morphisms) composing them to form larger tangles. The moves identify different such words into equivalence classes. Sometimes it is convenient to introduce special tangles replacing the coupons depicting them for simplicity and clarity. A selection of the graphical entities we are going to use is given as follows:

(2.2)

We may drop identity morphisms and their coupons. We may use a node to depict an algebra multiplication (or comultiplication in the inverted reading of the tangle). We also use the cross for the invertible involutive switch on tensors, and we drop usually the tensor signs on input and output labels, or even the labels if they are clear from the context. Creating and deleting elements depicts maps $a : I \to A$ or $f : A \to I$. Scalars (ring elements) do not have a graphical counterpart (void diagram) or are depicted by a tangle with no input and output lines (closed graph).

A braiding will be depicted by keeping over or under information as usually done in knot theory. This reads as follows:

$$\begin{array}{c}a\\\\A\end{array}\;;\;\begin{array}{c}A\\\\f\\\end{array}\;;\;\begin{array}{c}B\quad A\\\boxed{R}\\A\quad B\end{array}\;=\;\begin{array}{c}B\quad A\\\diagup\!\!\!\diagdown\\A\quad B\end{array}R\;;\;\begin{array}{c}B\quad A\\\boxed{R^{-1}}\\A\quad B\end{array}\;=\;\begin{array}{c}B\quad A\\\diagdown\!\!\!\diagup\\A\quad B\end{array}R^{-1}\quad(2.3)$$

As in (2.1) we sometimes use labelled circles and not triangles to depict creation and deletion of elements. If we want to distinguish a module A and its duals A^* (or $^*A$) we need either labels or *oriented tangles*. We use downward-oriented lines for A and upward-oriented lines for A^* (and $^*A$).

2.2.2 Basic rules of graphical calculus

We do note have space to formally introduce graphical calculus in full detail, so we restrict ourselves to presenting the basic facts on how to manipulate tangles.

2.2.3 Horizontal and vertical composition, sliding

We work in a rigid symmetric monoidal category $\mathcal{C}$, with a 'tensor' bifunctor $\otimes$ as monoidal structure. We have two types of composition of morphisms. The (partial) composition of morphisms in the category is called *vertical composition*, and it is depicted as 'stacking' (compatible) coupons of morphisms. We allow rewrites of coupons to be merged at their vertical boundary.

$$A \xrightarrow{f} B \xrightarrow{g} C \qquad\qquad \begin{array}{c}A\\\boxed{f}\\B\\\boxed{g}\\C\end{array} = \begin{array}{c}A\\\boxed{g\circ f}\\C\end{array} \qquad (2.4)$$

More delicate is the *horizontal composition* of morphisms on tensor products. We have

$$\begin{array}{c} & B\otimes C & \\ {}^{f\otimes C}\nearrow & & \searrow^{B\otimes g} \\ A\otimes C \xrightarrow{f\otimes g} & & B\otimes D \\ {}_{A\otimes g}\searrow & & \nearrow_{f\otimes D} \\ & A\otimes D & \end{array} \qquad \text{with identity maps denoted by } 1_A = A, \text{ etc.} \qquad (2.5)$$

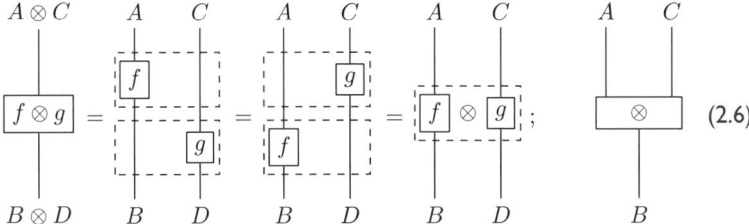

$$\tag{2.6}$$

From this we conclude that we are allowed to i) reduce boxes containing morphisms of the form $f \otimes g$ to two unconnected boxes f and g, and ii) that we are allowed to slide these boxes along each other. Note that we drop isomorphisms as shown in the rightmost tangle in (2.6). This generalizes in an obvious way to n inputs and m outputs of the tangles.

Warning: if we work in a braided monoidal category, these moves are no longer available, but need to be altered. Let $R : B \otimes A \to A \otimes B$ be a braiding, acting on elements as $R(b \otimes a) = a_R \otimes b_R = a_r \otimes b_r$ using a different Heyneman–Sweedler notation for 2-2-morphisms. The ordinary algebra structure of a symmetric monoidal tensor product on $A \otimes B$ is defined as $(a \otimes b)(c \otimes d) = ac \otimes bd$ or $m_{A \otimes B} = (m_A \otimes m_B)(A \otimes \sigma_{A,B} \otimes B)$ where σ is the switch map on tensors. Now, let $A \#_R B = A \otimes B$ as a $\Bbbk$-module, and define a new twisted multiplication, the smash product (recall that $A = 1_A$ etc.):

$$m_{A\#_R B} = (m_A \otimes m_B)(A \otimes R_{A,B} \otimes B) \qquad (a\#b)(c\#d) = ac_R \# b_R d. \tag{2.7}$$

Then the following holds.

Theorem 2.6 (Caenepeel, Militaru and Zhu, 2002, p. 50) *The triple (A, B, R) is a smash product structure if and only if*

$$R(b \otimes A) = A \otimes b \qquad R(B \otimes a) = a \otimes B$$
$$R(bd \otimes a) = a_{Rr} \otimes b_r d_R \qquad R(b \otimes ac) = a_R c_r \otimes b_{Rr} \tag{2.8}$$

for all $a, c \in A$ and $b, d \in B$.

The first two identities follow from the unit law, while the second two relate to (strict) associativity. In this case, we get $(f \otimes g) = (f \otimes 1)(1 \otimes g)$ but the other decomposition is not direct $((1 \otimes f)(g \otimes 1) = g_R \otimes f_r)$. Hence 'sliding' fails to be true and needs modification. More generally speaking the smash product is related to the question if $X \cong A \otimes B$ factorizes as an algebra, see *loc. cit.*. There exists a bijective correspondence between algebra structures on $A \otimes B$ such that the injections ι_A, ι_B are algebra maps and smash product structures (A, B, R). In what follows we mostly use the graphical language for the symmetric monoidal case only, hence assume the sliding move holds.

2.2.4 Closed structure, isotopy

A tangle is just a representative of an equivalence class of diagrams. The equivalence is isotopy of tangles, which is used in knot theory. We call a tangle an n-m tangle, or tangle with arity (n, m), if it has n input lines (starting at a discrete set of points on a top horizontal line) and m output lines (ending at a discrete set of points on a bottom horizontal line), e.g. a multiplication is a 2–1 tangle, a comultiplication is a 1–2 tangle and scalars will be denoted by 0–0 tangles or even by a void tangle. A *critical point* in a tangle is a point on the plane that has a vertical tangent. We are allowed to smoothly 'bend' lines in a tangle, provided that we keep their topology and do not introduce or destroy critical points or crossings of lines. Moves introducing further freedom to modify tangles impose additional conditions on the underlying category.

In what follows, we need closed structures on the underlying monoidal category. These come in a left and right version. In the case where we have a symmetry $\sigma : X \otimes Y \to Y \otimes X$ or a braid, then any two of left duality, right duality, and braiding defines the third. For an in-depth discussion see Kassel (1995).

Definition 2.7 *A monoidal category C is rigid, if for all $X \in C$ there exist $X^*, {}^*X$ such that the following universal morphisms exist:*

- *right duality (often denoted also b_X, d_X):*

$$ev_X : X^* \otimes X \to I_X, \quad cev_X : I_X \to X \otimes X^*,$$
$$\text{satisfying} \quad (I_X \otimes cev_X)(ev_X \otimes I_X) = I_X$$
$$(ev_X \otimes I_{X^*})(I_{X^*} \otimes cev_X) = I_{X^*}$$
(2.9)

- *left duality (often denoted also $\tilde{b}_X$, $\tilde{d}_X$):*

$$_X ev : X \otimes X^* \to I_X, \quad _X cev : I_X \to X^* \otimes X,$$
$$\text{satisfying} \quad (I_{{}^*X} \otimes {}_X cev)({}_X ev \otimes I_{{}^*X}) = I_{{}^*X}$$
$$({}_X ev \otimes I_X)(I_X \otimes {}_X cev) = I_X$$
(2.10)

- *symmetry (braiding) $\sigma_{X,Y} : X \otimes Y \to Y \otimes X$, satisfying the braid equation and invertibility*

$$(\sigma \otimes I)(I \otimes \sigma)(\sigma \otimes I) = (\sigma \otimes I)(\sigma \otimes I)(I \otimes \sigma); \qquad \sigma\sigma^{-1} = I \otimes I$$
(2.11)

The last equation as tangles represent the Reidemeister 3 move *R3* (2.14) and Reidemeister 2 move *R2* (2.15).

The conditions on ev, cev are depicted as *topological move* (or Reidemeister 0 move *R0*). The topological move allows deletion or introduction of two compatible extrema. This move is also ambiguously addressed as 'yanking', but we will see below that Frobenius bilinear forms also allow a 'yanking' of lines, so we reject this term.

We have introduced oriented lines to depict objects X, X^*, and $^*X$. If the distinction between left and right dual is vital, we need to apply labels. The topological move $R0_r$ for the right duality (the left dual tangles are obtained by inverting orientation) depicts the conditions in (2.9) (and, with inverted orientation, that of (2.10)).

$$R0_r: \quad \diagup\!\!\!\!\diagdown = | \quad ; \quad \diagdown\!\!\!\!\diagup = | \qquad (2.12)$$

The Reidemeister 2 move **R2** depends on what is encoded by the lines in a tangle, see Section 2.2.5. If the lines are assumed to be one-dimensional 'strings' (sic the name of the calculus), then straightening a loop introducing a twist θ in a string does not matter. In this case the Reidemeister 2 move is given by (graphically as lhs of (2.15)):

$$\theta := (I \otimes \text{ev})(\sigma \otimes I)(I \otimes \text{cev}) = I \qquad (2.13)$$

If we assume that lines in a tangle are more complicated objects, e.g. ribbons or cylinders (in TQFT, see Section 2.2.5), then we need to keep track of the twists. In this case $\theta \neq I$ and the Reidemeister 2 move needs another loop θ^{-1} with an inverse braid to compensate.

To summarize, we have the following isotopy moves on tangles

$$\text{Reidemeister 1} \quad ; \quad \text{Reidemeister 3} \qquad (2.14)$$

where the braiding is trivial for the switch map σ. The two alternative Reidemeister 2 moves (with and without a twist/braiding) read

$$\boxed{\theta} := \quad ; \quad \text{Reidemeister 2} \quad \text{Reidemeister 2'} \qquad (2.15)$$

In the symmetric monoidal case the twist morphism θ_A can be seen roughly as the composition $A \to {}^*A \to (^*A)^*$. If $^*A \cong A^*$ then θ_A is the canonical identification $A \cong (A^*)^*$, explaining why loops can be undone. The modified Reidemeister 2' move equals identity using the two identifications $A \cong (A^*)^*$ and $A \cong {}^*(^*A)$, explaining why two loops are necessary, and also the braiding, of course.

Moreover, we can move lines above and below extrema (evaluation and coevaluations). As an example look at

$$\text{[diagram]} = \text{[diagram]} = \text{[diagram]} \quad ; \text{ with } \quad \text{[diagram]} := \text{[diagram]} \qquad (2.16)$$

The other cases are similar.

To relate right and left duality, we need the braiding. If $\sigma^2 = I \otimes I$ is the ordinary switch, we get a symmetric monoidal category and can drop the left/right distinction, as they are related by the identity morphism, see left tangle equation in (2.17). If σ is a proper braid (2.11), then we can derive the left duality from the right duality. This is shown in the following equations for the 'cup' tangles ev

$$\underset{\text{if symmetric}}{\bigcup = \text{[diagram]}} \; ; \quad \bigcup = \text{[diagram]} = \text{[diagram]} = \text{[diagram]} \qquad (2.17)$$

The 'cap' tangles cev are related in a similar fashion. There are some subtleties in the interplay between dualities and twists, which we do not contemplate here further; see Joyal and Street (1991b), Turaev (1994), Kassel (1995), and Street (2007).

In a symmetric monoidal category we can assume **R0, R1, R2, R3** with a trivial twist. As we also get $^*A \cong A^*$ we can drop orientation too. This is called 'ambient isotopy'. If we work in a ribbon category we replace **R2** by its modified version **R2'** and keep track of twist and braid morphisms. This is called 'regular isotopy'; for terminology see Kauffman (1991). In what follows we will for simplicity work mainly in the symmetric monoidal setting.

2.2.5 Tangles not depicting strings

'String diagrams', related to symmetric monoidal categories, take their name from picturing them literally as strings, assumed to have zero radial extension. Such strings cannot be twisted (or twisting them is irrelevant). However, there are mathematical structures that are sensitive to twisting. They may be depicted, for example, by ribbons and lead accordingly to 'ribbon categories'. This is a reason to speak instead about 'tangle diagrams'. However, there are still more thickenings of tangles, such as cylinders in topological quantum field theory (TQFT). If we still use 'strings' to depict them, we are forced to adjust the allowed moves, for example to change the Reidemeister 2 move. This leads to different notions of isotopy; see Kauffman (1991). Here we just depict ribbons and the relevant tangles for TQFT for reference, but in the sequel we will just use strings. This reduces the artwork considerably and does not lose information if care is taken about using only allowed rewriting rules.

$$\text{[diagram]} = \text{[diagram]} \; ; \quad \text{[diagram]} \; ; \quad \text{[diagram]} \; ; \quad \text{[diagram]} \qquad (2.18)$$

The left equation shows how a loop, if straightened, produces a 2π twist in the ribbon. The three rightmost diagrams depict two-dimensional surfaces, cobordisms, which connect the (oriented, one-dimensional) circles at the top with (oriented) circles at the bottom. They are called the 'disk' (the unit, or counit if inverted), the 'trinion' (the Frobenius algebra product, or coproduct if inverted), and the 'cylinder' (identity map). We have depicted a twisted line on the cylinder, showing that one can also have a twist (Dehn twist) on cylinders. A TQFT based on these diagrams allows all two-dimensional Riemannian surfaces to be constructed and characterized; see Lawrence 1996.

2.3 Frobenius and Hopf algebras

In this section we recall some facts about, and most importantly some characterizations of, Frobenius algebras. We want to emphasize especially the difference between the multiplications in endormorphisms rings $\text{End}_R(M)$ over an R-module M and the left (right) action induced by an algebra multiplications $m_A : A \otimes A \to A$, where we look at the right (left) factor A as a left A-module $_AA$ (right A-module A_A). Our main sources are Yamagata (1996), Kadison (1999), Caenepeel, Militaru and Zhu (2002), Murray (2005), Lorenz (2011), and Lorenz and Fitzgerald Tokoly (2011). General texts on Hopf algebras and modules are Sweedler (1969), Abe (1980), Kasch (1982), Caenepeel (1998), and Street (2007).

Usually we work over a field, but we will recall here some more general notions where we also allow more general base rings R, examples being a residual field, a ring extension, or even a non-commutative ring.

2.3.1 Actions, coactions, representations, and two multiplications

To understand the similarities and differences between the Frobenius and endormorphism structures, we need to look briefly at algebra representations. We do that superficially only to the extent which is necessary for our purpose.

Let $_RM_S$ be a (p.f.) R, S-bimodule and $\mathcal{E}(M) = \text{End}(M) \cong M \otimes_S M^*$ be the endormorphism ring over M. Let $\{x_i\}_{i=1}^n$ be a set of generators for M and $\{f_i\}_{i=1}^n$ be a dual basis for M^*, i.e. $\text{ev}(f_i \otimes x_j) = f_i(x_j) = \delta_{ij}$. Given a non-degenerate bilinear form, $\beta : M \otimes M \to R$ with inverse $\overline{\beta} = \sum x_i \otimes y_i$ (see (2.45)). The bilinear form provides us with another set of generators $\{y_i\}_{i=1}^n$ for M, such that $\sum \beta(m, x_i) y_i = m$ for all $m \in M$. For simplicity we denote $\beta^* = \overline{\beta}$, as it is distinguished by its type signature; see (2.20). The next tangles describe a left A action on $_RM_S$, and how left-duality allows definition therefrom of a right A action on $_SM_R^*$. For the moment we use 2-tangles, where the area depicts the ring in question; for more details on such tangles see for example Luauda (2006) and Khovanov (2010). The rightmost tangle is a coaction for which similar results hold by tangle symmetry.

$$\underset{R M_S \quad sM_R^*}{\overset{A\ _R M_S \quad sM_R^*\ A}{\big| S\ ;\ S \big|}} := \underset{sM_R^* \quad A\ _R M_S}{\overset{sM_R^*\ A \quad _R M_S}{\big| S \quad ;\quad S \big|}} \qquad (2.19)$$

The Frobenius property induces an isomorphism between left modules $_AM$ and right modules M_A, which does not follow from duality alone. We can use the bilinear form β and its dual $\bar{\beta}$ to relate left and right actions.

$$\underset{\text{Frobenius}}{\cong} := \big|\beta\big|\big|\bar{\beta}\big|\ ;\qquad \begin{array}{c} M\otimes M \xrightleftharpoons[\bar{\beta}]{\beta} R \\ M^*\otimes M^* \xrightleftharpoons[\beta]{\bar{\beta}} R \end{array} \qquad (2.20)$$

We will study the properties of this isomorphism below in more detail.

A (finite) algebra A can be represented by a map into an endomorphism ring $\text{End}(M) \cong M \otimes M^*$. The algebra product is mapped homomorphically onto the natural product of endomorphisms given by the universal evaluation map; that is, composition of endomorphisms. For our purpose we use maps h, y, see (2.21), such that $y \circ h = 1_A$; that is, we use faithful representations. In the light of Wedderburn's theorem we may even assume, for simplicity, that $A \cong \text{End}(M)$, hence assuming A is simple such that $h \circ y = 1_{\text{End}(M)}$. Now we can look 'inside' the multiplication in A obtaining the left isomorphism in the next display.

$$\cong \quad \cong \quad ;\qquad \begin{array}{c} A \xrightleftharpoons[y]{h} M\otimes M^* \\ A \xrightleftharpoons[\sqcup]{\sqcap} M\otimes M \end{array} \qquad (2.21)$$

The second isomorphism is more subtle, as it involves the bilinear form. This multiplication is called β-multiplication and operates on $M \otimes M$. The choice of β has to be compatible with the morphisms $\sqcup, \sqcap$ in (2.21). With $a, b \in A$ such that $a = \sum a_{ij} x_i \otimes y_j$, $b = \sum b_{ij} u_i \otimes v_j$ one obtains the multiplication

$$ab = \sum a_{ij} x_i \otimes y_j \sum b_{ij} u_i \otimes v_j = \sum a_{ij} b_{lm} \beta(y_j, u_l) x_i \otimes v_m \qquad (2.22)$$

We remark here only that the information flow in the endomorphism ring situation (having an upwards/back in time flow) is different to β-multiplication (related to a Frobenius algebra), which has only downward information flow. This difference allows one in quantum teleportation to *choose* an entangled Bell state (related to $\bar{\beta}$) and make different Bell measurements (related to β up to unitary transformations), while the endomorphic situation does not allow one this freedom. This may have some implications in linguistic models of meaning, see Subsection 2.3.10.

2.3.2 Some notions from ring and module theory

All modules we are going to use are finitely generated projective (f.p.) over a base field or ring. This is implied by the invertibility of the Frobenius bilinear form (parastrophic matrix) hinging on a good duality theory. This enables one to dualize algebra structures providing a coalgebra structure, which fails in the general situation. Let A be an R-algebra with structure maps m_A, η_A. We denote by A^{op} the opposite algebra over the same R-module A, with the opposite multiplication $m_A^{op} = m_A \circ \sigma$. It is useful to introduce the *enveloping algebra* $A^e = A \otimes A^{op}$, which allows one to rewrite A, A-bimodules ${}_A M_A$ as A^e-left modules.

A *derivation* $D : A \to M$ is a linear operator from the A, A-bimodule A to the A, A-bimodule M, such that

$$D(ab) = D(a).b + a.D(b) \qquad (2.23)$$

where the module M is represented by a bold line. The bold–unbold 'multiplication'-like tangle is the right/left action of A on M. Let $\mathrm{Der}_R(A, M)$ be the R-module of derivations. A derivation D_m is called *inner derivation* if there exists an $m \in M$ such that $D_m(a) = am - ma$. Now define the space of A-invariants of M as $M^A := \{m \in M \mid am = ma\}$. It is obvious that for all $m \in M^A$ the inner derivation vanishes, $D_m = 0$. If $M = A$ as R-modules, the space of invariants is just the kernel of the multiplication map $I(A) = \mathrm{Ker}(m_A)$. One finds the following sequence to be exact:

$$0 \to M^A \to M \to \mathrm{Der}_R(A, M) \qquad (2.24)$$

Using the isomorphisms between A, A-bimodules and A^e-left modules shows that $M^A \cong \mathrm{Hom}_{A^e}(A, M)$ and $M \cong \mathrm{Hom}_{A^e}(A^e, M)$. It is also easy to see that $m_A : A^e \to A$ is an epimorphism. Hence the following sequence is exact:

$$0 \to I(A) = \mathrm{Ker}(m_a) \to A \otimes A^{op} \to A \to 0 \qquad (2.25)$$

A situation that is important in the Frobenius case is when this sequence is split. That is, there exists a map $\delta : A \to I(A) :: a \mapsto \delta(a) = a \otimes 1 - 1 \otimes a$ whose image $I(A)$ in A is an ideal $AI(A) = I(A) = I(A)A$. Then $A^e = A \otimes A^{op}$ decomposes as a direct sum $A^e = I(A) \oplus A$ and there is an idempotent pair $(e, 1 - e)$ projecting onto the two spaces. This gives by standard algebra arguments some structure results.

Lemma 2.8

$$\mathrm{Hom}_{A^e}(I(A), M) \cong \mathrm{Der}_R(A, M) \qquad (2.26)$$

Applying the functor $\mathrm{Hom}_{A^e}(-, A)$ to the exact sequence (2.25) shows that $HH^1(A, M) = \mathrm{Ext}^1_{A^e}(A, M) \cong \mathrm{Der}_R(A, M)/\mathrm{InnDer}_R(A, M)$, where HH^1 is the first Hochschild cohomology group. As an aside, having a Hopf algebra structure allows one to formalize several cohomology theories in a uniform manner; see Sweedler (1968). The graphical calculus is not (very) sensitive to the underlying ring structure, so we do not go deeper into ring theory here. The main result we quote establishes the existence of a splitting idempotent for the class of finite projective algebras we are interested in.

Theorem 2.9 *Let R be a commutative ring. For R-algebras A the following statements are equivalent.*

  (i) *A is projective as a left A^e-module.*
  (ii) *The exact sequence (2.25) for A^e-modules is split.*
  (iii) *There exists a splitting idempotent element $e = \sum e_{(1)} \otimes e_{(2)} \in A \otimes A$ such that for all $a \in A$, $ae = ea$ and $\sum e_{(1)} e_{(2)} = 1$ holds.*

An R-algebra A is *separable* iff A/R is a separable ring extension. That is, $m : A^e \to A$ is a split epimorphism of A^e-modules. By the above theorem this is equivalent to saying that A is A^e-projective or that there exists a splitting idempotent e as in (iii). For further details see Kadison (1999, Section 5.2). The conditions in (iii) translate into the following graphical statements ($\eta : I \to A$ is the unit map):

$$\underset{\sum e_{(1)} \otimes e_{(2)}}{\cap} \; : \; \cup\cap = \cap\cup =: \cap \; ; \; \bigcirc = \overset{\eta}{|} \qquad (2.27)$$

We defined the comultiplication map $\delta : A \to A \otimes A :: a \mapsto ae = ea$. Coassociativity follows from the symmetric definition of δ and from associativity of the product in A and the 'sliding' of morphisms. Try it! For more information about splitting idempotents and quadratic algebras see Hahn (1994) and Caenepeel, Militaru and Zhu (2002), as we want to avoid to discussing Azumaya and Taylor–Azumaya algebras. We will use generators and bases so we quote two more standard results from algebra, guaranteeing the existence of bases (generators).

Theorem 2.10 *Any projective separable algebra A over a commutative ring R is finitely generated. A separable algebra A over a field $\Bbbk$ is semisimple.*

Using this theorem, we find a finitely generated projective R-module M, with generators $\{x_i\}_1^n$ and a dual module M^* with dual basis $\{f_i\}_1^n$ such that $A \cong M \otimes_R M^*$. The Frobenius homomorphism will allow us to replace the dual module A^* by A and the dual basis by a *reciprocal basis* $\{y_i\}_1^n$.

2.3.3 Frobenius functors

Let A, S be rings and let $_A\mathcal{M}$, $_S\mathcal{M}$ be the categories of (f.p.) A-modules and S-modules. Let $i : S \to A$ be an injection, then any A-module can be turned into an S module. This defines the restriction functor R in the opposite direction

$$\text{R} : {}_A\mathcal{M} \to {}_S\mathcal{M} :: {}_A M \mapsto {}_S M :: sm = i(s)m \tag{2.28}$$

This is an instance of a change of base functor. Now R has a left adjoint T ⊣ R (induction functor) and a right adjoint R ⊣ H (coinduction functor) defined as follows:

$$\text{T} : {}_S\mathcal{M} \to {}_A\mathcal{M} :: \begin{cases} {}_S M \mapsto A \otimes_S {}_S M \\ f \mapsto A \otimes_S f \end{cases}$$

$$\text{H} : {}_S\mathcal{M} \to {}_A\mathcal{M} :: \begin{cases} {}_S M \mapsto \text{Hom}_S(A, {}_S M) \ni h \\ f \mapsto f \circ h \end{cases} \tag{2.29}$$

This means one has $\text{Hom}_A(\text{T}(M), N) \cong \text{Hom}_S(M, \text{R}(N))$ and $\text{Hom}_S(\text{R}(N), M) \cong \text{Hom}(N, \text{H}(M))$. The Frobenius property is captured by the following:

Definition 2.11 *A ring extension A/S is a Frobenius extension if and only if* H *and* T *are naturally equivalent as functors from* $_S\mathcal{M} \to {}_A\mathcal{M}$.

We get a Frobenius structure (Kadison, 1999; Caenepeel, Militaru and Zhu, 2002; Khovanov, 2010); that is, a triple $(\beta, \{x_i\}, \{y_i\})$ where $\lambda \in \text{Hom}_{S-S}(A, S)$ is the Frobenius homomorphism with $\beta(a, b) = \lambda(ab)$ (inverse $\bar{\beta} : S \to A \otimes A$), and $\{x_i\}, \{y_i\}$ are generators of A fulfilling the β-multiplication equations (see tangles (2.45) and (2.46)).

$$\sum_i x_i \beta(y_i, a) = a, \qquad \sum_i \beta(a, x_i) y_i = a \tag{2.30}$$

This is a generalization of the theorem 2.13, valid for non-commutative rings, with many interesting applications; see for example Khovanov (2006). Pairs of functors (T, H) with T ⊣ R ⊣ H such that T $\cong$ H is an isomorphism are called Frobenius pairs of Frobenius functors.

2.3.4 Graphical characterization of Frobenius algebras

Let $_R\mathcal{M}$ be a tensor category of (f.p.) R-modules and $A \in {}_R\mathcal{M}$. Let A be an R-algebra with structure maps (μ_A, η_A). If A is Frobenius then further structure maps exist, such as the Frobenius homomorphism $\lambda : A \to R$ or, equivalently, an associative bilinear form (see Section 2.3.7) $\beta = \lambda \circ \mu_A : A \otimes A \to R$. However, it is graphically more effective to use the dual Λ of the Frobenius homomorphism and the splitting idempotent element e to define a coalgebra structure $\delta : A \to A \otimes A$, as in (2.27). We define the coproduct as follows:

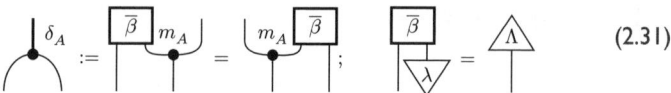 (2.31)

It is easy to show graphically that Λ is a unit for m_A and λ is a counit for δ_A. Hence we arrive at the following:

Definition 2.12 *A Frobenius algebra A is a quintuple $(A, m_A, \lambda, \delta_A, \Lambda)$ such that λ is a counit for δ_A, Λ is a unit for m_A, and the multiplication and comultiplication fulfill the compatibility law*

$$\quad (2.32)$$

together with the units and 'yanking' rules given in (2.31) and their duals.

If we interpret multiplication as an A action of the (left/right) A-module A, and comultiplication similarly as (left/right) coaction, then this compatibility relation reads '(left/right) actions and coactions commute'.

2.3.5 Algebraic characterizations of Frobenius algebras

Frobenius algebras can be characterized in a number of ways, emphasizing different aspects of this structure.

Theorem 2.13 (Caenepeel, Militaru and Zhu, 2002) *Let A be an n-dimensional $\Bbbk$-algebra. The following statements are equivalent:*

1. *A is Frobenius.*
2. *There exists a Frobenius isomorphism $\beta^r \in \mathrm{Hom}_{\Bbbk}({}_AA, {}_AA^*)$ for left A-modules in ${}_A\mathcal{M}$.*
3. *There exists a Frobenius isomorphism $\beta^l \in \mathrm{Hom}_{\Bbbk}(A_A, A^*_A)$ for right A-modules in $\mathcal{M}_A$.*
4. *The left regular representation $\mathbf{l}$ and right regular representation $\mathbf{r}$ are equivalent.*
5. *There exists $a_i \in \Bbbk$ such that the parastrophic matrix (2.1) is invertible.*
6. *There exists a non-degenerate associative bilinear form $\beta : A \times A \to \Bbbk$ (associativity : $\beta(ab, c) = \beta(a, bc)$ for all $a, b, c \in A$).*
7. *There exists a hyperplane in A that does not contain a non-zero right ideal of A.*
8. *There exists a pair $(\lambda, \overline{\beta})$, called a Frobenius pair, where $\lambda \in A^*$ is a Frobenius homomorphism and $\overline{\beta} : \Bbbk \to A \otimes A$ $(\overline{\beta} = \Delta(1) = \sum \overline{\beta}_{(1)} \otimes \overline{\beta}_{(2)})$ such that for all $a \in A$*

$$a\overline{\beta} = \overline{\beta}a \qquad \sum \lambda(\overline{\beta}_{(1)})\overline{\beta}_{(2)} = \sum \overline{\beta}_{(1)}\lambda(\overline{\beta}_{(2)}) = 1 \quad (2.33)$$

where we used Heyneman–Sweedler notation for the comultiplication.

2.3.6 Some properties of Frobenius and Hopf algebras

Frobenius algebras share some similarities with Hopf algebras, but also exhibit different features. We will discuss the relation between Frobenius and finite Hopf algebras in Section 2.3.9. Moreover, see Chapters 4 and 5 in this book.

In quantum information theory it is appropriate to distinguish two extremal cases.

Definition 2.14 *A Frobenius algebra A is called* special *or* trivially connected, *if the loop operator equals identity* $l = m_A \circ \delta_A = 1_A$ *(see (2.35)), or* connected *if* l *is invertible. A Frobenius algebra is called* totally disconnected *if the loop operator decomposes as* $l = m_A \circ \delta_A = \Lambda \circ \lambda$.

The special or connected case (left-hand side of eqn. (2.34)) shows that by 'yanking' one generates a tensor state that cannot be factored, while the right-hand side shows that the totally disconnected case produces a product state.

$$ \tag{2.34} $$

In Coecke, Pavlovic and Vicary (2011) and Coecke, Paquette and Perdrix (2008) it is shown that special Frobenius algebras are in one-to-one correspondence with a choice of an orthonormal basis. Moreover, one can characterize classical structures and complementarity on q-bits using special Frobenius algebras (Coecke and Duncan, 2011).

$$ \tag{2.35} $$

Let $l = m \circ \delta$ be a 'loop'. One has the following normal form (or spider) theorem:

Theorem 2.15 *Let A be a symmetric Frobenius algebra; that is,* $m_A^{op} = m_A \circ \sigma = m_A$. *Then any tangle 'foo' with arity* (n, m) *(n inputs, m-outputs) can be transformed using Frobenius moves and associativity to the normal form* $\delta^{m-1} \circ l^r \circ m^{n-1}$ *(with* $m^0 = l = \delta^0 = \text{Id}$) *for some non-negative integer r, see (2.35). If A is special* $l = 1_A$.

This theorem is proven by recursion. The Frobenius property (2.32) and associativity for m_A and δ_a allows us to interchange the order of multiplications and comultiplications. In this way the inputs can be multiplied together and the outputs produced by comultiplications. In this process a certain number of 'loops' occur, which vanish if A is special. Composing an n, m-tangle with k loops in its normal form with an m, p-tangle with l loops produces a n, p-tangle with $k + l + m$ loops. If A is not symmetric one needs to deal with the Nakayama automorphisms α, see Section 2.3.8. We note here that α can be constructed using the bilinear forms $\beta, \bar{\beta}$ and the trivial symmetry σ (switch) as in (2.35) along the lines on which we constructed the twist θ in (2.15) from left/right dualities and the braiding.

Theorem 2.16 *A special symmetric Frobenius algebra A is a bialgebra.*

$$\tag{2.36}$$

Another interesting property, which can be used for example in singular value decomposition (Fauser, 2006), is the following fact, true for any convolution algebra $\mathrm{Hom}(A, A)$, hence for Hopf and Frobenius.

Theorem 2.17 *The operators 'loop' $l = m_A \circ \delta_A$ of arity $(1, 1)$ and 'fork' $f = \delta_A \circ m_A$ of arity $(2, 2)$, see (2.35), fulfill the same minimal polynomial, hence have the same positive spectrum up to a null-space.*

In the case of a special Frobenius algebra we see that $l = 1_A$ and hence f is a projector onto a space isomorphic to A in $A \otimes A$.

Kuperberg ladders (Kuperberg, 1991; Fauser, 2002) are the counterparts in a Hopf algebra to the leftmost/rightmost tangles in (2.32). A Hopf algebra comes with an antipode S ((2.5) and Section 2.3.9) causing the ladder tangle to be invertible (left-hand side of (2.37)). These tangles play a role in invariant theory of 3-manifolds.

$$\tag{2.37}$$

The rightmost equation in (2.37) shows a Frobenius algebra homomorphism H : $A \to B$ such that $m_B \circ (\mathrm{H} \otimes \mathrm{H}) = \mathrm{H} \circ m_A$. In red/green-calculus the map in use is the Hadamard gate, which is invertible, allowing a 'color change'; that is, a change of algebra structure. As special Frobenius algebras encode bases, this is essentially an entangling operator changing the underlying classical structure. In the Hopf algebraic case Sweedler developed a powerful cohomology theory (Sweedler, 1968) with a Hopf algebra action, which provides a classification of such maps. Algebra homomorphisms fall into the trivial cohomology class.

As a last example in this subsection we consider a module A carrying a Hopf and a Frobenius algebra structure at the same time. To make this situation well behaved we demand the following *distributive laws* (also called the Laplace property (Rota and Stein, 1994)) to hold as compatibility relations. (White dots belong to the Hopf algebra, black to Frobenius.)

$$\text{[diagram]} \quad ; \quad \text{[diagram]} \tag{2.38}$$

In Fauser (2001) and Brouder *et al.* (2004) (see also Carroll, 2005), it was demonstrated how these 'Laplace Hopf algebras' produce, via twistings, all multiplicative structures in a perturbative quantum field theory (see also Fauser and Jarvis, 2006, and Fauser, 2008, for a fancy number theoretic application). In Fauser and Jarvis (2004), Fauser *et al.* (2006), and Fauser, Jarvis and King (2010), among others, it was demonstrated how this structure underlies invariant rings, providing powerful tools to shorten proofs and allowing solution of otherwise difficult problems. As a mental picture, the Hopf algebra operates on a tensor module as 'concatenation', while the Frobenius structure encapsulates the discrete permutation symmetries S_k on tensor terms $\sigma : \otimes^k A \to \otimes^k A$.

2.3.7 Associative regular bilinear forms and Frobenius homomorphisms

Frobenius algebras come with a second way, beside the closed structures, to identify f.p. modules with their duals. This is essentially using the parastrophic matrix from (2.1).

Definition 2.18 (Murray, 2005) *Let $\Bbbk$ be a residual field. A regular associative bilinear form is a $\Bbbk$-linear map $\beta \in \mathrm{Bil}^r_{\mathrm{ass}}(A, \Bbbk) : A \otimes_{\Bbbk} A \to \Bbbk$ such that*

$$\beta(ab, c) = \beta(a, bc) \qquad \text{associativity}$$
$$\forall a \in A \text{ with } a \neq 0, \exists b \in A \text{ such that } \beta(a, b) \neq 0 \qquad \text{non-degenerate} \tag{2.39}$$

The linear form $\lambda := \beta(-, 1) = \beta(1, -)$ is called Frobenius homomorphism. If $\lambda(ab) = \lambda(ba)$—that is the Nakayama automorphisms $\alpha = \mathrm{Id}$ (see Section 2.3.8)— it is called trace form; *see tangle in eqn (2.40)*.

Usually bilinear forms are depicted as cup-tangles. To avoid confusion with the evaluation maps from the closed structure we denote them as coupons.

$$\text{[diagram]} \tag{2.40}$$

Two bilinear forms are related by a *homothety*, $\beta \simeq \beta'$ if there exists a unit $k \in \Bbbk^{\times}$ and an automorphism $V \in \mathrm{Aut}_{\Bbbk}(A)$ such that

$$\beta(a, b) = k\beta'(Va, Vb) \tag{2.41}$$

Two bilinear forms related by a homothety are not essentially different and the set Bil^r_{ass} can be partitioned into homothety equivalence classes. The rightmost equation in (2.40) shows further that the Frobenius homomorphisms λ and $\lambda' = k\lambda V$ are related if V is an algebra homomorphism, as in the right-hand side of (2.37).

2.3.8 Nakayama automorphism

We call a bilinear form symmetric if $\forall a, b \in A \,.\, \beta(a, b) = \beta(b, a)$. Note that this does not imply that A is symmetric; in general $A \neq A^{op}$. The *Nakayama automorphism* $\alpha \in \text{Aut}_{k-alg}(A)$ measures the deviation from symmetry of β.

$$\beta(a, b) = \beta(b, \alpha(a)) \tag{2.42}$$

The Nakayama automorphism is unique up to inner automorphisms. First fix an isomorphism $\phi : \text{Hom}(_A A, _A A^*)$. Any other such isomorphism is given by first applying an automorphism to $_A A$ and then applying ϕ. This results in a transformation to a new Frobenius linear form $\lambda' = u\lambda : c \mapsto \lambda(ck)$ and a new bilinear form $\beta'(a, b) = \beta(a, bu)$, where $u \in A^\times$ is a unit in the algebra A. The corresponding Nakayama automorphism transforms as $\alpha' = I_u \circ \alpha$, where $I_u(a) = uau^{-1}$ is an inner automorphism.

The bilinear form β is symmetric iff $\alpha = \text{Id}$. Note that in a braided setting the ordinary switch can be addressed as a virtual crossing (Kauffman, 1999) not encoding over/under information.

$$\begin{array}{ccc} \text{symmetry} & \text{Nakayama automorphism} & \text{transposition} \end{array} \tag{2.43}$$

The non-degenerate bilinear form $\beta : A \otimes A \to \Bbbk$ induces the notion of an adjoint on $\text{End}(A)$ via $\beta(a, V^t b) := \beta(Va, b)$, which we call *transposition*. This transposition fulfills the usual properties such as linearity, $(UV)^t = V^t U^t$, and $(U^{-1})^t = (U^t)^{-1}$. In the presence of a non-trivial Nakayama automorphism transposition is in general *not* an involution.

$$\beta(a, V^{t^2} b) = \beta(V^t a, b) = \beta(\alpha^{-1} b, V^t a) = \beta(V\alpha^{-1} b, a) = \beta(a, \alpha V \alpha^{-1} b), \tag{2.44}$$

hence we get $V^{t^2} = \alpha V \alpha^{-1}$. If the Frobenius algebra is symmetric, then $\alpha = \text{Id}$ and transposition is an involution.

Lemma 2.19 (Murray, 2005) *Bilinear forms β and β' are homothetic iff $\exists k \in \Bbbk^\times$ and $V \in \text{Aut}_\Bbbk(A)$ such that $\rho_u = kV^t V \in \text{Aut}_\Bbbk(A)$.*

The reader may compare ρ_u to the definition of a positive operator in quantum mechanics. In the same line of thought, we notice the following. Let A be a Frobenius $\Bbbk$-algebra with bilinear form β and Nakayama automorphism α. As seen above, two homothetic forms $\beta'(a, b) = \beta(a, bu)$ are related by a unit $u \in A^\times$ with Nakayama's $\alpha' = I_u \circ \alpha$. This shows that the order of the Nakayama automorphism is independent of the choice of the form in the homothety equivalence class. This can be used to define the following norm function. Let $\alpha^n = 1$ be a Nakayama automorphism of finite order. Define the algebra norm $N_\alpha(a) := a\alpha(a)\ldots\alpha^{n-1}(a)$. This norm can be interpreted as the evaluation of a term t^n at a in the Ore ring of right twisted polynomials $A[t, \alpha]$ (see Lam and Leroy, 1988; Bueso, Gómes-Torrecillas and Verschoren, 2003; Abramov, Le and Li, 2005). If $\alpha^n = I_a$ then $\alpha(a) = a$ and one gets $(\alpha')^n = I_{N_\alpha(u)a}$. In the involutive case, important for quadratic algebras and Clifford algebras (Hahn, 1994), or more generally for $*$-algebras, one defines such norms using 'special elements' or directly using the involution. In quantum theory one usually assumes an involutive $*$-automorphism.

The closed structures allow us to relate morphisms f in $\mathrm{Hom}(A, B)$ to dualized morphisms f^* in $\mathrm{Hom}(B^*, A^*)$ etc. bending lines up or down; that is, using the topological move R0 (2.12) or 'yanking'. Having the Frobenius bilinear form β available, we have a second possibility to bend lines, which this time produces maps in $\mathrm{Hom}(A, B^*)$ etc. We need first to define the inverse map $\overline{\beta} : \Bbbk \to A \otimes A$, which exists due to non-degeneracy of β. We should write β_A and $\overline{\beta}_A$ for the components of the map on the category, but to unclutter the notation we drop these indices.

$$A \xrightarrow{A \otimes \overline{\beta}} A \otimes A \otimes A \xrightarrow{\beta \otimes A} A \qquad (2.45)$$

$\overline{\beta}$ is a right inverse. Using the Nakayama automorphism α we see that it is also a left inverse ($\overline{\alpha} = \alpha^{-1}$):

$$\qquad (2.46)$$

Here the crossings are 'virtual'; that is, the switch σ. The moves in (2.45) and (2.46) should be compared with the topological moves (2.12) for the cup/cap tangles of the closed structure. Here the left/right aspect is taken care of by the Nakayama automorphism.

A main characterization of a Frobenius algebra in Theorem 2.13, and one which generalizes, is given by the Frobenius isomorphism ${}_A A \cong {}_A A^*$ of the left A-modules. The Frobenius bilinear form and its dual allow us to construct such morphisms together with the closed structures (Fuchs, 2006; Fuchs and Stigner,

2008). We define left/right module maps $\beta^r \in \text{Hom}(A, A^*)$, $\beta^l \in \text{Hom}(A, {}^*A)$ and their inverses. The left/right aspect refers to the closed structures involved.

$$(2.47)$$

and

$$(2.48)$$

The proof that the identity holds in (2.47) and (2.48) requires *both* the inverse Frobenius bilinear form (2.46) and the topological move for right/left duality (2.12), and is left as an easy exercise.

We close this section about the Nakayama automorphism and its implications on 'yanking' moves by showing that the left/right duality imposed by the closed structures is related to the module homomorphisms $\beta^\bullet$ and $\overline{\beta}^\bullet$ (with $\bullet = r$ or l, Frobenius isomorphisms)

$$(2.49)$$

In Fuchs (2006) it is further graphically shown that the Nakayama automorphism $\alpha = \overline{\beta}^l \beta^r$ is actually an algebra automorphism. Compare this form of α with the form for α given in (2.35), which does not use the closed structure or the braid.

2.3.9 Finite Hopf algebras as Frobenius algebras

Hopf algebras are discussed at length in Chapters 4 and 5 of this book. We will provide here only the basic facts that relate them to Frobenius algebras. In defn 2.5 we saw that a Hopf algebra over H is at the same time an unital algebra (H, μ_H, η_H) and a counital coalgebra $(H, \Delta_H, \epsilon_H)$ that are compatible by the Hopf compatibility law, which includes the switch map σ. The multiplication μ_H extends to a multiplication $\mu_{H \otimes H}$.

$$\Delta_H \mu_H = \mu_{H\otimes H}(\Delta_H \otimes \Delta_H) \qquad \mu_{H\otimes H} := (\mu_H \otimes \mu_H)(\mathrm{Id} \otimes \sigma \otimes \mathrm{Id}) \qquad (2.50)$$

Hence Δ_H is an algebra morphism $(H, \mu_H) \to (H \otimes H, \mu_{H\otimes H})$. A Hopf algebra unifies the concept of a (Lie-)group and a (Lie-)algebra at the same time. An element $g \in H$ is called *group-like* if $\Delta(g) = g \otimes g$; that is, the diagonal action or 'copying'. An element $p \in H$ is called *primitive* (or algebra-like) if $\Delta(p) = p \otimes 1 + 1 \otimes p$. Primitive elements generate, for example, the (universal enveloping) Lie algebra of a Lie group seen as Hopf algebra. This analogy extends to the action of a Hopf algebra on a module M. Hence we have an H-action $H \otimes M \to M$ and an H-coaction $M \to H \otimes M$, which need to fulfill an analogue of the Hopf compatibility law. In graphical terms this reads, using white nodes for Hopf co/multiplications, as follows:

$$(2.51)$$

The multiplication map $\mu_{H\otimes H}$ is depicted as the dashed box, and modules receive bold lines. The graphical description makes it clear that one can interchange the role of multiplication and comultiplication and we see that μ_H is a morphism of coalgebras $(H \otimes H, \Delta_{H\otimes H})$ and (H, Δ_H).

$$\Delta_H \mu_H = (\mu_H \otimes \mu_H)\Delta_{H\otimes H} \qquad \Delta_{H\otimes H} := (\mathrm{Id} \otimes \sigma \otimes \mathrm{Id})(\Delta_H \otimes \Delta_H) \qquad (2.52)$$

just moving the dashed box in the tangle up. The compatibility law for a Hopf action on a left H-co/module M has a bold rightmost line in the left-hand side of (2.51).

The conditions under which a finite Hopf algebra over a commutative ring R is Frobenius were worked out by Larson and Sweedler (1969) for R, a principal ideal domain, and by Pareigis (1971) for general R with $\mathrm{Pic}[R] = 0$. (The abelian Picard group $\mathrm{Pic}[R]$ consists of the set of isoclasses $[X]$ of linebundles X over R (i.e. $X^* \otimes X \cong R$) with $\otimes$ as multiplication $[X] + [X'] = [X \otimes X']$ and $X \mapsto X^* = \mathrm{Hom}(X, R)$ as inverse.) These results have consequences for the existence and invertibility of the antipode, which in turn is relevant for Hopf algebra cohomology and invariants of 3-manifolds. So called integrals provide the main tool to prove these facts. Integrals allow one to construct Frobenius homomorphisms and equivalently the Frobenius bilinear form.

Let $\epsilon : H \to R$ be the augmentation map, which is an algebra homomorphism $\epsilon(ab) = \epsilon(a)\epsilon(b)$.

Definition 2.20 *A right (left) integral is a $\mu_r \in H$ ($\mu_l \in H$) satisfying for all $h \in H$ the relation $h\mu_r = \epsilon(h)\mu_r$ ($\mu_l h = \epsilon(h)\mu_l$). The space of all right (left) integrals is denoted as*

$$\int_H^r := \{\mu_r \in H \mid \forall h \in H. \, h\mu_r = \epsilon(h)\mu_r\} \tag{2.53}$$

$$(\int_H^l := \{\mu_l \in H \mid \forall h \in H. \, \mu_l = \epsilon(h)\mu_l\})$$

Graphically, integrals look like

$$\tag{2.54}$$

Let H be a finitely generated projective module over a commutative ring with $\text{Pic}[R] = 0$ underlying a Hopf algebra H. (The mild condition $\text{Pic}[R] = 0$ can be lifted by studying quasi-Frobenius rings, which we do not pursue here; see for example Nicholson and Yousif, 2003.) The crucial property we need to construct a Frobenius homomorphism is that $\int_H^r$ is a one-dimensional module over R

$$\int_H^l H \simeq H \simeq H \int_H^r \quad \text{with} \quad \int_H^l \cong R \cong \int_H^r \tag{2.55}$$

As $\int_H^r \cong {}_R R$ as an R-module is one dimensional, it is an invertible module. The following theorem, taken from Lorenz (2011), summarizes the work of Larson and Sweedler (1969), Pareigis (1971), and Oberst and Schneider (1973).

Theorem 2.21 *Let H be a finite projectively generated Hopf algebra over the commutative ring R. Then the following hold.*

- *The antipode S is bijective (has a linear inverse). This implies $\int_H^r = S(\int_H^l)$.*
- *H is a Frobenius R-algebra iff $\int_H^r \cong R$. This holds true if $\text{Pic}[R] = 0$. Moreover, if H is Frobenius, then the dual Hopf algebra H^* is Frobenius too.*
- *Let H be Frobenius. Then H is symmetric iff*
  (i) H is unimodular (i.e. $\int_H^r = \int_H^l$), and
  (ii) S^2 is an inner automorphism of H.

The existence of a one-dimensional R-module of right/left integrals entails the construction of the Frobenius isomorphism $\beta^\bullet : H \to H^*$ using the Hopf algebra structure on H. The following tangle diagrams explain how to construct the Frobenius homomorphism and Frobenius isomorphisms out of an unimodular integral $\Lambda = \mu_l = \mu_r$ and the Hopf algebra structure:

$$\tag{2.56}$$

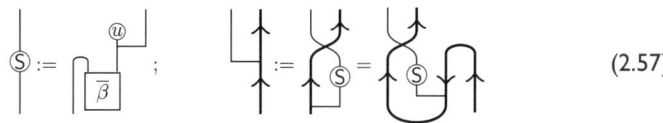

(2.57)

The first equation in (2.56) defines the (dual) bilinear form from the integral Λ. The second equation defines *right orthogonality* of the form (see Larson and Sweedler, 1969), and we need left orthogonality too. Orthogonality guarantees that the Frobenius isomorphism respects the module structure. The third equation defines the element u, which is needed to show the existence of the antipode in (2.57). Finally the right-hand side of equation (2.57) defines the left action of H on a dual module $_HM^*$ using the antipode and the right action on M_H^*, which by duality (2.19) comes from the left action $_HM$, establishing an Frobenius isomorphism.

2.3.10 Information flow: Frobenius versus closed structures

A difference we hit over and over again is how the 'yanking' is realized in the closed and Frobenius situations. If we interpret the orientation of tangles as information flow, then a module propagates information in time (downwards), while a dual module propagates information backwards in time (upwards). A further difference is that the closed structure is characterized by a universal property, while the Frobenius structure depends on a *choice* of a bilinear form.

If we look at teleportation protocols modelled by 'yanking', the ability of Alice to chose between different Bell measurements indicates that she is not dealing with a unique map, but with a Frobenius structure. Also the creation of a shared entangled state is not unique, as every Bell state does the trick. The need to communicate classical information to Bob emerges from the need to communicate this choice of one of the four Bell states in an a priori mutually agreed on classical basis, which can be described by a special Frobenius algebra. Hence the 'yanking' in teleportation should be thought of as a Frobenius algebra related property.

A similar situation emerges in canonical quantization of fields. The negative frequency parts are interpreted not as 'particles' (field modes) travelling backwards in time, but as 'antiparticles' propagating homochronos. This doubling of field modes resembles then the Frobenius-type information flow as discussed for teleportation. The β-multiplication is then given via the reproducing kernel property of the field propagators $\psi(x) = \int_Y g(x,y)\psi(y)$, but for a continuum. This calls for the extension of Frobenius structures to a non-unital situation, because on an infinite-dimensional space one cannot have a unit (see Abramsky and Heunen, 2012; Coecke and Heunen, 2011).

With regard to the topic of this book, it is worth looking at the vector space semantics of meanings of natural languages. There one has a set of grammatical types of vector spaces. The meaning of words are represented by vectors in different types of vector spaces—for nouns, transitive verbs, etc. On top of these types one has a Lambeck pregroup (see Lambek, 1999; Preller and Sadrzadeh, 2011a; Coecke, Sadrzadeh and Clark, 2010; and Chapters 9 and 13 in this volume), which is weakening the closed structure we are using here. Using a^l, a^r for left and right adjoints (duals), and having an order on the 'tensor monoid', one ends up with, among others, relations of the type $a^l a \leq 1 \leq aa^l$ and $aa^r \leq 1 \leq a^r a$. Preller and Sadrzadeh (2011a) and Coecke, Sadrzadeh and Clark (2010) show the striking similarity between this structure and categorical quantum mechanics, especially teleportation. With respect to our comment above, we think an investigation should be undertaken of how this pregroup approach relates to the Frobenius setup. First we note that any corpus of words is finite, hence we can assume a good duality or a Frobenius structure of its freely generated vector spaces of types. As the order of words in sentences is crucial for their meaning, one has to deal with non-symmetric Froebnius algebras; this also prevents left and right duality coinciding. Hence one needs to take the Nakayama automorphism into account. As we have seen in Section 2.3.8 the braiding and the left/right Frobenius isomorphisms can be transformed into one another (2.49). The pictures in Preller and Sadrzadeh (2011a, p. 150) and the preface of this book describing sentences such as 'John likes Mary' or 'John does not like Mary' would then be replaced by tree-like structures, where the multiplications are Frobenius pairings on the types, such as $V \otimes W \to J$ for a transitive verb. The order structure may call for lax-Frobenius structures. We close this speculation with the remark that in phylogenetic biomathematics one encounters very similar problems when reconstructing ancestral relationships from present-day gene sequences. This is an analogous process to the linguistic setup, reconstructing a tree which has as its 'words' the preset-day gene expression and as its 'semantic meaning' the ancestral relation including branching times (see for example Jarvis, Bashford and Sumner, 2005; Draisma and Kuttler, 2009; Jarvis and Sumner, 2012; Sumner, Holland and Jarvis, 2012).

2.4 A few pointers to further literature

Frobenius algebras emerge in a large number of situations, which we have had no opportunity to discuss here. Instead, we give a few hints as to where such developments are to be found, and what problems they are addressing. Our pick of the literature is subjective, with a bias towards graphically minded work.

Starting with Abramsky and Coecke (2004), compact closed (dagger) categories have been used to analyse quantum theory and especially quantum protocols. Graphical methods have been utilized very heavily and have led to 'picturalism' in quantum theory (Coecke and Duncan, 2011). The use of several Frobenius algebra structures at the same time (red-green-, black-white-, and rgb-calculi), which is discussed in Chapter 3, has captured, among other things, the concept of orthogonal bases (Coecke, Pavlovic and Vicary, 2011; Coecke, Paquette and Perdrix, 2008), complementary bases, the dagger structure, and such things as connectedness or disconnectedness of product states (e.g. entanglement for Bell, Greenberger–Horne–Zeilinger, and Werner states). Specialness of Frobenius algebras is not found in invariant rings, leading to non-trivial relations between reciprocal bases and dual bases, but still Frobenius structures can be employed there (Khovanov, 1997). The product of group characters can be understood along these lines, and the reader is invited to examine MacDonald (1979, 2nd edn. pp. 305–309) to see how it can be written in terms of the graphical calculus used here. A ring extension $\mathbb{Q}[q, t]$ leads to Macdonald polynomials. Below we will, however, give a pointer showing that it is not straightforward to generalize this 'cartesian' setup (also needed for canonical quantization) to manifolds and general coordinate invariance.

A theme that is amenable to graphical treatment is the relation between Frobenius structures of iterated ring extensions (or towers of algebras) and the Jones polynomial (see Kadison, 1999; Khovanov, 1997; Müger, 2003a). If seen as ring extensions $A_n/A_{n-1}/\ldots/A_0$ the various extensions provide separability idempotents, allowing the introduction of the relevant Markov traces and finally of the Jones polynomial.

A further theme is the relation of characters and certain trace modules with Frobenius structures. A *character* χ_V of a module V in A-mod$_R$ is given by

$$\chi_V(a) = \text{Tr}_{V/R}(a_V) \in R, \quad (a \in A) \tag{2.58}$$

where $a_V \in \text{End}(V)$ is given by $a_V(v) = av$. Hence characters form a subset of trace forms

$$\chi_v \in \frac{A^*}{[A^*, A^*]} \subseteq A^* \tag{2.59}$$

which vanish on $[A, A]$ (are constant on orbits). This allows one, for example, to define the Higman trace $\tau = \tau_\beta$ for a Frobenius algebra A as

$$\tau_\beta : A \to A :: a \mapsto \sum_i x_i a y_i \tag{2.60}$$

independent on the choice of generators. τ is $\mathcal{Z}(A)$ linear, with $\mathcal{Z}(A)$ the center of A. For a matrix algebra $A = M_n(R)$ one finds $\tau(a) = \text{trace}(a) 1_{n \times n}$, and for a group algebra $A := \mathbb{C}G$ one obtains the averaging or (up to normalization) Reynolds operator $\tau(a) = \sum_{g \in G} gag^{-1}$.

Furthermore, one can define the Casimir operator, equivalent to the Higman trace, if A is symmetric, as

$$c : A \to \mathcal{Z}(A) :: a \mapsto \sum_i y_i a x_i \tag{2.61}$$

which has deep connections to the Grothendieck groups, K-theory, and restriction and induction functors (for example see Lorenz and Fitzgerald Tokoly, 2011; Lorenz, 2011 and references therein). It is remarkable that the related diagrams, easily drawn, of the Higman trace, or the Casimir operator emerge rather naturally in graphical calculations with Frobenius algebras.

As we have already mentioned TQFT, we here only remark that Frobenius structures play a prominent role in that field (see Atiyah, 1989; Kock, 2003). The general idea of applying a functor with codomain a tensor monoid of vector spaces has proved to be very versatile. Similar constructions can be found in the theory of vertex operators and rational conformal quantum field theories (see for example Fuchs, Runkel and Schweigert, 2002; Fuchs, Runkel and Schweigert, 2007; and Barmeier et al., 2010, which make heavy use of graphical calculi and provide further references).

A theme related to classical physics, is that of Frobenius manifolds (Hitchin, 1997). If M is a manifold of dimension n, one can impose the existence of the following sections

$$\theta \in C^\infty(T^*M) \qquad g \in C^\infty(S^2 T^*M) \qquad c \in C^\infty(S^3 T^*M) \qquad (2.62)$$

on the (co)tangent space of M. Here g is a metric on M, with covariant derivative ∇; θ is a 1-form (related to the Frobenius homomorphism); and c is a symmetric rank-3 tensor. Let $\{e_i\}$ be an orthogonal basis of TM_x, with respect to the scalar product $(e_i, e_i) = \pm 1$, diagonalizing the left regular representation l_a for $a \in TM_x$. After rescaling, the e_i are mutually annihilating idempotents. Letting $e = \sum_i e_i$ be the unit of this algebra, then we get $\theta(v) = (e, v)$ and $(u, v) = c(e, u, v)$. Letting $\mu_i = \theta(e_i)$ and letting $f_i \in TM_x^*$ be the dual basis of the e_i, then the structure maps can be written as

$$\theta = \sum \mu_i f_i \qquad g = \sum \mu_i f_i f_i \qquad c = \sum \mu_i f_i f_i f_i \qquad (2.63)$$

The question that arises is whether this structure is compatible with the differential structure—that is, the covariant derivative—defined by the metric. This compatibility leads to Chazy's *non-linear* differential equation (Coecke, 2007). Chazy's equation provides a 'potential' for θ, g, and c resulting in the proper definition of Frobenius manifolds.

If the metric is an Egoroff metric, having a sort of potential form, $g = \sum_i \mu_i f_i f_i = \sum_i \frac{\partial \phi}{\partial x_i} dx_i^2$, then ∇c is symmetric, which implies $d_A c = 0$. This result relates orthogonal coordinates (bases) to Frobenius structures, as we have seen in the quantum information setting. However, on a manifold we need to be careful, as not all orthogonal frames are allowed. This can be seen by the following example. The metric $dx^2 + dy^2$ defines a Frobenius structure on $\mathbb{R}^2$, satisfying the condition above with $\phi = x_1 + x_2$. The same (in $\mathbb{R}^2 \setminus \{0\}$) metric in polar coordinates $dr^2 + r^2 d\phi^2$ does not carry a Frobenius structure with ∇c symmetric. This example shows clearly that general coordinate transformations, and hence general covariance, are not compatible with (symmetric) Frobenius structures. It sheds also some light on the usage of special Frobenius algebras in the semantics of quantum protocols as mentioned above.

Frobenius algebras can help to construct solutions of the Yang–Baxter equation (Beidar, Fong and Stolin, 1997) (which is related to our remark about the Jones polynomial, and Kadison's work). In a broader sense, one wants to study entwined modules; that is, modules $A \otimes C$ where A has an algebra structure and C has a coalgebra structure. It turns out that Frobenius functors play a crucial role in studying such entwined modules. This is developed at length in Caenepeel, Militaru and Zhu (2002). It turns out that these powerful algebraic tools allow one to attack non-linear differential equations and they also provide solutions to the Yang–Baxter equation. A structure prominently used in solid-state and high-energy physics problems is often termed there 'integrable models'.

We have in this work always assumed that the Frobenius algebra structure is associative. This can be relaxed to lax-Frobenius algebras or even general non-associative Frobenius algebras. To my knowledge little research has been done in this direction, but see Oziewicz and Wene (2011), where a first attempt is made to study non-associative Frobenius algebras, also using graphical methods. Even in the seemingly trivial case of complex numbers, which can be seen as a Frobenius algebra over the real numbers (Kock, 2003), one encounters different Frobenius algebras if associativity is dropped.

Acknowledgements

I would like to thank the organizers of the conference *The categorical flow of information in quantum physics and linguistics* for inviting me to this interesting event. Special thanks are due to Chris Heunen for helping me with the preparation of the TikZ graphics, and to Ronald C. King and Peter D. Jarvis for a critical reading of a draft of this chapter.

CHAPTER 3

A Graphical Approach to Measurement-Based Quantum Computing

ROSS DUNCAN

(Université Libre de Bruxelles)

3.1 Introduction

Quantum computation, at least for the finite-dimensional systems usually considered, lives in the setting of finite-dimensional Hilbert spaces. Even ignoring the possibly enormous dimension of the spaces involved, a Hilbert space is a very rich mathematical environment, which often hides the structure of the states and conceals the behaviour of their maps, making it difficult to analyse quantum programs. Can this difficulty be circumvented?

In this chapter, we will present an abstract formulation of quantum theory, based on algebraic features present in the Hilbert space theory, but making no reference to Hilbert spaces themselves. The reader will perhaps be unsurprised to learn that the tool of choice for this reformulation of quantum mechanics is category theory, and in particular the theory of symmetric monoidal categories (SMCs).

In a seminal paper, Abramsky and Coecke (2004) introduced the notions of †-symmetric monoidal category (†-SMC) and †-compact category and, exploiting the fact that the category of finite-dimensional Hilbert spaces and linear maps (henceforth called **fdHilb**) forms a †-compact category, gave a high-level proof of correctness of the quantum teleportation protocol (Bennett *et al.*, 1993). In so doing, they showed that quantum protocols do not necessarily rely upon the full apparatus of Hilbert spaces; a more abstract presentation of quantum mechanics can suffice. We will use such a high-level presentation to analyse measurement-based quantum programs.

As discussed earlier in this volume, †-compact categories admit a graphical notation where the morphisms of the category are represented by diagrams. Sequential composition of morphisms is represented by plugging together diagrams, and parallel composition (i.e. the tensor product) is represented by juxtaposition of diagrams. The crucial point, fully elaborated by Selinger (2011), is that the equations of the category are fully captured by homotopic transformation of diagrams. In other words, two morphisms are equal, according to the axioms of †-SMCs, if and only if the corresponding diagrams can be continuously deformed into each other. Hence, by transcribing a morphism into the graphical notation, a large amount of equational structure is incorporated directly into the syntax of the formal system.

The axioms of †-compact categories will not be sufficient, however, to study the quantum systems of interest in this chapter. We will introduce more structure by choosing specific generators for the category of diagrams, and imposing certain equations between diagrams involving those generators. These equations induce an equivalence relation between diagrams based on substitution. More concretely, we view each equation, say $L = R$, as a rewrite rule, and whenever we find L occurring as a subdiagram in some larger diagram D, we may perform the rewrite $D[L] \to D[R]$, as shown in Figure 3.1. Since all diagrams are typed this substitution is always possible. We will treat diagrams and rewriting informally here, but the interested reader can find a detailed account in the paper of Dixon and Kissinger (2010).

The additional equations imposed on diagrams are those corresponding to two different algebra structures found on the underlying Hilbert space. Thanks to the theorem of Coecke, Pavlovic and Vicary (2011b), there is a bijective correspondence between orthonormal bases for the Hilbert space—which from our point of view represent quantum observables—and special commutative †-Frobenius algebras. Therefore we encode each observable by an algebra, and its associated equations give the first collection of rewrite rules. For the purpose of analysing measurement-based quantum programs, we will only need to consider two different algebras, namely those corresponding to the X and Z spin observables. These observables have, in addition, a further property: they are complementary, and in a particularly strong sense. Intuitively, complementarity means that perfect knowledge of one observable implies complete ignorance of the other. In previous work, Coecke and the author

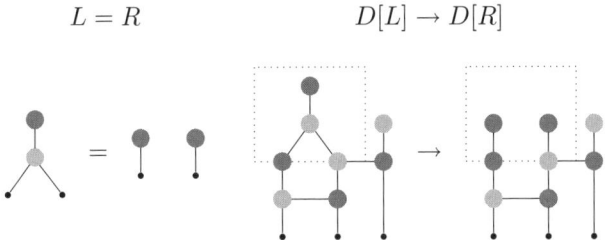

Fig 3.1 Rewriting as substitution

(Coecke and Duncan, 2011) showed that this kind of complementarity can also be formalized in terms of algebras. Strongly complementary observables form a bialgebra, in fact a Hopf algebra. We also impose the defining equations of these structures onto the diagrammatic language to get another family of rewrite rules.

In summary, the graphical calculus consists of diagrams generated by two Frobenius algebras for the X and Z observables; the equations imposed by the monoidal structure may be effectively forgotten because, thanks to the context of †-compact categories, the notion of equality of diagrams already contains all of them. We then impose rewrite rules corresponding to equations for the Frobenius structure, and then further equations stating that these generators, when combined appropriately, form a Hopf algebra. This setup, combined with additional elements to be introduced later, forms the zx-calculus, also known as the red–green calculus. Since this book is published in black and white, however, dark and light grey will take the place of red and green.

The zx-calculus is known to be weaker than the full theory of Hilbert spaces, but it is sound, meaning that any equation derived in it will also hold when translated back to the Hilbert space formalism. It replaces matrices with a structured and discrete notation that exposes the relationship between different parts of a quantum system. Furthermore, its graphical nature allows quantum circuits to be represented very easily, and more importantly, the ad hoc notation for graph states used in measurement-based quantum computation can be derived from algebraic considerations alone. Hence we can see the beautiful interplay between the structure of an entangled state and the algebraic objects that would represent this state. The rewrite rules then allow transformations between, e.g. the circuit model and the measurement-based model, and expose how information flows within the entangled state during the process of executing a measurement-based program. This last fact will be at the heart of our analysis: we will demonstrate how the nondeterminism induced by quantum measurements may be tracked through the graph structure of an entangled state and verify that a given computation is in fact deterministic.

3.1.1 Background and related work

We assume that the reader is familiar with the basics of monoidal category theory; aside from other chapters in this volume, the articles of Abramsky and Tzevelekos (2011) and Coecke and Paquette (2011) provide suitable introductions, while Mac Lane (1971) is the standard text. Compact closed categories were introduced by Kelly and Laplaza (1980), and the notion of dagger-compactness first arose in Abramsky and Coecke (2004). Diagrammatic notation for monoidal categories has a long history going back to work of Kelly (1972*a*, 1972*b*) and Penrose (1971); one can also view the proof-net syntax of linear logic in this light (Girard, 1996; Blute *et al.*, 1991). The essential reference on this subject is Selinger's survey (Selinger, 2011), which pulls together a great deal of material scattered throughout the literature.

The study of quantum mechanics through categorical eyes was initiated by the paper of Abramsky and Coecke (2004), and the explicit use of diagrams was emphasized by Coecke (2010). The notion of classical structure, here referred to as *observable structure* was introduced by Coecke and Pavlovic (2007) and further developed by those authors in collaboration with Paquette (Coecke and Paquette, 2006; Coecke, Paquette and Pavlovic, 2010a). The key theorem, that any classical structure in finite-dimensional Hilbert space is equivalent to a basis for that space, was shown by Coecke, Pavlovic and Vicary (2011b). The idea that complementarity could also be formalized in terms of interacting algebras was introduced by Coecke and the author (Coecke and Duncan, 2008; Coecke and Duncan, 2011); these papers introduced several fundamental ideas that have since found application in areas as diverse as quantum foundations (Edwards, 2009; Coecke and Edwards, 2011; Coecke, Edwards and Spekkens, 2011a), topological quantum computation (Horsman, 2011), and measurement-based quantum computing, which is our main concern here.

One can view the quantum teleportation protocol (Bennett *et al.*, 1993) as the first measurement-based quantum computation; albeit the program computes the identity function. Gottesman and Chuang (1999) showed later that this idea could be generalized to a universal computation model. The model of interest here is not the teleportation model, but rather the one-way model introduced by Raussendorf and Briegel (Raussendorf and Briegel, 2001; Raussendorf and Briegel, 2003; Raussendorf, Browne and Briegel, 2003). Our work here is based on the work carried out by Danos, Kashefi and Panangaden (2007) to provide this model with a formal syntax, the measurement calculus. Danos and Kashefi (2005) introduced the concept of *flow*, later renamed *causal* flow, to study the problem of determinism in the one-way model. Mhalla and Perdrix (2008) demonstrated an efficient algorithm for finding optimal flows, while Browne *et al.* (2007) introduced the notion of *generalized flow*.

This chapter mainly draws on the author's joint work with Perdrix (Duncan and Perdrix, 2010), although work relating the graphical/categorical approach to measurement-based quantum computing (MBQC) goes back rather further (Duncan, 2006).

3.1.2 Outline of the chapter

The next section is a primer on the basics of quantum mechanics. Section 3.3 introduces the algebraic framework of interacting observables (as presented in Coecke and Duncan, 2011) in full generality; Section 3.4 presents the zx-calculus, which is the specific instance of that framework as a formal graphical calculus based on qubits with the spin observables X and Z. Section 3.5 introduces the one-way model and the measurement calculus, and shows how to represent them in the zx-calculus. Finally, Section 3.6 examines determinism in the one-way model, and shows how to use the property of flow to generate rewrite sequences to prove the correctness of the measurement calculus programs.

3.1.3 Notation

We will use the Dirac notation throughout: vectors are denoted by *kets* $|\psi\rangle$, and their duals by *bras* $\langle\phi|$. The inner product is written $\langle\phi\,|\,\psi\rangle$, and is taken to be linear in the second component and anti-linear in the first. We will usually denote $\mathbb{C}^2$ by Q, since it is the state space of *qubits*. The vectors comprising the standard basis for Q, sometimes called the computational 'or Z-basis', are written $|0\rangle$ and $|1\rangle$; we denote the X-basis by

$$|+\rangle = \frac{1}{\sqrt{2}}(|0\rangle + |1\rangle) \qquad |-\rangle = \frac{1}{\sqrt{2}}(|0\rangle - |1\rangle)$$

When writing tensor products of qubits we will usually suppress the tensor symbol, and write e.g. $|00\rangle$ in place of $|0\rangle \otimes |0\rangle$. The base field $\mathbb{C}$ will often be written simply as I, since it is the unit of the monoidal category structure.

If u and v are vertices of an undirected graph, then $v \sim u$ means that they are adjacent.

When drawing diagrams, we use the pessimistic convention: diagrams should be read from top to bottom.

3.2 The rudiments of quantum computing

This section is necessarily rather brief; for a more complete treatment we suggest the excellent books by Mermin (2007), and Kaye, Laflamme and Mosca (2007).

Whereas a classical bit has only two values, its quantum analogue—the qubit—is a unit vector in a two-dimensional Hilbert space. It is impossible to distinguish two states that differ only by a global phase, so we quotient the state space by the relation $|\psi\rangle \sim e^{i\alpha}|\psi\rangle$. The state space of a compound quantum system, i.e. one formed by combining individual systems, is given by the tensor product of the constituent state spaces, so a state consisting of n qubits is a vector in a Hilbert space of dimension 2^n. Such a state space necessarily contains states that cannot be decomposed into a product of n individual qubits. For example, the Bell state,

$$|\Phi_+\rangle = \frac{|00\rangle + |11\rangle}{\sqrt{2}},$$

is a perfectly valid state of two qubits, but there are no single-qubit states such that $|\phi\rangle \otimes |\psi\rangle = |\Phi_+\rangle$. This simple mathematical fact underlies the phenomenon of *entanglement*. These indecomposable states reflect non-local correlations between the subsystems, and form the main building block of the paradigm called *measurement-based quantum computation*.

The evolution of an undisturbed quantum system is given by a unitary operator:

$$|\psi(t)\rangle = U_t|\psi(t_0)\rangle$$

A quantum computation is typically described as a quantum circuit: this is just a sequence of unitary operations acting on some number of qubits. Unitarity implies that quantum computations are reversible, since the unitaries form a group. The exception is quantum measurement.

Quantum measurements have two properties that run contrary to classical intuition. Firstly their outcomes are *probabilistic*: in almost all quantum states the outcome of a given measurement cannot be known with certainty. Secondly, they have *side-effects*, so that state after a measurement will usually be different to that before it. Mathematically speaking, we identify a quantum measurement with a self-adjoint operator,

$$M = \sum_i \lambda_i |v_i\rangle\langle v_i|.$$

The possible observed values are the eigenvalues λ_i. We are only concerned with non-degenerate measurements here, so we assume that all the λ_i are distinct and non-zero. Given a quantum state $|\psi\rangle$, the probability of observing λ_i is given by the inner product

$$p(\lambda_i) = |\langle v_i | \psi \rangle|^2.$$

Most importantly, the new state of the system after the measurement is the corresponding eigenvector $|v_i\rangle$. In other words, observing λ_i is effectively the same as acting on the state with the projection operator $|v_i\rangle\langle v_i|$ or, if the measured system is destroyed by the measurement, which will be the case for the systems of interest here, simply $\langle v_i|$. The actual values of the measurements are not important here, so we will regard them simply as labels for the outcomes.

If one part of an entangled quantum state is measured, the effect of that measurement can be observed in other parts of the state. For example, consider again the Bell state $|\Phi_+\rangle$. If its first qubit is measured in the $|0\rangle, |1\rangle$ basis then the two outcomes are equally likely; suppose that $|0\rangle$ is observed. The resulting effect is to act on the joint state with the operator $|0\rangle\langle 0| \otimes \mathrm{id}$. Ignoring normalization, we have

$$(|0\rangle\langle 0| \otimes \mathrm{id})|\Phi_+\rangle = (|0\rangle\langle 0| \otimes \mathrm{id})|00\rangle + (|0\rangle\langle 0| \otimes \mathrm{id})|11\rangle$$
$$= \langle 0 | 0\rangle|00\rangle + \langle 0 | 1\rangle|11\rangle = |00\rangle.$$

Hence, now performing the same measurement on the second qubit will produce outcome $|0\rangle$ with probability one. Despite acting on only one part of the system, we have produced a global change. (Notice also that the new joint state is no longer entangled.)

This, in a nutshell, is the concept behind measurement-based quantum computation: we begin with a large entangled state, and by performing carefully chosen measurements upon it, the unmeasured parts are driven toward the desired result. As a first approximation, we could say that the structure of entangled quantum states defines the desired computation, while the measurements themselves function more to 'push' information through this structure. From this point of of view, the measurements play a role similar to the evaluation maps in functional programming, effectively 'applying' their outcomes to the function defined by the rest of the state. (This is not entirely accurate; as we shall see, the choice of measurements does play a role in defining the computation.)

It is frequently useful to generalize the notion of quantum state to admit probabilistic mixtures of states. In this setting, states comprising a single state vector as described above are called *pure states*, while the others are called *mixed states*. These more general states are represented by *density operators*; that is, trace-one Hermitian matrices of the form

$$\rho = \sum_i p_i |\psi_i\rangle \langle \psi_i|$$

where $0 \leq p_i \leq 1$ and $\sum_i p_i = 1$. Note that the components $|\psi_i\rangle$ need not be orthogonal. For pure states we have $p_1 = 1$ and $p_i = 0$ for $i > 1$. The decomposition of a mixed state is not unique. For example, the maximally mixed qubit can arise by preparing either $|0\rangle$ or $|1\rangle$ with equal probability, or equivalent by preparing either $|+\rangle$ and $|-\rangle$:

$$\frac{|0\rangle\langle 0| + |1\rangle\langle 1|}{2} = \begin{pmatrix} 1/2 & 0 \\ 0 & 1/2 \end{pmatrix} = \frac{|+\rangle\langle +| + |-\rangle\langle -|}{2}.$$

The most general class of operations that are possible in the density matrix formalism are *completely positive maps*, also called *superoperators*. The action of such a map $\mathcal{E}$ on a state ρ is given by:

$$\rho' = \mathcal{E}\rho\mathcal{E}^\dagger.$$

Since $\mathcal{E}$ is positive ρ' is again a density matrix. The most basic examples of superoperators are unitary maps and quantum measurements. For example, given a measurement $M = \sum_i \lambda_i |v_i\rangle \langle v_i| = \sum_i \lambda_i P_i$, the effect of performing the measurement on state ρ is

$$\rho' = \sum_i P_i \rho P_i = \sum_{i,j} \text{Tr}[\rho P_i] |v_i\rangle \langle v_i|$$

where $\text{Tr}[\rho P_i]$ gives the probability of observing outcome i. While mixed states arise for a variety of reasons in quantum computation, in this chapter the randomness introduced by measurement will by the only source of uncertainty.

3.3 Observables and strong complementarity

3.3.1 Observables and observable structures

Given a quantum system whose state space is the Hilbert space A, we will assume that any orthonormal basis $\{|a_i\rangle\}_i$ for A defines an observable on that system. Furthermore, these will be the only observables of interest.

The no-cloning (Wootters and Zurek, 1982) and no-deleting (Pati and Braunstein, 2000) theorems state that it is impossible to perfectly copy or erase an unknown quantum state. However, it is possible to perform both of these operations if the state is guaranteed to be an outcome of a known observable; that is, if it is a member of some given basis. We can therefore view each quantum observable as determining a classical data type whose elements are possible outcomes, and whose operations are copying and deleting. For example, the copying and deleting operations for the standard basis are given by the linear maps

$$\delta_Z : Q \to Q \otimes Q \qquad \epsilon_Z : Q \to I$$
$$\delta_Z : |i\rangle \mapsto |ii\rangle \qquad \epsilon_Z : |i\rangle \mapsto 1$$

Note that ϵ_Z is an unnormalized bra, namely $\sqrt{2}\langle+|$. Graphically we will denote these operations by

$$\delta_Z = \qquad \epsilon_Z =$$

What axioms should such operations obey? Informally we may say that if we copy something, and then copy one of the copies, it should not matter which copy we copied; that if we copy something and immediately erase one of the copies, the combined operation should have no effect; and, if we copy something, the two copies may be exchanged without making any difference. The same thing stated formally is that (δ_Z, ϵ_Z) should form a cocommutative comonoid on Q. Presented graphically we have the following:

Since we operate in a †-category we may also consider the adjoint operations

$$\delta_Z^\dagger = \qquad \epsilon_Z^\dagger =$$

which automatically form a commutative monoid, and obey the same pictorial equations as (δ_Z, ϵ_Z) but flipped upside down:

Notice that $\epsilon_Z^\dagger$ is an unnormalized ket, $\sqrt{2}|+\rangle$.

Taken together, the 4-tuple $(\delta_Z, \epsilon_Z, \delta_Z^\dagger, \epsilon_Z^\dagger)$ forms a special commutative †-Frobenius algebra; this amounts to saying that in addition to the above, the following equations also hold:

These equations, the Frobenius law and the special condition respectively, will not be motivated here (see Coecke and Paquette, 2006; Coecke and Pavlovic, 2007).

The preceding discussion of algebras is motivated by the following theorem:

Theorem 3.1 (Coecke, Pavlovic and Vicary, 2011) *There is a bijective correspondence between orthonormal bases for a finite-dimensional Hilbert space A, and special commutative Frobenius algebras on A.*

Since we have assumed that all observables are non-degenerate, Theorem 3.1 permits us to treat Frobenius algebras of the above type as an abstract version of quantum observables. This definition, moreover, makes no reference to the fact that the underlying object is a Hilbert space, and hence it can be used in any †-SMC. For this reason, we will henceforward refer to special commutative †-Frobenius algebras by the term *observable structure*.[1]

Remark 3.2 Note that this representation of observables by algebras is not a representation of the *measurement* of an observable. The additional formal apparatus required to account for the non-deterministic aspect of the measurement will be introduced in Section 3.4.4.

3.3.2 Unbiasedness and the phase group

Given an orthonormal basis $\{|a_i\rangle\}_i$ for a d-dimensional Hilbert space A, a vector $|\psi\rangle$ is unbiased for $\{|a_i\rangle\}_i$ if, for all i, we have $|\langle a_i | \psi \rangle| = \frac{1}{\sqrt{d}}$. For example, $|+_\alpha\rangle = \frac{1}{\sqrt{2}}(|0\rangle + e^{i\alpha}|1\rangle)$ is unbiased for the standard basis on Q for all values of α; indeed these are the only unbiased states for the standard basis. Incorporating this concept into our diagrammatic language yields a surprising amount of power.

---

[1] The same object has also been called a *classical structure* (Coecke, Paquette and Pavlovic, 2010) and a *basis structure* (Edwards, 2009).

Recall that in a †-category a morphism $f : A \to B$ is unitary if $f^\dagger \circ f = \mathrm{id}_A$ and $f \circ f^\dagger = \mathrm{id}_B$; diagrammatically this is written,

$$\begin{array}{c}\boxed{f}\\\boxed{f}\end{array} = \Big| = \begin{array}{c}\boxed{f}\\\boxed{f}\end{array}$$

where the picture for $f^\dagger$ is obtained by flipping the picture for f upside down. This notion of unitarity agrees with usual one in **fdHilb**. Now we can make:

Definition 3.3 *Let (δ, ϵ) be an observable structure on A and let $\alpha : I \to A$ be a point of A. Define a map $\Lambda(\alpha) : A \to A$ by*

$$\Lambda(\alpha) := \delta^\dagger \circ (\alpha \otimes \mathrm{id}A) =$$

Definition 3.4 *Let $\alpha : I \to A$ be a point of A, and $\Lambda(\cdot)$ as in Definition 3.3. We say α is unbiased for (δ, ϵ) if $\Lambda(\alpha)$ is unitary.*

When α is unbiased, the map $\Lambda(\alpha)$ is called a *phase map* for (δ, ϵ). According to the unit law for the monoid structure, $\Lambda(\epsilon^\dagger)$ yields the identity map, which is unitary. Therefore every observable structure has at least one unbiased point, namely $\epsilon^\dagger$. Furthermore, since they are unitary the phase maps form a group, indeed an abelian group, as the following calculation shows:

Since all the phase maps commute, we make the following notational convention:

for which we have the equations

$$\left(\,\begin{array}{c} \bullet \\ \alpha \\ \bullet \end{array}\, \right)^{\dagger} = \begin{array}{c} \bullet \\ -\alpha \\ \bullet \end{array} \quad \text{and} \quad \begin{array}{c} \bullet \\ \alpha \\ \beta \\ \bullet \end{array} = \begin{array}{c} \bullet \\ \alpha+\beta \\ \bullet \end{array}.$$

The alert reader will have noted that the unbiased points themselves form an abelian group, isomorphic to the phase group, under the multiplication $\delta^{\dagger}$; one can equivalently define the phase group via this route.

Remark 3.5 While we will be exclusively interested in the case where $\alpha : I \to A$ is an unbiased point for some observable, most of the above still holds when α is an arbitrary point of A. In that case we get a commutative monoid rather than a group. Note especially that Theorem 3.6, below, still applies.

Returning to the example of (δ_Z, ϵ_Z) on the qubit, the vectors $|+_\alpha\rangle$, when multiplied by $\sqrt{2}$, yield the phase maps

$$\Lambda_Z(\sqrt{2}|+_\alpha\rangle) = Z_\alpha = \begin{pmatrix} 1 & 0 \\ 0 & e^{i\alpha} \end{pmatrix},$$

comprising rotations around the Z axis of the Bloch sphere.

We can now state the key theorem for the diagrammatic treatment of observable structures and their phase groups:

Theorem 3.6 (Coecke and Duncan, 2011) *Let D be a connected diagram generated by an observable structure (δ, ϵ) and its phase group; then D is determined completely by its number of inputs, its number of outputs, and the sum $\sum_i \alpha_i$ of phase group elements occurring in it.*

The above result is effectively a normal form theorem for observable structures, but we will use it instead to justify a new notational convention, and simply collapse any connected diagram down to a single vertex, which we refer to as a *spider*:

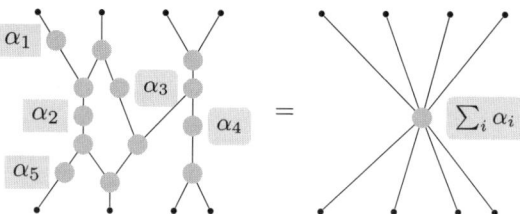

The label will be omitted when $\alpha = 0$. We can therefore adopt spiders as the generators of the diagrammatic language, governed by a single equational scheme, the spider rule:

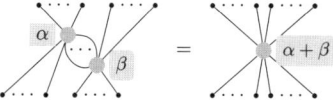

Example 3.7 What is the spider with one input, one output, and $\alpha = 0$? The answer is provided by the unit law of the observable structure: it must be the identity, as shown below.

Once we have made the above convention with respect to the identity, all the earlier equations are included in the spider rule, hence this formulation is equivalent to the definition in terms of δ, ϵ and $\Lambda(\alpha)$.

Example 3.8 For any given observable structure (δ, ϵ) we can produce a bipartite state $d : I \to A \otimes A$ by $d = \delta \circ \epsilon$:

If we partially compose this state with its adjoint we obtain, via the spider rule, the identity:

Hence, the object A bears a self-dual compact structure. It is straightforward to construct observable structures for $A \otimes A$ given one on A, so the monoidal category generated by A is compact closed.

3.3.3 Strong complementarity

In quantum theory, two observables are said to be *complementary* if measuring one of them reveals no information about the other, for example the X and Z spins. Notice that both elements of X basis, $|+\rangle$ and $|-\rangle$, are unbiased with respect to the Z basis, and vice versa; these bases are said to be *mutually unbiased*. Mutually unbiased bases correspond to complementary observables: given an eigenstate of Z, the inner product with either eigenstate of X has the same absolute value, and hence both outcomes are equiprobable when an X measurement is performed. In this section we will present, though not justify, a characterization of complementarity in terms of observable structures rather than bases. In fact, we will present the axioms for observable structures that are *strongly complementary*, a property enjoyed by well-behaved pairs of observables. While the observables we are most interested in—the X and Z spins—are strongly complementary, the material of this section is completely general; the special features of the X and Z observables are treated in the next section.

Since we are now dealing with two observables, we will have two observable structures, (δ_g, ϵ_g) and (δ_r, ϵ_r), which are represented by light and dark grey spiders.

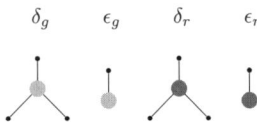

The map δ was originally introduced as a copying operation, but it was axiomatized without any reference to the objects it copies. We now correct this.

Definition 3.9 *A point $k : I \to A$ is called* classical *for an observable structure (δ, ϵ) if it satisfies $\delta \circ k = k \otimes k$.*

In the language of coalgebras, classical points are called *set-like elements*.

Lemma 3.10 *Let k denote a classical point for (δ_g, ϵ_g), and let α be any unbiased point for (δ_g, ϵ_g); k is an eigenpoint of the corresponding phase map $\Lambda_g(\alpha)$.*

Note the appearance of a scalar element here—the eigenvalue of k.

Remark 3.11 Any diagram with no inputs or outputs represents an arrow of type $I \to I$. When interpreted in **fdHilb** these are simply complex numbers. Since quantum mechanics does not distinguish states that differ by a scalar factor we will ignore these whenever they appear. Furthermore, *many of the equations presented below hold only up to scalar normalizing factor*. We omit these in order to simplify the presentation; if needed they can easily be reconstructed.

Definition 3.12 *Two observable structures (δ_g, ϵ_g) and (δ_r, ϵ_r) on A are called* strongly complementary *if*

1. for every point $k : I \to A$, if k is classical for (δ_g, ϵ_g) then it is unbiased for (δ_r, ϵ_r), and vice versa
2. $\epsilon_r^\dagger$ is classical for (δ_g, ϵ_g) and $\epsilon_g^\dagger$ is classical for (δ_r, ϵ_r), i.e.:

3. the equation $(\delta_r^\dagger \otimes \delta_r^\dagger) \circ (\mathrm{id}A \otimes \sigma \otimes \mathrm{id}A) \circ (\delta_g \otimes \delta_g) = \delta_g \circ \delta_r^\dagger$ holds, i.e.:

$$ \qquad (3.1)$$

where σ denotes the symmetry of the monoidal structure.

Since we are operating in †-SMC, conditions 2 and 3 also imply flipped versions of the same equations. Given this, these two conditions could be replaced by a unified condition:

- The 4-tuple $(\delta_g, \epsilon_g, \delta_r^\dagger, \epsilon_r^\dagger)$ forms a *bialgebra* on A.

In fact, as well being a bialgebra, a pair of strongly complementary observable structures is in addition a Hopf algebra. Recall $d = \delta \circ \epsilon^\dagger$, and define $s : A \to A$ by $s = (d_r^\dagger \otimes \mathrm{id}A) \circ (\mathrm{id}A \otimes d_g)$. We introduce a new element of the graphical notation for s:

Now we have:

Lemma 3.13 *The 5-tuple $(\delta_g, \epsilon_g, \delta_r^\dagger, \epsilon_r^\dagger, s)$ forms a Hopf algebra on A, i.e.:*

$$\delta_r^\dagger \circ (s \otimes \mathrm{id}A) \circ \delta_g = \epsilon_r^\dagger \circ \epsilon_g.$$

$$ \qquad (3.2)$$

Remark 3.14 We have stated Lemma 3.13 as a consequence of the bialgebra structure; in fact, under a mild side condition, eqn (3.2) can be shown to be equivalent to condition 1 of defn 3.12. See Coecke and Duncan (2011) for full details.

The classical points have some useful additional properties, which we will now state; the reader can find proofs in Coecke and Duncan (2011).

Thanks to defn 3.12, if k is classical for (δ_g, ϵ_g) then $\Lambda_r(k)$ is an element of the phase group for the strongly complementary observable (δ_r, ϵ_r). We draw the classical points in the colour of the observable with respect to which they are unbiased, and rely on the label to indicate that it is in fact a classical point: Latin letters will indicate classical points, while Greek letters will denote arbitrary unbiased points.

Proposition 3.15 *Let k, k' be classical points for (δ_g, ϵ_g), and let h be classical for (δ_r, ϵ_r); then:*

1. *the phase map $\Lambda_r(k)$ is a comonoid homomorphism of (δ_g, ϵ_g):*

2. *the phase maps $\Lambda_g(h)$ and $\Lambda_r(k)$ commute, up to a scalar factor:*

3. *the point $\delta_r^\dagger(k \otimes k')$ is also classical for (δ_g, ϵ_g):*

Corollary 3.16 *If (δ_g, ϵ_g) has finitely many classical points, then they form a subgroup among the group of unbiased points of (δ_r, ϵ_r).*

When we consider observable structures over Hilbert spaces, having finitely many classical points is just the statement that the underlying space is finite dimensional. Since this is the case for all the situations of interest for this chapter we will henceforth assume that the *classical phases* always form a subgroup.

There is also an important interaction between the classical points and the phase group.

Proposition 3.17 *Let k be a classical point for (δ_g, ϵ_g), let α be an unbiased point for (δ_g, ϵ_g), and define $k \bullet \alpha := \Lambda_r(k) \circ \alpha$; then:*

- *$k \bullet \alpha$ is again unbiased for (δ_g, ϵ_g);*
- *$\Lambda_r(k) \circ \Lambda_g(\alpha) = \Lambda_g(k \bullet \alpha) \circ \Lambda_r(k)$.*

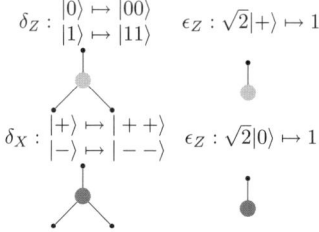

- *$\Lambda_r(k)$ is a group automorphism of the unbiased points of (δ_g, ϵ_g), and conjugation by $\Lambda_r(k)$ is an automorphism of the corresponding phase group.*

To restate some of the preceding: the classical points of one observable structure always form a subgroup among the unbiased points of the strongly complementary observable; this subgroup in turn acts as an automorphism group upon the unbiased points of the first observable structure.

While all the preceding results will be (sometimes implicitly) used in the subsequent sections, we will be able to make this all rather more concrete by focusing on the specific case of the Z and X spin observables.

3.4 The zx-calculus

3.4.1 The Z and X observables

In this section, and in the rest of the chapter, we will represent the Z and X spin observables by following two strongly complementary observable structures on $\mathbb{C}^2$:

$$\delta_Z : \begin{array}{l}|0\rangle \mapsto |00\rangle \\ |1\rangle \mapsto |11\rangle\end{array} \qquad \epsilon_Z : \sqrt{2}|+\rangle \mapsto 1$$

$$\delta_X : \begin{array}{l}|+\rangle \mapsto |++\rangle \\ |-\rangle \mapsto |--\rangle\end{array} \qquad \epsilon_Z : \sqrt{2}|0\rangle \mapsto 1$$

One of the most significant simplifications that occurs when working with the Z and X observables is that they both generate the same compact structure; that is, we have the equation

$$d_Z \quad = \quad d_X \quad = \quad |00\rangle + |11\rangle$$

Since there is no need to distinguish between a light or dark cup (or cap), we will drop the dot from diagrammatic notation whenever possible. Since the category bears a single compact structure, we can treat the internal structure of any diagram as an undirected graph and appeal to the principle of diagrammatic equivalence described earlier: if two diagrams are isomorphic as labelled graphs, they are equal.

In direct consequence, the antipode map of the Hopf algebra structure is trivial:

$$s = (d_Z^\dagger \otimes \mathrm{id} A) \circ (\mathrm{id} A \otimes d_X) = \mathrm{id} Q$$

The defining equation of the Hopf algebra structure can therefore be simplified:

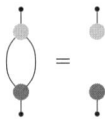

Recall that a point $|\alpha_Z\rangle$ is unbiased for the standard basis $|0\rangle, |1\rangle$ if and only if it has the form $\frac{1}{\sqrt{2}}(|0\rangle + e^{i\alpha}|1\rangle)$. Hence the phase group for the Z observable is just the circle group, i.e. the interval $[0, 2\pi)$ under addition modulo 2π. The X observable's phase group is isomorphic, so we represent the unbiased points and phase maps as shown below.

$$\alpha, \alpha', \beta, \beta' \in [0, 2\pi)$$

The phase maps are unitary, so the dagger sends each element to its inverse, i.e. it negates the angle:

Since we are operating in dimension 2, there are two classical points, corresponding to the angles 0 and π. The action of the non-trivial classical map is to negate the phase:

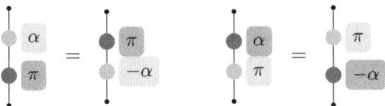

Given any two bases for a Hilbert space there is a unitary isomorphism that maps one basis to the other; in the case of the Z and X bases this map is the familiar Hadamard matrix. We will introduce an extra element into the diagrammatic language to represent this map:

$$H = \frac{1}{\sqrt{2}} \begin{pmatrix} 1 & 1 \\ 1 & -1 \end{pmatrix} = \boxed{H}$$

The Hadamard is a self-adjoint unitary, hence we have the equations:

Since the Hadamard maps one basis to the other, we could use it to *define* one observable structure in terms of the other:

Strictly speaking, it is redundant to continue with both the Z and X observable structures: we could eliminate, for example, the X vertices. However, for some purposes it is more convenient to use both Z and X, while for other purposes it is easier to work with just Z and H, so we maintain all three elements in the syntax, and endorse the following *colour duality principle*.

Proposition 3.18 *Every statement made in the diagrammatic language also holds with the colours reversed.*

We will switch freely between the two-coloured presentation, and the one-colour and Hadamard view depending on which is most convenient at any given time.

3.4.2 Syntax and semantics I

The zx-calculus is a formal graphical notation, based on the notion of an *open graph*.

Definition 3.19 *An* open graph *is a triple* (G, I, O) *consisting of an undirected graph* $G = (V, E)$ *and distinguished subsets* $I, O \subseteq V$ *of input and output vertices I and O. The set of vertices* $I \cup O$ *is called the* boundary *of G, and* $V \setminus (I \cup O)$ *is the* interior *of G.*

A term of the zx-calculus is called a *diagram*; this is an open graph with some additional properties and structure.

Definition 3.20 *A* diagram *is an open graph* (G, I, O), *where (i) all the boundary vertices are of degree one; (ii) the set of inputs I and the set of outputs O are both totally ordered; and (iii) whose interior vertices are restricted to the following types:*

- *Z vertices with m inputs and n outputs, labelled by an angle* $\alpha \in [0, 2\pi)$; *these are denoted* $Z_n^m(\alpha)$ *and shown graphically as light circles,*
- *X vertices with m inputs and n outputs, labelled by an angle* $\alpha \in [0, 2\pi)$; *these are these are denoted* $X_n^m(\alpha)$ *and shown graphically as dark circles,*
- *H (or Hadamard) vertices, restricted to degree 2; shown as squares.*

If a X or Z vertex has $\alpha = 0$ *then the label is entirely omitted. The allowed vertices are shown in Figure 3.2.*

Since the inputs and outputs of a diagram are totally ordered, we can identify them with natural numbers and speak of the *k*th input, etc.

Remark 3.21 When a vertex occurs inside the graph, the distinction between inputs and outputs is purely conventional; one can view them simply as vertices of degree $n + m$. However, this distinction allows the semantics to be stated more directly, see below.

The collection of diagrams forms a compact category in the obvious way: the objects are natural numbers and the arrows $m \to n$ are those diagrams with m inputs and n outputs; composition $g \circ f$ is formed by identifying the inputs of g with the outputs of f and erasing the corresponding vertices; $f \otimes g$ is the diagram formed by the disjoint union of f and g with I_f ordered before I_g, and similarly for the outputs. This is basically the free (self-dual) compact category generated by the arrows shown in Figure 3.2.

We can make this category †-compact by specifying that $f^\dagger$ is the same diagram as f, but with the inputs and outputs exchanged, and all the angles negated.

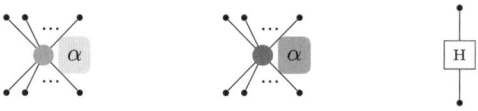

Fig 3.2 Interior vertices of diagrams

This construction yields a category that does not incorporate the algebraic structure of strongly complementary observables. To obtain the desired category we must quotient by the equations shown in Figure 3.3. We denote the category so-obtained by $\mathbb{D}$.

Remark 3.22 The equations shown in Figure 3.3 are not exactly those described in Sections 3.3 and 3.4.1, but they are equivalent to them. We shall therefore, on occasion, use properties discussed earlier as derived rules in computations.

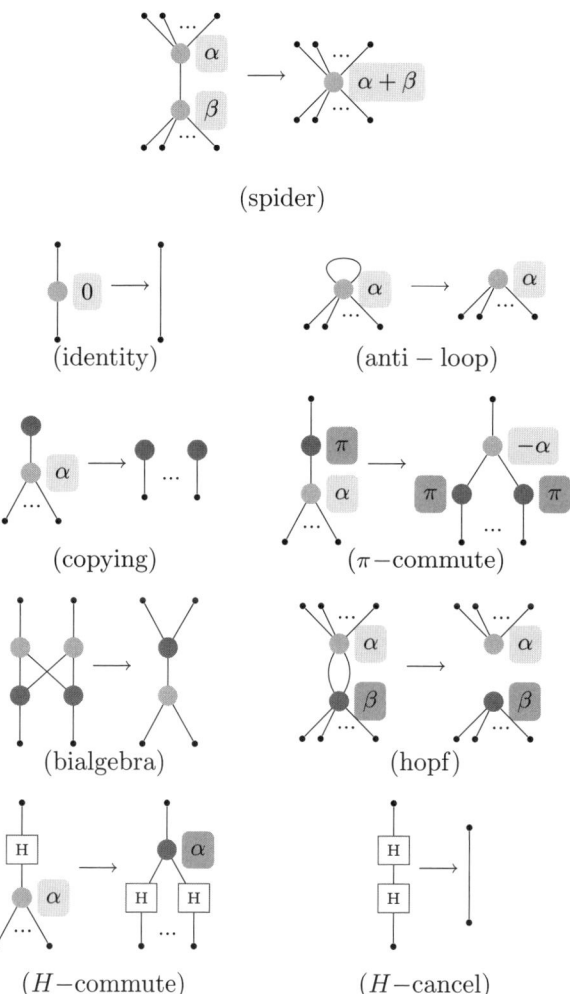

Fig 3.3 Rewrite rules for the zx-calculus. We present the rules for the Z subsystem; to obtain the complete set of rules exchange the colours in the rules shown above.

THE ZX-CALCULUS | 69

Since $\mathbb{D}$ is a monoidal category we can assign an interpretation to any diagram by providing a monoidal functor from $\mathbb{D}$ to any other monoidal category. Since we are interested in quantum mechanics, the obvious target category is **fdHilb**.

Definition 3.23 *Let $[\![\cdot]\!] : \mathbb{D} \to$ **fdHilb** be a symmetric monoidal functor defined on objects by*

$$[\![1]\!] = \mathbb{C}^2$$

and on the generators by:

$$[\![Z_n^m(\alpha)]\!] = [\![\ \begin{array}{c}\vdots\\ \alpha \\ \vdots\end{array}\ ]\!] = \begin{cases} |0\rangle^{\otimes m} \mapsto |0\rangle^{\otimes n} \\ |1\rangle^{\otimes m} \mapsto e^{i\alpha}|1\rangle^{\otimes n} \end{cases},$$

$$[\![X_n^m(\alpha)]\!] = [\![\ \begin{array}{c}\vdots\\ \alpha \\ \vdots\end{array}\ ]\!] = \begin{cases} |+\rangle^{\otimes m} \mapsto |+\rangle^{\otimes n} \\ |-\rangle^{\otimes m} \mapsto e^{i\alpha}|-\rangle^{\otimes n} \end{cases},$$

$$[\![H]\!] = [\![\ \boxed{H}\ ]\!] = \tfrac{1}{\sqrt{2}}\begin{pmatrix} 1 & 1 \\ 1 & -1 \end{pmatrix}.$$

The value of $[\![\cdot]\!]$ on all other objects and arrows is then fixed by the requirement that it be a symmetric monoidal functor.[2]

Theorem 3.24 (Soundness) *For any diagrams D and D' in $\mathbb{D}$, if $D = D'$ then $[\![D]\!] = [\![D']\!]$ in **fdHilb**.*

Proof *Notice that the compact closed structure is preserved automatically because $[\![\cdot]\!]$ is a monoidal functor. It just remains to check that all the equations of Figure 3.3 hold in the image of $[\![\cdot]\!]$.* □

Remark 3.25 While Theorem 3.24 shows that every equation provable in the zx-calculus is true in Hilbert spaces, the converse does not hold: there are diagrams D and D' such that $[\![D]\!] = [\![D']\!]$ but the equation $D = D'$ cannot be derived from the rules of the calculus. See Duncan and Perdrix (2009) for details.

3.4.3 Quantum circuits

The quantum circuit model is simple and intuitive quantum computational model. Analogous to traditional Boolean circuits, a quantum circuit consists of a register of qubits to which quantum logic gates—that is one- or two-qubit unitary

---

[2] The full details of this construction regarding cyclic graphs and traces can be found in Duncan (2006).

$$Z = \begin{pmatrix} 1 & 0 \\ 0 & -1 \end{pmatrix} \qquad X = \begin{pmatrix} 0 & 1 \\ 1 & 0 \end{pmatrix}$$

$$Z_\alpha = \begin{pmatrix} 1 & 0 \\ 0 & e^{i\alpha} \end{pmatrix} \qquad H = \tfrac{1}{\sqrt{2}} \begin{pmatrix} 1 & 1 \\ 1 & -1 \end{pmatrix}$$

$$\wedge X = \begin{pmatrix} 1 & 0 & 0 & 0 \\ 0 & 1 & 0 & 0 \\ 0 & 0 & 0 & 1 \\ 0 & 0 & 1 & 0 \end{pmatrix} \qquad \wedge Z = \begin{pmatrix} 1 & 0 & 0 & 0 \\ 0 & 1 & 0 & 0 \\ 0 & 0 & 1 & 0 \\ 0 & 0 & 0 & -1 \end{pmatrix}$$

Fig 3.4 Quantum logic gates

operations—are applied, in sequence and in parallel.[3] A fairly typical set of logic gates is shown in Figure 3.4, but these are not all necessary, as the following theorem states.

Theorem 3.26 (Barenco et al., 1995) *The set $\{Z_\alpha, H, \wedge X\}$ suffices to generate all unitary matrices on Q^n.*

Corollary 3.27 *The zx-calculus can represent all unitary matrices on Q^n.*

Proof It suffices to show that there are zx-calculus terms for the matrices Z_α, H, and $\wedge X$. We have

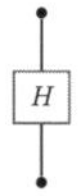

$$[\![\; \boxed{H} \;]\!] = H, \qquad [\![\; \alpha \;]\!] = Z_\alpha \quad \text{and} \quad [\![\; \bullet\!-\!\bullet \;]\!] = \wedge X$$

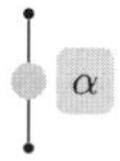

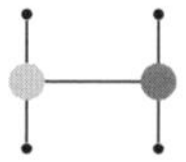

which can be verified by direct calculation. Note that

$$[\![\; \text{⋏} \;]\!] = [\![\; \text{⋎} \;]\!]$$

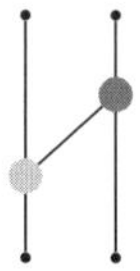

so the presentation of $\wedge X$ is unambiguous. □

Example 3.28 (The $\wedge Z$-gate) The $\wedge Z$-gate can be obtained by using a Hadamard (H) gate to transform the second qubit of a $\wedge X$ gate. We obtain a simpler representation using the colour-change rule

---

[3] The circuit model usually incorporates measurements too, but this will not be necessary here. See e.g. Chapter 4 of Nielsen and Chuang (2000).

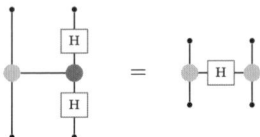

From the presentation of $\wedge Z$ in the zx-calculus, we can immediately read off that it is symmetric in its inputs. Furthermore, we can prove one of the basic properties of the $\wedge Z$ gate, namely that it is self-inverse.

Example 3.29 (Bell state) The following is a zx-calculus term representing a quantum circuit that produces a Bell state, $|00\rangle + |11\rangle$. We can verify this fact by the equations of the calculus.

The corresponding zx-calculus derivation is a proof of the correctness of this circuit.

The zx-calculus can represent many things that do not correspond to quantum circuits. We now present a criterion to recognize which diagrams do correspond to quantum circuits.

Definition 3.30 *A diagram is called* circuit-like *if:*

1. *all of its vertices can be covered by a set $\mathcal{P}$ of disjoint directed paths, each of which ends in an output*
2. *for every oriented cycle γ in the diagram, if γ contains at least 2 edges from different paths in $\mathcal{P}$, then it traverses at least one of them in the direction opposite to that induced by the path, and*
3. *it is a simple graph and is three-coloured.*

Intuitively, the paths of $\mathcal{P}$ represent the trajectories of the individual qubits through the circuit, whereas those edges not included in any path represent two-qubit gates. Condition 2 guarantees that the diagram has a causally consistent temporal order, while condition 3 forces the diagram to be minimal with respect to the spider, anti-loop, and Hopf rules.

Remark 3.31 Definition 3.30 requires each path to end in an output vertex, but does not demand that the initial vertex is an input. This allows the representation of quantum circuits with some or all inputs fixed; see Example 3.29 above.

Example 3.32 The following diagram is not circuit-like, since condition 2 fails.

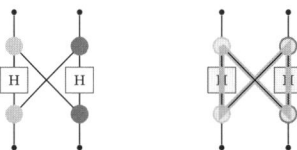

There is only one possible path covering of this diagram; the indicated cycle runs contrary to the path.

Example 3.33 The following diagram is circuit-like and, as shown, is equivalent to something which clearly *looks* like a circuit.

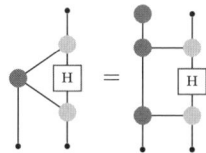

Of course, the right-hand diagram is *not* circuit-like because it does not satisfy condition 3; indeed, it reduces to the left-hand diagram by two applications of the spider rule.

As the preceding example shows, the definition of circuit-like—in particular the technical third condition—is rather stronger than strictly necessary to ensure that a diagram corresponds to a valid circuit. For example, it forces the two-qubit gates to be $\wedge X$ rather than $\wedge Z$. However it is not hard to prove the following result:

Proposition 3.34 *If $D : n \to m$ is a circuit-like diagram, then $[\![D]\!]$ is a unitary embedding $Q^n \to Q^m$; conversely, if $[\![D]\!]$ is a unitary embedding, then there exists some circuit-like D' such that $D = D'$ by the rules of the zx-calculus.*

3.4.4 Syntax and semantics II: measurements

While the version of the zx-calculus we have presented so far can represent the projection onto some measurement outcome, for example we have $[\![\,\begin{smallmatrix}\bullet\\|\\\bullet\end{smallmatrix}\,]\!] = \langle +|$, the physical process of measurement, including its non-deterministic aspect, cannot be represented. To address this we now introduce the notion of a $\mathcal{V}$-labelled diagram. This will require a modification to the syntax, and a new interpretation based on superoperators rather than **fdHilb**.

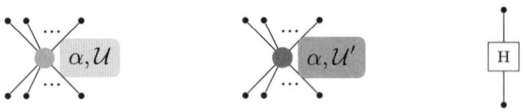

Fig 3.5 Interior vertices of $\mathcal{V}$-labelled diagrams

Definition 3.35 *Let $\mathcal{V}$ be some set of variables. A* conditional diagram *is a diagram (cf defn 3.20) where each Z or X vertex with $\alpha \neq 0$ is additionally labelled by a set $\mathcal{U} \subseteq \mathcal{V}$.*

If a vertex is labelled by a $\mathcal{U} \neq \varnothing$ then it is called *conditional*, otherwise it is *unconditional*. A diagram with no conditional vertices is called unconditional. The allowed vertices of $\mathcal{V}$-labelled diagrams are shown in Figure 3.5.

The equational rules must also be modified to take account of the labels: certain rewrites are only allowed when the variable sets agree, in a sense that will be made clear below. The updated rules are shown in Figure 3.6.

For any given $\mathcal{V}$, the $\mathcal{V}$-labelled diagrams, quotiented by the equations of Figure 3.6, form a †-compact category denoted $\mathbb{D}(\mathcal{V})$; $\mathbb{D}(\varnothing)$ is exactly the category $\mathbb{D}$ defined earlier.

Definition 3.36 *A function $v : \mathcal{V} \to \{0,1\}$ is called a* valuation *of $\mathcal{V}$; for each valuation v, we define a functor $\hat{v} : \mathbb{D}(\mathcal{V}) \to \mathbb{D}$ which produces a new diagram by relabelling the Z and X vertices. If a vertex z is labelled by α and $\mathcal{U}$, then $\hat{v}(z)$ is labelled by 0 if $\sum_{s \in \mathcal{U}} v(s) = 0$ and α otherwise.*

The modified rewrite rules for $\mathcal{V}$-labelled diagrams are simply the original equations, with the constraint that they should be true in all valuations.

Definition 3.37 *Let D be a diagram in $\mathbb{D}(\mathcal{V})$ such that every variable of $\mathcal{V}$ occurs in D. Define a symmetric monoidal functor $[\![\cdot]\!]_\mathcal{V} : \mathbb{D}(\mathcal{V}) \to$ **SuperOp** by setting $[\![1]\!]_\mathcal{V} = \mathbb{C}^2 \times \mathbb{C}^2$ and, for every diagram D, defining $[\![D]\!]_\mathcal{V}$ as the superoperator constructed by summing over all the valuations of $\mathcal{V}$:*

$$\rho \mapsto \sum_{v \in 2^\mathcal{V}} [\![\hat{v}(D)]\!] \rho [\![\hat{v}(D)]\!]^\dagger.$$

Proposition 3.38 *For any diagram D in $\mathbb{D}(\mathcal{V})$, we have $[\![D^\dagger]\!]_\mathcal{V} = [\![D]\!]_\mathcal{V}^\dagger$.*

Example 3.39 The following diagram represents the measurement of a single qubit in the basis $|\pm_\alpha\rangle = |0\rangle \pm e^{i\alpha}|1\rangle$.

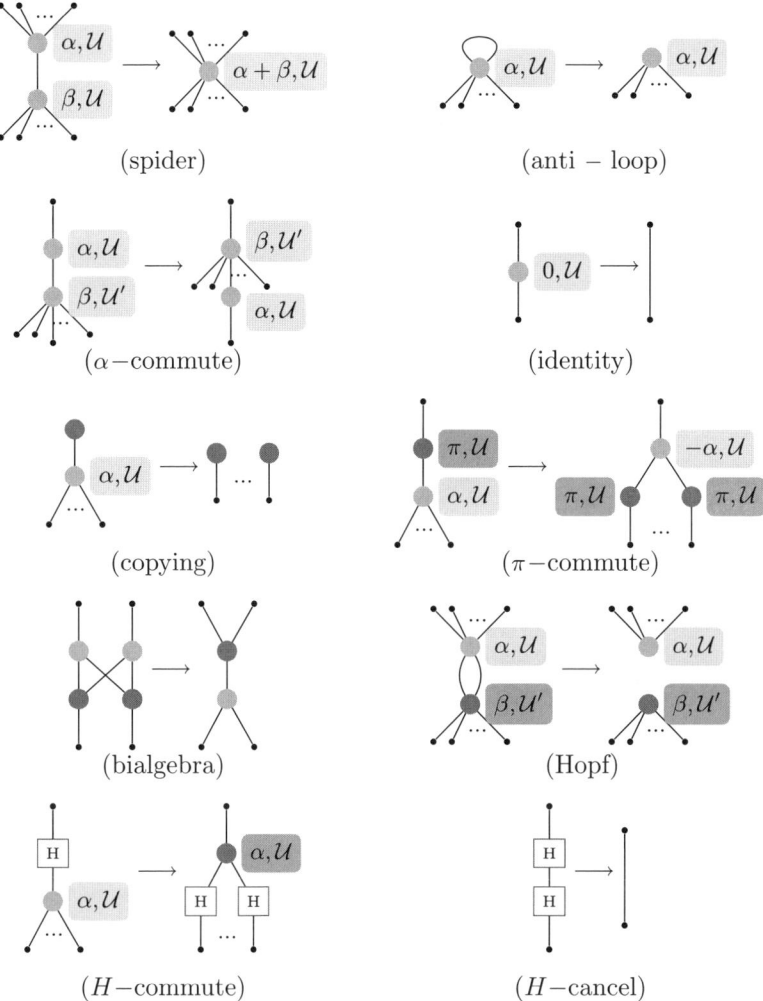

Fig 3.6 Rewrite rules for the zx-calculus with conditional vertices. We present the rules for the Z subsystem; to obtain the complete set of rules exchange the colours in the rules shown above.

The variable i encodes which of the two possible outcomes occurred, as can be seen by computing the denotation:

$$\rho \mapsto \sum_{v=0,1} [\![\hat{v}(PIC)]\!] \rho [\![\hat{v}(PIC)]\!]^{\dagger} = \langle +_\alpha | \rho | +_\alpha \rangle + \langle +_\alpha | Z \rho Z | +_\alpha \rangle$$

$$= \langle +_\alpha | \rho | +_\alpha \rangle + \langle -_\alpha | \rho | -_\alpha \rangle$$

Example 3.40 A classically controlled Pauli-Z operation is represented by the following diagram:

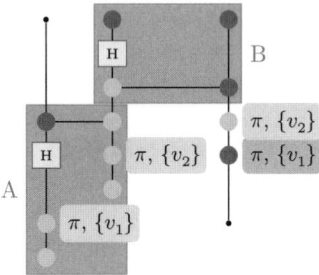

Example 3.41 Combining the two previous examples, we present the teleportation protocol (Bennett *et al.*, 1993).

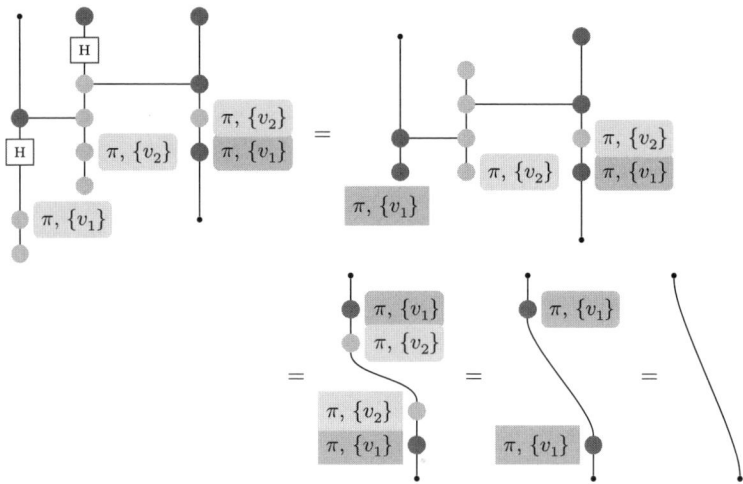

The block labelled B is the circuit to prepare a Bell state, while A represents the Bell basis measurement of two qubits. Notice that the same variables label the measurements as the corresponding correction operators, indicating that these operations must be correlated. We can rewrite this diagram to prove the correctness of the protocol.

Since all the conditional vertices are removed by the final step, we can conclude that the teleportation protocol is *deterministic*.

Example 3.42 It is also possible to write down diagrams that correspond to quite unphysical operations. For example the diagram

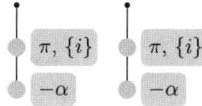

represents two one-qubit measurements whose outcomes are always perfectly correlated, regardless of the input. This is of course plainly impossible in quantum mechanics.

To avoid such situations, each single-qubit measurement must be labelled by a fresh variable; any other vertex labelled by the same variable must be interpreted as an operation that is classically controlled by the outcome of that measurement.

3.5 The measurement calculus

Now we turn our attention to the details of MBQC. While there are several approaches to MBQC, we will be concerned only with the *one-way model* (1WQC) introduced by Raussendorf and Briegel (Raussendorf and Briegel, 2001; Raussendorf and Briegel, 2003; Briegel *et al.*, 2009).

Whereas the quantum circuit model consists of reversible unitary gates, the 1WQC carries out computation via the irreversible state changes induced by quantum measurements. The computation begins with some input qubits coupled to a large entangled resource state, called a cluster state or *graph state*. Single qubit measurements are performed upon the state. Since the measured qubits are no longer entangled with the rest of the resource we can view the measurement process as consuming the resource, hence the name *one-way*. The computation proceeds by a number of rounds of measurement, where the choice of measurement performed in later rounds may depend on the observed outcomes of earlier measurements, until finally only the output qubits remain unmeasured. It may then be necessary to apply some single-qubit unitary corrections to obtain the desired result. Aside from the initial creation of the resource state, all the operations of the 1WQC act locally on a single qubit, so it is perhaps surprising that the 1WQC is universal for quantum computing: any unitary operation on n qubits may be computed by the 1WQC.

We shall formally describe the 1WQC using the syntax of the *measurement calculus* of Danos, Kashefi and Panangaden (2007). Measurement calculus programs, called *patterns*, consist of a (finite) set of qubits, a (finite) set of Boolean variables called *signals*, and a sequence of *commands* $C_n \ldots C_1 C_0$, read from right to left. The possible commands are:

- N_i : initialize qubit i in the state $|+\rangle$
- E_{ij} : entangle qubits i and j by apply applying a $\wedge Z$ operation
- M_i^α : measure qubit i in the basis $|\pm_\alpha\rangle := \frac{1}{\sqrt{2}}(|0\rangle \pm e^{i\alpha}|1\rangle)$; we assume the measurements are destructive, so qubit i will play no further role in the computation
- X_i and Z_i : apply a 1-qubit Pauli X (resp. Z) operator to qubit i; these are called *corrections*.

The measurement and correction commands may be classically controlled by the value of one or more signals.

- ${}^s[M_i^\alpha]^t$: perform the measurement $M_i^{(-1)^s\alpha + t\pi}$
- X_i^s and Z_i^s : if $s = 1$ perform the command X_i (resp. Z_i), otherwise do nothing.

In principle the signals could obtain their values from any source, but they will always be associated to the outcome of a measurement already performed in the same pattern.[4] If the $|+_\alpha\rangle$ outcome is obtained, the corresponding signal is set to zero; otherwise it set to one. This introduces the third of three determinacy conditions.

1. The initialization of a qubit is the first command acting on it.
2. The measurement of a qubit is the last command acting on it.
3. No command depends upon a measurement not already performed.

Any qubits not initialized are *inputs*; any qubit not measured is an *output*.

Example 3.43 Consider the two-qubit pattern

$$\mathfrak{P}_H := X_2^{s_1} M_1^0 E_{12} N_2.$$

Since qubit 1 is not initialized, it must be an input; similarly qubit 2 is an output. The only signal is associated to the measurement of qubit 1. Suppose that the input qubit is in state $|\psi\rangle = a|0\rangle + b|1\rangle$. The execution proceeds as follows:

$$|\psi\rangle \xrightarrow{N_2} |\psi\rangle \otimes |+\rangle = a|00\rangle + a|01\rangle + b|10\rangle + b|11\rangle$$
$$\xrightarrow{E_{12}} a|00\rangle + a|01\rangle + b|10\rangle - b|11\rangle$$
$$= |+\rangle(a|+\rangle + b|-\rangle) + |-\rangle(a|+\rangle - b|-\rangle)$$

So far all we have done is construct the initial entanglement. The next step is the measurement M_1^0. Suppose that the result of the measurement is 0, i.e. the projection

---

[4] It is easy to generalize the conditional commands to allow classical control by some arithmetic expression of signals, but this is not required here, and indeed does not increase the expressiveness of the measurement calculus.

onto $|+\rangle$; then qubit 1 is eliminated and we have the new state $a|+\rangle + b|-\rangle$. Since the signal s_1 is zero, there is no need to perform the final correction. On the other hand, should the outcome of the measurement be 1, the resulting state will be $a|+\rangle - b|-\rangle$, and since $s_1 = 1$ the X_2 correction must be applied, again producing a final state of $a|+\rangle + b|-\rangle$. Hence the overall effect of $\mathfrak{P}_H$ is independent of the outcome of the measurement; in either case the computer applies a Hadamard gate upon its input.

This example illustrates a key feature of the 1WQC. The measurement introduces a branch in the execution where one outcome corresponds to the 'correct' behaviour, in the sense that the projection achieves the desired computational effect, and one branch contains an 'error' that must be corrected later in the pattern. More generally, a pattern with n measurements potentially performs 2^n different linear maps on its input. These are called the *branch maps*; the branch where all the measurements output 0 is called the *positive branch*, and we use the convention that this branch is the computation that we intend to carry out. The pattern is deterministic if all the branch maps have the same effect on the input as the positive branch.

Remark 3.44 The concept of branch map is used in Danos, Kashefi and Panangaden (2007) to define the semantics of the measurement calculus. We omit this, because we shortly provide a semantics via a translation into the zx-calculus. The interpretation presented here is equivalent to that of Danos, Kashefi and Panangaden (2007).

Theorem 3.45 *For any pattern $\mathfrak{P}$, there exists an equivalent pattern—in the sense of having the same semantics—whose command sequence has the form*

$$\mathfrak{P}^* = CMEN$$

where C, M, E, and N are sequences of commands consisting exclusively of corrections, measurements, entangling operations, and initializations respectively.

The pattern $\mathfrak{P}^*$ is said to be in *standard form*; any pattern may be transformed into an equivalent by a rewrite procedure presented in Danos, Kashefi and Panangaden (2007). There are two main ingredients to this procedure. The first are the commuting relations between the Pauli matrices and the $\wedge Z$. The second is the following identity, allowing corrections to be absorbed into conditional measurements:

$$^s[M_i^\alpha]^t = M_i^\alpha Z_i^t X_i^s. \tag{3.3}$$

However, this will also function as the *definition* of the conditional measurement, in which corrections are only conditional commands. Needless to say, with this convention Theorem 3.45 no longer applies. We will adopt the following convention

TABLE 3.1 Translation from pattern to diagram

N_i	E_{ij}	M_i^α	X_i^s	Z_i^s
●	●—H—●	● $\pi, \{i\}$ $-\alpha$	● $\pi, \{s\}$	● $\pi, \{s\}$

for the rest of the chapter: patterns are always assumed to be in standard form, including conditional measurements, which are then replaced with unconditional measurements, as per eqn (3.3).

Definition 3.46 *Let $\mathfrak{P}$ be a pattern with qubits ranged over by V, and let $I, O \subseteq V$ denote its input and output qubits respectively. We define a diagram $D(\mathfrak{P}) : |I| \to |O|$ in $\mathbb{D}(V \setminus I)$, by composing the subdiagrams corresponding to the command sequence of $\mathfrak{P}$, as shown in Table 3.1.*

Since the semantics of the diagram $D(\mathfrak{P})$ capture the meaning of the pattern $\mathfrak{P}$, reasoning in the zx-calculus allows properties of the pattern to be derived by graphical manipulations of $D(\mathfrak{P})$. Let us reconsider Example 3.43.

Example 3.47 The pattern $\mathfrak{P}_H$ produces the following diagram:

$$D(\mathfrak{P}_H) := \quad \pi, \{s_1\} \quad \text{—H—} \quad \pi, \{s_1\}$$

Note that the vertical wires correspond to the physical qubits; we shall stick to this convention whenever convenient. Now we can deduce that $\mathfrak{P}_H$ computes the Hadamard gate using purely diagrammatic reasoning:

$$\pi, \{s_1\} \; \text{—H—} \; \pi, \{s_1\} \;=\; \pi, \{s_1\} \; \text{—H—} \; \pi, \{s_1\} \;=\; \pi, \{s_1\} \; \text{—H—} \; \pi, \{s_1\}$$

$$=\; \pi, \{s_1\} \; \text{—H—} \; \pi, \{s_1\} \;=\; \boxed{H}$$

80 | MEASUREMENT-BASED QUANTUM COMPUTING

Example 3.48 The ubiquitous CNOT operation can be computed by the pattern $\mathfrak{P}_{\wedge X} = X_4^3 Z_4^2 Z_1^2 M_3^0 M_2^0 E_{13} E_{23} E_{34} N_3 N_4$ (Danos, Kashefi and Panangaden, 2007). This yields the diagram

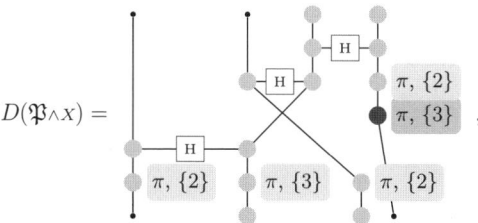

with qubit 1 the leftmost, and qubit 4 is the rightmost. Now, we can prove the correctness of the pattern by rewriting:

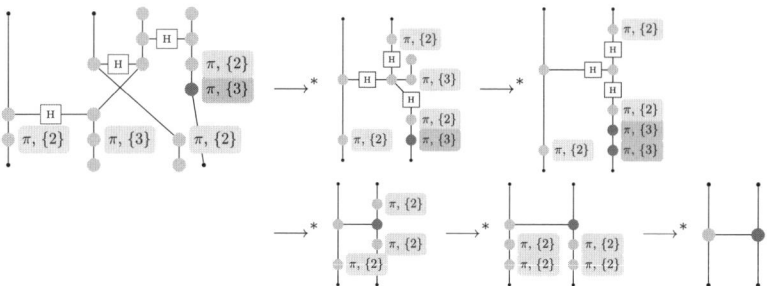

One can clearly see in this example how the non-determinism introduced by measurements is corrected by conditional operations later in the pattern. The possibility of performing such corrections depends on the *geometry* of the pattern, the entanglement graph implicitly defined by the pattern. This will be the main concern of the next section.

3.6 Determinism and flow

Consider the pattern $\mathfrak{N} = Z_1^{s_2} M_2^0 M_3^\alpha E_{12} E_{23} N_2 N_3$. Working in the zx-calculus, it can be rewritten as follows:

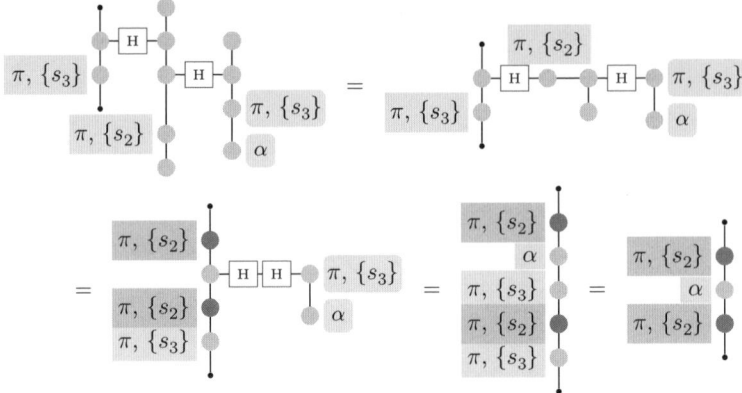

Hence we have a pattern that is non-deterministic, acting as a Z-rotation by either α or $-\alpha$ depending on the outcome of measurement 2. Furthermore, unlike the previous examples, there is no way to remove the dependence on s_2 by correction at qubit 1, or by conditional measurement at qubit 3.

As this example shows, not every pattern performs a deterministic computation, and this possibility depends not just upon the conditional operations introduced by the programmer, but also the structure of the entangled resource used in the computation. This structure is called the *geometry* of the pattern. A pattern can perform a deterministic computation if its geometry has a graph-theoretic property called *flow*. Examples 3.43 and 3.48 have flow, while the pattern $\mathfrak{N}$ above does not.

Flow is a sufficient property for determinism, but not a necessary one: it guarantees *strong* and *uniform* determinism. Strong determinism means that all the branch maps of the pattern are equal, while uniformity means that the pattern is deterministic for all possible choices of its measurement angles. Before moving on, we make the following obvious observation:

Theorem 3.49 *If $D(\mathfrak{P})$ can be rewritten to an unconditional diagram then $\mathfrak{P}$ is strongly deterministic.*

Hence to show that a pattern is deterministic, it suffices to find some rewrite sequence that removes all the conditional vertices. Typically this is done by 'pushing' the conditional vertex introduced by measurement through the diagram until it meets a matching corrector. The two conditional vertices then cancel each other out. The rest of this section will explore when this is possible.

Definition 3.50 *The* geometry *of a pattern $\mathfrak{P}$, denoted $\Gamma_{\mathfrak{P}}$, is the open graph $((V, E), I, O)$ defined by taking the qubits of $\mathfrak{P}$ as vertices V, the input and output* qubits as the sets I and O, and defining the edge relation by $v \sim u$ if and only if the command E_{vu} occurs in $\mathfrak{P}$.

Example 3.51 Consider $\mathfrak{P}_{\wedge X} = X_4^3 Z_4^2 Z_1^2 M_3^0 M_2^0 E_{13} E_{23} E_{34} N_3$ as in Example 3.48. We then have

$$\Gamma_{\mathfrak{P}_{\wedge X}} = $$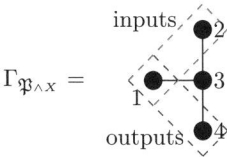

Definition 3.52 Let $G = ((V, E), I, O)$ be an open graph; a flow on G is a pair $(f, \prec)$, where f is a function $V \setminus O \to V \setminus I$ and $\prec$ is a partial order on V, satisfying

(F1). $f(u) \sim u$
(F2). $u \prec f(u)$
(F3). if $f(u) \sim v$ and $u \neq v$ then $u \prec v$.

Intuitively, the function f specifies a causal successor for every measured qubit. Should the measurement at qubit i give the 'wrong' answer, then a conditional operation at qubit $f(i)$ can be used to correct the resulting error. The partial order ensures that no causal loops can form; that is, it is not necessary to apply a correction to a qubit that has already been measured. We have the following:

Theorem 3.53 (Danos and Kashefi, 2006) If G is an open graph with flow then there exists a strongly and uniformly deterministic pattern $\mathfrak{P}$ such that $G = \Gamma_{\mathfrak{P}}$.

It must be emphasized that flow is a property of the geometry *not* the pattern itself. In order for a pattern to be deterministic, correct placement of the conditional operations is still required. The pattern $\mathfrak{P}$ is given explicitly in Danos and Kashefi (2006); we will reconstruct this later.

Definition 3.54 Let $\Gamma = ((V, E), I, O)$ be an open graph; we define an unconditional diagram $D(\Gamma)$ as follows.

- The vertices of $D(\Gamma)$ are given by the disjoint union $V + E + I + O$. Should Γ have vertex v contained in both I and O then $D(\Gamma)$ contains three corresponding vertices in $D(\Gamma)$; we use subscripts v_V, v_I, v_O to disambiguate.
- The vertices are typed depending which disjoint subset they originate in: those from V (the original vertices of Γ) have type Z, without any label; those from E have type H; and those from $I + O$ are boundary vertices, with I providing the inputs and O the outputs.
- If e is an edge in Γ connecting vertices u and v, then we have $u_V \sim e$ and $e \sim v_V$ in $D(\Gamma)$. For the boundary vertices we have $v_I \sim v_V$ and $v_O \sim v_V$.

Example 3.55 Consider again the $\wedge X$ pattern, or rather its geometry.

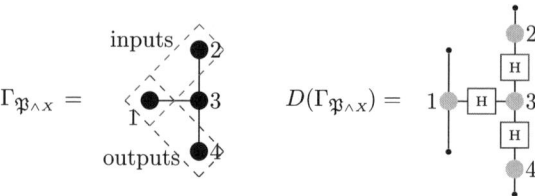

Theorem 3.56 *Let $\mathfrak{P}$ be a pattern; if $\Gamma_\mathfrak{P}$ has flow then $D(\Gamma_\mathfrak{P})$ is equivalent to a circuit-like diagram.*

Proof Suppose that $\Gamma_\mathfrak{P}$ has a flow $(f, \prec)$. Let J be the vertices of $\Gamma_\mathfrak{P}$ that are minimal with respect to $\prec$. For each vertex $j \in J$ we can define a finite sequence $p_j = j, f(j), f^2(j), \ldots, f^n(j)$ where the last element of the sequence is an output qubit. By definition of f, the collection $\cup_{j \in J} p_j$ contains all the vertices of $\Gamma_\mathfrak{P}$. Each p_j defines a path in $D(\Gamma_\mathfrak{P})$, and the collection of these paths covers all the Z vertices of $D(\Gamma_\mathfrak{P})$; we can trivially extend these paths to include the boundary vertices adjacent to their end points. The collection $\{p_j\}_{j \in J}$ provides the path covering required by the definition of circuit-like (cf. defn 3.30). $D(\Gamma_\mathfrak{P})$ satisfies condition 2, because of the partial order structure of the flow, and condition 3 by construction, however it does not satisfy condition 1, since some vertices are not covered by the path. Specifically, those H vertices e where $v \sim e \sim u$ and $f(u) \neq v$ are not covered by the path. At each such vertex we perform the following rewrite:

Now the H vertices can be removed via the rewrite shown below:

…after which any pairs of adjacent H vertices may be cancelled, and the spider rule can be used to guarantee that the diagram is three coloured. □

The converse to Theorem 3.56 does not hold; in Duncan and Perdrix (2010) it is shown that geometries that have *generalized flow*, discussed below, can be rewritten to circuit-like diagrams. However, we can give a weaker result that holds for flow.

Definition 3.57 *Let D be a diagram, and let $U \subseteq V$ be a set of its vertices satisfying:*

- *$u \in U$ implies that u has type H*
- *$v_1 \sim u \sim v_2$ implies that v_1 and v_2 are either both of type Z or both of X.*

Then D is called weakly circuit-like *if the following conditions hold.*

1. *The vertices $V \setminus U$ can be covered by a set $\mathcal{P}$ of disjoint directed paths, each of which ends in an output.*
2. *For every oriented cycle γ in the diagram, if γ contains at least two edges from different paths in $\mathcal{P}$, then it traverses at least one of them in the direction opposite to that induced by the path; and,*
3. *It is a simple graph and is three-coloured.*

Weakly circuit-like diagrams correspond to circuits where the two-qubit gates may be $\wedge Z$ gates as well as $\wedge X$ gates.

Theorem 3.58 *Let $\mathfrak{P}$ be a pattern; if $D(\Gamma_\mathfrak{P})$ is weakly circuit-like then $\Gamma_\mathfrak{P}$ has flow.*

Proof By construction, the Z vertices are in bijective correspondence with the qubits of $\mathfrak{P}$; hence we need only define a flow (f, $\prec$) over the Z vertices. Let p be one of the paths of the path covering of $D(\Gamma_\mathfrak{P})$; p then defines a linear order over the Z vertices it covers. Define f by $f(u) = v$ whenever v is the successor of u in this order. Since the paths are disjoint, the same procedure can be carried out for every path, and since the paths cover all the Z vertices this defines the required function f. The union of these linear orders gives a partial order over the Z vertices; to obtain the required $\prec$ this order can be completed by imposing condition (F3). The acyclicity condition 2 guarantees that this is possible. $\square$

Evidently, the circuit-like diagram produced from $D(\Gamma_\mathfrak{P})$ is not equivalent to $D(\mathfrak{P})$, but they are closely related.

Definition 3.59 *Let $\mathfrak{P}$ be pattern; construct a new diagram $D(\Gamma_\mathfrak{P})^*$ from $D(\Gamma_\mathfrak{P})$ as follows.*

- *If ${}^s[M_i^\alpha]^t$ occurs in $\mathfrak{P}$ then modify $D(\Gamma_\mathfrak{P})$ by adjoining the subdiagram corresponding to the measurement, as shown below:*

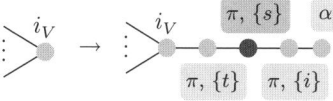

- *If X_i^s or Z_i^s occurs in $\mathfrak{P}$ then adjoin the subdiagram corresponding to the correction as shown below:*

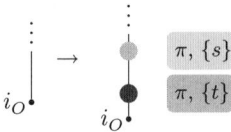

Note that since the $\mathfrak{P}$ is in standard form, corrections can only appear at an output qubit.

Example 3.60 Recall the pattern $\mathfrak{P}_{\wedge X} = X_4^3 Z_4^2 Z_1^2 M_3^0 M_2^0 E_{13} E_{23} E_{34} N_3 N_4$. We have:

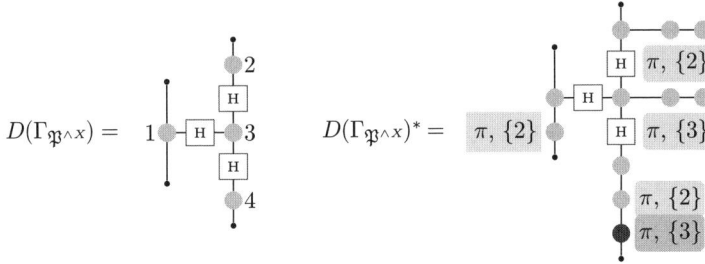

Lemma 3.61 *For any pattern $\mathfrak{P}$ we have $D(\mathfrak{P}) \to D(\Gamma_\mathfrak{P})^*$*

Proof All the Z vertices in $D(\mathfrak{P})$ that are introduced by N_i and E_{ij} commands form a connected subgraph, hence they can all be contracted together via the spider rule; this gives $D(\Gamma_\mathfrak{P})^*$. □

Corollary 3.62 *If $\mathfrak{P}$ has flow then the subgraph of $D(\mathfrak{P})$ excluding the measurements rewrites to a circuit-like diagram.*

If $\mathfrak{P}$ has a circuit-like geometry then Theorem 3.53 shows that it could be deterministic, if there are corrections in the appropriate places. The zx-calculus can be used to determine where the corrections must be placed. Suppose that we have the configuration shown below. (The dotted boxes represent either measurements or outputs.)

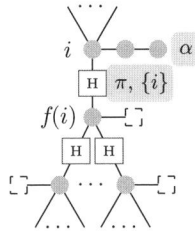

The measurement at qubit *i* introduces an error term that must be cancelled at a later qubit. We can perform the following rewrite sequence:

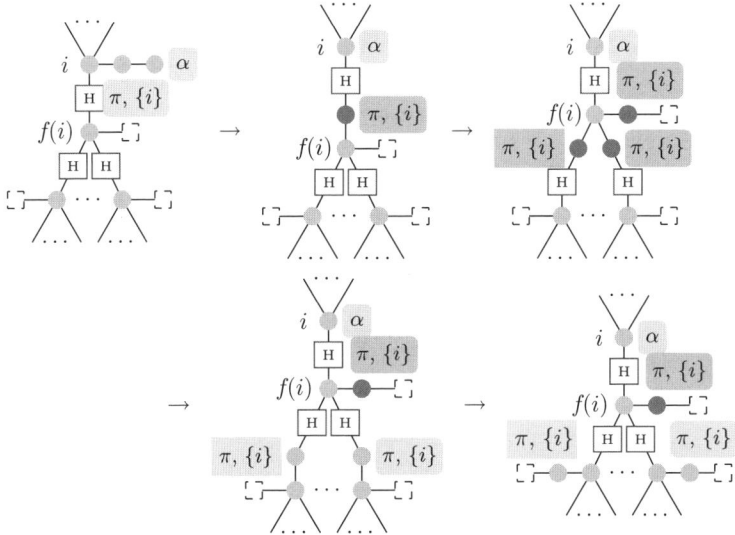

Hence in order to correct for the measurement at qubit *i*, we must perform a conditional X at qubit $f(i)$, and a condition Z at all its neighbours except i itself. Since the geometry is circuit-like all of these qubits are later in the execution of the pattern than the measurement of *i* itself. Hence we can conclude as follows.

Theorem 3.63 *Let $\mathfrak{P}$ be a pattern such that $\Gamma_{\mathfrak{P}}$ has flow and, for every measured qubit i, the command sequence contains correctors $X_{f(i)}^i$ and Z_j^i for all $i \neq j \sim f(i)$; then $D(\mathfrak{P})$ rewrites to an unconditional circuit-like diagram.*

Of course, if *j* is a measured qubit the corrections can be absorbed into the measurements.

The *Z* correction at qubit *j* is effectively the same as an error introduced by measuring *j*, hence this operation can be deferred by another step, and the correction performed at $f(j)$ and its neighbours. However, the same is not true for the corrector *X*. Consider the following diagrams:

In this case we have $[\![D_1]\!] \neq [\![D_2]\!]$ unless $\alpha = 0$ or $\alpha = \pi$, hence these diagrams cannot be rewritten to one another. Therefore uniform determinism requires that the X correction be performed at $f(i)$ and not later in the computation. On the other hand, if $\alpha \in \{0, \pi\}$ then these diagrams are equal. This points to an important advantage of the zx-calculus.

The diagram $D(\mathfrak{P})$ contains all the information of the pattern itself, not just its geometry. Therefore, by rewriting as discussed above, the correctness of $\mathfrak{P}$ can be verified directly and, in particular, should the MBQC programmer have made an error in the placement of the corrections this error will be revealed. Furthermore, since the zx-calculus is sensitive to the values of the angles, it can also be used to show that a pattern is deterministic even when it is not uniformly so. Consider the pattern $\mathfrak{P} = M_3^0 M_2^\alpha E_{23} E_{12} N_2 N_3$. This pattern does not have flow. However it is deterministic:

The disconnected component is just a scalar factor, which we drop since it has no bearing on the computation.

In this chapter we have focused on flow, but flow is not a necessary condition for strong and uniform determinism. With flow, each measurement has a single successor where the correction must be performed. If each qubit has instead a set of correcting qubits this yields the notion of *generalized flow* (Browne et al., 2007). A computation is called *stepwise* deterministic if after each measurement the non-determinism can be removed by correction. Generalized flow is both necessary and sufficient for strong, stepwise, uniform determinism. The zx-calculus can also be used to handle generalized flow; in fact using the rewrite rules, especially the bialgebra rule, any geometry that has generalized flow can be rewritten to an equivalent pattern that has flow. The details can be found in Duncan and Perdrix (2010).

3.7 Conclusions

Complementarity has long been recognized as one of the fundamental ingredients of quantum mechanics, although it is usually understood negatively, as the failure of certain classical properties. Here we have demonstrated a positive characterization of complementarity in terms of the existence of certain algebraic structures. The zx-calculus provides a very rich language for reasoning about quantum systems that fully exploits this algebraic structure.

Due to its graphical nature, the zx-calculus is extremely legible, exposing the close parallels between quantum circuits and 1WQC patterns with flow. In the preceding section we have seen how the almost metaphorical property of flow actually defines a trajectory within a diagram along which information—in this the outcome of some quantum measurement—must travel in order to produce a deterministic computation. This analysis can be taken further in the analysis of *generalized flow* (Duncan and Perdrix, 2010), where the role of the bialgebra rule is crucial, functioning as a kind 'interference' principle, where different paths cancel out.

The graphical syntax we have employed here is not simply a gimmick. Since the zx-calculus is based on algebraic first principles, it can unify differing computational paradigms such quantum circuits and the 1WQC in a single setting. Furthermore, since it is based on graphs it is amenable to automation, opening the door to mechanized reasoning about quantum programs (see Quantomatic homepage at https://sites.google.com/site/quantomatic/).

Acknowledgements

This chapter is based on work originally carried in collaboration with Bob Coecke (Coecke and Duncan, 2008; Coecke and Duncan, 2011) and Simon Perdrix (Duncan and Perdrix, 2009; Duncan and Perdrix, 2010). The author is supported financially by the FRS-FNRS.

CHAPTER 4

Quantum Groups and Braided Algebra

SHAHN MAJID
(Queen Mary, University of London)

4.1 Introduction

What *is* a quantum group? Here is a non-technical explanation for a very general audience. One answer is that it is a new, more general notion of symmetry called a 'quantum symmetry'. This means that it is an algebra but with enough other structure so as to to be able to play the role of a symmetry in a generalized sense. This additional structure is what I call a 'xerox map' Δ, which allows, in the case of a classical group G, to duplicate elements in the sense $\Delta g = g \otimes g$ for each element $g \in G$. This duplication is needed because when you say that an element is a symmetry of a picture, for example, what you really mean is that if x, y in the picture enjoy a certain relation then the transformed points $g.x, g.y$ enjoy the same relation. There is also a map ϵ needed to express when something is unchanged under the action and a map S that plays the role of inverse. The other ingredient—as in quantum computing—is to extend things linearly. Thus an actual group is a trivial example of a quantum group built on the vector space kG over a field k with basis $\{g \in G\}$, forming an algebra by extending the group product linearly, and additional structures

$$\Delta g = g \otimes g, \quad \epsilon g = 1, \quad Sg = g^{-1}$$

for all $g \in G$. In general, a quantum group is an associative unital algebra H equipped with algebra-preserving maps $\Delta : H \to H \otimes H, \epsilon : H \to k$ (the 'coproduct' and 'counit' respectively) and a linear map $S : H \to H$ (the 'antipode') subject to

$$(\Delta \otimes \mathrm{id})\Delta = (\mathrm{id} \otimes \Delta)\Delta, \quad (\epsilon \otimes \mathrm{id})\Delta = (\mathrm{id} \otimes \epsilon)\Delta$$

and

$$\cdot(\mathrm{id} \otimes S)\Delta = \cdot(S \otimes \mathrm{id})\Delta = 1\epsilon$$

Although proposed in the 1940s as 'Hopf algebras', the subject of quantum groups took off in the 1980s with the appearance *en masse* of truly more general examples beyond classical groups and Lie algebras. These remain the two main classes of true quantum groups known so far.

1. The bicrossproduct ones, due to the author, and associated to a local Lie group factorization (or a linear splitting of a Lie algebra) (Majid, 1988; Majid, 1991*d*). They are relevant to quantum gravity as quantizations of the Poincaré group or Euclidean group of motions and are behind the prediction (Amelino-Camelia and Majid, 2000) that the speed of light is not quite constant due to quantum gravity effects.
2. The q-deformation quantum groups $U_q(\mathfrak{g})$ of Drinfeld and Jimbo, associated to complex simple Lie algebras (Drinfeld, 1987). The importance of these is a 'quasitriangular structure' whereby they generate braided categories as their representation category. As such, they have revolutionized the theory of knots and three-manifolds as well as being objects of study in their own right.

There are by now many books on quantum groups but we refer particularly to Majid (1995*a*, 2002), in which we took a 'braided algebra' approach that allows, for example, the construction of $U_q(\mathfrak{g})$ by means of braid and tangle diagrams. Key to this was a theory of 'braided groups' or Hopf algebras in braided categories and their 'wiring up', much as one wires up the chips in a computer (Majid, 1993*a*; Majid, 1991*a*; Majid, 1994*a*). This was my own 'slant' on the Drinfeld–Jimbo quantum groups developed over about 60 papers in the 1990s and is our topic here. I have also proposed such objects as the correct syntax for certain types of 'topological quantum computer' based in braided categories although this has yet to be demonstrated.

We start in Section 4.2 with a preliminary discussion of braided categories, within which we shall work, and a central theorem due to myself and independently to V.G. Drinfeld about the 'centre' of a monoidal category. We then illustrate how to do group and quantum group theory in such a braided category in Section 4.3 (Majid, 1995*a*; Majid, 2002; Majid, 1994*a*), including the examples of the braided line and quantum-braided plane. Section 4.4 outlines the more advanced theory of braided groups covering the most important known constructions and some applications. This section is necessarily more technical than the others. Section 4.5 explains how braided algebra solved the Lie algebra problem for quantum groups $U_q(\mathfrak{g})$—a finite-dimensional object that generates the quantum group and which is subject to self-contained axioms (Majid, 1994*c*). The theory also includes an application (Lopez Pena, Majid and Rietsch, 2010) to the theory of finite simple groups where any finite group can be given a (trivially braided) braided-Lie algebra structure. Section 4.6 concludes with a taste of how one can do geometry in the form of gauge theory and, ultimately, Riemannian geometry at this level of braid and tangle diagrams. This discussion is taken from Majid (1999c, 1999a).

4.2 Braided categories

Categories feature in many of the chapters in this volume so suffice it to say that a category $\mathcal{C}$ for our purposes is just the following.

1. A collection of *objects* $V, W, Z, U, \ldots$
2. A specification of a set $\mathrm{Mor}(V, W)$ of allowed 'maps' $V \to W$ for each V, W.
3. A composition operation $\circ : \mathrm{Mor}(W, Z) \times \mathrm{Mor}(V, W) \to \mathrm{Mor}(V, Z)$ with properties analogous to the composition of maps.
4. A requirement that every set $\mathrm{Mor}(V, V)$ should contain an identity element id_V such that $\phi \circ \mathrm{id} = \phi$, $\mathrm{id} \circ \phi = \phi$ for any morphism for which $\circ$ is defined.

In our case all objects will be concrete sets with structure and morphisms will often be certain classes of maps between sets, although neither assumption is required. In particular, we indicate objects as $V \in \mathcal{C}$ by an abuse of set theory notations. Informally, a 'map' $F : \mathcal{C} \to \mathcal{V}$ between categories is called a *functor*—but note that this does not just map over objects but also morphisms between them in a compatible way so that

$$F(\phi \circ \psi) = F(\phi) \circ F(\psi),$$

for all composable morphisms ϕ, ψ. A natural transformation $\theta : F \to G$ or $\theta \in \underline{\mathrm{Nat}}(F, G)$ between two such functors means a collection $\{\theta_V : F(V) \to G(V) \mid V \in \mathcal{C}\}$ of morphisms in $\mathcal{V}$ that are 'functorial' in the sense that

$$\theta_W \circ F(\phi) = G(\phi) \circ \theta_V, \quad \forall \phi : V \to W$$

The natural transformation θ is called a 'natural isomorphism' if each θ_V is an isomorphism. There are similar notions for contravariant functors.

Definition 4.1 *A monoidal category is* $(\mathcal{C}, \otimes, \underline{1}, \Phi, l, r)$, *where:*

1. $\mathcal{C}$ *is a category*
2. $\otimes : \mathcal{C} \times \mathcal{C} \to \mathcal{C}$ *is a functor*
3. *a natural isomorphism* $\Phi : (\otimes) \otimes \to \otimes(\otimes)$, *i.e. a collection of functorial isomorphisms*

$$\Phi_{V,W,Z} : (V \otimes W) \otimes Z \cong V \otimes (W \otimes Z), \quad \forall V, W, Z \in \mathcal{C},$$

obeying the 'pentagon condition' in Figure 4.1.

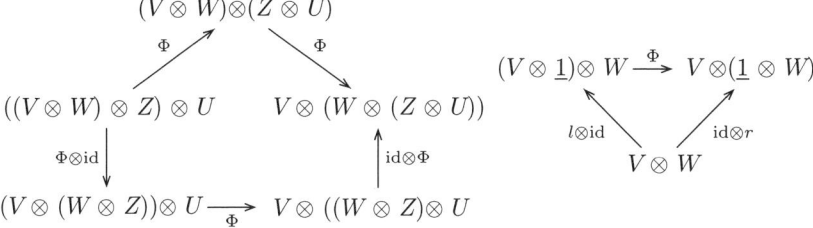

Fig 4.1 Pentagon condition for Φ and triangle condition for compatibility with *l, r*.

4. *a unit object* $\underline{1}$ *and associated natural isomorphisms* $l : \mathrm{id} \to (\ ) \otimes \underline{1}$, $r : \mathrm{id} \to \underline{1} \otimes (\ )$, *i.e. a collection of functorial isomorphisms, obeying the 'triangle condition' in Figure 4.1.*

In practice we can just omit brackets and write expressions such as $V \otimes W \otimes Z \otimes U$ quite freely. There will be several ways to fill in the brackets and Φ in our expressions, but Mac Lane's coherence theorem (Mac Lane, 1971) says that all the different ways will coincide. The maps l, r associated to the unit object also take care of themselves once the consistency condition stated is satisfied. We will therefore soon suppress all these maps.

My favourite construction for monoidal categories is the notion of a 'representation', which was introduced in Majid (1991g, 1992a). The general setting is for objects of Mon/$\mathcal{V}$ consisting of a monoidal category $\mathcal{C}$ and a monoidal functor $F : \mathcal{C} \to \mathcal{V}$ to a fixed monoidal category $\mathcal{V}$. A monoidal functor by definition comes with a natural isomorphism $f \in \underline{\mathrm{Nat}}(\otimes F^2, F(\otimes))$, i.e. a functorial collection of isomorphisms

$$f_{X,Y} : F(X) \otimes F(Y) \cong F(X \otimes Y), \quad \forall X, Y \in \mathcal{C}$$

subject to an obvious compatibility with the associators and units of $\mathcal{C}, \mathcal{V}$. We say that $\mathcal{C}$ is 'fibred' over $\mathcal{V}$.

In this context a right-representation of $\mathcal{C}$ in $\mathcal{V}$ is a pair (V, λ_V) where $V \in \mathcal{V}$ and $\lambda_V \in \underline{\mathrm{Nat}}(V \otimes F, F \otimes V)$, i.e. a functorial collection of morphisms

$$\lambda_{V,X} : V \otimes F(X) \to F(X) \otimes V, \quad \forall X \in \mathcal{C}$$

such that

$$\lambda_{V,\underline{1}} = \mathrm{id}, \quad (\mathrm{id} \otimes \lambda_{V,Y})(\lambda_{V,X} \otimes \mathrm{id}) = (f_{X,Y}^{-1} \otimes \mathrm{id})\lambda_{V,X\otimes Y}(\mathrm{id} \otimes f_{X,Y})$$

We denote the category of all such representations by $\mathcal{C}^\circ$ (with an obvious notion of morphism), and a similar notion with left-representation by $^\circ\mathcal{C}$.

Theorem 4.2 (Majid, 1991g; Majid, 1992a) $\mathcal{C}^\circ$ *and* $^\circ\mathcal{C}$ *are monoidal categories fibred over* $\mathcal{V}$ *by the forgetful functor. Moreover, there is a canonical inclusion* $\mathcal{C} \subseteq {}^\circ(\mathcal{C}^\circ)$ *as fibred monoidal categories.*

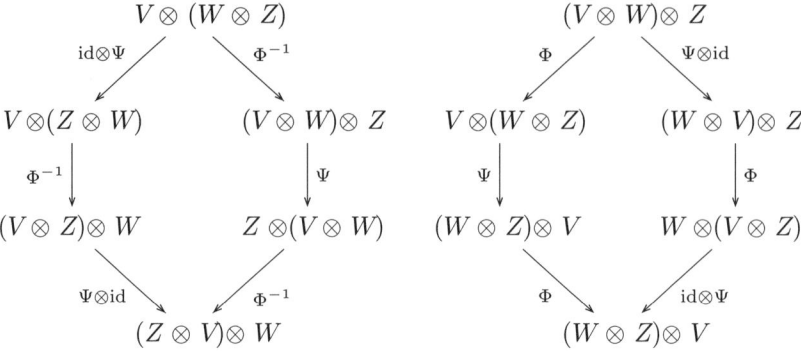

Fig 4.2 Hexagon conditions for Ψ.

Proof The heart of the proof is that if $(V.\lambda_V)$ and (W, λ_W) are such representations then

$$(V \otimes W, \lambda_{V \otimes W}), \quad \lambda_{V \otimes W, X} = (\lambda_{V,X} \otimes \mathrm{id})(\mathrm{id} \otimes \lambda_{W,X}), \quad \forall X \in \mathcal{C}$$

is a representation. The trivial representation is $\lambda_{\underline{1},X} = \mathrm{id}$. □

When objects have duals (are 'rigid') the λ are invertible, or one can just define $\mathcal{C}^\circ$ with isomorphisms from the outset. In this case we can use the inverse of a left-representation to give a right-representation and work only with these. It is my favourite theorem because it is a vast generalization of the Pontryagin theorem for abelian groups and I have taken the view (Majid, 1991e) that such duality ideas are at the heart of the notion of quantum gravity as well as of the notion of physical reality in general (see my answer to Plato's cave in Majid, 2007).

Definition 4.3 *A braided monoidal category $(\mathcal{C}, \otimes, \underline{1}, \Phi, l, r, \Psi)$ is:*

1. *a monoidal category $(\mathcal{C}, \otimes, \underline{1}, \Phi, l, r)$*
2. *a natural isomorphism $\Psi : \otimes \to \otimes^{\mathrm{op}}$, i.e. a collection of functorial isomorphisms*

$$\Psi_{V,W} : V \otimes W \cong W \otimes V, \quad \forall V, W \in \mathcal{C},$$

obeying the 'hexagon condition' in Figure 4.2; here $\otimes^{\mathrm{op}}(V, W) = W \otimes V$.

The model is the usual flip or transposition map $V \otimes W \cong W \otimes V$ for vector spaces. Note that we do not assume that $\Psi_{V,W} = \Psi_{W,V}^{-1}$ (if this holds for all V, W then we have a 'symmetric' monoidal category). The braided case is due to Joyal and Street (1993).

All of the above leads to the following convenient notation. We write morphisms pointing generally downwards (say) and denote tensor product by horizontal juxtaposition. Instead of the usual arrow for Ψ, Ψ^{-1} we use the shorthand

$$\Psi_{V,W} = \underset{W\quad V}{\overset{V\quad W}{\diagdown\mkern-14mu\diagup}}\,, \quad (\Psi_{W,V})^{-1} = \underset{W\quad V}{\overset{V\quad W}{\diagup\mkern-14mu\diagdown}}$$

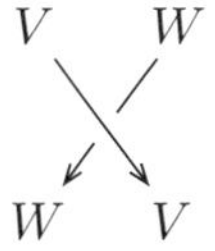

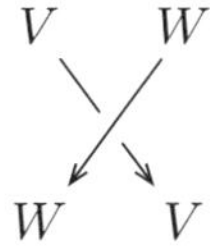

to distinguish them. We denote any other morphisms as nodes on a string with the appropriate number of input and output legs. In this notation, the hexagon conditions and the functoriality of the 'braiding' Ψ appear as shown in Figure 4.3, where the doubled lines in part (a) refer to the composite objects $V \otimes W$ and $W \otimes Z$ in a convenient extension of the notation. The functoriality of Ψ is expressed in part (b) as the assertion that a morphism $\phi : V \to Z$ can be pulled through a braid crossing (and similarly for Ψ^{-1}).

The coherence theorem for braided categories (written out formally in Joyal and Street (1993)) asserts that if two composite morphisms

$$V_1 \otimes V_2 \otimes \cdots \otimes V_n \to V_{\sigma(1)} \otimes V_{\sigma(2)} \otimes \cdots \otimes V_{\sigma(n)}$$

(for some permutation σ and some bracketings on each side) that are built from $\Psi, \Psi^{-1}, \Phi, \Phi^{-1}$ correspond to the same braid, then they coincide as morphisms. If we do not know that all the Ψ are invertible we say that the category is 'prebraided'.

Proposition 4.4 (Majit, Drinfeld) *Let C be a monoidal category viewed as fibred over itself by the identity functor. Then the invertible version of C° is braided with*

$$\Psi_{(V,\lambda_V),(W,\lambda_W)} = \lambda_{V,W}$$

This special case is also called the 'centre' $Z(C)$.

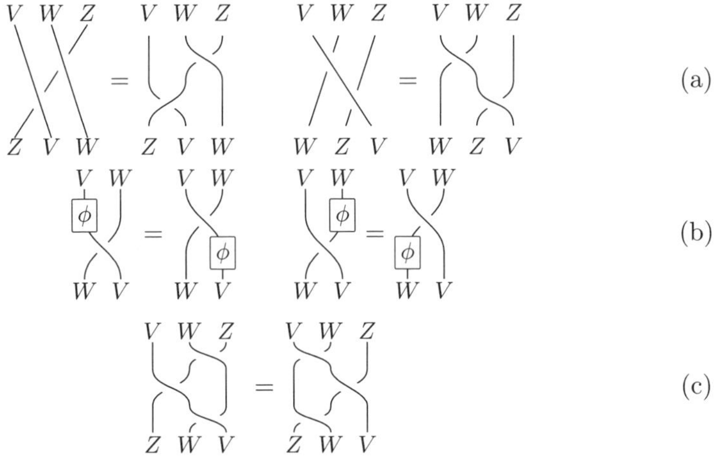

Fig 4.3 Hexagons (a) and functoriality (b) in the diagrammatic notation. They imply the Yang–Baxter or braid relations (c).

This special case of my duality construction was pointed out as being braided, together with the following example, in a letter to me from V.G. Drinfeld as a comment on the preprint version of Majid (1991g). The proof appeared in my later book (Majid, 1995a).

Example 4.5 If H is a quantum group with invertible antipode then its category of modules ${}_H\mathcal{M}$ is a monoidal category and $Z({}_H\mathcal{M})$ is the category of crossed (or Drinfeld–Radford–Yetter) H-modules. If H is finite-dimensional then the latter category is isomorphic to ${}_{D(H)}\mathcal{M}$, where $D(H)$ is Drinfeld's 'quantum double' quantum group.

Not every quantum group has its category of modules ${}_H\mathcal{M}$ braided. Those that do are ones with an additional *quasitriangular structure* (Drinfeld, 1987). There is a dual notion of *comodules* of a quantum group and a *coquasitriangular structure* making the category ${}^H\mathcal{M}$ of comodules braided. In general if $F: \mathcal{C} \to$ Vec is a monoidal functor then it factors through the category of comodules of a Hopf algebra with its forgetful functor, and if $\mathcal{C}$ is braided then this Hopf algebra is coquasitriangular (see Majid 1995a). So for the purposes of this article the most important thing about quantum groups is that they generate braided categories either directly or via the quantum double. The latter in the finite-dimensional case is of the factorization form $D(H) = H \bowtie H^{*op}$ as shown in Majid (1990) but an action of H^{*op} is the same thing as a coaction of H and in that formulation the notion of a crossed H-module or Drinfeld–Radford–Yetter module makes sense for any Hopf algebra—it is basically just equivalent to Drinfeld's quantum double construction and can be taken as a definition of that in the infinite-dimensional case.

For a different kind of result, if $\mathcal{C}$ is braided and fibred over a monoidal category $\mathcal{V}$ then there is a functor $R : \mathcal{C} \to \mathcal{C}^\circ$ (defined by Majid (1992a)):

$$R(X) = (F(X), \lambda_{F(X)}), \quad \lambda_{F(X),Y} = f_{Y,X}^{-1} F(\Psi_{X,Y}) f_{X,Y}.$$

This specializes the case of Example 4.5 when H is quasitriangular to a functor ${}_H\mathcal{M} \to {}_{D(H)}\mathcal{M}$ in the finite-dimensional case or ${}_H\mathcal{M} \to Z({}_H\mathcal{M})$ in the general case, an important functor introduced by Majid (1991b).

4.3 Braided groups

Now consider an algebra *in* a braided category $\mathcal{C}$. A usual algebra B means, of course, a map $\cdot : B \otimes B \to B$ and a unit element, which we denote as a map $\eta : k \to B$ sending 1 to the unit element, subject to axioms

$$\cdot (\cdot \otimes \mathrm{id}) = \cdot (\mathrm{id} \otimes \cdot), \quad \cdot (\eta \otimes \mathrm{id}) = \mathrm{id} = \cdot (\mathrm{id} \otimes \eta)$$

as maps. Usually in the category of vector spaces we do not explicitly write the associator Φ nor the maps r, l that afford the identifications $B = k \otimes B = B \otimes k$, but the obvious maps here are implicit, so as to make sense of algebraic expressions. We require just the same in a braided category except that:

- the role of k is played by $\underline{1}$, the unit object for the tensor product
- all maps are required to be morphisms
- we may need more non-trivial associator Φ and morphisms l, r to be inserted for compositions to make sense, i.e.

$$\cdot (\cdot \otimes \mathrm{id}) = \cdot (\mathrm{id} \otimes \cdot) \Phi_{B,B,B}, \quad \cdot (\eta \otimes \mathrm{id}) r_B = \mathrm{id} = \cdot (\mathrm{id} \otimes \eta) l_B.$$

In our diagrammatic language of Figure 4.3 the product morphism becomes a 'processing box' with two legs coming in at the top and one coming out at the bottom. Since we will use this a lot we will just denote it by an annotated blob and sometimes not even write the blob or its annotation (just as one does not always write the product of an algebra explicitly). The above axioms of an algebra are then as in Figure 4.4(a) and say that the two different branchings for two products are the same (one could then write just one node with three lines to express this) and

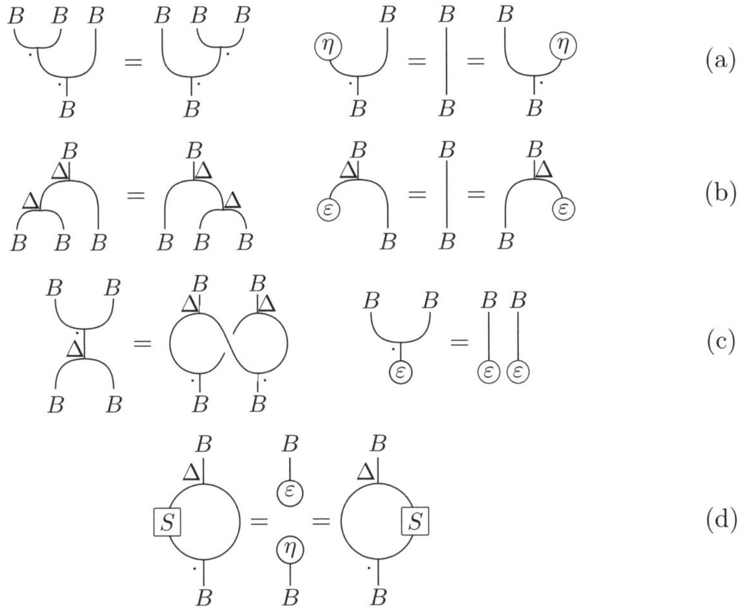

Fig 4.4 Axioms (a) of an algebra in a braided category, (b) of a coalgebra, (c) further axioms for a bialgebra, and (d) of a quantum group in a braided category, or 'braided group'. Read down the page.

that we may 'prune' an identity element branch. As before, the unit object $\underline{1}$ is denoted by omission in the diagrammatic notation, as are Φ, l, r, but they should all be understood.

Example 4.6 Let G be an abelian group and $F: G \times G \to k$ a nowhere vanishing function that is 1 if either argument is the group identity. Then the category of G-graded vector spaces over k is a symmetric monoidal category with

$$\Psi_{V,W}(v \otimes w) = \frac{F(|v|, |w|)}{F(|w|, |v|)} w \otimes v$$

$$\Phi_{V,W,Z}(v \otimes (w \otimes z)) = \frac{F(|v|, |w| + |z|)F(|w|, |z|)}{F(|v|, |w|)F(|v| + |w|, |z|)}(v \otimes w) \otimes z$$

on elements of homogeneous degree $|\ |$. We are writing the group additively. Moreover, $B = \mathrm{span}_k\{e_x \mid x \in G\}$ with a new product $e_x \bullet e_y = F(x, y)e_{x+y}$ is an algebra in this category with grading $|e_x| = x$ (see Albuquerque and Majid, 1999).

Here F is a cochain and the coefficients of Φ are its 3-cocycle boundary $\phi = \partial F$. Any 3-cocycle ϕ on a group makes the category of group-graded spaces non-trivially monoidal, and the 'dual' or 'centre' construction of Proposition 4.4 applied to this gives a braided category, computed in Majid (1998b) as the modules of the double quasi-Hopf algebra.

Example 4.7 The octonions $\mathbb{O}$ are an associative commutative algebra in the category of $\mathbb{Z}_2^3$-graded spaces (Albuquerque and Majid, 1999) with symmetry and associator

$$\Psi(e_a \otimes e_b) = \psi_2(a, b)e_b \otimes e_a, \quad \Phi((e_a \otimes e_b) \otimes e_c) = \psi_3(a, b, c)e_a \otimes (e_b \otimes e_c)$$

where on $a_i \in \mathbb{Z}_2^n$ and $m \geq 1$ we define

$$\psi_m(a_1, a_2, \cdots, a_m) = \begin{cases} 1 & \text{if } \{a_i\} \text{ linearly dependent} \\ -1 & \text{if not} \end{cases}$$

so that

$$\psi_1(a) = (-1)^{\delta_{a,0}}, \quad \psi_n(a_1, \cdots, a_n) = (-1)^{|a_1, a_2, \cdots a_n|}, \quad \psi_m = 1, \quad \forall m > n$$

For general n one can show that

$$\psi_m = \partial \psi_{m-1}, \quad \forall m \text{ even}$$

so that all even ψ_m are coboundaries. It also follows that when m is odd ψ_m is not a cocycle when $m < n - 1$ but ψ_n is. This ψ_n for n odd could still be a coboundary and indeed, when $n = 3$, one can write $\psi_3 = \partial F$ (Albuquerque and Majid, 1999), where

$$F(a,b) = (-1)^{\sum_{i \geq j} a_i b_j + a_0 b_1 b_2 + b_0 a_1 b_2 + b_0 b_1 a_2} = \begin{pmatrix} 1 & 1 & 1 & 1 & 1 & 1 & 1 & 1 \\ 1 & 1 & 1 & 1 & 1 & 1 & -1 & -1 \\ 1 & 1 & 1 & 1 & -1 & 1 & -1 \\ 1 & 1 & 1 & 1 & 1 & -1 & -1 & 1 \\ 1 & 1 & 1 & -1 & 1 & 1 & 1 & -1 \\ 1 & 1 & 1 & -1 & 1 & 1 & -1 & 1 \\ 1 & 1 & 1 & -1 & 1 & -1 & 1 & 1 \\ 1 & 1 & 1 & -1 & 1 & -1 & -1 & -1 \end{pmatrix}$$

Here $a = (a_0, a_1, a_2)$ and $b = (b_0, b_1, b_2)$, which on the right we read in binary as $a_2 a_1 a_0 \in 0, 1, \cdots, 7$. This F has a remarkable property (Majid, 2005) of being invariant under Fourier transform of both arguments up to a cyclic rotation of the copies of $\mathbb{Z}_2^3$, and the algebra with product given by F, as an example of the preceding general construction, is the octonions. We speculate that for $n = 4$ the functions ψ_4 can similarly be used to define some kind of braided 2-category and ψ_3 an algebraic object in it, but this remains to be worked out.

Note that in this example we are actually interested in coboundaries, which is the exact opposite of the more usual consideration of the cohomology ring, which in the above case is, additively,

$$H^*((\mathbb{Z}_2)^n, \mathbb{Z}_2) = \mathbb{Z}_2[x_1, \cdots, x_n]$$

where x_i are degree 1. Thus there are plenty of other $\mathbb{Z}_2^n$-graded monoidal categories. For example, when $n = 3$ we have a cousin of the octonion category with 3-cocycle $\phi(a, b, c) = (-1)^{a_0 b_1 c_2}$.

For our first bit of diagrammatic algebra we prove the following fundamental lemma in the case where the category is braided. Note that in the diagrammatic notation the product and coproduct, being morphisms, can be taken through a braid crossing much as in Figure 4.3(b). I used these methods from 1989 but my first publications with them are (Majid, 1991a; Majid, 1993d; Majid, 1994b) dating from 1990–91. My review (Majid, 1994a) contained a convenient collection.

Lemma 4.8 *(SM, 1989) If B, C are two algebras in a braided category then there is a new algebra $B \underline{\otimes} C$ in the same category, built on $B \otimes C$ with product*

$$\cdot_{B \underline{\otimes} C} = (\cdot_B \otimes \cdot_C) \Phi^{-1}_{B,B,C \otimes C} (\mathrm{id} \otimes \Phi_{B,C,C})(\mathrm{id} \otimes (\Psi_{C,B} \otimes \mathrm{id}))(\mathrm{id} \otimes \Phi^{-1}_{C,B,C}) \Phi_{B,C,B \otimes C}$$

and tensor product unit.

Proof This is done in Figure 4.5 using the diagrammatic notation following Majid (1994a). We use functoriality to push the product of C down and over to the right, We then use associativity in B, C and then functoriality to push the product of B up and under to the right. □

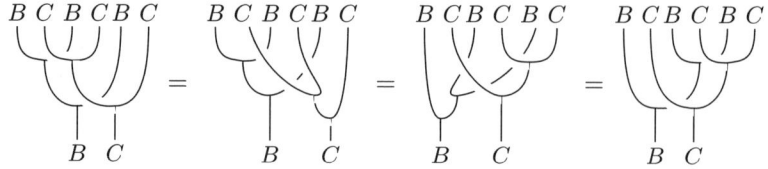

Fig 4.5 Proof that the tensor product of two braided algebras is associative. Read down the page—or up the page for a result about coalgebras.

This means, for example, that you can tensor product $\mathbb{O} \otimes \mathbb{O} \cdots \otimes \mathbb{O}$ to obtain new algebras in the same category.

Similarly, a coalgebra in the braided category just means the same as an algebra but with all structure morphisms going the other way. This means an object B and morphisms $\Delta : B \to B \otimes B$, $\epsilon : B \to \underline{1}$ and the coassociativity and counity axioms

$$(\Delta \otimes \mathrm{id})\Delta = \Phi^{-1}_{B,B,B}(\mathrm{id} \otimes \Delta)\Delta, \quad r_B^{-1}(\epsilon \otimes \mathrm{id})\Delta = \mathrm{id} = l_B^{-1}(\mathrm{id} \otimes \epsilon)\Delta.$$

The diagrammatic form of this is shown in Figure 4.4(b) and for brevity we again write the 'box' for the coproduct as a blob or node. These are just the 'turning upside down' of the diagrams for an algebra. Turning the statement in Lemma 4.8 and its proof upside down gives a result about the tensor product of two coalgebras—an example of a 'two for the price of one' principle. Actually one can also reflect diagrams horizontally, so that one actually has 'four for the price of one' in the theory, except that in the present case the diagrams are already symmetric so we do not get a new result.

The notions of algebra and coalgebra make sense in any monoidal category. This braiding is needed in the Lemma and hence to define the notion of a bialgebra—a compatibility between the structures of an algebra and coalgebra on the same object—shown in Figure 4.4(c). From the lemma we see that this condition has the meaning that $\Delta : B \to B \otimes B$ is an algebra homomorphism. For a quantum group in the braided category or a *braided group* we also need an antipode as shown in Figure 4.4(d) (this axiom itself does not require a braiding and so makes sense more generally).

The amazing thing is that all the main theorems of Hopf algebras go through at this level of diagrammatic algebra and hence apply in any braided category. As an example, recall that for a usual Hopf algebra the inversion or antipode is antimultiplicative in the sense $S(ab) = (Sb)S(a)$ for all elements a, b. This in turn generalizes the anti-automorphism property of group inversion in a group. We have the same in any braided category:

Proposition 4.9 (Majid, 1993d) *If B is a braided group, the antipode obeys*

$$S \cdot = \cdot \Psi_{B,B}(S \otimes S), \quad \Delta S = (S \otimes S)\Psi_{B,B}\Delta$$

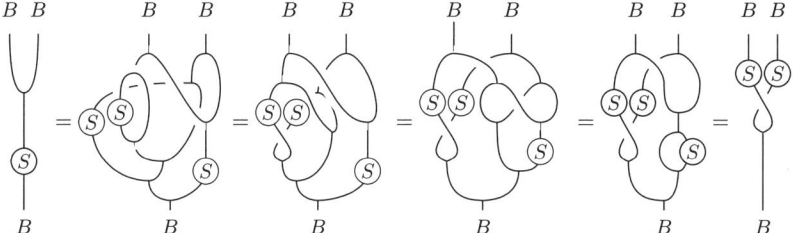

Fig 4.6 Diagrammatic proof of antipode property Proposition 4.9. Read down the page—or up the page for a different result.

Proof One half of this is shown in Figure 4.6 (Majid, 1993d; Majid, 1994a). The insertion of the 'antipode loops' in the first equality does not change anything by the co/unity or 'pruning' axioms of co/algebras. We then use functoriality to push one of the S morphisms to the left and co/associativity to rebranch the trees. We then then cancel an antipode loop to arrive at the right-hand side. The other half is free by the two for one principle—turn the page upside down or read everything up the page with different annotations (swap the product with a coproduct). □

What this amounts to then is a deep generalization of group theory in which algebra is 'wired up', much like wires in a computer except that when wires cross we need to use the braiding or inverse braiding and we have to read the diagrams down the page, or possibly up the page for the dual computation, but not, for example, sideways. For example, one can see that every braided group acts on itself by the adjoint action. This is a map $B \otimes B \to B$ defined by

and such diagram methods were used by the author to prove that this is indeed an action of the algebra of B on B, and that it respects the product of B (B becomes a B-module algebra). We refer readers to Majid (1994a) for details. The adjoint action generalizes the action by conjugation of a group on itself—just read the diagram with the diagonal coproduct and antipode given by inversion, as in Section 4.1, and trivial braiding.

There is also a duality operation for braided groups at least when B is rigid. For a left dual we suppose an evaluation map $\mathrm{ev} : B^* \otimes B \to \underline{1}$, denoted as a 'cup', and also a coevaluation $\mathrm{coev} : B \to B \otimes B^*$, denoted by a 'cap' subject to two 'bend-straightenning axioms'. The relation between the product of B^* and the coproduct of B is then given by adjointing in the category and makes sense for algebras and coalgebras in a monoidal category without reference to a braiding. In the braided group case this makes $B^\star$ another braided group, as shown for the coproduct axiom in Figure 4.7 (Majid, 1994a). An application is that $B^\star$ acts canonically on B and respects its product (B becomes a $B^\star$-module algebra). This generalises the left regular representation of a group on the algebra of functions on the group by left translation, generated by left-invariant vector fields. Thus it underlies the notion of 'braided differentiation'.

An application of these ideas is to the Fourier transform. Typically one has a unique-up-to-scale translation-invariant integration $\int : B \to \underline{1}$ except that this will not necessarily be a morphism (a categorical way to deal with this is for the integral to be valued in a fixed invertible object (Bespalov et al., 2000)) and using this we define the *braided Fourier transform* by

$$\mathcal{S} = (\int \otimes \mathrm{id})(\cdot \otimes \mathrm{id})(\mathrm{id} \otimes \mathrm{coev}) : B \to B^\star$$

as shown in the lower box on the left of Figure 4.8. The exp here is simply the coevaluation $\underline{1} \to B \otimes B^\star$. The upper box is a right-handed version of the regular action of $B^\star$ on B and Figure 4.8 shows that Fourier transform intertwines this action with the product of $B^\star$, i.e. turns differential equations on B into algebraic equations on $B^\star$ (see Kempf and Majid (1994)).

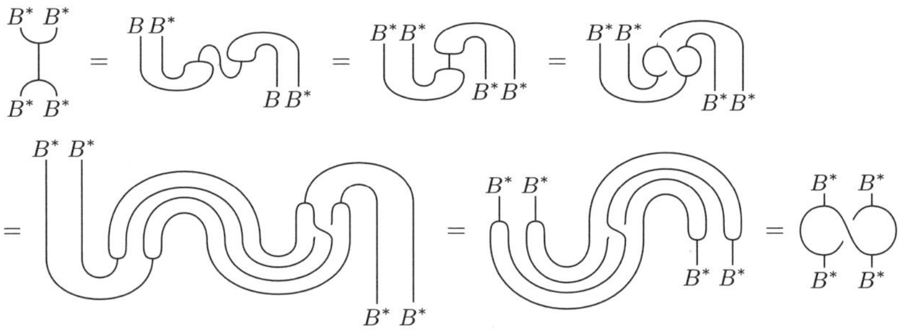

Fig 4.7 Main part of proof that the dual $B^\star$ is again a braided group if B is. Read down the page.

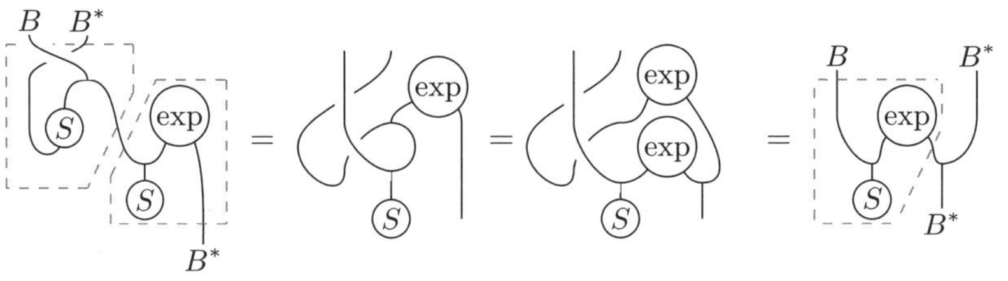

Fig 4.8 Proof that Fourier transform (lower and right boxes) intertwines the left regular action (upper box) with the product of $B^\star$. Read down the page.

Example 4.10 (Braided line) Let q be a primitive r-th root of unity and $\mathcal{C}$ the category of $\mathbb{Z}_r$-graded vector spaces with braiding $\Psi_{V,W}(v \otimes w) = q^{|v||w|} w \otimes v$ on homogeneous elements $v \in V, w \in W$ of degrees $|v|, |w|$. Let $B = c_q[\mathbb{C}] = \mathbb{C}[x]/\langle x^r \rangle$ with $\Delta x = x \otimes 1 + 1 \otimes x$, $\epsilon x = 0$, $Sx = -x$ extended as a Hopf algebra in $\mathcal{C}$ with $|x| = 1$. This implies, for example

$$\Delta x^n = \sum_{m=0}^n \begin{bmatrix} n \\ m \end{bmatrix}_q x^m \otimes x^{n-m}, \quad Sx^n = q^{\frac{n(n-1)}{2}}(-x)^n.$$

The first follows from the q-binomial theorem since $(1 \otimes x)(x \otimes 1) = \Psi(x \otimes x) = qx \otimes x = q(x \otimes 1)(1 \otimes x)$ and the second from Proposition 4.9. The q-binomial coefficients here are defined as usual but with q-intergers $[t]_q = (1 - q^t)/(1 - q)$ in place of usual integers. The dual braided group has the same form and the regular action in one convention is generated by $\partial_q : c_q[\mathbb{C}] \to c_q[\mathbb{C}]$ defined (Majid, 1993c) as the linear part of the Taylor expansion

$$\Delta f = 1 \otimes f + x \otimes \partial_q f + x^2 \otimes \cdots$$

This turns out to be the q-difference operator

$$\partial_q(f) = \frac{f(x) - f(qx)}{x(1-q)}$$

which turns up in q-analysis. In terms of this, the adjoint action of the generator x comes out (Majid, 1995b) as the braided vector field

$$\mathrm{Ad}_x(f) = (1-q)x^2 \partial_q f$$

which has no classical analogue (classically the adjoint action is trivial). Also, the naive (non-morphism) integration comes out as $\int x^m = \delta_{m,r-1}$ (there is also a braided improper integration and a Gaussian-weighted integral in Kempf and Majid (1994) as well as in Bespalov et al. (2000)) and, as the braided group is basically self-dual, the Fourier transform comes out as an operator

$$\mathcal{S} : c_q[\mathbb{C}] \to c_q[\mathbb{C}], \quad \mathcal{S}(f)(x) = \int f(x) e_q^{x|p}, \quad e_q^{x|p} := \sum_{m=0}^{r-1} \frac{x^m \otimes p^m}{[m]_q!}.$$

Here $|p| = -1$ and the pairing between p^n and x^n comes out as $[n]_q!$, which is the reason for the exponential.

One can do the same 'braided affine line' $\mathbb{A}_q = k[x]$ over any field k and $q \in k^*$, giving a braided group in the category of $\mathbb{Z}$-graded spaces with braiding and formulae essentially the same as the above, except that the exponential needs to be treated formally because the algebra is not finite-dimensional. Even the humble braided line for generic q has applications to mathematics (e.g. Cameron and Majid (2003)). Similarly in two dimensions one has the affine braided plane $\mathbb{A}_q^{2|0} = k\langle x, y\rangle/yx = qxy$. The algebra here is sometimes called a Manin quantum plane but its full structure as a braided group is due to the author in the framework of Majid (1993c). Here

$$\Delta(x^m y^n) = \sum_{r=0,s=0}^{r=m,s=n} \begin{bmatrix} m \\ r \end{bmatrix}_{q^2} \begin{bmatrix} n \\ s \end{bmatrix}_{q^2} q^{s(m-r)} x^r y^s \otimes x^{m-r} y^{n-s}$$

with braiding

$$\Psi(x \otimes x) = q^2 x \otimes x, \quad \Psi(y \otimes y) = q^2 y \otimes y$$
$$\Psi(x \otimes y) = qy \otimes x, \quad \Psi(y \otimes x) = qx \otimes y + (q^2 - 1)y \otimes x$$

The braided category here is that of certain types of $U_q(gl_2)$-modules or of $k_q[GL_2]$-comodules for a certain quantum group, or in some contexts a central extension of the more familiar quantum groups $U_q(sl_2)$ and $k_q[SL_2]$. One similarly has braided differentials and, formally or in a finite-dimensional quotient, braided Fourier theory etc.

4.4 Applications of braided groups

So far we have covered the basic theory of braided groups established by me at the end of the 1980s and early 1990s. It would be too technical to cover everything beyond that but here I will sketch some of the more advanced theory and mention some new trends.

For any braided group B in any braided category $\mathcal{V}$ there is a notion of module $(V, \alpha_V : B \otimes V \to V)$, where $V \in \mathcal{V}$ and α_V obeys a condition that polarizes the notion of an associative product. We have already given the example of the adjoint action. The first lemma here is that any two modules can be tensor producted via the coproduct Δ, and as a result the category $_B\mathcal{V}$ of B-modules in $\mathcal{V}$ becomes a monoidal one fibred over $\mathcal{V}$ via the forgetful functor. This is shown in Figure 4.9 (Majid, 1993d; Majid, 1994a). One similarly has a monoidal category $^B\mathcal{V}$ of braided comodules in $\mathcal{V}$, which is the same concept with arrows reversed. In the rigid case this can be identified with $_{B^*}\mathcal{V}$. We also have a reconstruction theorem dating from 1989:

Theorem 4.11 (Majid, 1991a, 1991f, 1992b, 1993a) *Let $F : \mathcal{C} \to \mathcal{V}$ be a monoidal category $\mathcal{C}$ fibred over a cocomplete braided monoidal category $\mathcal{V}$ with the image of $\mathcal{C}$ rigid. Then F factors $\mathcal{C} \to {}^B\mathcal{V} \to \mathcal{V}$ through the forgetful functor for a certain braided group $B \in \mathcal{V}$ universal with this property. Moreover, $\mathcal{C}^\circ$ can be identified with $_B\mathcal{V}$.*

The braided group B here is denoted $\mathrm{Aut}^F(\mathcal{C}, \mathcal{V})$ and is constructed as a certain coend (Majid, 1991a, 1992b, 1993a). All that we really require is that this coend exists (that a certain functor is representable). Assuming representability

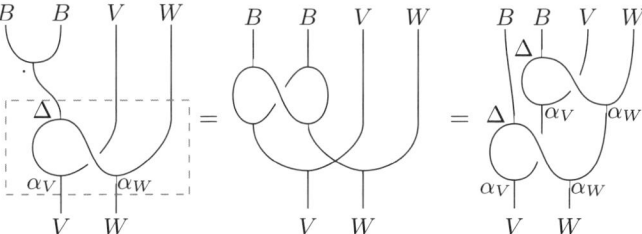

Fig 4.9 Diagrammatic proof that tensor product (boxed) of two modules of a braided group B is another module. Read down the page—or up the page for the same result for comodules.

we also have a module version of the reconstruction theorem with factorization $\mathcal{C} \to {}_B\mathcal{V} \to \mathcal{V}$ and giving $\mathrm{Aut}_F(\mathcal{C}, \mathcal{V})$ essentially dual to the one before. We will write $\mathrm{Aut}^{\cdot}(\mathcal{V}) = \mathrm{Aut}^{\mathrm{id}}(\mathcal{V}, \mathcal{V})$ and $\mathrm{Aut}.(\mathcal{V})$ in the module version. There is a canonical homomorphism $\mathrm{Aut}^{\cdot}(\mathcal{V}) \to \mathrm{Aut}.(\mathcal{V})$, which is essentially an isomorphism when $\mathcal{V}$ is maximally far from symmetric. A more explicit version of $\mathrm{Aut}^{\cdot}(\mathcal{V})$ in some cases was found independently by Lyubashenko, along with an operation of 'Fourier transform' on such objects. A version of these findings appeared in Lyubashenko (1995). The Fourier theory as related to our point of view appeared in Lyubashenko and Majid (1994).

Corollary 4.12 Majid (1991a, 1992b, 1993a) *Every coquasitriangular Hopf algebra H has a braided version $B(H) \in {}^H\mathcal{M}$, its 'transmutation'.*

Proof We take the identity functor of $\mathcal{C} = \mathcal{V} = {}^H\mathcal{M}$. One similarly has the result for any quasistriangular H with $B(H) \in {}_H\mathcal{M}$ (see Majid, 1993d). □

This key result ensured that there was a plentiful supply of braided groups. Similarly if $H \to H_1$ is a map of Hopf algebras with H quasistriangular, we have a functor ${}_{H_1}\mathcal{M} \to {}_H\mathcal{M}$ and now transmutation gives a braided group $B(H, i, H_1) \in {}_H\mathcal{M}$.

Example 4.13 (q-Minkowski space) Consider the quantum group $H = \mathbb{C}_q[SU_2]$. Its transmutation $B_q[SU_2]$ by Corollary 4.12 is the algebra $\mathbb{C}\langle \alpha, \beta, \gamma, \delta \rangle$ modulo the relations (Majid, 1991c; Majid, 1992b)

$$\beta\alpha = q^2\alpha\beta, \quad \gamma\alpha = q^{-2}\alpha\gamma, \quad \delta\alpha = \alpha\delta, \quad \alpha\delta - q^2\gamma\beta = 1$$

$$[\beta, \gamma] = (1 - q^{-2})\alpha(\delta - \alpha), \quad [\delta, \beta] = (1 - q^{-2})\alpha\beta, \quad [\gamma, \delta] = (1 - q^{-2})\gamma\alpha,$$

$$\Delta \begin{pmatrix} \alpha & \beta \\ \gamma & \delta \end{pmatrix} = \begin{pmatrix} \alpha & \beta \\ \gamma & \delta \end{pmatrix} \otimes \begin{pmatrix} \alpha & \beta \\ \gamma & \delta \end{pmatrix}, \quad \epsilon \begin{pmatrix} \alpha & \beta \\ \gamma & \delta \end{pmatrix} = \begin{pmatrix} 1 & 0 \\ 0 & 1 \end{pmatrix}$$

If we drop the third 'braided-determinant' relation then we have the braided matrices $B_q[M_2]$ as a transmutation of the quantum matrices $\mathbb{C}_q[M_2]$. There is also a natural $*$-involution $\alpha^* = \alpha, \delta^* = \delta, \beta^* = \gamma$ making this the space of 2×2 'braided hermitian matrices' and $B_q[SU_2]$ a q-hyperboloid. This is analogous to Minkowski space as the space of 2 × 2 hermitian matrices with Lorentzian distance given by the determinant.

In physics nowadays there is a lot of interest in the idea that the real world is actually better described by a quantum space-time; by a non-commutative coordinate algebra. The above model, worked out in the 1990s by me and independently the group of Wess in Munich, remains one of the most non-trivial but also the least explored due to its complexity, in contrast to other quantum space-times related to the other (bicrossproduct) type of quantum group. It is also interesting that the Aut. construction equipped with the Fourier operator in the special case of fusion categories plays a key role in the modern approach to topological quantum field theories and hence to topological quantum computing, knot and 3-manifold invariants, and 2+1 quantum gravity.

What happens if we start with a braided group $B \in \mathcal{V}$ and reconstruct from the forgetful functor ${}^B\mathcal{V} \to \mathcal{V}$? The automorphism braided group comes out as a cross coproduct $B {>\!\!\blacktriangleleft}\, \mathrm{Aut}(\mathcal{V})$. Similarly we have a cross product braided group $B {>\!\!\triangleleft}\, \mathrm{Aut}.(\mathcal{V})$ in the module reconstruction setting.

Corollary 4.14 (Majid, 1994b) *If H is a quasitriangular Hopf algebra and $B \in {}_H\mathcal{M}$ then there is an ordinary Hopf algebra $B{>\!\!\triangleleft}\, H$, its 'bosonization', where we make a cross product by the given action of H and a cross coproduct by a certain induced coaction of H.*

Proof Here $\mathcal{V} = {}_H\mathcal{M}$ is braided and we can identify $B{>\!\!\triangleleft}\, B(H)$ as a transmutation of an ordinary Hopf algebra $B{>\!\!\triangleleft}\, H$ by an inclusion $H \hookrightarrow B{>\!\!\triangleleft}\, H$. Alternatively, we can reconstruct from the forgetful functor all the way to Vec rather than to ${}_H\mathcal{M}$. □

The situation is similar in the comodule setting with H coquasitriangular. Bosonization and transmutation are not inverse: $B(H){>\!\!\triangleleft}\, H$ is a Hopf algebra canonically associated to H, in the nicest 'factorizable' case isomorphic to the quantum double of H. Bosonization can also be understood as an instance of a more general 'biproduct' $B{>\!\!\triangleleft}\, H$ for any braided group $B \in Z({}_H\mathcal{M})$ by the given action and coaction. Conversely, if $H \overset{\leftarrow}{\hookrightarrow} H_1$ is a Hopf algebra H_1 with an inclusion from and projection to a Hopf algebra H then $H_1 \cong B{>\!\!\triangleleft}\, H$ for some braided group $B \in Z({}_H\mathcal{M})$. This theory was anticipated by Radford (1985) before braided categories were invented. This modern form was due to the author (Majid, 1993b).

Here is another general construction. Let B be a braided group with dual B^* in a braided category $\mathcal{V}$. We can similarly consider the category ${}_{B^*}\mathcal{V}_B$ of *braided crossed bimodules* consisting of objects that are modules of B, B^* in a compatible way as shown in Figure 4.10 (Majid, 2002). This is the braided version of the same ideas as for the quantum double of a quantum group and makes sense more generally as compatible B-modules and B-comodules (as first observed in Bespalov (1997)). Either way the category is itself a braided monoidal one. So for every braided group B in a braided category $\mathcal{V}$ we get a new braided category.

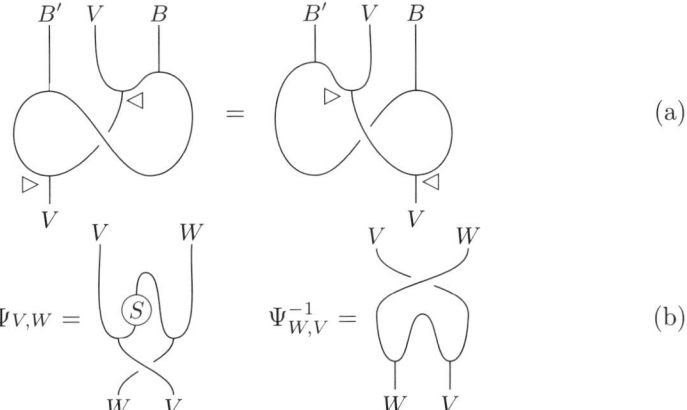

Fig 4.10 Compatibility condition and braiding for the category of ${}_{B'}\mathcal{V}_B$ modules. Here B' is dually paired to B, e.g. $B^\star$. Read down the page.

Example 4.15 Let $\mathcal{V}_n$ be the category of comodules of the quantum group $\mathbb{C}_q[SL_n]$. We let $\tilde{\mathcal{V}}_n$ be the extension to comodules of the centrally extended quantum group and $\mathbb{A}_q^{n|0} \in \tilde{\mathcal{V}}_n$ the braided plane. The above construction gives $\mathcal{V}_{n+1}$. Thus all quantum group representation categories for this family of quantum groups can be built up inductively (Majid, 1999b; Majid, 2002). The approach generalizes to the construction of all $\mathbb{C}_q[G]$-comodule categories by repeated adjunction of braided groups (each step increases the rank of the underlying Lie group by one).

We also have a forgetful functor ${}_{B^\star}\mathcal{V}_B \to \mathcal{V}$ and applying our module reconstruction theorem in the representable case we have a certain braided group $B^\star {>}\triangleleft \mathrm{Aut}.(\mathcal{V}) \triangleright{<} B^{op}$.

Corollary 4.16 (Majid, 1999b) *Let H be a quasitriangular Hopf algebra and B a braided group with dual in ${}_H\mathcal{M}$. Then there is a new quasitriangular Hopf algebra $B^\star {>}\triangleleft H \triangleright{<} B^{op}$, the 'double bosonization' of B.*

Proof We use the module version of Theorem 4.11 for the reconstruction from ${}_B\mathcal{V}_{B^\star}$, which gives a braided group $B^\star {>}\triangleleft B(H) \triangleright{<} B^{op}$ in ${}_H\mathcal{M}$. One can then recognize this as the result of transmutation from a quantum group $B^\star {>}\triangleleft H \triangleright{<} B^{op}$ with respect to the inclusion of H. Alternatively, which is the line taken in Majid 1999b, we can use ordinary reconstruction all the way to Vec rather than to ${}_H\mathcal{M}$ to give an ordinary quantum group rather than a braided one. □

One can apply this theory to explicitly construct the reduced finite-dimensional quantum groups $u_q(\mathfrak{g})$ at roots of unity. A slight generalization in Majid 1999b similarly gives the quantum groups $U_q(sl_{n+1}) = \mathbb{A}_q^{n|0} {>}\triangleleft \tilde{U}_q(sl_n) \triangleright{<} \mathbb{A}_q^{n|0\ op}$ etc, i.e. it inductively constructs all quantum groups $U_q(\mathfrak{g})$.

We conclude with a general construction for braided groups in an abelian braided category $\mathcal{V}$ that includes the braided planes $\mathbb{A}_q^{n|0}$. It is normally done in the strictly associative case but should apply more generally with care. Let $V \in \mathcal{V}$ be an object, then

$$TV = \underline{1} \oplus V \oplus (V \otimes V) \oplus \cdots$$

forms a braided group with $\Delta v = v \otimes 1 + 1 \otimes v$ in the case where there are elements, or $\Delta = l_V + r_V$ in degree 1 in terms of maps. We define 'braided integers' and 'braided factorials' (Majid, 1993c, 1995a):

$$[n, \Psi] = \mathrm{id} + \Psi_{12} + \Psi_{12}\Psi_{23} + \cdots + \Psi_{12}\cdots\Psi_{n-1,n} : V^{\otimes n} \to V^{\otimes n}$$

$$[n, \Psi]! = ([n-1, \Psi]! \otimes \mathrm{id})[n, \Psi] : V^{\otimes n} \to V^{\otimes n}$$

The factorial can also be written as a 'braided symmetrizer' $\sum_{\sigma \in S_n} \Psi_{i_1} \cdots \Psi_{i_{l(\sigma)}}$, where we write a permutation in terms of simple reflections and replace these by $\Psi_i = \Psi$ acting in the corresponding position. This is analogous to an example of an antisymmetrizer construction used by Woronowicz (1989) in another context and which in our approach corresponds to $[n, -\Psi]!$. We then define

$$S(V) = TV/\oplus_n \mathrm{Ker}\,[n, \Psi]!, \quad \Lambda(V) = TV/\oplus_n \mathrm{Ker}\,[n, -\Psi]!$$

as the braided symmetric and braided antisymmetric algebras. Both are braided groups. Another point of view on this construction is as follows, in the case where V is rigid. The pairing between V^*, V extends as a braided-Hopf algebra pairing between $T(V^*)$ and TV, which on tensor powers is given by $[n, \Psi]!$ (Majid, 1993c, 1995a). Hence the coradical of the pairing on the TV side is $S(V)$ (we have seen all this already in Example 4.10). The situation is similar for the exterior algebra case. Both of these points of view and, importantly, the braided group structure were introduced by me and used notably in Majid (1999b) and Majid (2004) (but it should be noted that some authors call $S(V)$ the 'Nichols–Woronowciz algebra').

Example 4.17 (Majid, 1999b) Let l be the rank of a complex simple Lie algebra $\mathfrak{g}$ and $a = (a_{ij})$ the $l \times l$ Cartan matrix that defines the Lie algebra. A *Cartan datum* in the sense of Lusztig (1993) is a symmetric bilinear on $\mathbb{Z}^l$, which we denote as a dot product, such that in the basis directions $i, j = 1, 2, \cdots l$,

$$i \cdot i \in \{2, 4, 6, \cdots\}, \quad a_{ij} = \frac{2i \cdot j}{i \cdot i} \in \{0, -1, -2, \cdots\}, \quad \forall i \neq j$$

(when this exists one says that the Cartan matrix is symmetrizable; this includes all complex simple $\mathfrak{g}$). From this data we build a braided category of $\mathbb{Z}^l$-graded vector spaces with braiding

$$\Psi(v \otimes w) = q^{i \cdot j} w \otimes v, \quad \forall |v| = g_i, |w| = g_j$$

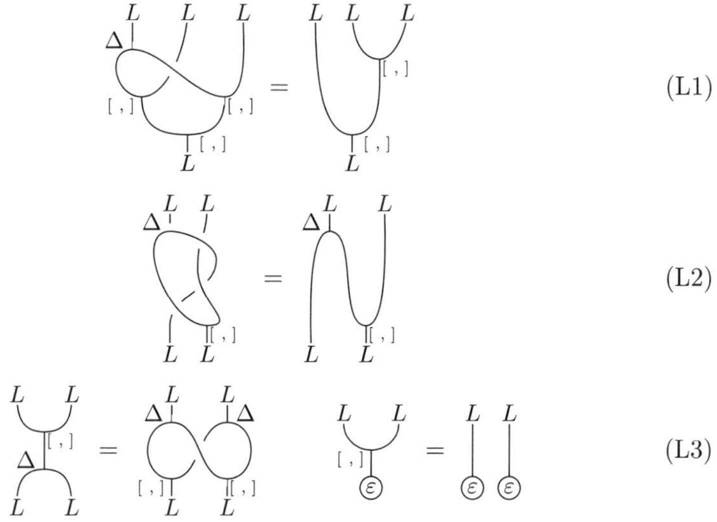

$$\text{(L1)}$$

$$\text{(L2)}$$

$$\text{(L3)}$$

Fig 4.11 Axioms of a braided-Lie algebra. Read down the page.

where $\{g_i\}$ are the multiplicative group generators of $\mathbb{Z}^l$. We consider V a vector space with basis $\{x_i\}$ and degrees $|x_i| = g_i$ and $|y_i| = g_i^{-1}$ to make it an object in the braided category. Then $TV = \mathbb{C}\langle x_1, \cdots, x_l \rangle$ is the free non-associative algebra on generators $\{x_i\}$ and $S(V) = U_q(n_+)$ the positive root subalgebra of the quantum group $U_q(\mathfrak{g})$. On the other side $S(V^\star) = U_q(n_-)$. These are now non-degenerately paired or in some sense dual braided groups in our braided category. The entire quantum group can be recovered by a version of Corollary 4.16 (Majid, 1999b) using an additional 'root datum'. This gives an interpretation of some of Lusztig's approach to quantum groups (Lusztig, 1993).

4.5 Braided-Lie algebras

Sometimes diagrammatic or braided algebra as we have been doing requires a different picture that is useful even in the unbraided case. An example of this is the notion of *braided-Lie algebra* (Majid, 1994c) in a braided category $\mathcal{V}$. Usually a Lie algebra requires a vector space with a bracket obeying antisymmetry and a Jacobi identity. However, the kind of 'cyclic rotation' that defines the latter makes no sense in a strictly braided category where $\Psi^2 \neq \text{id}$ and one needs a new concept. Now, we have seen in the Section 4.1 that the notion of 'symmetry' has at its core the idea of replication or a 'xerox map', so let us start with an object $\mathcal{L} \in \mathcal{V}$ that is a coalgebra, i.e. with morphisms $\Delta : \mathcal{L} \to \mathcal{L} \otimes \mathcal{L}$ and counit $\epsilon : \mathcal{L} \to \underline{1}$. We also need a bracket morphism $[\,,\,] : \mathcal{L} \otimes \mathcal{L} \to \mathcal{L}$ and we impose on it the axioms shown diagrammatically in Figure 4.11 (Majid, 1994c).

Lemma 4.18 (Majid, 1995b) *For any braided-Lie algebra $\mathcal{L} \in \mathcal{V}$ the morphism $\tilde{\Psi}$: $\mathcal{L} \otimes \mathcal{L} \to \mathcal{L} \otimes \mathcal{L}$ defined by*

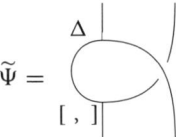

obeys the braid relations on $\mathcal{L}^{\otimes 3}$.

In view of this it is natural to require that this $\tilde{\Psi}$ is invertible. We call this a *regular* braided-Lie algebra.

An actual Lie algebra $\mathfrak{g}$ can be seen as a braided-Lie algebra of the form $\mathcal{L} = k \oplus \mathfrak{g}$ in Vec (so with trivial braiding of the underlying category, although with a non-trivial $\tilde{\Psi}$ even in this case, provided the Lie bracket is non-zero). Here

$$[c, v] = v, \quad [v, c] = 0, \quad [c, c] = c, \quad \Delta v = v \otimes c + c \otimes v, \quad \Delta c = c \otimes c$$

where c spans the copy of k. The axioms of a braided-Lie algebra then amount to the bracket $[\,,\,] : \mathfrak{g} \otimes \mathfrak{g} \to \mathfrak{g}$ obeying

$$[[v, w], z] + [w, [v, z]] = [v, [w, z]], \quad \forall v, w, z \in \mathfrak{g}$$

while regularity is automatic. We do not require antisymmetry of the bracket, which means that a braided-Lie algebra of this form is the same as saying that $\mathfrak{g}$ is a Leibniz algebra (Loday and Pirashvili, 1993), a slightly more general notion than that of a Lie algebra but including it. In fact the theory of braided-Lie algebras arose independently at about the same time and can be considered as a different route to Leibniz algebras.

Similarly, we can consider $\mathcal{L} = kX$ where X is a set, and $\Delta x = x \otimes x$ for all $x \in X$. Writing $[x, y] = {}^x y$ as a notation, the axioms boil down in this case to

$${}^{({}^x y)}({}^x z) = {}^x({}^y z), \quad \forall x, y, z \in X$$

The regularity condition amounts to the requirement that for every x, z there is a unique y such that ${}^x y = z$. Such a structure is variously called a 'rack' or 'quandle', a more general notion than a group or conjugacy class in a group with $[x, y] = xyx^{-1}$ (Fenn and Rourke, 1992). This concept in its modern form dates from the 1980s with earlier roots. In fact a quandle as opposed to a general rack has the further condition ${}^x x = x$, $\forall x \in X$, and this is expressed in braided-Lie algebra terms as the further condition

$$[\,,\,]\Delta = \mathrm{id}$$

This holds automatically in the Leibniz algebra case. Thus at the least our theory uniformly applies to two otherwise very disparate structures already of interest.

Theorem 4.19 (Majid, 1994c) *When $\mathcal{V}$ is abelian there is a braided enveloping bialgebra $U(\mathcal{L}) \in \mathcal{V}$ and, when $\mathcal{L}$ is rigid, a braided Killing form defined as*

$$U(\mathcal{L}) = T\mathcal{L}/\ \vcenter{\hbox{[diagram]}}$$

Here $U(\mathcal{L})$ has coproduct extending the one of $\mathcal{L}$ and K is braided-symmetric in the sense $K\tilde{\Psi} = K$.

This is non-trivial even in the Vec case. Then $U(k \oplus \mathfrak{g})$ is a quadratic ordinary bialgebra associated to any Leibniz algebra with relations $xy - yx = c[x,y]$ and c central. In the Lie algebra case there is a bialgebra homomorphism $U(k \oplus \mathfrak{g}) \to U(\mathfrak{g})$ sending $c = 1$. The Killing form restricts to the usual Killing form and in addition

$$K(c,c) = 1, \quad K(c,x) = K(x,c) = 0.$$

For a conjugacy class or ad-stable subset $X \subset G$ of a group not containing the group identity we have a quadratic bialgebra $U(kX)$ with quadratic relations $xy = [x,y]x$ and a bialgebra map $U(kX) \to kG$, which is a surjection if X generates the group. The Killing form here comes out as $K(x,y) = |Z(xy) \cap X|$ where $Z(xy)$ is the centralizer of xy.

Moreover, braided-Lie algebras are finite-dimensional objects that generate $U_q(\mathfrak{g})$ from a self-contained set of axioms for the maps above and the braiding $\Psi_{\mathcal{L},\mathcal{L}}$:

Theorem 4.20 (Majid, 1994c; Gomez and Majid, 2003) *For $U_q(\mathfrak{g})$ with generic q, there is a braided-Lie algebra $\mathcal{L}$ associated to each irrep of $\mathfrak{g}$, of square dimension and living in the braided category of finite-dimensional $U_q(\mathfrak{g})$-modules. There is a standard one for which $U(\mathcal{L}) \to U_q(\mathfrak{g})$ is essentially a surjection (more precisely $U(\mathcal{L}) \to B_q[G]$ is a surjection).*

Proof One can reduce the braided-Lie algebra structure to R-matrix identities (Majid, 1994c), which gives the standard calculus using the defining representation, but in principle can also be used in other representations. Alternatively, differential forms $\Omega^1(\mathbb{C}_q[G])$ in the context of non-commutative geometry have been classified under the assumption of left–right translation invariance of the differential calculus and are known to correspond in the generic case and for calculi with classical limit to irreps (Majid, 1998a). The left-invariant 1-forms Λ^1 then define a braided-Lie algebra by $\mathcal{L} = \Lambda^{1*}$ (see Gomez and Majid, 2003). We should explain also that $U_q(\mathfrak{g})$ has the same algebra as $B(U_q(g))$ and the latter is almost self-dual so almost isomorphic to $B_q(G)$; more precisely it is a localization of the latter. Either one should work in a formal power-series setting or it is to $B_q(G)$ that the braided-enveloping algebra more naturally surjects. □

Also, in some cases, with more structure, one can quotient $U(\mathcal{L})$ to a braided group. This is the case for instance for $U(kX)$ if $X \subset G$ is closed under group inversion, in which case we can add the relations $xx^{-1} = 1$ for all $x \in X$ to give a cocommutative Hopf algebra. When X generates we have a surjection of groups $G_X \to G$ where G_X is generated by X with the same relations we have mentioned for $U(kX)$ along with $xx^{-1} = e$ where e is the group identity. This is reminiscent of the connected simply connected Lie group associated to a Lie group and with the same Lie algebra.

More recently we have looked (Lopez Pena, Majid and Rietsch, 2010) at the Killing form in this context and find—experimentally—that when G is simple the Killing form appears to be non-degenerate for all braided-Lie algebras associated to $X \subseteq G \setminus \{e\}$ closed under inversion. We also prove it in some cases, notably the case of $X = G \setminus \{e\}$ with G obeying the Roth property (this applies to sporadic groups, indeed for all finite simple groups other than certain members of one family). The Killing form in this case can also be regarded as a map $\mathcal{L} \to \mathcal{L}$ by composing with the inverse of the Euclidean inner product and we find—again experimentally—the remarkable property that for X a conjugacy class the eigenspace decomposition typically decomposes kX under conjugation into irreps of G. If one plots the eigenvalue against the irrep one has a kind of 'bar code' for every conjugacy class in which, typically, the different multiples of an irrep occurring in kX are now separated. We do not yet have conditions by which to associate a particular irrep to a conjugacy class (which if we succeeded would be remarkable) nor to single out a particular conjugacy class as most suitable, but these are all ideas for 'transfer' of Lie theory to finite groups suggested by the theory above.

4.6 Diagrammatic geometry

Finally, we show how one can begin to do differential geometry in an abelian monoidal category $\mathcal{V}$. The theory of principal bundles, Riemannian connections, and gauge fields has been worked out at this diagrammatic level at least with the universal calculus (Majid, 1999a) but as this requires a fair bit of classical differential geometry to appreciate, I will limit myself here to the sector of the theory where the bundles are trivial. In this case the main datum is an algebra $A \in \mathcal{V}$ equipped with a calculus Ω^1, which for the sake of concreteness we take to be the universal one $\Omega^1 = \mathrm{Ker}(\cdot : A \otimes A \to A)$ with $\mathrm{d} = \mathrm{id} \otimes \underline{1} - \underline{1} \otimes \mathrm{id}$ as the 'differential' $\mathrm{d} : A \to \Omega^1$. This extends to an acyclic complex (Ω, d) with $\mathrm{d}^2 = 0$ and a super-Leibniz rule. Here $\Omega^n \subset A^{\otimes n+1}$ is the joint kernel of the product of adjacent factors. All of this makes sense in an abelian monoidal category. Useful quotients leading to non-universal calculi are likely to require a braided category or other further structure.

We also require a 'gauge group' and at our level we need only a coalgebra B. We are thinking of B as functions on the group and A as functions on the base space, so that a gauge transformation is a morphism $\gamma : B \to A$ and we require this to be invertible in the convolution sense

$$\cdot(\gamma \otimes \gamma^{-1})\Delta = \mathrm{id} = \cdot(\gamma^{-1} \otimes \gamma)\Delta.$$

A gauge field is similarly a map $B \to \Omega^1$ rather than a group or Lie algebra-valued 1-form as usually. Its curvature $F(A) = \mathrm{d}A + A * A$ in a convolution sense is shown in Figure 4.12 (Majid, 1999a) and changes by conjugation in the sense shown in the second line of the figure when A transforms as shown. This is the 'fundamental lemma' of non-abelian gauge theory and it is remarkable that it makes sense at a planar level of diagrams without braid crossings, i.e. in any abelian monoidal category. One needs more structure such as a braiding and a braided group or braided-Lie coalgebra for deeper aspects of the geometry and for non-trivial bundles.

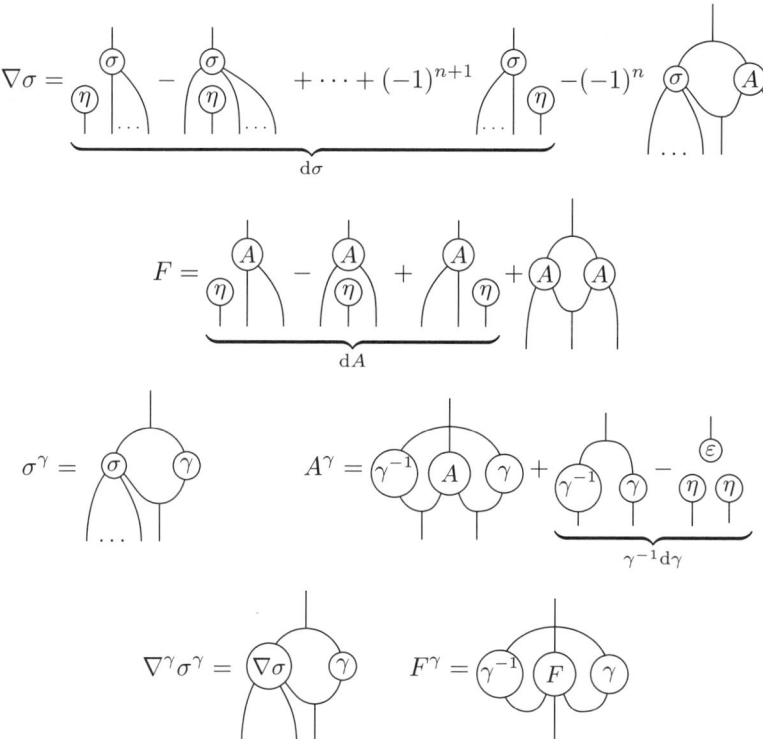

Fig 4.12 Local form of gauge theory in a monoidal category. Here a gauge field is a map $B \to \Omega^1 \subset A \otimes A$ and matter field $\sigma : V \to \Omega^n \subset A^{\otimes n+1}$ where V is a B-comodule. Read down the page.

Finally, matter fields are maps $V \to \Omega^n$ where V is a right-handed B-comodule. This is the analogue of forms with values in a representation space of the gauge group classically. The gauge field defines a covariant derivative on such matter fields, taking the form $\nabla \sigma = d\sigma - (-1)^n \sigma * A$ in a convolution sense as shown on the top left in Figure 4.12. It transforms covariantly if σ transforms as shown. We are denoting the coaction here by the same notation as for the coproduct. One significance of the planar nature of these diagrams may be that they can be realized as processes in a quantum computer. In quantum computing one may not easily be able to 'wire up' as for a regular computer but rather it may make sense for the output of one process to feed into another literally by concatenation.

CHAPTER 5

Hopf Algebras—Variant Notions and Reconstruction Theorems

JOOST VERCRUYSSE
(Université Libre Bruxelles)

5.1 Introduction

Since the discovery of Hopf algebras in the 1940s, they have appeared as useful tools in various fields of mathematics, such as number theory (the group ring of a formal group is a Hopf algebra), algebraic geometry (the algebra of regular functions on an algebraic group is a Hopf algebra, or more generally Hopf algebras are constructed from affine group schemes), Lie theory (the universal enveloping algebra of a Lie algebra is a Hopf algebra), Galois theory and separable field extensions (in relation with so-called Hopf–Galois theory), graded ring theory (one constructs a Hopf algebra from a graded ring and there is a strong relationship between graded modules and Hopf modules), locally compact group theory (quantum group theory), combinatorics (see e.g. Aguiar and Mahajan, 2006, and the references therein), quantum mechanics and so on.

It should be of no surprise that because of the large variety of applications of the theory of Hopf algebras, a wide variety of mutations of the original definition has also occurred in the literature. Some of these variations are *quasi Hopf algebras* (Drinfeld, 1989), which weaken certain (co)associativity constraints, *weak Hopf algebras* (Böhm, Nill and Szlachányi, 1999), which weaken certain compatibility conditions on the (co)unit, *Hopf algebroids* (Böhm and Szlachányi, 2004), which allow the transition to a non-commutative base, *multiplier Hopf algebras* (van Daele, 1994), which are non-unital as algebra, *group Hopf co-algebras* (Turaev, 2000), which are a dual version of group graded of Hopf-algebras, and *Hopfish algebras* (Tang,

Weinstein and Zhu, 2007), which is a Morita invariant notion of Hopf algebras where structure maps are replaced by bimodules.

Although the motivation to introduce these alternative notions was often quite diverse, it is surprisingly impressive how many features of the initial Hopf algebra theory can be transferred to each of the generalized versions. An explanation of this striking fact might be offered by the use of monoidal categories. First of all, Hopf algebras, originally defined over a base field, can be defined in any braided monoidal category. Most of the theory can be quite easily lifted over to this setting, sometimes under additional assumptions such as the existence and preservation of certain (co)limits. Several, but not all, of the above-mentioned generalizations can be understood as particular cases of Hopf algebras in a suitably chosen braided monoidal category. Perhaps a more subtle approach, which we will use as a starting point in this survey, is not to treat Hopf algebras directly as objects with certain properties *in* a braided monoidal category, but to study them as monads *on* a monoidal category and characterize them by means of the Tannaka reconstruction (also known as Tannaka duality or Tannaka–Krein duality).

The Tannaka reconstruction theorem originally stated that a compact topological group is completely determined by its finite-dimensional representations. This result has been generalized to Hopf algebras and large classes of quantum groups. Let k be a field and Vect_k^f the category of finite-dimensional k-vector spaces. In this setting, the Tannaka reconstruction theorem can be stated as follows.

Theorem 5.1 *There is a bijective correspondence between the following objects:*

(i) *k-Hopf algebras H*
(ii) *autonomous monoidal categories $\mathcal{V}$, together with a strict monoidal autonomous functor $U : \mathcal{V} \to \text{Vect}_k^f$.*

Under this correspondence, $\mathcal{V} \simeq \mathsf{M}_H^f$ (the category of finite-dimensional representations of H) and U is the usual forgetful functor.

After recalling some necessary notions from monoidal categories and basic Hopf algebra theory in Sections 5.2 and 5.3, it is our aim is to recall in Section 5.4 the basic ideas of the Tannaka reconstruction theorem and some variations of it. We make a distinction between:

- what we call 'simple reconstruction', where the algebra (or, dually, coalgebra) object is already given by the onset, and the reconstruction concerns only the bialgebra or Hopf algebra structure in relation to the monoidal structure on the category of (co)representations (see Section 5.4.1 for the case of monads and Section 5.4.2 for the algebra case)
- what we could call 'the difficult part' of the reconstruction, which concerns the 'real' Tannaka reconstruction, and allows reconstruction of the algebra (or coalgebra) object itself from its category of (usually only finite-dimensional) (co)representations (see Section 5.4.3).

We will then study in Section 5.5 how the variations on the notion of a Hopf algebra are related to variations of the Tannaka theorem—by changing the base category, the properties of the 'forgetful' functor, or both—and show that the different generalizations of Hopf algebras can be re-obtained in a natural way. Of course, as this is a survey, the results that will be stated are not meant to be exclusively new. All (or most) theorems have appeared before and we will try to be as precise as possible with references. We will refer for most proofs to these references as well.

There already exist several excellent surveys about Tannaka reconstruction, such as those by Hewitt and Ross (1970), Joyal and Street (1991c), and Schauenburg (1992). It makes no sense to repeat or copy this work here. We do not intend to explain precisely how the reconstruction is obtained explicitly, and what the motivation originally was to do this (this is done perfectly in the references above), but we will use this construction to (hopefully) provide some insight into the 'zoo' of Hopf-algebra-like structures.

Another survey that treats the different recent generalizations of Hopf algebras is Karaali (2008). The difference to the present paper is, firstly, the central role that we give to the Tannaka reconstruction theorem and secondly the fact that (for example) multiplier Hopf algebras and group Hopf co-algebras are not considered in Karaali's paper. Nevertheless, we certainly can recommend this paper for a different point of view, especially for the treatment of quasi-Hopf algebras, weak Hopf algebras and Hopf algebroids from the point of view of the dynamical quantum Yang–Baxter equation, which is not considered here.

5.2 Preliminaries

We suppose that the reader is familiar with the technicalities of monoidal categories, as this is one of the main subjects of this book. This section is only meant to fix the notation and terminology. For more details, we refer to, for example, Mac Lane (1971, Chapters VII and IX) and Chapter 4 of this volume.

First of all, for an object X in a category $\mathcal{C}$, we denote the identity morphism on X by id_X or just by X.

$$X \xrightarrow{id_X = X} X.$$

Monoidal categories

Recall that a *monoidal category* $(\mathcal{C}, \otimes, I, a, \ell, r)$ consists of a category $\mathcal{C}$, a functor $\otimes : \mathcal{C} \times \mathcal{C} \to \mathcal{C}$, a monoidal unit object $I \in \mathcal{C}$, an associativity constraint (a natural isomorphism) $a_{X,Y,Z} : X \otimes (Y \otimes Z) \to (X \otimes Y) \otimes Z$, and unit constraints (natural isomorphisms) $\ell_X : I \otimes X \to X$ and $r_X : X \otimes I \to X$, satisfying suitable compatibility conditions. By Mac Lane's coherence theorem, every monoidal category $(\mathcal{C}, \otimes, I, a, l, r)$ is monoidally equivalent to a strict monoidal category

$(C', \otimes', I', a', l', r')$, i.e. a', l' and r' are identities, so there is no need to write them. As a consequence of this theorem, we will also omit writing the terms a, l, r hereafter (unless mentioned explicitly otherwise), and a monoidal category will be denoted for short by $(C, \otimes, I)$. We will make computations and definitions as if C was strict monoidal but, by coherence, everything we do and prove remains valid in the non-strict setting (this will have important implications for quasi-Hopf algebras).

BRAIDINGS AND SYMMETRIES

We call a monoidal category $(C, \otimes, I, a, l, r)$ *braided* if there exist natural isomorphisms $\gamma_{X,Y} : X \otimes Y \to Y \otimes X$, for all $X, Y \in C$ satisfying appropriate compatibility conditions with a, l and r. If $\gamma_{X,Y}^{-1} = \gamma_{Y,X}$ for all $X, Y \in C$, then C is said to be a *symmetric monoidal category*.

RIGIDITY

An object X in a monoidal category is called *left rigid* if there exists an object X^* together with morphisms $\eta : I \to X \otimes X^*$ and $\epsilon : X^* \otimes X \to I$ such that

$$X \otimes \epsilon \circ a^{-1} \circ \eta \otimes X = X, \quad \epsilon \otimes X^* \circ a \circ X^* \otimes \eta = X^*$$

A *right rigid* object is defined symmetrically. A monoidal category is said to be *left rigid* (resp. right rigid) if every object is left (resp. right) rigid. Another name for a rigid monoidal category is an *autonomous (monoidal) category*. If C is braided, then it is right rigid if and only if it is left rigid.

A right *closed monoidal category* is a monoidal category C such that each endofunctor $X \otimes - : C \to C$ associated to an object $X \in C$ has a right adjoint $[X, -]$. A monoidal category is left closed if each endofunctor of the form $- \otimes X$ has a right adjoint. Braided monoidal categories are left closed if and only if they are right closed, that is they are closed for short. If a category is right (resp. left) rigid, then it is right (resp. left) closed and $[X, -] \simeq X^* \otimes -$.

MONOIDAL FUNCTORS

We warn the reader that our terminology of monoidal functors differs slightly from the one used in Chapter 4 of this volume. What is called a monoidal functor in Chapter 4 is called a *strong* monoidal functor here. If one uses the terminology of Chapter 4, then what is called a monoidal functor below, should be referred to as a *lax* monoidal functor.

A functor $F : C \to D$ between the monoidal categories $(C, \otimes, I)$ and $(D, \odot, J)$ is called a *monoidal functor* if there exists a D-morphism $\phi_0 : J \to F(I)$ and a natural transformation $\phi_{X,Y} : F(X) \odot F(Y) \to F(X \otimes Y)$, $X, Y \in C$ satisfying suitable compatibility conditions with relation to the associativity and unit constraints of C and D. Furthermore, we make a distinction between the notions of a *strong* monoidal functor (F, ϕ_0, ϕ), where ϕ_0 is an isomorphism and ϕ is a natural isomorphism, and a *strict* monoidal functor (F, ϕ_0, ϕ), where ϕ_0 is the identity morphism and ϕ is the identity natural transformation. Dually, an *op-monoidal functor* $F : \mathcal{C} \to \mathcal{D}$ is a functor for which there exists a morphism $\psi_0 : F(I) \to J$ in $\mathcal{D}$ and morphisms $\psi_{X,Y} : F(X \otimes Y) \to F(X) \odot F(Y)$ in $\mathcal{D}$ that are natural in $X, Y \in \mathcal{C}$, satisfying suitable compatibility conditions. A strong monoidal functor (F, ϕ_0, ϕ) is automatically op-monoidal. Indeed, one can take $\psi_0 = \phi_0^{-1}$ and $\psi = \phi^{-1}$. If $\mathcal{C}$ and $\mathcal{D}$ are braided monoidal categories, then a *braided monoidal functor* $F : \mathcal{C} \to \mathcal{D}$ is a monoidal functor such that $F\gamma_{X,Y} \circ \phi_{X,Y} = \phi_{Y,X} \circ \gamma_{FX,FY} : FX \odot FY \to F(Y \otimes X)$. Of course, one has a canonical definition of a *monoidal natural transformation* between monoidal functors.

Monoidal functors behave nicely with respect to adjuctions, as can be seen from the following classical theorem.

Theorem 5.2 *Let* $F : \mathcal{C} \to \mathcal{D}$ *be a left adjoint to* $G : \mathcal{D} \to \mathcal{C}$, *where* $(\mathcal{C}, \otimes, I)$ *and* $(\mathcal{D}, \odot, J)$ *are monoidal categories. Then F is op-monoidal if and only if G is a monoidal functor.*

If (F, G) is an adjoint pair of monoidal functors such that the unit and counit of this adjunction are monoidal natural transformations then we call (F, G) a *monoidal adjunction*. By the above theorem, the left adjoint of a monoidal adjucntion is also op-monoidal, in fact it is even strong monoidal.

In a natural way, one defines a monoidal functor between rigid monoidal categories to be rigid (or autonomous) if it preserves dual objects. A monoidal functor between (right, left) closed monoidal categories is said to be (right, left) closed if it commutes with the adjoints of the endofunctors of type $X \otimes -$.

5.3 Bialgebras and Hopf algebras in monoidal categories

5.3.1 Algebras and coalgebras with their representations

Let $\mathcal{C} = (\mathcal{C}, \otimes, I)$ be a monoidal category. An *algebra* or *monoid* in $\mathcal{C}$ is a triple $A = (A, m, u)$, where $A \in \mathcal{A}$ and $m : A \otimes A \to A$ and $u : I \to A$ are morphisms in $\mathcal{C}$ such that the following diagrams commute:

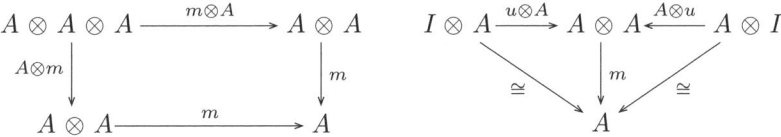

A *coalgebra* or *comonoid* in a monoidal category $\mathcal{A}$ is an algebra in the opposite category $\mathcal{C}^{\mathrm{op}}$.

A right *module* or *representation* over an algebra A in $\mathcal{C}$ is a pair (M, ρ_M) where M is an object of $\mathcal{C}$ and $\rho_M : M \otimes A \to M$ is a morphism in $\mathcal{C}$ satisfying the following associativity and unitary condition

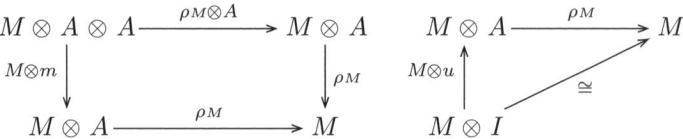

A left module (M, λ_M) is defined in a symmetric way by means of a left action $\lambda : A \otimes M \to M$.

A right *comodule* (M, ρ^M) over a coalgebra C in $\mathcal{C}$ is a right module over C, considered as an algebra in $\mathcal{C}^{\text{op}}$. In particular, the morphism $\rho^M : M \to M \otimes C$ in $\mathcal{C}$ is called the coaction.

Morphisms of algebras, coalgebras, modules, and comodules are defined in an obvious way as structure-preserving morphisms from $\mathcal{C}$. This leads to the introduction of the categories $\mathsf{Alg}(\mathcal{C})$, $\mathsf{Coalg}(\mathcal{C})$, $\mathcal{C}_A$, and $\mathcal{C}^C$ of respectively algebras in $\mathcal{C}$, coalgebras in $\mathcal{C}$, right modules over a fixed algebra A in $\mathcal{C}$, and right comodules over a fixed coalgebra in $\mathcal{C}$.

Example 5.3 Let k be a commutative ring. Modules (resp. comodules) over an algebra A (resp. coalgebra C) in $\mathcal{M}_k$ are exactly the classical modules over A as k-algebra (resp. the classical comodules over C as k-coalgebra). Modules and comodules are the classical examples.

Example 5.4 An algebra in $\underline{\mathsf{Set}}$ is a monoid (G, m, u). A module over G is a G-set. Every set X has a unique structure of a coalgebra in $\underline{\mathsf{Set}}$. The comultiplication $\Delta : X \to X \times X$ is the diagonal map $\Delta(x) = (x, x)$, and the counit is the unique map $\varepsilon : X \to \{*\}$. A comodule over X is then a set Y together with a (any) map $f : Y \to X$. The comultiplication $\rho_f : Y \to Y \times X$ is given by $\rho_f(y) = (y, f(y))$.

Example 5.5 Consider the category of categories (where of course some care has to be taken with respect to the kind of 'largeness' of the categories one is considering, so as to overcome set-theoretic problems, but we omit this discussion here), with cartesian product of categories as tensor product. An algebra in this category is a (strict) monoidal category. A module over a monoidal category $\mathcal{C}$ is a category $\mathcal{M}$ together with a bifunctor $\otimes_{\mathcal{M}} : \mathcal{M} \times \mathcal{C} \to \mathcal{M}$, such that we have natural transformations $M \otimes_{\mathcal{M}} (X \otimes Y) \cong (M \otimes_{\mathcal{M}} X) \otimes_{\mathcal{M}} Y$ and $M \otimes_{\mathcal{M}} I \cong M$ that satisfy a suitable collection of compatibility conditions. We call such a category a right $\mathcal{C}$-category.

Example 5.5 provides us with a tool to consider a wider range of modules and comodules. Let (A, m, u) be an algebra in the monoidal category $\mathcal{C}$ and let $(\mathcal{M}, \otimes_{\mathcal{M}})$ be a right $\mathcal{C}$-category. Then a right A-module in $\mathcal{M}$ is an object $M \in \mathcal{M}$ endowed with a morphism $\rho_M : M \otimes_{\mathcal{M}} A \to M$ in $\mathcal{M}$ satisfying the following associativity and unitary constraints

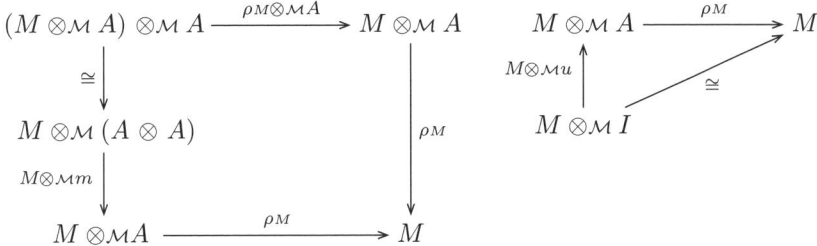

Example 5.6 Let A be an associative ring with unit. The category of (unital) A-bimodules ${}_A\mathcal{M}_A$ is a monoidal category; the monoidal product is given by the tensor product $\otimes_A$ over A, and the unit object by A. An algebra in ${}_A\mathcal{M}_A$ is just a ring morphism $\iota : A \to R$. A coalgebra in ${}_A\mathcal{M}_A$ is called an A-coring (see e.g. Brzezinski and Wisbauer (2003)). One can easily check that $\mathcal{M}_A$ is a right ${}_A\mathcal{M}_A$-category, where we use again the tensor product over A. So we can compute (co-)modules over an A-(co)ring in $\mathcal{M}_A$.

Example 5.7 Monads and comonads are algebras and coalgebras in a monoidal category of endofunctors $\mathsf{EndoFun}(\mathcal{C})$ on a category $\mathcal{C}$. I.e. a monad is an endofunctor $\mathbb{A} : \mathcal{C} \to \mathcal{C}$ with natural transformations $m : \mathbb{AA} \to \mathbb{A}$ and $u : id_\mathcal{C} \to \mathbb{A}$ satisfying suitable unitary and associativity conditions. Obviously, $\mathcal{C}$ is a left $\mathsf{EndoFun}(\mathcal{C})$-category. A module over a monad $(\mathbb{A}, m, u)$ in $\mathcal{C}$ is called an Eilenberg–Moore object. Explicitly, this is a pair (X, ρ_X), where $X \in \mathcal{C}$ and $\rho_X : \mathbb{A}(X) \to X$ is a morphism in $\mathcal{C}$ satisfying the obvious associativity and unitary conditions. We denote the category of Eilenberg–Moore objects (the EM category for short) by $\mathcal{C}_\mathbb{A}$. If $\mathbb{B}$ is a comonad, then we denote the category of comodules over $\mathbb{B}$ in $\mathcal{C}$ (also called the category of Eilenberg–Moore objects) by $\mathcal{C}^\mathbb{B}$.

If (A, m, u) is an algebra in a monoidal category $\mathcal{C}$, then this induces canonically a monad $(\mathbb{A}, M, U)$ on $\mathcal{C}$ by defining $\mathbb{A} = - \otimes A$, $M = - \otimes m : \mathbb{AA} = - \otimes A \otimes A \to \mathbb{A} = - \otimes A$ and $U \simeq u$ by putting for any $X \in \mathcal{C}$, $U_X = X \cong X \otimes I \xrightarrow{X \otimes u} X \otimes A = \mathbb{A}X$. The EM category of $\mathbb{A}$ is exactly the category of right A-modules.

5.3.2 Bialgebras and Hopfalgebras with their representations

Let $(\mathcal{C}, \otimes, I, \gamma)$ be a braided monoidal category. Then $\mathsf{Alg}(\mathcal{C})$ and $\mathsf{Coalg}(\mathcal{C})$ are monoidal categories with a strict monoidal forgetful functor to $\mathcal{C}$. Indeed, for any two algebras (A, m_A, u_A) and (B, m_B, u_B), we can construct a new algebra

$$(A \otimes B, m = (m_A \otimes m_B) \circ (A \otimes \gamma_{B,A} \otimes B), u_A \otimes u_B).$$

A *bialgebra* in $\mathcal{C}$ is a coalgebra in the monoidal category $\mathsf{Alg}(\mathcal{C})$ of algebras or equivalently, an algebra in the monoidal category of coalgebras $\mathsf{Coalg}(\mathcal{C})$. Explicitly, a bialgebra in $\mathcal{C}$ is an object B enriched with an algebra structure (B, m, u), and a coalgebra structure (B, Δ, ϵ) such that any of the following sets of equivalent conditions hold

- Δ and ϵ are algebra maps
- m and u are coalgebra maps
- the following diagrams are commutative:

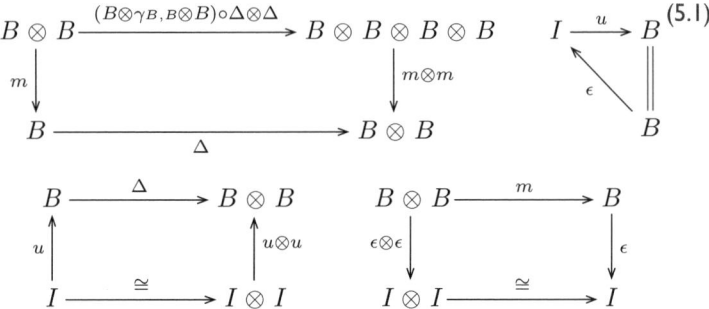
(5.1)

An *antipode* for a bialgebra H in $\mathcal{C}$ is a $\mathcal{C}$-morphism $S : H \to H$ such that

$$m \circ (H \otimes S) \circ \Delta = u \circ \epsilon = m \circ (S \otimes H) \circ \Delta.$$

In fact, this means that S is an inverse for the identity map on H in the monoid $\mathsf{End}_\mathcal{C}(H)$ for the convolution product $*$ that is defined by

$$f * g = m \circ (f \otimes g) \circ \Delta \in \mathsf{End}_\mathcal{C}(H)$$

for any $f, g \in \mathsf{End}_\mathcal{C}(H)$. A *Hopf algebra* in $\mathcal{C}$ is a bialgebra that possesses an antipode. An immediate property of the antipode tells us that it is an anti-algebra morphism and an anti-coalgebra morphism. Some authors only consider Hopf algebras with invertible antipodes; here we will mention explicitly when the antipode is invertible (with respect to usual composition).

Morphisms of bialgebras are $\mathcal{C}$-maps that preserve both the algebra and coalgebra structure. Morphisms of Hopf algebras are morphisms of the underlying bialgebras. It can be proven that Hopf algebra morphisms also preserve the antipode. Hence we construct the category $\mathsf{Bialg}(\mathcal{C})$ of bialgebras in $\mathcal{C}$ and its full subcategory $\mathsf{Hpfalg}(\mathcal{C})$ of Hopf algebras in $\mathcal{C}$.

For a bialgebra (hence also for a Hopf algebra), we can consider its category of right (resp. left) modules and right (resp. left) comodules. Another interesting category is the category of Hopf modules. A (right,right) *Hopf module* over a bialgebra

H is a triple (M, ρ_M, ρ^M), where (M, ρ_M) is a right H-module, (M, ρ^M) is a right H-comodule, and the following compatibility condition holds

$$(\rho_M \otimes m) \circ (M \otimes \gamma_{H,H} \otimes H) \circ (\rho^M \otimes \Delta) = \rho^M \circ \rho_M : M \otimes H \to M \otimes H.$$

Morphisms of H-Hopf modules are $\mathcal{C}$-morphisms that are at the same time H-module morphisms and H-comodule morphisms. The category of (right,right) H-Hopf modules is denoted by $\mathcal{C}_H^H$. For any bialgebra, we can consider the functor

$$- \otimes H : \mathcal{C} \to \mathcal{C}_H^H \tag{5.2}$$

If $\mathcal{C}$ admits equalizers, then this functor has a right adjoint that we denote by $(-)^{coH}$, the functor that takes H-coinvariants and that is computed by the following equalizer at any $M \in \mathcal{C}_H^H$:

$$M^{coH} \longrightarrow M \underset{M \otimes u}{\overset{\rho^M}{\rightrightarrows}} M \otimes H.$$

If H is a Hopf algebra, the fundamental theorem for Hopf modules says that this functor is an equivalence of categories (Takeuchi, 2000). The fact that this functor is an equivalence of categories is even equivalent with the bialgebra H being a Hopf algebra.

Theorem 5.8 (Fundamental theorem for Hopf modules) *Let H be a bialgebra in a braided monoidal category $\mathcal{C}$ with equalizers, then the following statements are equivalent.*

(i) *the functor $(-)^{coH} : \mathcal{C}_H^H \to \mathcal{C}$ is fully faithful;*
(ii) *the pair $(- \otimes H, (-)^{coH})$ is an equivalence of categories between $\mathcal{C}$ and $\mathcal{C}_H^H$;*
(iii) $\underline{\mathrm{can}} = (m \otimes H) \circ (H \otimes \Delta) : H \otimes H \to H \otimes H$ *is a $\mathcal{C}$-isomorphism;*
(iv) H *admits an antipode, i.e. H is a Hopf algebra in $\mathcal{C}$.*

It is generally known and easily verified that a monoidal functor $F : \mathcal{C} \to \mathcal{D}$ sends an algebra A in $\mathcal{C}$ to an algebra $F(A)$ in $\mathcal{D}$. Similarly, an op-monoidal functor sends a coalgebra C in $\mathcal{C}$ to a coalgebra $F(C)$ in $\mathcal{D}$. Finally, a strong monoidal functor sends a bialgebra B (respectively a Hopf algebra H) to a bialgebra $F(B)$ (respectively a Hopf algebra $F(H)$) in $\mathcal{D}$.

Example 5.9 As in $\underline{\mathrm{Set}}$, all objects are in a unique way coalgebras, $\mathrm{Coalg}(\underline{\mathrm{Set}}) = \underline{\mathrm{Set}}$. Consequently every monoid, i.e. every algebra in $\underline{\mathrm{Set}}$, is also an algebra in $\mathrm{Coalg}(\underline{\mathrm{Set}})$, hence a bialgebra in $\underline{\mathrm{Set}}$. A Hopf algebra in $\underline{\mathrm{Set}}$ is nothing other than a group.

Example 5.10 A bialgebra (resp. a Hopf algebra) in $\mathcal{M}_k$ is a classical k-bialgebra (resp. Hopf k-algebra). As the linearization functor $k- : \underline{\mathrm{Set}} \to \mathcal{M}_k$ is a strong monoidal functor, every group algebra kG is in a canonical way a (cocommutative)

Hopf algebra. On the other hand, the contravariant functor $\mathrm{Fun}(-,k): \underline{\mathrm{Set}} \to \mathcal{M}_k$, which sends a set X to the vector space of k-valued functions on $\overline{\overline{X}}$, is only a monoidal functor (if we consider it as a covariant functor from $\underline{\mathrm{Set}}^{op} \to \mathcal{M}_k$). Hence $\mathrm{Fun}(X, k)$ has the structure of a (commutative) algebra for any set X. If we consider only finite sets, then $\mathrm{Fun}(-,k): \underline{\mathrm{Set}}^f \to \mathcal{M}_k$ again becomes strong monoidal and hence $\mathrm{Fun}(G, k)$ Hopf algebra is a (commutive) Hopf algebra if G is a finite group, but in general $\mathrm{Fun}(G, k)$ is no longer a Hopf algebra when G is an infinite group. However, if we consider the more restrictive functor of regular functions on affine sets (spaces) $\mathcal{O}: \mathrm{Aff} \to \mathcal{M}_k$, then we again obtain a strong monoidal functor. As the algebra structure is obtained from the unique coalgebra structure on the affine set given by the diagonal map, this 'explains' why regular functions on an affine space give rise to a commutative algebra. Moreover if G is an affine group, then $\mathcal{O}(G)$ is a Hopf algebra that is commutative as algebra. By deforming the multiplicative structure of these algebras, one can construct non-commutative non-cocommutative Hopf algebras that are called quantum groups (see e.g. Kassel (1995)). (Often, quantum groups are considered from the point of view of differential geometry rather than from algebraic geometry, considering Lie groups rather than affine groups.)

Example 5.11 Let $\mathcal{M}_k^{\mathbb{Z}_2}$ be the category of $\mathbb{Z}_2$-graded k-modules. This is a symmetric monoidal category with a symmetry given by

$$\sigma_{X,Y}: X \otimes Y \to Y \otimes X, \quad \sigma_{X,Y}(x \otimes y) = (-1)^{|x||y|} y \otimes x,$$

where $|\cdot|$ denotes the degree of an element. A Hopf algebra in this category is a Hopf superalgebra.

5.3.3 Bimonads and Hopf monads

The word bimonad (and Hopf monad) is (are) in use for different notions that are all meant to generalize bialgebras (or Hopf algebras) in the setting of monads. The idea of a bimonad in the sense of Bruguières, Lack and Virelizier (2011) and Moerdijk (2002)—in the latter paper the author uses, somewhat misleadingly, the term 'Hopf monad' for what is called a bimonad in the former and also below—is to transfer the notion of a bialgebra from the setting of braided monoidal categories to normal monoidal categories. This is similar to the case of normal monads, where it became possible to transfer the notion of an algebra from the setting of monoidal categories to arbitrary categories. Mesablishvili and Wisbauer (2011) even try to take the notion of a bimonad a step further, attempting to free the notion of a bialgebra even of the monoidal structure. The price to pay is that reconstruction theorems are not as easily obtained in this last setting, but some other Hopf-algebraic features, such as the fundamental theorem, arise more naturally. So both approaches have their advantages and disadvantages. Because this paper is greatly motivated by the reconstruction theorems, we will advocate the approach of Bruguières *et al.* and Moerdijk.

A *bimonad* $\mathbb{B} = (\mathbb{B}, M, U)$ on a monoidal category $\mathcal{C}$ is an opmonoidal monad; that is, $\mathbb{B}$ is a monad on $\mathcal{C}$ such that the underlying functor $\mathbb{B}$ is an op-monoidal functor and the natural transformations $M : \mathbb{BB} \to \mathbb{B}$ and $U : id_\mathcal{C} \to \mathbb{B}$ are opmonoidal natural transformations. In particular, $\mathbb{B}$ is endowed with a natural transformation $\phi^\mathbb{B}_{X,Y} : \mathbb{B}(X \otimes Y) \to \mathbb{B}X \otimes \mathbb{B}Y$ and a morphism $\phi : \mathbb{B}I \to I$ satisfying suitable compatibility conditions (see e.g. Bruguières, Lack and Virelizier (2011, Definition 2.4), but we will omit the technicalities in this review). A *bicomonad* is defined as a monoidal comonad.

The *right fusion operator* of a bimonad $\mathbb{B}$ is defined as the natural transformation

$$\underline{\text{can}}^r_{X,Y} = (M_X \otimes \mathbb{B}Y) \circ \phi^\mathbb{B}_{\mathbb{B}X,Y} : \mathbb{B}(\mathbb{B}X \otimes Y) \to \mathbb{B}X \otimes \mathbb{B}Y.$$

Similarly one defines a left fusion operator $\underline{\text{can}}^\ell : \mathbb{B}(1_\mathcal{C} \otimes \mathbb{B}) \to \mathbb{B} \otimes \mathbb{B}$.

A right (resp. left) *Hopf monad* is a bimonad such that its right (resp. left) fusion operator is a natural isomorphism. A Hopf monad is at the same time a left and right Hopf monad.

Example 5.12 Any bialgebra $(B, m, u, \Delta, \epsilon)$ in a braided monoidal category $\mathcal{C}$ induces bimonad $\mathbb{B} = - \otimes B$ on $\mathcal{C}$. Let us just give the formula for the structural natural transformation $\phi^\mathbb{B}$ and the morphism ϕ,

$$\phi^\mathbb{B}_{X,Y} = (X \otimes \gamma_{Y,B} \otimes B) \circ X \otimes Y \otimes \Delta : (X \otimes Y) \otimes B \to (X \otimes B) \otimes (Y \otimes B);$$

$$\phi = \epsilon : I \otimes B \cong B \to I.$$

Moreover, any Hopf algebra H induces right Hopf monad $\mathbb{H} = - \otimes H$ on $\mathcal{C}$. In this situation, the right fusion operator is given explicitly by

$$\underline{\text{can}}^{\mathbb{H},r}_{X,Y} = (X \otimes \gamma_{Y,H} \otimes H) \circ (X \otimes Y \otimes \underline{\text{can}}) \circ (X \otimes \gamma^{-1}_{Y,H} \otimes H) :$$

$$X \otimes H \otimes Y \otimes H \to X \otimes H \otimes Y \otimes H$$

where $\underline{\text{can}}$ is the canonical map that appears in Theorem 5.8. In a similar way, the left fusion operator is in correspondence with the morphism $\underline{\text{can}} \circ \gamma_{H,H}$, where $\gamma_{H,H}$ denotes the braiding on the Hopf algebra H. If the antipode is invertible, then $\mathbb{H}$ is also a left Hopf monad, hence a Hopf monad.

Recall that an object $G \in \mathcal{C}$ is called a generator if and only if the functor $\text{Hom}_\mathcal{C}(G, -) : \mathcal{C} \to \underline{\text{Set}}$ is fully faithful. If the category $\mathcal{C}$ has coproducts, this is furthermore equivalent with the fact that for any object $X \in \mathcal{C}$ there is a canonical epimorphism $f_X : H = \coprod_{f:G \to X} G \to X$, where the coproduct takes over a number of copies of G. Therefore, we find a fork

$$\coprod_{\substack{(g,h):G \to H, \, st \\ f_X \circ g = f_X \circ h}} G \xrightarrow[h_X]{g_X} \coprod_{f:G \to X} G \xrightarrow{f_X} X \quad (5.3)$$

In general this diagram is not a coequalizer, but G is called a *regular generator* if eqn (5.3) is a coequalizer for every $X \in \mathcal{C}$ (see e.g. Kelly, 2005, p. 81). If we work, moreover, in a category where the endofunctors $- \otimes X$ and $X \otimes -$ preserve colimits, it is not hard to proof that for any two objects X and Y one obtains a canonical epimorphism of the form $f_X \otimes f_Y : \coprod G \otimes G \to X \otimes Y$, hence two morphisms $h, h' : X \otimes Y \to Z$ are identical in $\mathcal{C}$ if and only if $h \circ (f \otimes g) = h' \circ (f \otimes g)$ for all $f : G \to X$ and $g : G \to Y$.

We now state a first simple 'reconstruction-type' theorem.

Theorem 5.13 *Let $\mathcal{C}$ be a cocomplete braided monoidal category such that I is a regular generator of $\mathcal{C}$ and that $- \otimes X$ and $X \otimes -$ preserve colimits in $\mathcal{C}$ for all $X \in \mathcal{A}$. Let (B, m, u) be an algebra in $\mathcal{C}$ and $(\mathbb{B}, M, U)$ the associated monad on $\mathcal{C}$. Then there is a bijective correspondence between*

  (i) *bialgebra structures on B (in $\mathcal{C}$) with underlying algebra (B, m, u)*
  (ii) *bimonad structures on $\mathbb{B}$ (on $\mathcal{C}$) with underlying monad $(\mathbb{B}, M, U)$.*

Moreover, under the above correspondence, there is furthermore a bijective correspondence between antipodes turning B into a Hopf algebra and inverses of the right fusion operator turning $\mathbb{B}$ into a right Hopf monad. Moreover, there is a bijective correspondence between invertible antipodes on B and inverses for both the left and right fusion operators turing $\mathbb{B}$ into a Hopf monad.

Proof We already know from Example 5.12 that a bialgebra B induces a bimonad $\mathbb{B} = - \otimes B$. Conversely, if the monad $\mathbb{B} = - \otimes B$ is a bimonad, then one can easily verify that the morphisms $\Delta = \phi^\mathbb{B}_{I,I} : B \cong (I \otimes I) \otimes B \to (I \otimes B) \otimes (I \otimes B) \cong B \otimes B$ and $\epsilon = \phi : B \cong I \otimes B \to I$ define a comultiplication and counit on B that turn B into a bialgebra.

Let us prove that these constructions give a bijective correspondence. For any two objects $X, Y \in \mathcal{C}$, and any two morphisms $f : I \to X$ and $g : I \to Y$, we find by naturality of $\phi^\mathbb{B}$

$$\phi^\mathbb{B}_{X,Y} \circ (f \otimes g \otimes B) = (f \otimes B \otimes g \otimes B) \circ \phi^\mathbb{B}_{I,I}$$

Moreover, by the properties of I as monoidal unit in $\mathcal{C}$, we can consider $\phi^\mathbb{B}_{I,I} : B \to B \otimes B$ and it follows furthermore that

$$(f \otimes B \otimes g \otimes B) \circ \phi^\mathbb{B}_{I,I} = (X \otimes \gamma_{Y,B} \otimes B) \circ (X \otimes Y \otimes \phi^\mathbb{B}_{I,I}) \circ (f \otimes g \otimes B)$$

As this holds for all morphisms f, g and I is a regular generator, we obtain that (see remark above)

$$\phi^\mathbb{B}_{X,Y} = (X \otimes \gamma_{Y,B} \otimes B) \circ (X \otimes Y \otimes \phi^\mathbb{B}_{I,I}).$$

In the same way, one proves that $\epsilon = \phi$.

For the last statement, we know from Theorem 5.8 that antipodes on the bialgebra B are in correspondence with the bijectivity of the canonical map $\underline{\text{can}} : B \otimes B \to B \otimes B$. Furthermore, as mentioned in Example 5.12, the fusion operator is, up to natural isomorphism, completely determined by the canonical map. Hence the fusion operator is invertible if and only if the canonical map is invertible if and only if B has an antipode. Finally, it is known that both can and can $\circ\, \gamma_{H,H}$ are bijective if and only if the antipode of a Hopf algebra is invertible. □

Remark 5.14 In Hopf algebra theory (over a base field), one sometimes considers Hopf algebras with a one-sided antipode; that is, a *right* antipode for a bialgebra H is a morphism $S : H \to H$ that is a *right* inverse of the identity with relation to the convolution product. A right Hopf algebra is then a bialgebra that possesses a right antipode. Left antipodes and left Hopf algebras are defined in the same way. Let us remark that this terminology does not fully correspond to the case of Hopf monads. As we have seen, a *right* Hopf monad is a bimonad, such that the *right* fusion operator is a natural isomorphism. By the above theorem, if the monad $\mathbb{H}$ is of the form $- \otimes H$ for an algebra H, then right Hopf monad structures on $\mathbb{H}$ correspond exactly to bialgebra structures on H with a usual antipode; that is, ordinary Hopf algebra structures on H (not just right). Similarly, a left Hopf monad of the form $H \otimes -$ also corresponds to a Hopf algebra structure on H. So $H \otimes -$ is a left Hopf monad if and only if $- \otimes H$ is a right Hopf monad if and only if H is a Hopf algebra. In a similar way as Theorem 5.13, one proves that a (left and right) Hopf monad of the form $- \otimes H$ corresponds exactly to a Hopf algebra H with invertible antipode.

Remark 5.15 As mentioned above, Mesablishvili and Wisbauer (2011) introduced an alternative notion of bimonad and Hopf monad. In their work a bimonad on a (not necessarily monoidal) category $\mathcal{C}$ consists of a monad $(\mathbb{B}, m, u)$ on $\mathcal{C}$ such that the underlying functor is endowed at the same time with the structure of a comonad $(\mathbb{B}, \Delta, \epsilon)$ on $\mathcal{C}$, and there is a mixed distributive law $\lambda : \mathbb{B}\mathbb{B} \to \mathbb{B}\mathbb{B}$ relating the monad and comonad structure. This setting allows consideration of a suitable kind of Hopf module $\mathcal{C}_\mathbb{B}^\mathbb{B}$ along with a comparison functor $\mathcal{C} \to \mathcal{C}_\mathbb{B}^\mathbb{B}$, similar to the functor given in eqn (5.2). An antipode can now be introduced, similar to the case of usual Hopf algebras, as a natural transformation $S : \mathbb{B} \to \mathbb{B}$ satisfying $m * (SH) * \Delta = u * \epsilon = m * (HS) * \Delta$ (where $*$ is the Godement product). Theorem 5.8 has a very natural generalization in this setting. (A version of the fundamental theorem has also been given in the setting of Hopf monads by Bruguières, Lack and Virelizier.)

5.4 Reconstruction theorems

5.4.1 Monoidal structure on the representation categories of bimonads and Hopf monads

As mentioned earlier, it is classically known that bialgebras can be characterized as those algebras whose category of modules is a monoidal category with a strict monoidal forgetful functor. The monadic version of this theorem is of folkloristic knowledge. For a slightly different formulation of the following theorem, see Moerdijk (2002, Theorem 7.1) or Szlachányi (2003, Section 2.4).

Theorem 5.16 *Let $(\mathbb{B}, M, U)$ be a monad on a monoidal category C. Then there is a bijective correspondence between:*

(i) *bimonad structures on the functor $\mathbb{B}$ with underlying monad $(\mathbb{B}, M, U)$; and*
(ii) *monoidal structures on the EM category of the monad $(\mathbb{B}, M, U)$ such that the forgetful functor to C is a strict monoidal functor.*

The following theorem follows now by a duality argument.

Theorem 5.17 *Let $(\mathbb{B}, M, U)$ be a monad on a monoidal category C. Then there is a bijective correspondence between:*

(i) *stuctures of a monoidal monad on the functor $\mathbb{B}$ with underlying monad $(\mathbb{B}, M, U)$; and*
(ii) *monoidal structures on the Kleisli category of the monad $(\mathbb{B}, M, U)$ such that the forgetful functor to C is a strict monoidal functor.*

As will become clear when we proceed, to characterize Hopf algebras, we need rigidity conditions on our underlying category. A first theorem that already fully characterizes Hopf monads in a very beautiful and general way is the following theorem, a reformulation of Bruguières, Lack and Virelizier (2011, Theorem 3.6) where we only ask for a closed monoidal category (such as a module category). Although it is not as strong as the original Tannaka theorem from the reconstruction point of view, it explains very nicely the internal meaning of Hopf monads, and therefore Hopf algebras, in terms of monoidal categories.

Theorem 5.18 (Bruguières *et al.*, 2011, Theorem 3.6) *Let $(\mathbb{B}, M, U)$ be a monad on a right closed monoidal category C. Then there is a bijective correspondence between*

(i) *right Hopf monad structures on $\mathbb{B}$ with underlying monad $(\mathbb{B}, M, U)$; and*
(ii) *right closed monoidal structures on the EM category of the monad $(\mathbb{B}, M, U)$ such that the forgetful functor to C is right closed and strict monoidal.*

In the classical theory of Hopf algebras, rather than considering a closed structure on the category of (right) H-modules over a Hopf algebra H, one often uses the antipode to put a right H-module structure on the dual space M^* of any right H-module M as follows:

$$f \cdot h = f(e_i \cdot S(h))f_i,$$

where $f \in M^*$, $h \in H$ and $\{e_i, f_i\} \in M \times M^*$ is a finite dual basis. This property follows now directly from Theorem 5.18.

Corollary 5.19 *Let C^f be the right rigid full subcategory of the closed monoidal category C. If $\mathbb{B}$ is a Hopf monad on C, then there is a right rigid full subcategory $C^f_{\mathbb{B}}$ of the EM-category of $\mathbb{B}$ such that the forgetful functor restricts and corestricts to a functor $U^f : C^f_{\mathbb{B}} \to C^f$ that is strict monoidal (and hence right rigid).*

Proof One can construct the category $\mathcal{C}_\mathbb{B}^f$ as the pullback of the functor $U : \mathcal{C}_\mathbb{B} \to \mathcal{C}$ and the embedding functor $\mathcal{C}^f \hookrightarrow \mathcal{C}$, as in the following diagram.

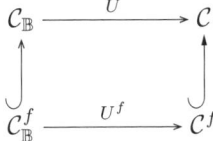

Explicitly, $\mathcal{C}_\mathbb{B}^f$ consists of all objects $(X, \rho) \in \mathcal{C}_\mathbb{B}$ such that $U(X, \rho) = X$ is rigid. We denote $[X, -]$ for the right adjoint of $X \otimes - : \mathcal{C} \to \mathcal{C}$ and $[(X, \rho), -]_\mathbb{B}$ for the right adjoint of $(X, \rho) \otimes - : \mathcal{C}_\mathbb{B} \to \mathcal{C}_\mathbb{B}$. Then we have, $U \circ [(X, \rho), -]_\mathbb{B} = [X, -] \simeq X^* \otimes -$. Moreover, $[(X, \rho), I]_\mathbb{B} \in \mathcal{C}_\mathbb{B}$, so X^* can be endowed with a $\mathbb{B}$-module structure ρ^*, and we find that (X, ρ) is right rigid with dual (X^*, ρ^*). □

As Theorem 5.16 and Theorem 5.18 are 'if and only if' theorems, where Corollary 5.19 only works in one direction, the natural question arises whether the statement of Corollary 5.19 also has an inverse; that is: 'Can we reconstruct a Hopf monad structure on $\mathbb{B}$ by only knowing the rigid monoidal structure of $\mathcal{C}_\mathbb{B}^f$?'. The first problem is that in general the monad $\mathbb{B}$ is not a functor defined on $\mathcal{C}^f$. If one thinks about (Hopf) monads that arise from ordinary k-(Hopf) algebras, this is only the case if the underlying space of the (Hopf) algebra is itself finite-dimensional. In this case we can apply (Bruguières, Lack and Virelizier, 2011, Theorem 3.10) and find the following.

Theorem 5.20 *Let $\mathcal{C}$ be a rigid monoidal category and $(\mathbb{B}, M, U)$ a monad on $\mathcal{C}$. There is a bijective correspondence between:*

(i) *Hopf monad structures on $\mathbb{B}$ with underlying monad $(\mathbb{B}, M, U)$; and*
(ii) *rigid monoidal structures on the EM category of the monad $(\mathbb{B}, M, U)$ such that the forgetful functor to $\mathcal{C}$ is strict monoidal (and hence rigid).*

5.4.2 Simple reconstruction of bialgebras and Hopf algebras

In this section, we will apply the results of the previous section on the case where the bimonad is obtained from a bialgebra. We give a reconstruction-type theorem for bialgebras and Hopf algebras that are not yet of the type of the Tannaka reconstruction theorem; that is, we do not reconstruct a the whole algebra from only pieces of finite-dimensional information, rather we reconstruct the coalgebraic structure on a given algebra from the monoidal structure on its category of representations. This is, however, the right step up towards a full Tannaka reconstruction theorem, as it fully characterizes bialgebras and Hopf algebras, as one sees from Theorem 5.21 and Theorem 5.22 respectively.

Theorem 5.21 Let $B = (B, m_B, u_B)$ be an algebra in a braided monoidal category $\mathcal{A} = (\mathcal{A}, \otimes, I, \gamma)$. Suppose that I is a regular generator of $\mathcal{A}$ and that $- \otimes X$ and $X \otimes -$ preserve colimits in $\mathcal{A}$, for all $X \in \mathcal{A}$. There is a bijective correspondence between:

(i) monoidal structures on $\mathcal{A}_B$ such that the forgetful functor $\mathcal{A}_B \to \mathcal{A}$ is strict monoidal, and
(ii) bialgebra structures $(B, m_B, u_B, \Delta_B, \varepsilon_B)$ on B.

Proof Follows from Theorem 5.13 and Theorem 5.16. □

Theorem 5.22 Let $B = (B, m_B, u_B)$ be an algebra in a closed braided monoidal category $\mathcal{A} = (\mathcal{A}, \otimes, I, \gamma)$. Suppose that I is a regular generator of $\mathcal{A}$ and that $- \otimes X$ and $X \otimes -$ preserve colimits in $\mathcal{A}$ for all $X \in \mathcal{A}$. There is a bijective correspondence between:

(i) right closed monoidal structures on $\mathcal{A}_B$ such that the forgetful functor $\mathcal{A}_B \to \mathcal{A}$ is strict monoidal and right closed, and
(ii) Hopf algebra structures $(B, m_B, u_B, \Delta_B, \varepsilon_B)$ on B.

Proof Follows from Theorem 5.13 and Theorem 5.18. □

Notice that the last two theorems are in particular applicable to the case $\mathcal{C} = \mathcal{M}_k$, i.e. they characterize classical bialgebras and Hopf algebras.

5.4.3 Tannaka reconstruction of bialgebras and Hopf algebras

We will not discuss the Tannaka reconstruction theorem in terms of Hopf monads or bimonads, of which an explicit description is not known to the author (although we can refer the interested reader to the recent work by Booker and Street (2011) for an even more general approach; see also Section 5.5.3.6). In recent years, several variations and generalizations of the theorem in different settings have been formulated. See for example Joyal and Street (1991c), Majid (1995a), McCurdy (2012), Schäppi (2011b), Schäppi (2011a), Schauenburg (1992), and Szláchanyi (2009).

Our formulation is taken from McCurdy (2012). Let $\mathcal{C}$ be a complete braided monoidal category such that all endofunctors of the form $- \otimes X$ and $X \otimes -$ for $X \in \mathcal{C}$ preserve limits. The category strmon $/ \mathcal{C}$ is defined as the category whose objects are pairs $(\mathcal{V}, F)$, where $\mathcal{V}$ is a monoidal category and $F : \mathcal{V} \to \mathcal{C}$ is a strict monoidal functor $F : \mathcal{V} \to \mathcal{C}$ such that FX is right rigid for any object $X \in \mathcal{V}$. A morphism $H : (\mathcal{V}, F) \to (\mathcal{W}, G)$ in this category is a strict monoidal functor $H : \mathcal{V} \to \mathcal{W}$ such that $F = G \circ H$. The category strmon* $/ \mathcal{C}$ is the subcategory of strmon $/ \mathcal{C}$ that consists of functors $F : \mathcal{V} \to \mathcal{C}$, where $\mathcal{V}$ is a right rigid category.

Let $\mathcal{C}$ be a braided monoidal category. As we know from Theorem 5.21, every bialgebra B in $\mathcal{C}$ induces a strict monoidal functor $U_B : \mathcal{C}_B \to \mathcal{C}$. If B is moreover a Hopf algebra, then we have by Corollary 5.19 also functor $U_B^f : \mathcal{C}_B^f \to \mathcal{C}^f$. Therefore, we find functors $\tilde{U} : \mathsf{Bialg}(\mathcal{C}) \to \mathsf{strmon}\swarrow \mathcal{C}$, $\tilde{U}(B) = (\mathcal{C}_B, U_B)$, and $\tilde{U}^* : \mathsf{Hpfalg}(\mathcal{C}) \to \mathsf{strmon}^* \swarrow \mathcal{C}$, $\tilde{U}^*(B) = (\mathcal{C}_B^f, U_B^f)$. The Tannaka reconstruction allows in first place a left adjoint for these functors.

Theorem 5.23 *Let $\mathcal{C}$ be a complete braided monoidal category such that all endofunctors of the form $-\otimes X$ and $X \otimes -$ for $X \in \mathcal{C}$ preserve limits.*

(i) *The functor $\tilde{U} : \mathsf{Bialg}(\mathcal{C}) \to \mathsf{strmon} \swarrow \mathcal{C}$ has a left adjoint* tan.
(ii) *The functor $\tilde{U}^* : \mathsf{Hpfalg}(\mathcal{C}) \to \mathsf{strmon}^* \swarrow \mathcal{C}$ has a left adjoint* tan*.

The existence of the functors tan and tan* is based on the so-called end-construction, which is in fact a particular limit (see e.g. Kelly, 2005). This construction allows us to build up an algebra out of its category of representations, or dually a coalgebra out of its category of corepresentations (in the latter case this coalgebra is sometimes called the *coendomorphism coalgebra*, or *coend* for short). As in our situation, the representation categories posses an additional monoidal structure, and the reconstructed algebra will inherit an additional structure as well, leading to a bialgebra or Hopf algebra. We refer for a full proof of Theorem 5.23 to for example Street (2007, Section 16) or McCurdy (2012, Section 6).

The next question that arises is whether the above theorem completely determines bialgebras and Hopf algebras. That is, when are the functors $\tilde{U}$, $\tilde{U}^*$ or their adjoints fully faithful. In particular, if H is a bialgebra (or a Hopf algebra), one can wonder if H is isomorphic to the reconstructed algebra $\tan\tilde{U}H$; this is the so-called *reconstruction problem*. A second problem is termed the *recognition problem* and refers to the question of whether the pair $\tilde{U}\tan(\mathcal{V}, F)$ is isomorphic to $(\mathcal{V}, F)$ in $\mathsf{strmon} \swarrow \mathcal{C}$; that is, whether the functor F is essentially unique.

It turns out that in many of the cases of interest, such as when $\mathcal{C} = \mathsf{Vect}(k)$, the category of vector spaces over a fixed field k, both problems have a positive answer, leading in particular to Theorem 5.1. Generalizations to a general categorical setting often become highly technical, and we omit them here. Let us briefly summarize some results.

- Day (1996) solved both problems for finitely presentable, complete and cocomplete symmetric monoidal closed categories for which the full subcategory of objects with duals is closed under finite limits and colimits.
- McCrudden (2002) proved the reconstruction problem for so-called Maschkean categories, which are certain abelian monoidal categories in which all monomorphisms split.
- Probably the most general (symmetric) setting can be found in Schäppi (2011b), where the author deals with complete and cocomplete symmetric monoidal and closed categories (called cosmoi).

5.5 Variations on the notion of Hopf algebra

As we have seen, the reconstruction theorems (Theorem 5.1 and Theorem 5.22) fully characterize Hopf algebras over a field. By varying the base category or the properties of the forgetful functor, we will recover in this section the different variations on the notion of a classical Hopf algebra that were defined over the last decades.

5.5.1 Variations on the properties of the forgetful functor

5.5.1.1 QUASI-HOPF ALGEBRAS

Let $(\mathcal{C}, \otimes, I)$ and $(\mathcal{D}, \odot, J)$ be monoidal categories. We will say that a functor $F : \mathcal{C} \to \mathcal{D}$ is a quasi-monoidal functor if there is a natural isomorphism $\psi_{X,Y} : F(X) \odot F(Y) \cong F(X \otimes Y)$ and a $\mathcal{C}$-isomorphism $\psi_0 : F(I) \cong J$ (without any further conditions). Clearly any strong monoidal functor is a quasi-monoidal functor. Then we can introduce quasi-bialgebras by postulating the following characterization:

Let (H, m, u) be a k-algebra. Then there is a bijective correspondence between:

(i) *quasi-bialgebra* structures on H with underlying algebra (H, m, u); and
(ii) monoidal structures on the category of right H-modules $\mathcal{M}_H$ such that the forgetful functor $U : \mathcal{M}_H \to \mathcal{M}_k$ is a quasi-monoidal functor.

Clearly, every (usual) k-bialgebra is a quasi-bialgebra, but the converse is not true. The main difference with usual bialgebras is that the comultiplication in a quasi-bialgebra is not necessarily coassociative. Quasi-bialgebras were introduced by Drinfeld (1989) using the following more explicit description. Let H be a k-algebra with an invertible element $\Phi \in H \otimes H \otimes H$, and endowed with a comultiplication $\Delta : H \to H \otimes H$ and a counit $\varepsilon : H \to k$ satisfying the following conditions for all $a \in H$:

$$(H \otimes \Delta) \circ \Delta(a) = \Phi[(\Delta \otimes H) \circ \Delta(a)]\Phi^{-1},$$
$$(\varepsilon \otimes H) \circ \Delta = H = (H \otimes \varepsilon) \circ \Delta.$$

Furthermore, Φ has to be a normalized 3-cocycle, in the sense that

$$[(H \otimes H \otimes \Delta)(\Phi)] [(\Delta \otimes H \otimes H)(\Phi)] = (1 \otimes \Phi) [(H \otimes \Delta \otimes H)(\Phi)] (\Phi \otimes 1)$$
$$(H \otimes \varepsilon \otimes H)(\Phi) = 1 \otimes 1.$$

Then H is a quasi-bialgebra. The correspondence with the characterization above follows from the fact that the associativity constraint in the monoidal category of right H-modules $\mathcal{M}_H$ over a quasi-bialgebra H can be constructed as

$$a_{X,Y,Z} : X \otimes (Y \otimes Z) \to (X \otimes Y) \otimes Z, \; a_{X,Y,Z}(x \otimes (y \otimes z)) = ((x \otimes y) \otimes z) \cdot \Phi$$

Conversely, if $\mathcal{M}_H$ is monoidal, then we recover $\Phi = \alpha_{H,H,H}(1 \otimes 1 \otimes 1)$, which satisfies the conditions of a normalized 3-cocycle as a consequence of the constraints in a monoidal category.

If moreover, there exist elements $\alpha, \beta \in H$ and an anti-algebra morphism $S : H \to H$ such that

$$S(a_{(1)})\alpha a_{(2)} = \epsilon(a)\alpha; \quad a_{(1)}\beta S(a_{(2)}) = \epsilon(a)\beta,$$

for all $a \in H$ and

$$X^1 \beta S(X^1)\alpha X^3 = 1 = S(x^1)\alpha x^2 \beta S(x^3).$$

where we have denoted $\Phi = X^1 \otimes X^2 \otimes X^3$ and $\Phi^{-1} = x^1 \otimes x^2 \otimes x^3$, then H is called a quasi-Hopf algebra. By means of Tannaka reconstruction, one then finds the following characterization.

There is a bijective correspondence between:

(i) quasi-Hopf k-Hopf algebras H; and
(ii) right rigid monoidal categories $\mathcal{V}$ together with a right rigid quasi-monoidal functor $\mathcal{V} \to \mathcal{M}_k$.

Usual group algebras give rise to usual Hopf algebras, similarly examples of quasi-Hopf algebras can be constructed by deforming group algebras with normalized 3-cocycles on this group in classical sense. As quasi-Hopf algebras are in bijective correspondence with certain classes of monoidal categories, these results can be used to compute all possible monoidal structures on certain (small) categories. This is, for example, done in Bulacu, Caenepeel and Torrecillas (2011) and see also references therein.

In contrast to usual bialgebras and Hopf algebras, which are self-dual objects in a braided monoidal category, the quasi-version is not self-dual. Nevertheless, it is possible to define 'dual quasi-bialgebras' or 'co-quasi-bialgebras' (and Hopf algebras). In this dual setting, the objects are usual coalgebras, but posses a non-associative algebra structure, which is governed by a 'reassociator' $\phi \in \mathrm{Hom}(H \otimes H \otimes H, k)$. More precisely, a co-quasi-bialgebra is a coalgebra H such that its category of comodules $\mathcal{M}^H$ is monoidal and the forgetful functor to k-modules is quasi-monoidal. Hence one can consider algebras in the category $\mathcal{M}^H$, called H-comodule algebras. Let A be an H-comodule algebra. Then A is a right H-comodule equipped with H-comodule morphisms $\mu : A \otimes A \to A, \mu(a \otimes b) = a \cdot b$ and $\eta : k \to A$. However, the triple (A, μ, η) is not an associative k-algebra. In contrast, A satisfies the following quasi-associativity condition:

$$(a \cdot b) \cdot c = a_{[0]} \cdot (b_{[0]} \cdot c_{[0]}) \phi(a_{[1]}, b_{[1]}, c_{[1]}).$$

It should be remarked that different from the classical case, H with regular multiplication is not an H-comodule algebra. Since the forgetful functor $U : \mathcal{M}^H \to \mathcal{M}_k$ is a not a strong monoidal functor, the underlying k-vector space of the H-comodule algebra A in general no longer possesses the structure of an associative k-algebra.

Recall that two co-quasi-bialgebras H and H' are called gauge-equivalent iff there exists a monoidal equivalence $F : \mathcal{M}^H \to \mathcal{M}^{H'}$ that commutes with the forgetful functors. In particular, if a co-quasi-Hopf algebra H is gauge equivalent with a usual Hopf algebra H_F, this means that we have a monoidal functor $F : \mathcal{M}^H \to \mathcal{M}^{H_F}$ such that $U_H = U_{H_F} \circ F : \mathcal{M}^H \to \mathcal{M}_k$. As we know that the forgetful functor $U_{H_F} : \mathcal{M}^{H_F} \to \mathcal{M}_k$ is strict monoidal, we find that the forgetful functor U_H is again monoidal, although not necessarily strict monoidal. Consequently, we obtain that any (initial non-associative) H-comodule algebra also possesses the structure of an associate k-algebra by deforming its multiplication by means of the functor U_H in the following way:

$$\mu' : A \otimes A = U_H A \otimes U_H A \longrightarrow U_H(A \otimes A) \xrightarrow{\mu} U_H A = A,$$

where μ is the multiplication of the H-comodule algebra A. Using the converse argument, certain associative k-algebras can be deformed into non-associative ones by means of a gauge-transform. This idea is explored in a very elegant way in Albuquerque and Majid (1999), where it is shown that the octonions arise as a deformation of the group algebra $k[\mathbb{Z}_2 \times \mathbb{Z}_2 \times \mathbb{Z}_2]$ by a 2-cochain, and in this way can be interpreted as a comodule algebra over a co-quasi-Hopf algebra.

5.5.1.2 QUASI-TRIANGULAR HOPF ALGEBRAS

Similar to the previous section, here we can introduce quasi-triangular Hopf algebras by postulating the following characterization:

Let (H, m, u) be a k-algebra. Then there is a bijective correspondence between

(i) *quasitriangular Hopf algebra* structures on H with underlying algebra (H, m, u); and
(ii) right closed braided monoidal structures on the category of right H-modules $\mathcal{M}_H$ such that the forgetful functor $U : \mathcal{M}_H \to \mathcal{M}_k$ is a right closed braided strict monoidal functor.

Quasi-triangularity is a in fact a property of a Hopf algebra, not a true variation on the axioms. Every quasi-triangular Hopf algebra is a Hopf algebra, but not conversely. Explicitly, a Hopf algebra $(H, m, u, \Delta, \epsilon, S)$ is quasi-triangular if there exists an invertible element (called the R-matrix) $R = r^1 \otimes r^2 = R^1 \otimes R^2 \in H \otimes H$ such that:

$$r^1 x_{(1)} \otimes r^1 x_{(2)} = x_{(2)} r^1 \otimes x_{(1)} r^2,$$
$$r^1_{(1)} \otimes r^1_{(2)} \otimes r^2 = r^1 \otimes R^1 \otimes r^2 R^2,$$
$$r^1 \otimes r^2_{(1)} \otimes r^2_{(2)} = r^1 R^1 \otimes R^2 \otimes r^1;$$

for all $x \in H$.

As a consequence of the properties of quasi-triangularity, the R-matrix is a solution of the Yang–Baxter equation. Therefore H-modules over a quasi-triangular Hopf algebra are, for example, studied to determine quasi-invariants of braids and knots. In fact, quasi-triangularity can already be considered for bialgebras, leading to the expected versions of reconstruction and Tannaka theorems.

5.5.1.3 WEAK HOPF ALGEBRAS

Let $(\mathcal{C}, \otimes, I)$ and $(\mathcal{D}, \odot, J)$ be monoidal categories. A functor $F : \mathcal{C} \to \mathcal{D}$ is called *Frobenius monoidal* if F has a monoidal structure (ϕ, ϕ_0) and opmonoidal structure (ψ, ψ_0) such that the following diagrams commute for all $A, B, C \in \mathcal{C}$:

$$\begin{array}{ccc}
F(A \otimes B) \odot FC & \xrightarrow{\phi_{A \otimes B, C}} & F(A \otimes B \otimes C) \\
{\scriptstyle \psi_{A,B} \odot FC} \downarrow & & \downarrow {\scriptstyle \psi_{A, B \otimes C}} \\
FA \odot FB \odot FC & \xrightarrow{FA \odot \phi_{B,C}} & FA \odot F(B \otimes C)
\end{array}$$

$$\begin{array}{ccc}
FA \odot F(B \otimes C) & \xrightarrow{\phi_{A, B \otimes C}} & F(A \otimes B \otimes C) \\
{\scriptstyle FA \odot \psi_{B,C}} \downarrow & & \downarrow {\scriptstyle \psi_{A \otimes B, C}} \\
FA \odot FB \odot FC & \xrightarrow{\phi_{A,B} \odot FC} & F(A \otimes B) \odot FC
\end{array}$$

A Frobenius monoidal functor is called *separable Frobenius monoidal* if, moreover,

$$\phi_{A,B} \circ \psi_{A,B} = F(A \otimes B)$$

for all $A, B \in \mathcal{C}$. Any strong monoidal functor is separable Frobenius monoidal.

Again, we introduce the next notion by postulating the following characterization.

Let (H, m, u) be a k-algebra. Then there is a bijective correspondence between:

(i) *weak bialgebra* structures on H with underlying algebra (H, m, u); and
(ii) monoidal structures on the category of right H-modules $\mathcal{M}_H$ such that the forgetful functor $U : \mathcal{M}_H \to \mathcal{M}_k$ is a separable Frobenius monoidal functor.

The classical definition of a weak bialgebra and a weak Hopf algebra was given in Böhm, Nill and Szlachányi (1999). A k-algebra (H, m, u) is a weak bialgebra if H has a k-coalgebra structure (H, Δ, ϵ), such that Δ is a multiplicative map (i.e. the first diagram of eqn (5.1) commutes) and the following weaker compatibility conditions hold:

$$(\Delta(1) \otimes 1)(1 \otimes \Delta(1)) = (\Delta \otimes H)\Delta(1) = (1 \otimes \Delta(1))(\Delta(1) \otimes 1)$$

$$\epsilon(b1_{(1)})\epsilon(1_{(2)}b') = \epsilon(bb') = \epsilon(b1_{(2)})\epsilon(1_{(1)}b'),$$

for $b, b' \in H$. A weak Hopf algebra is a weak bialgebra that is equipped with a k-linear map $S : H \to H$ satisfying

$$h_{(1)} S(h_{(2)}) = \epsilon(1_{(1)}h)1_{(2)}$$

$$S(h_{(1)})h_{(2)} = 1_{(1)}\epsilon(h1_{(2)})$$

$$S(h_{(1)})h_{(2)}S(h_{(3)}) = S(h).$$

The Tannaka reconstruction gives us the following characterization (see e.g. McCurdy (2012)).

There is a bijective correspondence between:

(i) weak Hopf k-Hopf algebras H; and
(ii) right rigid monoidal categories $\mathcal{V}$ together with a right rigid separable Frobenius monoidal functor $\mathcal{V} \to \mathcal{M}_k$.

Weak Hopf algebras are related to bimonads and Hopf monads, as we will discuss in more detail in Section 5.5.2.1. To study the particularities of weak Hopf algebra theory, however, weak monads, weak bimonads, and weak Hopf monads were introduced in a series of papers (see e.g. Böhm, 2010, and Böhm, Lack and Street, 2011b).

5.5.2 Variations on the monoidal base category

5.5.2.1 HOPF ALGEBROIDS

It took quite a long time to establish the correct Hopf-algebraic notion over a non-commutative base. The reasons for the difficulties are quite clear. First of all, if R is a non-commutative ring then the category of right R-modules $\mathcal{M}_R$ is no longer monoidal (in general). Therefore we have to look instead to the category of R-bimodules ${}_R\mathcal{M}_R$, which is monoidal but in general still not braided. So bialgebras and Hopf algebras cannot be computed *inside* this category. However, we can compute bimonads and Hopf-monads *on* this category. This is how the theory of bialgebroids was developed. Historically, bialgeboids and Hopf algebroids were constructed first in a more direct way, and the interpretion via bimonads and Hopf monads is only very recent. However, in order to make the relation between the different variations on Hopf-algebra-like structures more prominent, we take this new approach in this explanation and introduce bialgebroids by postulating the following characterization, a reformulation of Schauenburg (1998, Theorem 5.1):

Let B be an $R \otimes R^{op}$-algebra (B, m, u) (i.e. an algebra in the monoidal category ${}_R\mathcal{M}_R$). Then there is a bijective correspondence between:

(i) (right) *R-bialgebroid* structures on B, with underlying algebra (B, m, u); and
(ii) monoidal structures on the category of right B-modules such that the forgetful functor $U : \mathcal{M}_B \to {}_R\mathcal{M}_R$ is strict monoidal.

A particular feature of bialgebroids is that rather than a unit map $u : R \otimes R^{op} \to B$, one considers the source and target maps $s : R \to B$ and $t : R^{op} \to B$, which are the combination of the unit map u with the canonical injections $R \to R \otimes R^{op}$ and $R^{op} \to R \otimes R^{op}$ (respectively).

As one can see, in contrast to the case over a commutative base, the notion of a bialgeboid is not left–right symmetric. A left bialgebroid is introduced symmetrically, as an $R \otimes R^{op}$-algebra with monoidal structure on its category of left modules.

Due to this asymmetry, several different notions of a Hopf algebroid were introduced in the literature. Some of these were shown to be equivalent, although this was far from being trivial. We omit this discussion here, but refer to the review by Böhm (2009). The presently overall accepted notion of a Hopf algebroid (introduced in Böhm and Szlachányi (2004) for bijective antipodes) consists of a triple (H_L, H_R, S), where H_L is a left L-algebroid, H_R is a right R-algebroid, such that H_L and H_R share the same underlying k-algebra H. The structure maps have to satisfy several compatibility conditions, for which we refer to Böhm (2009, Definition 4.1), and $S : H \to H$ is the k-linear antipode map that satisfies the following axiom:

$$\mu_L \circ (S \otimes_L H) \circ \Delta_L = s_R \circ \epsilon_R, \qquad \mu_R \circ (H \otimes_R S) \circ \Delta_R = s_L \circ \epsilon_L,$$

where $\mu_L, \Delta_L, \epsilon_L$, and s_L are respectively the multiplication map, the comultiplication map, the counit map, and the source map of the left bialgebroid H_L.

The way we defined a right bialgebroid B over R, tells us immediately that $- \otimes_R B$ is a bimonad on ${}_R\mathcal{M}_R$. It turns out that the bimonad $- \otimes_R H_R$ associated to a Hopf algebroid is a right Hopf monad on ${}_R\mathcal{M}_R$ and the bimonad $H_L \otimes_L -$ becomes a left Hopf monad on ${}_L\mathcal{M}_L$. The converse is not always true, but the alternative notion of a $\times_R$-Hopf algebra introduced by Schauenburg (2000) is defined as a right R-bialgebroid such that the canonical morphism $\underline{\text{can}} : H \otimes_{R^{op}} H \to H \otimes_R H$, $\underline{\text{can}}(h \otimes_{R^{op}} h') = h_{(1)} \otimes_R h_{(2)} h'$ is a bijection. This leads to the following characterization.

Let B be an $R \otimes R^{op}$ algebra. Then there is a bijective correspondence between:

(i) $\times_R$-Hopf algebra structures on B, and
(ii) right closed monoidal structures on the category of right B-modules such that the forgetful functor $U : \mathcal{M}_B \to {}_R\mathcal{M}_R$ is right closed.

It should be remarked that weak Hopf algebras (resp. weak bialgebras) are strongly related to Hopf algebroids (resp. bialgebroids). To some extent, weak Hopf algebras have been a key motivation for recent developments in the theory of Hopf algebroids. Let H be a weak bialgebra. Then $\mathcal{M}_H$ is a monoidal category, but the forgetful functor $\mathcal{M}_H \to \mathcal{M}_k$ is not strict monoidal, nor even strong monoidal; it is only separable Frobenius monoidal. However, if we consider the H-subalgebra

$$R = \mathrm{Im}\, \pi_R, \quad \text{where} \quad \pi_R : H \to H, \ \pi_R(b) = 1_{(1)}\epsilon(b1_{(2)}),$$

which is called the target space, then we do obtain a strict monoidal functor $\mathcal{M}_H \to {}_R\mathcal{M}_R$. In this way, the weak bialgebra H becomes a right R-bialgebroid, and in a similar way a weak Hopf algebra becomes a weak Hopf algebroid (see Böhm, 2009, Sections 3.2.2 and 4.1.2). For a detailed discussion of the monoidal properties of the representation categories of weak bialgebras, refer to Böhm, Caenepeel and Janssen (2011a).

5.5.2.2 HOPF GROUP COALGEBRAS

Hopf group coalgebras were introduced by Turaev in his work on homotopy quantum field theories (see Turaev, 2010, and the earlier preprint Turaev, 2000). The purely algebraic study of these objects was initiated by Virelizier (2002). Explicitly, a Hopf group coalgebra is a family of algebras $(H_g, \mu_g, \eta_g)_{g \in G}$ indexed by a group G with unit e, together with a family of algebra maps

$$\Delta_{g,h} : H_{gh} \to H_g \otimes H_h, \quad \forall g, h \in G$$

and an algebra map $\epsilon : H_e \to k$ and a family of k-linear maps $S_g : H_{g^{-1}} \to H_g$, $\forall g \in G$ such that the following compatibility conditions hold for all $g, h, f \in G$

$$(\Delta_{g,h} \otimes H_f) \circ \Delta_{gh,f} = (H_g \otimes \Delta_{h,f}) \circ \Delta_{g,hf}$$
$$(H_g \otimes \epsilon) \circ \Delta_{g,e} = H_g = (\epsilon \otimes H_g) \circ \Delta_{e,g}$$
$$\mu_g \circ (S_g \otimes H_g) \circ \Delta_{g^{-1},g} = \eta_g \circ \epsilon = \mu_g \circ (H_g \otimes S_g) \circ \Delta_{g,g^{-1}}$$

In Caenepeel and De Lombaerde (2006) the nice observation was made that these objects can be understood as Hopf algebras in a particular symmetric monoidal category.

Let us first recall a general construction in category theory. Let $\mathcal{C}$ be a category. Then we can construct a new category $\mathsf{Fam}(\mathcal{C})$, the category of families in $\mathcal{C}$, as follows:

- an object in $\mathsf{Fam}(\mathcal{C})$ is a pair $(I, \{C_i\}_{i \in I})$, where $I \in \underline{\mathsf{Set}}$
- a morphism in $\mathsf{Fam}(\mathcal{C})$ is a pair $(f, \phi) : (I, \{C_i\}_{i \in I}) \to (J, \{D_j\}_{j \in J})$ consisting of a map $f : I \to J$ and a family of $\mathcal{C}$-morphisms $\phi_i : C_i \to D_{f(i)}$.

Dually, we define $\mathsf{Maf}(\mathcal{C}) = \mathsf{Fam}(\mathcal{C}^{op})^{op}$. This is the category with the same objects, but morphisms are pairs $(f, \phi) : (I, \{C_i\}_{i \in I}) \to (J, \{D_j\}_{j \in J})$ consisting of a map $f : J \to I$ and a family of $\mathcal{C}$-morphisms $\phi_j : C_{f(j)} \to D_j$. If $\mathcal{C}$ is monoidal, braided monoidal, or closed than $\mathsf{Fam}(\mathcal{C})$ and $\mathsf{Maf}(\mathcal{C})$ are as well, in a canonical way.

In Caenepeel and De Lombaerde (2006), the category Fam($\mathcal{C}$) was called the Zunino category and Maf($\mathcal{C}$) was called the Turaev category associated with $\mathcal{C}$. The reason for introducing these names is the observation that was made in the above cited paper, that relates the algebras, coalgebras, bialgebras, and Hopf algebras in these categories to structures that were earlier studied by Turaev and Zunino.

- Algebras in Fam($\mathcal{C}$) are nothing other than algebras graded by a monoid; coalgebras in Fam($\mathcal{C}$) are just a family of coalgebras indexed by a set.
- Algebras in Maf($\mathcal{C}$) are nothing other than families of algebras indexed by a set; coalgebras in Maf($\mathcal{C}$) are coalgebras that are a kind of co-graded coalgebras, called G-coalgebras in Turaev (2000).
- Bialgebras and Hopf algebras in Fam($\mathcal{C}$) are Hopf algebras graded by a group, and bialgebras and Hopf algebras in Maf($\mathcal{C}$) are 'co-graded' versions, called group Hopf coalgebras in Turaev (2000).

As Hopf group coalgebras are just Hopf algebras, a particular braided monoidal category, the reconstruction theorems, Theorem 5.21 and Theorem 5.22, can be directly applied to this situation.

5.5.2.3 MULTIPLIER HOPF ALGEBRAS

Let A be a non-unital algebra. A (right) A-module is called *firm* if the multiplication map induces an isomorphism $M \otimes_A A \cong M$. The algebra A is called a firm algebra if it is firm as left or equivalently right regular A-module. In this situation, the category of firm A-bimodules is again a monoidal category with monoidal unit A. Examples of this kind of non-unital algebra are so-called algebras with *local units*, that is, algebras such that for any $a \in A$, there exists an element $e \in A$ such that $ae = a = ea$. If S is an infinite set, then the algebra of functions with finite support fHom(S, k) is a non-unital algebra with local units.

Multiplier Hopf algebras are a generalization of Hopf algebras in the setting of non-unital algebras. They were motivated by study of non-compact quantum groups in the setting of C^*-algebras, but have been studied in a purely algebraic setting since their introduction. The non-compactness of their underlying space is directly related to the fact that the algebra is non-unital. However, it was proven that multiplier Hopf algebras always have local units.

If A is a non-unital k-algebra, then the multiplier algebra of A is the k-module $M(A)$ that is defined by the following pullback:

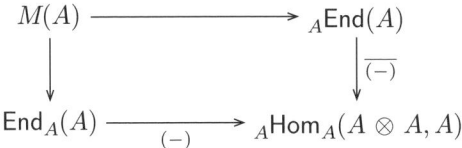

where we have used the linear maps

$$\overline{(-)} : {}_A\mathrm{End}(A) \to {}_A\mathrm{Hom}_A(A \otimes A, A), \quad \bar{\rho}(a \otimes b) = \rho(a)b, \text{ for } \rho \in {}_A\mathrm{End}(A); \quad (5.4)$$

$$\underline{(-)} : \mathrm{End}_A(A) \to {}_A\mathrm{Hom}_A(A \otimes A, A), \quad \underline{\lambda}(a \otimes b) = a\lambda(b), \text{ for } \lambda \in \mathrm{End}_A(A). \quad (5.5)$$

Notice that if A is unital then $A \cong \mathrm{End}_A(A) \cong {}_A\mathrm{End}(A) \cong {}_A\mathrm{Hom}_A(A \otimes A, A)$ in a canonical way, hence also $M(A) \cong A$. We can understand $M(A)$ as the set of pairs (λ, ρ) where $\lambda \in \mathrm{End}_A(A)$ and $\rho \in {}_A\mathrm{End}(A)$ such that

$$a\lambda(b) = \rho(a)b, \qquad (5.6)$$

for all $a, b \in A$. Elements of $M(A)$ are called *multipliers*. For any $x \in M(A)$, we will represent this element as (λ_x, ρ_x). Moreover, for any $a \in A$, we will denote

$$a \cdot x = \rho_x(a), \quad x \cdot a = \lambda_x(a),$$

then eqn (5.6) reads as $a(xb) = (ax)b$. One obtains in a canonical way that $M(A)$ is a unital algebra and there is a canonical algebra morphism $\iota : A \to M(A)$. Moreover, if $f : A \to M(B)$ is a morphism of algebras and the maps

$$A \otimes B \to B, \quad a \otimes b \mapsto f(a) \cdot b$$
$$B \otimes A \to B, \quad b \otimes a \mapsto b \cdot f(a)$$

are surjective, one can always extend this map to a morphism $\bar{f} : M(A) \to M(B)$. Such morphisms are called non-degenerate.

Let A be an algebra with local units. Consider a non-degenerate algebra morphism $\Delta : A \to M(A \otimes A)$, such that for all $a, b \in A$, $\Delta(a)(1 \otimes b) \in A \otimes A$ and $(b \otimes 1)\Delta(a) \in A \otimes A$. Then we can express the following coassociativity condition for all $a, b, c \in A$,

$$(c \otimes 1 \otimes 1)(\Delta \otimes A)(\Delta(a)(1 \otimes b)) = (A \otimes \Delta)((c \otimes 1)\Delta(a))(1 \otimes 1 \otimes b).$$

Now consider the following 'fusion maps' or canonical maps

$$T_1 : A \otimes A \to A \otimes A, \quad T_1(a \otimes b) = \Delta(a)(1 \otimes b)$$
$$T_2 : A \otimes A \to A \otimes A, \quad T_1(a \otimes b) = (a \otimes 1)\Delta(b).$$

Following van Daele (1994), we say that A is a multiplier Hopf algebra if there is a non-degenerate coassociative comultiplication $\Delta : A \to M(A \otimes A)$ as above such that the maps T_1 and T_2 are bijective.

It can be shown that A is a multiplier Hopf algebra if and only if there exists a counit $\epsilon : A \to k$ and an antipode $S : A \to M(A)$, which satisfy conditions similar to the classical case, but which have to be formulated with the required care.

The full categorical description of multiplier Hopf algebras is not settled yet. A first attempt was made by Janssen and Vercruysse (2010). In this paper a reconstruction theorem for multiplier bialgebras was given. By a multiplier bialgebra we mean a non-unital algebra A (with local units) that has a coassociative non-degenerate comultiplication $\Delta : A \to M(A \otimes A)$ and a counit $\epsilon : A \to k$. Then according to Janssen and Vercruysse (2010, Theorem 2.9), we have the following characterization:

Given an algebra with local units A, there is a bijective correspondence between

(i) multiplier bialgebra structures on A; and
(ii) monoidal structures on the categories $\mathcal{M}_A$ of firm right A-modules, $_A\mathcal{M}$ of firm left A-modules, and the category $A - \text{Ext}$ of ring extensions $A \to A'$, where A' is again a ring with local units, such that the following diagram of forgetful functors is strict monoidal

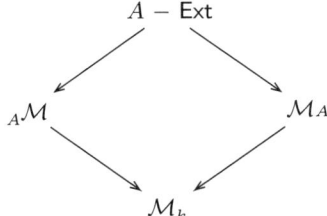

In a forthcoming study by Janssen and Vercruysse (2012), some classes of multiplier Hopf algebras, such as multiplier Hopf algebras with a complete set of central local units including discrete quantum groups, are studied as Hopf algebras in a particular monoidal category. This monoidal category is closely related to the category $\text{Maf}(\mathcal{C})$ of the previous section, which also indicates the close relationship between Hopf group coalgebras and multiplier Hopf algebras. In particular, for a group Hopf coalgebra $(H_g)_{g \in G}$, we have that $\oplus_{g \in G} H_g$ is a multiplier Hopf algebra, whose multiplier algebra is given by $M(\oplus_{g \in G} H_g) = \prod_{g \in G} H_g$.

5.5.2.4 HOM-HOPF ALGEBRAS

Another variation of the classical notion of Hopf algebra, which has received a lot of attention recently, are so-called Hom-Hopf algebras (see Makhlouf and Silvestrov, 2009). This concerns non-associative algebras H, whose non-associativity is ruled by a k-linear endomorphism $\alpha \in \text{End}(H)$. It was proven in Caenepeel and Goyvaerts (2011) that Hom-Hopf algebras can be viewed as Hopf algebras in a monoidal category whose objects are pairs (X, f), where X is a k-module and f is a k-automorphism of X, and where the associativity constraint is non-trivial. Moreover, a Hom-Hopf algebra is nothing else than a usual Hopf algebra, together with a Hopf algebra automorphism.

5.5.3 More Hopf algebra-type structures and applications

5.5.3.1 Yetter–Drinfel'd modules

For any monoidal category $\mathcal{C}$, one can construct its *center*, which is the braided monoidal category $\mathcal{Z}(\mathcal{C})$ whose objects are pairs (A, u), where A is an object of $\mathcal{C}$ and $u_X : A \otimes X \to X \otimes A$ is a natural transformation that satisfies

$$u_{X \otimes Y} = (X \otimes u_Y) \circ (u_X \otimes Y), \quad u_I \simeq A$$

An arrow from (A, u) to (B, v) in $\mathcal{Z}(\mathcal{C})$ consists of an arrow $f : A \to B$ in $\mathcal{C}$ such that

$$v_X(f \otimes 1_X) = (1_X \otimes f) u_X.$$

The category $\mathcal{Z}(\mathcal{C})$ becomes a braided monoidal category with the tensor product on objects defined as

$$(A, u) \otimes (B, v) = (A \otimes B, w)$$

where $w_X = (u_X \otimes 1)(1 \otimes v_X)$, and the obvious braiding. Let H be a k-bialgebra. As we have seen, the category of right H-modules is a monoidal category. We can now define the category of Yetter–Drinfel'd modules as the centre of the monoidal category H-modules, $\mathcal{YD}_H^H = \mathcal{Z}(\mathcal{M}_H)$. Moreover, as the tensor product is preserved, we obtain the following diagram of monoidal forgetful functors:

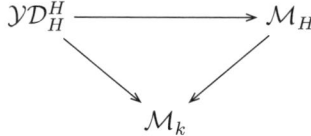

Hence there exists a quasitriangular bialgebra $\mathcal{D}(H)$ such that $\mathcal{YD}_H^H \cong \mathcal{M}_{\mathcal{D}(H)}$. We call $\mathcal{D}(H)$ the Drinfel'd double of H.

Moreover, if H is a finite-dimensional Hopf algebra, then one can consider the rigid category $\mathcal{M}_H^f$ and it can be checked that $\mathcal{Z}(\mathcal{M}_H^f)$ is rigid as well. This allows us to construct a Hopf algebra structure on $\mathcal{D}(H)$, which can in this case be computed explicitly as a crossed product on H and H^*.

Furthermore, as $\mathcal{YD}_H^H$ is again a braided monoidal category, one can consider bialgebras and Hopf algebras as being *inside* this category. These objects are called Yetter–Drinfeld Hopf algebras and are studied, for example, in Sommerhäuser (2002). As explained by Takeuchi (2000), this is (almost) equivalent to considering braided Hopf algebras—that is Hopf algebras in the category of k-modules—but where one uses a 'local braiding' instead of the usual twist maps.

5.5.3.2 MONOIDAL STRUCTURES ON CATEGORIES OF RELATIVE HOPF MODULES

Let H be a bialgebra. Then we know that the category of (right) H-modules $\mathcal{M}_H$ and the category of (right) H-comodules $\mathcal{M}^H$ are monoidal categories. Hence we can consider bimonads on these categories, and Hopf monads on these categories if H is a Hopf algebra. Suppose for example that A is an H-comodule algebra. Than we have a monad

$$\mathbb{A} = - \otimes A : \mathcal{M}^H \to \mathcal{M}^H.$$

Moreover, the category of $\mathbb{A}$-modules coincides with the category of relative Hopf modules $\mathcal{M}_A^H$. By Theorem 5.16 $\mathbb{A}$ is a bimonad if and only if $\mathcal{M}_A^H$ is a monoidal category and the forgetful functor $\mathcal{M}_A^H \to \mathcal{M}^H$ is a strict monoidal forgetful functor. For example, if A is an bialgebra in the monoidal centre of $\mathcal{M}^H$, that is A is a bialgebra in the category of Yetter–Drinfel'd modules over H, then the conditions are fulfilled and $\mathcal{M}_A^H$ is a monoidal category (see also Bulacu and Caenepeel, 2012).

5.5.3.3 HOPFISH ALGEBRAS

Hopfish algebras were introduced in Tang, Weinstein and Zhu (2007) and were motivated by Poisson geometry, where there was a need to describe the structure on irrational rotation algebras (see Blohmann, Tang and Weinstein, 2008). From the purely algebraic point of view, Hopfish algebras can be motivated by the very nice feature that they provide a Morita invariant notion of Hopf-algebra-like structure. It is clear that the notion of a usual Hopf algebra is not at all Morita invariant: if an algebra A is Morita equivalent with a Hopf algebra H, then A has not necessarily the structure of a Hopf algebra. It follows from the theory of Tang, Weinstein and Zhu that A, however, possesses a structure that they call a Hopfish algebra. In contrast to the usual Hopf algebras, Hopfish algebras should be understood using higher category theory. As the usual Hopf algebras live in a braided monoidal base category, Hopfish algebras live in a monoidal bicategory. The theory for Hopfish algebras, and Tannaka theory in particular, is far from having been fully explored.

Suppose that A is an algebra that is Morita equivalent to the Hopf algebra H. Then there is a strict Morita context (A, H, D, E, μ, τ) and we have functors

$$\mathcal{M}_A \xrightarrow{F} \mathcal{M}_H \xrightarrow{U} \mathcal{M}_k.$$

Here $F \simeq - \otimes_A D$ is an equivalence of categories, and U is the strict monoidal forgetful functor. Using the equivalence between $\mathcal{M}_H$ and $\mathcal{M}_A$ we can put a monoidal structure on the category $\mathcal{M}_A$ by

$$X \odot Y := ((X \otimes_A D) \otimes (Y \otimes_A D)) \otimes_H E$$

In particular, we see that the forgetful functor $U' : \mathcal{M}_A \to \mathcal{M}_k$ is no longer a (strict) monoidal functor, which explains that A is no longer a Hopf algebra, not even a bialgebra. The 'comultiplication' is now a bimodule rather than a morphism. Indeed, if we calculate $A \odot A = (D \otimes D) \otimes_H E =: \Delta_A$ then this is naturally a left $A \otimes A$-module and a right A-module. On the other side $\epsilon_A := k \otimes_H E$ is a k-A bimodule. The triple $(A, \Delta_A, \epsilon_A)$ then becomes a *sesquiunital sesquialgebra*, which satisfies the following coassociativity and counitality axioms:

$$(A \otimes \Delta_A) \otimes_{A \otimes A} \Delta_A \cong (\Delta_A \otimes A) \otimes_{A \otimes A} \Delta_A,$$

as $A \otimes A \otimes A$-A bimodules and

$$(\epsilon_A \otimes A) \otimes_{A \otimes A} \Delta_A \cong A \cong (A \otimes \epsilon_A) \otimes_{A \otimes A} \Delta_A$$

as A-bimodules. Conversely, the category of right A-modules of a sesquiunital sesquialgebra can be endowed with a monoidal structure, but without having a strict monoidal forgetful functor to the underlying category of k-modules.

Furthermore, A is a Hopfish algebra if it has an antipode, which is defined as a left $A \otimes A$-module S such that the k-dual of S is isomorphic with the right $A \otimes A$-module $\mathrm{Hom}_A(\epsilon_A, \Delta_A)$, and such that S is free of rank one if it is considered as an A-A^{op} bimodule.

5.5.3.4 TOPOLOGICAL HOPF ALGEBRAS AND LOCALLY COMPACT QUANTUM GROUPS

As is generally known, the theory of Hopf algebras has undergone a great revival thanks to the discovery of quantum groups. In contrast to pure Hopf algebras, the theory of quantum groups is, however, not always purely algebraic but often involves topological and analytical features. This has lead to generalizations of the notion of a Hopf algebra.

First, one can consider a monoidal category of well-behaving topological vector spaces with completed tensor product. A Hopf algebra in this category is called a topological Hopf algebra (see e.g. Larson, 1998, and Bonneau and Sternheimer, 2005).

Motivated by the Pontryagin duality for locally compact topological commutative groups, Kustermans and Vaes (2000) introduced locally compact quantum groups. At the time of writing, a full Tannaka reconstruction theory for these locally compact quantum groups is not known to the author.

5.5.3.5 COMBINATIONS

Of course, it is possible to make combinations of two or more of the structures that we have reviewed above. For example, weak quasi-Hopf algebras, quasitriangular quasi-Hopf algebras, weak group Hopf coalgebras, and weak multiplier Hopf algebras have been investigated by several authors.

5.5.3.6 FURTHER GENERALIZATIONS AND PROSPECTS

It became clear in recent years that the proper setting for combining all generalizations of Hopf algebras into a single and unifying framework, will not be braided monoidal categories, not even Hopf monads (as treated above). Rather one will have to move to a higher categorical setting, considering monoidal bicategories, 2-monoidal categories and possibly even more involved structures (see e.g. Chikhladze, Lack and Street, 2010, and Booker and Street, 2011). These recent developments are indicating that the theory of Hopf algebras and the interrelation with (higher) category have an exciting future ahead.

Acknowledgement

The author would like to thank Gabi Böhm for useful comments on an earlier version of the paper and Stef Caenepeel for discussions.

CHAPTER 6

Modular Categories

MICHAEL MÜGER

(Radboud University, Nijmegen)

6.1 Introduction

Modular categories, as well as the (possibly) more general non-degenerate braided fusion categories, are braided tensor categories that are linear over a field and satisfy some natural additional axioms, like existence of duals, semisimplicity, and an important non-degeneracy condition. Precise definitions will be given later. There are several reasons to study modular categories.

- As will hopefully become clear, they are rather interesting mathematical structures in themselves, well worth being studied for intrinsic reasons. For example, there are interesting number-theoretic aspects.
- Among the braided fusion categories, modular categories are the opposite extreme of the symmetric fusion categories, which are well known to be closely related to finite groups. Studying these two extreme cases is also helpful for understanding and classifying those braided fusion categories that are neither symmetric nor modular.
- Modular categories serve as input datum for the Reshetikhin–Turaev (RT) construction of topological quantum field theories (TQFTs) in $2+1$ dimensions and therefore give rise to invariants of smooth 3-manifolds. This goes some way towards making Witten's interpretation of the Jones polynomial via Chern–Simons quantum field theory (QFT) rigorous. (But since there still is no complete rigorous non-perturbative construction of the Chern–Simons QFTs by conventional QFT methods, there is also no proof of their equivalence to the RT-TQFTs constructed using the representation theory of quantum groups.)
- Modular categories arise as representation categories of loop groups and, more generally, of rational chiral conformal QFTs (CQFTs). In chiral CQFT

the field theory itself, its representation category, and the conformal characters form a remarkably tightly connected structure.
- Certain massive QFTs and quantum spin systems in two spatial dimensions lead to modular categories, e.g. Kitaev's 'toric code'.
- The recent topological approaches to quantum computing, while differing in details, all revolve around the notion of modular categories.

As the above list indicates, modular categories—and related mathematical subjects such as representation theory of loop groups and of quantum groups at root-of-unity deformation parameter—represent one of the most fruitful places of interaction of 'pure' mathematics and mathematical physics. (While modular categories play a certain role in string theory via their importance for rational conformal field theories, the author believes that their appearance in massive QFTs and spin models may ultimately turn out to be of larger relevance for realistic physics.)

This article assumes a certain familiarity with category theory, including monoidal (=tensor) categories. Concerning braided categories, only the definition, for which we refer to Chapter 4, will be assumed. Our standard reference for category theory is Mac Lane (1971). For a broader survey of some related matters concerning tensor categories, cf. also Müger (2010c).

6.2 Categories

We limit ourselves to recalling some basic definitions. A category is called an Ab-category if it is enriched over (the symmetric tensor category Ab of) abelian groups, i.e. all hom-sets come with abelian group structures and the composition ∘ of morphisms is a homomorphism with respect to both arguments. An Ab-category is called additive if it has a zero object and every pair of objects has a direct sum. If k is a field (or more generally, a commutative unital ring) then a category $\mathcal{C}$ is called k-linear if all hom-sets are k-vector spaces (or k-modules) and ∘ is bilinear. An object X is called simple if every monic morphism $Y \hookrightarrow X$ is an isomorphism and absolutely simple if $\text{End}\, X \cong k\text{id}_X$. If k is an algebraically closed field, as we will mostly assume, the two notions coincide. A k-linear category is semisimple if every object is a finite direct sum of simple objects. A semisimple category is called finite if the number of isomorphism classes of simple objects is finite. (There is a notion of finiteness for non-semisimple categories (cf. Etingof and Ostrik, 2004), but we will not need it.) A positive $*$-operation on a $\mathbb{C}$-linear category $\mathcal{C}$ is a contravariant endofunctor of $\mathcal{C}$ that acts like the identity on objects, is involutive ($** = \text{id}$) and anti-linear. (Dropping the $\mathbb{C}$-linearity, one arrives at the notion of a dagger-category.) A $*$-operation is called positive if $s^* \circ s = 0$ implies $s = 0$. A category equipped with a (positive) $*$-operation is called hermitian (unitary). A unitary category with finite-dimensional hom-spaces and splitting idempotents is semisimple (since finite-dimensional algebras with positive $*$-operation are semisimple, thus multimatrix, algebras).

6.3 Tensor categories

We assume familiarity with the basics of tensor (=monoidal) categories, including symmetric ones. We therefore limit the discussion to (i) issues of duality in not necessarily braided tensor categories, (ii) fusion categories and (iii) module categories versus categories of modules over an algebra in a tensor category.

6.3.1 Duality

In the following, we will state definitions and results for strict tensor categories, but everything can easily be adapted to the non-strict case.

Definition 6.1 Let $\mathcal{C}$ be a tensor category and let $X, Y \in \mathcal{C}$. If there exist morphisms $e : Y \otimes X \to \mathbf{1}$, $d : \mathbf{1} \to X \otimes Y$ satisfying

$$(\mathrm{id}_X \otimes e) \circ (d \otimes \mathrm{id}_X) = \mathrm{id}_X, \qquad (e \otimes \mathrm{id}_Y) \circ (\mathrm{id}_Y \otimes d) = \mathrm{id}_Y,$$

then Y is called a left dual of X and X is a right dual of Y. If every object has a left (right) dual, we say that $\mathcal{C}$ has left (right) duals.

It is easy to prove that duals are unique up to isomorphism: If Y, Y' are both left (or both right) duals of X then $Y \cong Y'$. This justifies writing $^{\vee}X$ ($X^{\vee}$) for a left (right) dual of X. However, if X admits a left dual $^{\vee}X$ and a right dual $X^{\vee}$, it may or may not be the case that $^{\vee}X \cong X^{\vee}$. Clearly, a left dual $^{\vee}X$ of X is also a right dual of X if and only if X is a left dual of $^{\vee}X$, thus if and only if X is isomorphic to its second left dual $^{\vee\vee}X$. If $^{\vee}X \cong X^{\vee}$ holds, we say that X has a two-sided dual and mostly write $\overline{X}$ rather than $^{\vee}X$ or $X^{\vee}$. If all objects have two-sided duals, we say that $\mathcal{C}$ has *two-sided duals*.

There are three situations where duals, to the extent that they exist, are automatically two-sided:

(i) $\mathcal{C}$ is hermitian: if $(^{\vee}X, e_X, d_X)$ defines a left dual of X, one finds that $(^{\vee}X, d_X^*, e_X^*)$ is a right dual (in this situation, some authors talk about conjugates rather than duals and give slightly different axioms; see Longo and Roberts, 1997)
(ii) $\mathcal{C}$ is braided, cf. Section 6.4.5
(iii) $\mathcal{C}$ is semisimple with simple unit (cf. Etingof, Nikshych and Ostrik, 2005, Proposition 2.1).

If $\mathcal{C}$ is linear over a field k and the unit $\mathbf{1}$ is absolutely simple, we can and will use the bijection $k \to \mathrm{End}\,\mathbf{1}$, $c \mapsto c\,\mathrm{id}_\mathbf{1}$ to identify $\mathrm{End}\,\mathbf{1}$ with k. If $X \in \mathcal{C}$ is absolutely simple and has a two-sided dual $\overline{X}$, one defines 'squared dimension'

$d^2(X) = (e' \circ d)(e \circ d')$, which is easily seen to be independent of the choices of the duality morphisms e, d, e', d' (cf. Müger, 2003a). If X, Y and $X \otimes Y$ are absolutely simple, one has $d^2(X \otimes Y) = d^2(X)d^2(Y)$. If $\mathcal{C}$ is semisimple and finite, we define $\dim \mathcal{C} = \sum_i d^2(X_i)$, where $\{X_i\}$ is a complete family of simple objects.

In the k-linear case, one would like to have a dimension function $X \mapsto d(X)$ that is additive and multiplicative. If $\mathcal{C}$ is semisimple and finite, this can be done using Perron–Frobenius theory (see Section 6.3.2). Another approach strengthens the requirement of existence of duals by introducing a new piece of structure:

Definition 6.2 *A left duality on a tensor category is an assignment $X \mapsto (^\vee X, e_X, d_X)$, where $^\vee X$ is a left dual of X, the morphisms $e_X : {}^\vee X \otimes X \to \mathbf{1}$, $d_X : \mathbf{1} \to X \otimes {}^\vee X$ satisfying the above identities. Similarly, a right duality is an assignment $X \mapsto (X^\vee, e'_X, d'_X)$ where the morphisms $e'_X : X \otimes X^\vee \to \mathbf{1}$, $d'_X : \mathbf{1} \to X^\vee \otimes X$ establish $X^\vee$ as a right dual of X.*

If $\mathcal{C}$ is a tensor category with a chosen left duality $X \mapsto (^\vee X, e_X, d_X)$ and $f : X \to Y$ then

$$^\vee f = (e_Y \otimes \mathrm{id}_{^\vee X}) \circ (\mathrm{id}_{^\vee Y} \otimes f \otimes \mathrm{id}_{^\vee X}) \circ (\mathrm{id}_{^\vee X} \otimes d_X) : {}^\vee Y \to {}^\vee X$$

extends $X \mapsto {}^\vee X$ to a contravariant endofunctor $^\vee -$ of $\mathcal{C}$. Since $^\vee Y \otimes {}^\vee X$ is a left dual of $X \otimes Y$, $^\vee -$ can be considered as a (covariant) tensor functor $\mathcal{C} \to \mathcal{C}^{\mathrm{op},\mathrm{rev}}$, where $\mathcal{C}^{\mathrm{rev}}$ coincides with $\mathcal{C}$ as a category, but has the reversed tensor product $X \otimes^{\mathrm{rev}} Y = Y \otimes X$. (Similarly for $-^\vee$.) If $^\vee -$ is a left duality and $\mathcal{C}$ has two-sided duals, one can find a natural isomorphism $\gamma : \mathrm{id} \to {}^{\vee\vee}-$. Since this is of little use unless γ is monoidal, one defines:

Definition 6.3 *A pivotal category (Freyd and Yetter, 1989; Freyd and Yetter, 1992) is a tensor category together with a left duality $^\vee -$ and a monoidal natural isomorphism $\gamma : \mathrm{id} \to {}^{\vee\vee}-$.*

Remark 6.4

1. Categories equipped with left and/or right dualities are often called rigid or autonomous, and pivotal categories are also called sovereign. We will avoid all these terms. The categories we consider will either have the property of possessing two-sided duals (without given duality structures) or be equipped with pivotal or spherical structures.
2. For a general tensor category with two-sided duals, there is little a priori reason to expect the existence of a pivotal structure, but see Theorems 6.6 and 6.7 below.

Definition 6.5 *Let $\mathcal{C}$ be a strict pivotal category with left duality $X \mapsto (\overline{X}, e_X, d_X)$ and monoidal natural isomorphism $(\alpha_X : X \to \overline{\overline{X}})_{X \in \mathcal{C}}$. For $X \in \mathcal{C}$ and $s \in \mathrm{End}\, X$, define the left and right traces of s by*

$$tr_X^L(s) = e_X \circ (\text{id}_{\overline{X}} \otimes (s \circ \alpha_X^{-1})) \circ d_{\overline{X}},$$
$$tr_X^R(s) = e_{\overline{X}} \circ ((\alpha_X \circ s) \otimes \text{id}_{\overline{X}}) \circ d_X.$$

(Both traces take values in End **1**.*) The left and right dimensions of X are defined by* $d_{L/R}(X) = tr_X^{L/R}(\text{id}_X)$.

The traces satisfy $\text{tr}_X(s \circ t) = \text{tr}_Y(t \circ s)$ for all $t : X \to Y$ and $s : Y \to X$, as well as $\text{tr}_{X \otimes Y}(s \otimes t) = \text{tr}_X(s)\text{tr}_Y(t)$ for $s \in \text{End}\, X$, $t \in \text{End}\, Y$. If X is absolutely simple, the dimensions defined in terms of the traces are connected to the intrinsic squared dimension by $d_L(X)d_R(X) = d^2(X)$.

A pivotal category is called spherical (Barrett and Westbury, 1999) if $tr_X^L = tr_X^R$ for all $X \in \mathcal{C}$. In this case we have $d_L(X) = d_R(X)$ and simply write $d(X)$. In fact, a pivotal category that is semisimple over a field is spherical if and only if $d_L(X) = d_R(X)$ for every X. To this day, all known examples of pivotal fusion categories are spherical. For example, if H is a finite dimensional semisimple Hopf algebra in characteristic zero, one automatically has $S^2 = \text{id}$ and therefore sphericity of the module category.

As noted before, in a hermitian category, duals (if they exist) are always two-sided. In a unitary category there is a canonical way of defining traces of endomorphisms, giving rise to positive dimensions, that makes no use of spherical structures (cf. Longo and Roberts, 1997). But as shown by Yamagami, a unitary category always admits a unique spherical structure that gives rise to these traces. Note, however, there are $\mathbb{C}$-linear categories that do not admit a unitary structure (cf. Rowell, 2005): essentially, quantum group categories at odd root of unity.

6.3.2 Fusion categories

Let k be an algebraically closed field. A k-linear tensor category is called a fusion category if it has finite-dimensional hom-spaces, is semisimple with finitely many isomorphism classes of simple objects, the unit **1** is absolutely simple, and all objects have duals (which are automatically two-sided by semisimplicity). A fusion subcategory of a fusion category is a full tensor subcategory that is again fusion (i.e. closed under direct sums and duals).

Even for fusion categories it is unknown whether pivotal structures always exist (see Douglas, Schommer-Pries and Snyder, 2011, for work currently in progress), but there has been a result in this direction.

Theorem 6.6 (Etingof, Nikshych and Ostrik, 2005) *Let $\mathcal{C}$ be a fusion category over $\mathbb{C}$ and let* $^\vee -$ *be an arbitrary left dual structure. Then the tensor functor* $^{\vee\vee\vee\vee}-$ *is naturally monoidally isomorphic to the identity functor.*

In a fusion category, we can find mutually non-isomorphic simple objects X_i, indexed by a finite set I, such that every object is a finite direct sum of the objects $X_i, i \in I$. There is a distinguished index $0 \in I$ such that $X_0 = \mathbf{1}$. Now we can define non-negative integers $N_{ij}^k \in \mathbb{N}_0$ via $X_i \otimes X_j \cong \oplus_{k \in I} N_{ij}^k X_k$. These numbers have various obvious properties, such as $N^j_{i0} = N^j_{0i} = \delta_{i,j}$. We also have $N^0_{ij} = \delta_{i,\bar{j}}$, where $i \mapsto \bar{i}$ is the involution on I defined by $\overline{X_i} \cong X_{\bar{i}}$. The structure $(I, 0, i \mapsto \bar{i}, N^{\cdot}_{\cdot,\cdot})$ is known as a discrete hypergroup. (In particular, every group G gives rise to a hypergroup, taking $I = G$, $0 = e$, $\bar{g} = g^{-1}$ and $N^k_{g,h} = \delta_{gh,k}$.)

For every discrete hypergroup $(I, \cdots)$ there is a unique group $G(I)$ equipped with a map $\partial : I \to G(I)$ satisfying

$$\partial 0 = e, \quad \partial\bar{i} = (\partial i)^{-1}, \quad \text{and} \quad N^k_{i,j} > 0 \Rightarrow \partial k = \partial i \cdot \partial j$$

and being universal for such maps. (Thus for every map $\partial' : I \to H$, where H is a group satisfying the same axioms as ∂, there is a unique group homomorphism $\alpha : G \to H$ such that $\partial' = \alpha \circ \partial$.) In view of this property, $G(I)$ is called the (universal) grading group of the hypergroup I (also 'groupification' might be a good term, in analogy to abelianization); see Baumgärtel and Lledó (2004) and Gelaki and Nikshych (2008). Notice that $G(I)$ is abelian when $\mathcal{C}$ is braided. If H is a compact group, let $\widehat{H}$ be the hypergroup corresponding to the semisimple category Rep H. Then the (discrete abelian) grading group $G(\widehat{H})$ is canonically isomorphic to Pontrjagin dual of the (compact abelian) centre $Z(H)$ of H (see Müger, 2004b).

Each matrix $N_i = (N^k_{ij})_{j,k}$ is irreducible and has non-negative entries. Thus it has a unique positive Perron–Frobenius eigenvalue, denoted FPdim(X_i), the Frobenius–Perron (FP) dimension of X_i (Etingof, Nikshych and Ostrik, 2005). The Frobenius–Perron dimension (Etingof, Nikshych and Ostrik, 2005) of the fusion category $\mathcal{C}$ is defined by

$$\text{FP dim}(\mathcal{C}) = \sum_{i \in I} \text{FP dim}(X_i)^2.$$

By definition, FPdim(X) and FPdim$(\mathcal{C})$ live in $\mathbb{R}_{>0}$ rather than in the ground field k, but if $\mathbb{R} \subset k$ it makes sense to compare FPdim(X) with $d^2(X)$ (which is canonically defined for all simple objects) and with $d(X)$ (if $\mathcal{C}$ has a spherical structure). It is classical that the FP dimension is the unique positive dimension function on a finite hypergroup I. In particular the FP dimension coincides with the positive dimension function defined on unitary categories with duals (Longo and Roberts, 1997). (Recall that the latter arises from a unique spherical structure.) Somewhat more generally, if $\mathcal{C}$ is spherical over $k \supset \mathbb{R}$ and $d(X) > 0$ for all X then $d(X) = \text{FP dim}(X)$ for all X. Since not every fusion category over $\mathbb{C}$ is unitarizable (in the sense of admitting a positive $*$-operation) it is important that there are the following remarkable results.

Theorem 6.7 (Etingof, Nikshych and Ostrik, 2005) *Let $\mathcal{C}$ be a fusion category over $\mathbb{C}$. Then:*

(i) *for every simple X, one has $0 < d^2(X) \leq \text{FP dim}(X)^2$ and thus $1 \leq \dim \mathcal{C} \leq \text{FP dim} \mathcal{C}$*

(ii) if $\dim \mathcal{C} = \operatorname{FP}\dim(\mathcal{C})$ (equivalent to $d^2(X) = \operatorname{FP}\dim(X)^2$ for every simple X) $\mathcal{C}$ admits a unique spherical structure for which $d(X) = \operatorname{FP}\dim(X) > 0$ for every X; such categories are called pseudo-unitary.

6.3.3 Algebras in tensor categories and their modules

A considerable part of (commutative) algebra can be generalized from the symmetric categories Ab and Vect$_k$ to arbitrary (braided) tensor categories. This plays an important role in the structural study of such categories and in particular of braided fusion categories. In this subsection we discuss some facts that do not require a braiding.

Definition 6.8 Let $\mathcal{C}$ be a strict tensor category. An algebra (or monoid) in $\mathcal{C}$ is a triple (A, m, η), where $A \in \mathcal{C}$, $m : A \otimes A \to A$ and $\eta : \mathbf{1} \to A$ satisfy $m \circ (m \otimes \operatorname{id}_A) = m \circ (\operatorname{id}_A \otimes m)$ (associativity) and $m \circ (\eta \otimes \operatorname{id}_A) = \operatorname{id}_A = m \circ (\operatorname{id}_A \otimes \eta)$ (unit property). If $\mathcal{C}$ is non-strict, one inserts associativity isomorphisms at the appropriate places.

Remark 6.9 At least in categories that are not linear over a field, it would be more appropriate to speak of monoids rather than algebras, but the latter term is used much more in the recent literature.

Definition 6.10 If $\mathcal{C}$ is a tensor category and (A, m, η) an algebra in $\mathcal{C}$ then a left A-module (in $\mathcal{C}$) is a pair (X, μ) where $X \in \mathcal{C}$ and $\mu : A \otimes X \to X$ satisfies $\mu \circ (m \otimes \operatorname{id}_X) = \mu \circ (\operatorname{id}_A \otimes \mu)$ and $\mu \circ (\eta \otimes \operatorname{id}_X) = \operatorname{id}_X$.

The left A-modules in $\mathcal{C}$ form a category, denoted A-Mod$_\mathcal{C}$ or $_A\mathcal{C}$, with hom-sets

$$\operatorname{Hom}_{A\mathcal{C}}((X, \mu), (X', \mu')) = \{t \in \operatorname{Hom}_\mathcal{C}(X, X') \mid t \circ \mu = \mu' \circ (\operatorname{id}_A \otimes t)\}.$$

Right modules and bimodules are defined analogously.

There is a functor $F_A : \mathcal{C} \to {}_A\mathcal{C}$, the free module functor, defined by $F_A : X \mapsto (A \otimes X, m \otimes \operatorname{id}_X)$. Notice that for every $(X, \mu) \in {}_A\mathcal{C}$ we have $\mu \in \operatorname{Hom}_{A\mathcal{C}}(F(X), (X, \mu))$. The functor F_A is faithful provided $s \mapsto \operatorname{id}_A \otimes s$ is injective, which usually is the case. Since the maps

$$\operatorname{Hom}_{A\mathcal{C}}(F_A(X), F_A(Y)) \to \operatorname{Hom}_\mathcal{C}(A \otimes X, Y), \quad s \mapsto s \circ (\eta \otimes \operatorname{id}_Y),$$

$$\operatorname{Hom}_\mathcal{C}(A \otimes X, Y) \to \operatorname{Hom}_{A\mathcal{C}}(F_A(X), F_A(Y)), \quad t \mapsto (m \otimes \operatorname{id}_Y) \circ (\operatorname{id}_A \otimes t)$$

are inverses of each other, we have a bijection $\mathrm{Hom}_{AC}(F_A(X), F_A(Y)) \cong \mathrm{Hom}_C(A \otimes X, Y)$. Thus F_A will in general not be full, and it can happen that F_A trivializes an object $X \in C$ in the sense of mapping it to a multiple of the unit object (A, m) of $_AC$.

The free A-modules form a full subcategory of $_AC$, but in order to say more about the module category, one needs a descent-type assumption such as the following.

Definition 6.11 *An algebra (A, m, η) in a tensor category is called separable if the multiplication morphism admits a splitting that is a morphism of A–A bimodules, i.e. a morphism $\tilde{m} : A \to A \otimes A$ satisfying*

$$(m \otimes \mathrm{id}_A) \circ (\mathrm{id}_A \otimes \tilde{m}) = \tilde{m} \circ m = (\mathrm{id}_A \otimes m) \circ (\tilde{m} \otimes \mathrm{id}_A), \qquad m \circ \tilde{m} = \mathrm{id}_A.$$

If C is k-linear and $\dim \mathrm{Hom}(\mathbf{1}, A) = 1$ then the algebra (A, m, η) is called connected.

Lemma 6.12 *If C is a tensor category and (A, m, η) a separable algebra, then every module $(X, \mu) \in {}_AC$ is a quotient of the free module $F(X)$.*

Proof Let $\tilde{m} : A \to A \otimes A$ be a splitting of m. Defining

$$\gamma = (\mathrm{id}_A \otimes \mu) \circ (\tilde{m} \otimes \mathrm{id}_X) \circ (\eta \otimes \mathrm{id}_X) : X \to A \otimes X,$$

an easy computation shows that $\gamma \in \mathrm{Hom}_{AC}((X, \mu), F(X))$ and $\mu \circ \gamma = \mathrm{id}_X$. Thus $\mu \in \mathrm{Hom}_{AC}(F(X), (X, \mu))$ is a split epimorphism. □

Remark 6.13 A notion similar to separability appeared in Bruguières (2000), where a morphism $\beta : \mathbf{1} \to A \otimes A$ was required, satisfying axioms following from the above ones if one takes $\beta = \tilde{m} \circ \eta$. (Notice that β is what is actually used in the lemma.) A related concept is that of a special Frobenius algebra, which is a quintuple $(A, m, \eta, \tilde{m}, \varepsilon)$ where $(A, m, \eta, \tilde{m})$ is a separable algebra and $(A, \tilde{m}, \varepsilon)$ is a coalgebra. (In particular, $\tilde{m}$ must be coassociative.)

Lemma 6.14 (Ostrik, 2003; Etingof et al., 2005; Davydov et al., 2012) *Let C be a fusion category and (A, m, η) an algebra in C. Then the following are equivalent.*

(i) *A is separable.*
(ii) *The category $_AC$ is semisimple.*
(iii) *The category C_A is semisimple.*
(iv) *The category $_AC_A$ is semisimple.*

Remark 6.15 For C spherical, (i)⇒(iv) was already shown by Müger (2003a), as part of the following.

Theorem 6.16 (Müger, 2003a) *If $\mathcal{C}$ is a spherical fusion category and $A \in \mathcal{C}$ a separable and connected algebra then $\mathcal{D} = {}_A\mathcal{C}_A$ is a spherical fusion category and it contains a separable connected algebra B such that $\mathcal{C} \simeq {}_B\mathcal{D}_B$. Calling fusion categories that are related in this way 'weakly monoidally Morita equivalent', weak monoidal Morita equivalence is an equivalence relation, denoted by $\approx$. If $\mathcal{C} \approx \mathcal{D}$ then $\dim \mathcal{C} = \dim \mathcal{D}$.*

Remark 6.17

1. If H is a finite dimensional semisimple and co-semisimple Hopf algebra then $H - \mathcal{M}od \approx \widehat{H} - \mathcal{M}od$.
2. The results of the theorem have been generalized to fusion categories in Etingof, Nikshych and Ostrik (2005). In that generality, one must use FPdim instead of dim.

6.3.4 Module categories and categories of modules

Definition 6.18 *If $\mathcal{M}$ is a category. $\text{End}\,\mathcal{M}$ denotes the category whose objects are the endofunctors of $\mathcal{M}$, i.e. functors $\mathcal{M} \to \mathcal{M}$, and whose morphisms are the natural transformations of these endofunctors. With $\otimes$ defined as composition of functors, $\text{End}\,\mathcal{M}$ is a strict tensor category. An alternative name for $\text{End}\,\mathcal{M}$ is $Z_0(\mathcal{M})$, the monoidal centre of $\mathcal{M}$.*

Definition 6.19 *Let $\mathcal{C}$ be a tensor category and $\mathcal{M}$ a category. A left, (resp. right), $\mathcal{C}$-module structure on $\mathcal{M}$ is a tensor functor $\mathcal{C} \to \text{End}\,\mathcal{M}$ (resp. $\mathcal{C}^{rev} \to \text{End}\,\mathcal{M}$). A left (right) $\mathcal{C}$-module category is a category $\mathcal{M}$ together with a left (right) $\mathcal{C}$-module structure. A module category is indecomposable if it is not the direct sum of two non-trivial module categories.*

Left (right) $\mathcal{C}$-module categories form a 2-category, whose 1-morphisms are functors between module categories intertwining the $\mathcal{C}$-actions and whose 2-mor-phisms are natural transformations between module functors.

Remark 6.20 Unpacking the above definition, one finds that an action of a tensor category $\mathcal{C}$ on a category $\mathcal{M}$ is the same as a functor $\mathcal{C} \times \mathcal{M} \to \mathcal{M}$ satisfying certain properties like the existence of natural isomorphisms $(C \otimes D) \otimes M \to C \otimes (D \otimes M)$, etc. For such an approach see e.g. Ostrik (2003).

If $\mathcal{C}$ is a tensor category and (A, m, η) an algebra in $\mathcal{C}$ then the left module category ${}_A\mathcal{C}$ has a structure as a right $\mathcal{C}$-module category: $(X, \mu) \otimes Y = (X \otimes Y, \mu \otimes \text{id}_Y)$. Similarly, $\mathcal{C}_A$ has a left $\mathcal{C}$-module structure. The question now arises whether every $\mathcal{C}$-module category is of this form. For fusion categories this is the case:

Theorem 6.21 (Ostrik, 2003) *Let $\mathcal{C}$ be a fusion category and $\mathcal{M}$ an indecomposable semisimple left module category over $\mathcal{C}$. Then there is a connected separable algebra (A, m, η) in $\mathcal{C}$ and an equivalence $\mathcal{M} \to \mathcal{C}_A$ of module categories.*

Thus (indecomposable, semisimple) module categories over a fusion category $\mathcal{C}$ and categories of modules for a separable algebra in $\mathcal{C}$ are essentially the same thing. This allows us to discuss the weak monoidal Morita equivalence of Müger (2003a) in terms of module categories: if $\mathcal{M}$ is a good module category over the fusion category $\mathcal{C}$ then the category $\mathrm{Hom}_\mathcal{C}(\mathcal{M}, \mathcal{M})$ of $\mathcal{C}$-module functors, which clearly is monoidal, actually is fusion and is called dual to $\mathcal{C}$ with respect to $\mathcal{M}$ (Etingof, Nikshych and Ostrik, 2005). The connection with Müger (2003a) is as follows: when $\mathcal{M} = {}_A\mathcal{C}$ then $\mathrm{Hom}_\mathcal{C}(\mathcal{M}, \mathcal{M}) \simeq {}_A\mathcal{C}_A$, where ${}_A\mathcal{C}_A$ acts on ${}_A\mathcal{C}$ by tensoring from the left.

6.4 Braided tensor categories

6.4.1 Centralizers in braided categories and the symmetric centre Z_2

In this section, we quickly discuss some aspects of braided tensor categories (BTCs) that do not require additional axioms like existence of duals or linearity over a ring or field. A braided tensor category is a tensor category $\mathcal{C}$ equipped with a braiding c. For the definition of the latter, we refer to Majid's contribution to the present volume. In principle, a braided tensor category should be written as $(\mathcal{C}, c)$, where $\mathcal{C}$ is a tensor category and c a braiding on $\mathcal{C}$, but we will suppress the c. (After all, one usually does the same with the various items of the monoidal structure.)

Definition 6.22 *Let $\mathcal{C}$ be a braided tensor category with braiding c. The opposite braiding $\tilde{c}$ is defined by $\tilde{c}_{X,Y} = (c_{Y,X})^{-1}$. The tensor category $\mathcal{C}$ equipped with the braiding $\tilde{c}$ is denoted $\tilde{\mathcal{C}}$.*

In the graphical calculus (where we draw morphisms going upwards) the braidings c and $\tilde{c}$ are represented by

$$c_{X,Y} = \begin{array}{c}Y \quad X \\ \diagup\hspace{-6pt}\diagdown \\ X \quad Y\end{array} \qquad \tilde{c}_{X,Y} = \begin{array}{c}Y \quad X \\ \diagdown\hspace{-6pt}\diagup \\ X \quad Y\end{array}$$

which is consistent in view of the isotopy

$$c_{Y,X} \circ \tilde{c}_{X,Y} = \quad = \quad = \mathrm{id}_{X \otimes Y}.$$

Remark 6.23

1. Clearly $\tilde{\tilde{c}} = c$ always holds. The stronger statement $\tilde{c} = c$ is equivalent to

$$c_{X,Y} \circ c_{Y,X} = \mathrm{id}_{Y \otimes X} \quad \forall X, Y \in \mathcal{C}. \tag{6.1}$$

Recall that a BTC satisfying this additional condition is called a symmetric tensor category (STC). For this reason the braiding (=symmetry) of an STC is depicted by a crossing with unbroken lines. Conversely, the definition of BTCs is obtained from the usual definition of STCs by dropping the condition in eqn (6.1) that the braiding be involutive. (In doing so, one must add a second hexagon axiom which follows from the first in the presence of eqn (6.1).)

2. Since examples of STCs abound, e.g. the category of sets, categories of modules and vector spaces, representation categories of groups, categories of sheaves, etc., it is quite astonishing that they were formalized only in 1963 (Benabou, 1963; Mac Lane, 1963). Considering that it is much harder to find interesting examples of non-symmetric BTCs, it is less surprising that their formalization took place only around 1985/6 in the first preprint versions of Joyal and Street (1993). To some extent, this was inspired by the surge of activity around new 'quantum' invariants in low-dimensional topology (Jones, 1985; Freyd et al., 1985), quantum groups (Drinfeld, 1987), conformal field theory (Moore and Seiberg, 1989a) and algebraic quantum field theory (Fredenhagen, Rehren and Schroer, 1989; Fredenhagen, Rehren and Schroer, 1992) in the second half of the 1980s. But there are also considerations intrinsic to (higher) category theory as well as in 'old-fashioned' (non-'quantum') algebraic topology that lead to BTCs (Joyal and Street, 1993).

3. In a BTC, any two objects commute up to isomorphism: $X \otimes Y \cong Y \otimes X$. Looking for a stronger statement to be attached to the expression 'X and Y commute', one could think of $c_{X,Y} = \mathrm{id}_{X \otimes Y}$ or the slightly weaker $X \otimes Y = Y \otimes X$. But these are hardly ever satisfied in interesting BTCs. This essentially leaves us with only the following option.

Definition 6.24 *Let C be a BTC.*

(a) *Two objects $X, Y \in C$ are said to commute if $c_{X,Y} = \tilde{c}_{X,Y}$, equivalently $c_{X,Y} \circ c_{Y,X} = \mathrm{id}_{Y \otimes X}$.*

(b) *Let $\mathcal{D} \subset \mathcal{C}$ be a subcategory. The centralizer $C_C(\mathcal{D})$ is the full subcategory of $\mathcal{C}$ defined by*

$$\mathrm{Obj}\, C_C(\mathcal{D}) = \{X \in \mathcal{C} \mid c_{X,Y} \circ c_{Y,X} = \mathrm{id}_{Y \otimes X} \,\forall Y \in \mathcal{D}\}.$$

(c) *The symmetric centre $Z_2(\mathcal{C})$ is defined as $C_C(\mathcal{C})$, i.e. the full subcategory defined by*

$$\mathrm{Obj}\, Z_2(\mathcal{C}) = \{X \in \mathcal{C} \mid c_{X,Y} \circ c_{Y,X} = \mathrm{id}_{Y \otimes X} \,\forall Y \in \mathcal{C}\}.$$

Remark 6.25

1. The definition of $Z_2(\mathcal{C})$ is due to Bruguières (2000) and Müger (2000), but the concept first appeared ten years earlier in the context of algebraic quantum field theory (Rehren, 1990). The centralizer $C_C(\mathcal{D})$ seems first to have appeared in Müger (2003c).

2. In most of the literature, $\mathcal{D}'$ is written instead of $C_\mathcal{C}(\mathcal{D})$. When the ambient category $\mathcal{C}$ is fixed, there is no risk of confusion.

3. $C_\mathcal{C}(\mathcal{D})$ depends only on the objects of $\mathcal{D}$. One easily sees, for any $\mathcal{D}$, that the monoidal unit $\mathbf{1}$ lies in $C_\mathcal{C}(\mathcal{D})$ and that $C_\mathcal{C}(\mathcal{D})$ is closed under tensor products, and thus is is a full monoidal subcategory. Furthermore, it is closed under isomorphisms (i.e. replete) and under direct sums, if they exist. The braiding that $C_\mathcal{C}(\mathcal{D})$ inherits from $\mathcal{C}$ in fact is a symmetry when $\mathcal{D} = \mathcal{C}$, thus $Z_2(\mathcal{C})$ is an STC. In fact, a BTC $\mathcal{C}$ is symmetric if and only if $\mathcal{C} = Z_2(\mathcal{C})$.

4. The objects of $Z_2(\mathcal{C})$ are called *transparent* (Bruguières, 2000), since for braidings involving them it there is no difference between over- and under-crossings, or *central*.

5. The name 'centre' for $Z_2(\mathcal{C})$ is amply justified by considerations from higher category theory (see for example Baez and Dolan, 1998, or Baez and May, 2010), some of which have already played a role in Joyal and Street (1993). In higher category theory, an infinite family of centre constructions is considered. In the more limited context of 1-categories, one deals with the bicategories consisting of categories, tensor categories, BTCs, and STCs, respectively. Notice that moving rightwards in this list adds a piece of structure (tensor structure, braiding) except in the last step, where a condition is added; namely eqn (6.1). It is clear that there are forgetful (2-)functors moving leftwards in this list. More interestingly, there are constructions, called centres, in the opposite direction. We have just defined the symmetric centre $Z_2(\mathcal{C})$ of a BTC $\mathcal{C}$, and the monoidal centre $Z_0(\mathcal{C}) = \text{End}\,\mathcal{C}$ of a category $\mathcal{C}$ was given in defn 6.18. The braided centre $Z_1(\mathcal{C})$ of a tensor category $\mathcal{C}$ will be defined below. (These constructions are categorifications of their simpler analogues for 0-categories, i.e. sets, where we deal with the 1-categories of sets, monoids, and commutative monoids. The centre $Z_1(M)$ of a monoid is well known, whereas the centre $Z_0(S)$ of a set S is the monoid of endomaps of S.) Notice that the centre constructions are compatible with equivalences of categories, but not with more general functors. Thus they are functorial only on the sub-bicategories of CAT, $\otimes$-CAT, etc., whose 1-morphisms are equivalences of categories, tensor categories, etc.

6. In view of the above, one realizes that STCs, which play a rather prominent role in large parts of mathematics, are but one extreme case of BTCs, singled out by the condition that they coincide with their symmetric centres. This makes it natural to ask whether interesting things can be said in the opposite extreme case, namely when the centre of a BTC $\mathcal{C}$ is trivial in the sense of containing only what it must contain, namely the unit object and its direct sums. (Compare with the theory of von Neumann algebras, where the commutative ones and those with trivial centre ('factors'), play distinguished roles.) This is indeed the case, modular categories just being braided categories having a bit more structure and having trivial symmetric centre $Z_2(\mathcal{C})$. Since this section is devoted to results requiring no additional axioms, the study of modular categories will begin later.

6.4.2 Rambling remarks on the construction of proper BTCs

So far, we have not given any example of a non-symmetric BTC. The simplest one, the free braided tensor category $\mathcal{B}$ generated by one object, or just the braid category, is constructed from the braid groups B_n and could have been found long before (Joyal and Street, 1993). Its objects are the non-negative integers $\{0, 1, 2, \ldots\}$ with addition as tensor product. The category is discrete, i.e. $\text{Hom}_\mathcal{B}(n, m) = \emptyset$ when $n \neq m$, the endomorphisms given by $\text{End}_\mathcal{B}(n) = B_n$ with composition as in B_n (i.e. concatenation of braids). The tensor product of morphisms is given by horizontal juxtaposition of braids, and the braiding is defined as in

$c_{3,2} =$

This category clearly is not symmetric. (For example, under the isomorphism $B_2 \to \mathbb{Z}$, the braid $c_{1,1} \circ c_{1,1}$ is mapped to $2 \neq 0$.)

An attempt to systematize the known constructions of braided categories was made in Müger 2010c, where three types of constructions were distinguished.

1. Braided deformations of STCs.
2. Free (=topological) constructions.
3. The braided centre $Z_1(\mathcal{C})$ of a tensor category.

While the philosophies behind these three approaches are quite different, they are by no means mutually exclusive. In fact, the most interesting braided categories, namely the representation categories of quantum groups, can be understood in terms of all three constructions!

Space constraints do not allow us to say much about the deformation approach. While it is usually formulated in terms of a 'q-deformation' of the universal enveloping algebra $U(\mathfrak{g})$ of a simple Lie algebra, giving rise to a quasi-triangular Hopf algebra $U_q(\mathfrak{g})$ (see for example Majid, 1995a; Kassel, 1995; Chari and Pressley, 1995) one may argue that (as always?) a categorical perspective provides additional insight. In other words, the representation categories $\mathcal{C}(\mathfrak{g}, q) = U_q(\mathfrak{g}) - \mathcal{M}od$ can be obtained directly by deformation of the STCs $\mathcal{C}(\mathfrak{g}) = U(\mathfrak{g}) - \mathcal{M}od$. Such deformations are controlled by the third Davydov-Yetter cohomology (Davydov, 1998; Yetter, 1998; Yetter, 2001), and one can show (Etingof, Nikshych and Ostrik, 2005) that $H^3(D(\mathfrak{g}) - \mathcal{M}od)$ is one-dimensional for a simple Lie algebra $\mathfrak{g}$, explaining the one-parameter family of q-deformations. Actually constructing the braided deformed category $\mathcal{C}(\mathfrak{g}, q)$ from $\mathcal{C}(\mathfrak{g})$ can be done formally using Drinfeld associators (Kassel and Turaev, 1998) or analytically (i.e. non-formally) using the Knizhnik–Zamolodchikov connection (Kazhdan and Lusztig, 1994). The categories thus obtained can be shown to be equivalent to the representation categories of the corresponding quantum groups.

The approach via 'free constructions' generalizes the construction of the braid category $\mathcal{B}$ given above. Given a tensor category $\mathcal{C}$, there is a free BTC $F\mathcal{C}$ over $\mathcal{C}$, which reduces to $\mathcal{B}$ when $\mathcal{C}$ is the trivial tensor category $\{1\}$ (Joyal and Street, 1993). This construction provides a left adjoint to the forgetful 2-functor from the bicategory of braided tensor categories to the bicategory of tensor categories. There are analogous versions of this construction in case one studies categories with additional structures, such as duals or linearity over a field. Most important, at least as far as connections with low dimensional topology are concerned, are the various categories of tangles, which can be considered as free rigid braided category, free ribbon category, etc., generated by one object (Turaev, 1989; Freyd and Yetter, 1989; Turaev, 1994; Kassel, 1995; Yetter, 2001). The tangle categories are not linear over a field, but can easily be linearized using the free vector space functor from sets to vector spaces. The categories thus obtained are still too generic and too big (in the sense of having infinite-dimensional hom-spaces) to be really interesting, but quotienting them by a suitable ideal defined, for example, in terms of a link invariant, one can obtain rigid braided categories with finite-dimensional hom-spaces. For appropriate choices of the link invariant (HOMFLY or Kauffman polynomials), one actually obtains the representation categories of the quantum groups of types A–D (Turaev and Wenzl, 1997; Blanchet, 2000; Beliakova and Blanchet, 2001a; Beliakova and Blanchet, 2001b), including the most interesting (and difficult) root-of-unity case. Again, there is no space to go into this any further.

However, beginning in the next subsection, we will have more to say about the third approach to the construction of modular categories, the braided centre construction, since it is of considerable relevance for the structure theory of modular categories. This construction will play a central role in most of what follows.

6.4.3 The braided centre $Z_1(\mathcal{C})$

For simplicity, we give the following definition only for strict tensor categories, but the generalization is straightforward.

Definition/Proposition 6.26 *Let $\mathcal{C}$ be a strict tensor category.*

(a) *Let $X \in \mathcal{C}$. A half braiding e_X for X is a family $\{e_X(Y) : X \otimes Y \xrightarrow{\cong} Y \otimes X\}_{Y \in \mathcal{C}}$ of isomorphisms, natural with respect to Y, satisfying $e_X(1) = \text{id}_X$ and*

$$e_X(Y \otimes Z) = \text{id}_Y \otimes e_X(Z) \circ e_X(Y) \otimes \text{id}_Z \quad \forall Y, Z \in \mathcal{C}.$$

(b) Let $Z_1(\mathcal{C})$ be the category whose objects are pairs (X, e_X) consisting of an object and a half-braiding, the hom-sets being given by

$$\mathrm{Hom}_{Z_1(\mathcal{C})}((X, e_X), (Y, e_Y))$$
$$= \{t \in \mathrm{Hom}_\mathcal{C}(X, Y) \mid \mathrm{id}_X \otimes t \circ e_X(Z) = e_Y(Z) \circ t \otimes \mathrm{id}_X \quad \forall Z \in \mathcal{C}\}.$$

Now a tensor product of two objects is defined by $(X, e_X) \otimes (Y, e_Y) = (X \otimes Y, e_{X \otimes Y})$, where

$$e_{X \otimes Y}(Z) = e_X(Z) \otimes \mathrm{id}_Y \circ \mathrm{id}_X \otimes e_Y(Z).$$

The tensor unit is $(\mathbf{1}, e_1)$ where $e_1(X) = \mathrm{id}_X$. Defining composition and tensor product of morphisms to be inherited from $\mathcal{C}$, one verifies that $Z_1(\mathcal{C})$ is a strict tensor category. Finally,

$$c_{(X,e_X),(Y,e_Y)} = e_X(Y)$$

defines a braiding. The BTC $Z_1(\mathcal{C})$ will be called the braided centre of $\mathcal{C}$.

Remark 6.27

1. Usually, $Z_1(\mathcal{C})$ is denoted by $Z(\mathcal{C})$. We wrote Z_1 to avoid confusion with the symmetric centre Z_2 (if $\mathcal{C}$ is braided, both $Z_1(\mathcal{C})$ and $Z_2(\mathcal{C})$ are defined), but later we will drop the subscript and identify $Z = Z_1$.
2. In the same way as $Z_2(\mathcal{C}) = C_\mathcal{C}(\mathcal{C})$ is a special case of the centralizer $C_\mathcal{C}(\mathcal{D})$, there is a 'relative centre' $Z_1(\mathcal{C}, \mathcal{D})$ for a pair $\mathcal{D} \subset \mathcal{C}$. Its objects are pairs (X, e_X) where $X \in \mathcal{C}$ and e_X is a family of isomorphisms $X \otimes Y \to Y \otimes X$ for all $Y \in \mathcal{D}$. $Z_1(\mathcal{C}, \mathcal{D})$ is always monoidal, but not necessarily braided, and we have $Z(\mathcal{C}, \mathcal{C}) = Z_1(\mathcal{C})$.
3. The definition of Z_1 appeared in Joyal and Street (1991d) and Majid (1991g). The second reference also gave $Z(\mathcal{C}, \mathcal{D})$ and attributed $Z_1(\mathcal{C})$ to unpublished work of Drinfeld, which led many authors to call $Z_1(\mathcal{C})$ the 'Drinfeld centre'.
4. Despite its being somewhat involved, the definition of $Z_1(\mathcal{C})$ is quite natural. It is nevertheless instructive to give an interpretation in terms of bicategories. The point is that a (strict) tensor category is 'the same' as a (strict) 2-category with one object, and similarly, BTCs correspond to monoidal 2-categories with one object (Joyal and Street, 1993). Now, let $\mathcal{E}$ be the 2-category with one object corresponding to the tensor category $\mathcal{C}$ and let $\mathcal{F}$ be the monoidal 2-category $Z_0(\mathcal{E})$ of endo-2-functors of $\mathcal{E}$. If $\mathcal{F}_1 \subset \mathcal{F}$ is the full sub-2-category retaining only the unit object $\mathbf{1} = \mathrm{id}_\mathcal{E}$, it turns out that the braided category corresponding to $\mathcal{F}_1$ is nothing but $Z_1(\mathcal{C})$.
5. The construction of $Z_1(\mathcal{C})$ was preceded and probably motivated by Drinfeld's definition of the quantum double $D(H)$, which is a quasi-triangular Hopf algebra, of a Hopf algebra H. The two constructions are closely related, for one can construct, at least if H is finite-dimensional, an equivalence

$$D(H) - \mathcal{M}od \simeq Z_1(H - \mathcal{M}od)$$

of BTCs (Kassel, 1995). Since it is not true that all tensor categories arise from Hopf algebras, the construction of the braided centre $Z_1(\mathcal{C})$ can be considered a generalization of the Hopf algebraic quantum double. (One might also find the definition of $Z_1(\mathcal{C})$ more natural than that of $D(H)$.)

6. If $\mathcal{C}$ is k-linear, spherical or a $*$-category (=unitary category), the same holds for $Z_1(\mathcal{C})$. Other properties are much harder to show. In situations where a tensor category $\mathcal{C}$ is not of the form $H - \mathcal{M}od$ for some Hopf algebra H, it can actually be quite difficult to construct objects of $Z_1(\mathcal{C})$ different from $\mathbf{1}_{Z_1(\mathcal{C})}$. There are situations where $\mathcal{C}$ is quite big, but $Z_1(\mathcal{C})$ is 'trivial' (in the sense that $X \cong \mathbf{1} \; \forall X$): This happens if $\mathcal{C}_0$ is a category and $\mathcal{C}_1 = Z_0(\mathcal{C}_0) = \text{End}(\mathcal{C}_0)$, categorifying the simple fact that the centre (in the usual sense) $Z_1(M)$ of the monoid $M = Z_0(S)$ of endomaps of any set S is trivial in the sense of $Z_1(Z_0(S)) = \{id_S\}$. As we will see later, the situation is much better if $\mathcal{C}$ is a fusion category.

In view of its construction, it is clear that there is a forgetful tensor functor

$$K : Z_1(\mathcal{C}) \to \mathcal{C}, \quad (X, e_X) \mapsto X.$$

In general, there is no natural functor, in particular not an inclusion, from $\mathcal{C}$ to $Z_1(\mathcal{C})$. There are two exceptions: if $\mathcal{C}$ is a fusion category, K has a 2-sided (non-monoidal) adjoint $I : \mathcal{C} \to Z_1(\mathcal{C})$. This will be discussed later. The other exception is the case where $\mathcal{C}$ comes with a braiding c. While the definition of $Z_1(\mathcal{C})$ makes no reference to c, its existence has many consequences (Müger, 2003b):

1. There are two tensor functors $F_1, F_2 : \mathcal{C} \to Z_1(\mathcal{C})$, given by

$$F_1(X) = (X, e_X) \text{ with } e_X(Y) = c_{X,Y},$$
$$F_2(X) = (X, \widetilde{e}_X) \text{ with } \widetilde{e}_X(Y) = \widetilde{c}_{X,Y}.$$

Both functors are braided tensor functors from $\mathcal{C} = (\mathcal{C}, c)$ and $\widetilde{\mathcal{C}} = (\mathcal{C}, \widetilde{c})$, respectively, to $Z_1(\mathcal{C})$.

2. F_1 and F_2 are full and faithful, i.e. give embeddings $\mathcal{C} \hookrightarrow Z_1(\mathcal{C})$, $\widetilde{\mathcal{C}} \hookrightarrow Z_1(\mathcal{C})$. For example,

$\text{Hom}_{Z_1(\mathcal{C})}(F_1(X), F_1(Y))$
$= \text{Hom}_{Z_1(\mathcal{C})}((X, c_{X,\bullet}), (Y, c_{Y,\bullet}))$
$= \{t \in \text{Hom}_\mathcal{C}(X, Y) \mid id_X \otimes t \circ c_{X,Z} = c_{Y,Z} \circ t \otimes id_X \; \forall Z \in \mathcal{C}\}$
$= \text{Hom}_\mathcal{C}(X, Y),$

due to the naturality of c with respect to both arguments.

3. The full subcategories $F_1(\mathcal{C}) \subset Z_1(\mathcal{C})$ and $F_2(\widetilde{\mathcal{C}}) \subset Z_1(\mathcal{C})$ commute with each other, By the definitions of $Z_1(\mathcal{C})$ and of F_1, F_2, we have

$$c_{F_1(X),F_2(Y)} \circ c_{F_2(Y),F_1(X)} = c_{X,Y} \circ \widetilde{c}_{Y,X} = c_{X,Y} \circ c_{X,Y}^{-1} = \mathrm{id}_{Y \otimes X}.$$

An almost equally simple argument shows that $F_1(\mathcal{C}), F_2(\widetilde{\mathcal{C}})$ are each other's centralizers:

$$F_1(\mathcal{C})' = C_{Z_1(\mathcal{C})}(F_1(\mathcal{C})) = F_2(\widetilde{\mathcal{C}}), \qquad F_2(\widetilde{\mathcal{C}})' = C_{Z_1(\mathcal{C})}(F_2(\widetilde{\mathcal{C}})) = F_1(\mathcal{C}).$$

Consequentially,

$$F_1(\mathcal{C})'' = F_1(\mathcal{C}), \qquad F_2(\widetilde{\mathcal{C}})'' = F_2(\widetilde{\mathcal{C}}).$$

This is nice since a priori we only know that $\mathcal{D} \subset \mathcal{D}''$ for a tensor subcategory $\mathcal{D} \subset \mathcal{C}$. Again, for a fusion category $\mathcal{C}$, one can prove much stronger results.

4. Since $F_1(\mathcal{C})$ and $F_2(\widetilde{\mathcal{C}})$ are full subcategories of $Z_1(\mathcal{C})$, so is their intersection, and one finds

$$F_1(\mathcal{C}) \cap F_2(\widetilde{\mathcal{C}}) = F_1(Z_2(\mathcal{C})) = F_2(Z_2(\mathcal{C})).$$

5. In view of item 4, F_1 and F_2 combine to a braided tensor functor

$$H : \mathcal{C} \times \widetilde{\mathcal{C}} \to Z_1(\mathcal{C}), \qquad (X, Y) \mapsto F_1(X) \otimes F_2(Y).$$

In view of 5, this functor will be neither full nor faithful in general: if $X \in Z_2(\mathcal{C})$ and $X \not\cong 1$ then $(X, 1) \not\cong (1, X)$ but $H((X, 1)) = F_1(X) = F_2(X) = H((1, X))$, thus H does not reflect isomorphisms. However, we will see that the linearized version $\mathcal{C} \boxtimes \widetilde{\mathcal{C}} \to Z_1(\mathcal{C})$ of H actually is an equivalence when $\mathcal{C}$ is modular.

Another definition, due to Bezrukavnikov (2004), involving the braided centre will be useful later:

Definition 6.28 Let $\mathcal{C}$ be a BTC, $\mathcal{D}$ a tensor category and $F : \mathcal{C} \to \mathcal{D}$ a tensor functor. A central structure on F is a braided tensor functor $\widehat{F} : \mathcal{C} \to Z_1(\mathcal{D})$ such that $K \circ \widehat{F} = F$. Here $K : Z_1(\mathcal{D}) \to \mathcal{D}$ is the tensor functor that forgets the half-braiding, thus $\widehat{F}$ is a lift of F from $\mathcal{C}$ to $Z_1(\mathcal{C})$.

Remark 6.29 From the point of view of the Baez–Dolan picture of 'k-tuply monoidal n-categories' (Baez and Dolan, 1998), it is interesting to note the close analogy between central functors $\widehat{F} : \mathcal{C} \to Z_1(\mathcal{D})$ (where $\mathcal{C}$ is braided and $\mathcal{D}$ just monoidal) and actions $F : \mathcal{C} \to Z_0(\mathcal{D})$ (with $\mathcal{D}$ a category and $\mathcal{C}$ monoidal). In both cases, the centre Z_0 (resp. Z_1) on the r.h.s. serves to create the piece of structure (monoidal, braiding) that is needed in order to talk about a monoidal functor F or braided functor $\widehat{F}$.

6.4.4 Algebras and modules in BTCs and module categories of BTCs

Definition 6.30 *Let C be a BTC. An algebra (A, m, η) in C is called commutative if $m \circ c_{A,A} = m$. A commutative separable algebra is called étale.*

In the case where C is braided and the algebra (A, m, η) in C is commutative, we would like ${}_A C$ to be a (braided) tensor category. In order for this to hold we need an additional assumption:

From now on, we require without further mention that C has coequalizers (see Mac Lane 1971 for the definition.)

Later on, all categories we consider will be at least abelian and therefore satisfy this assumption.

Definition/Proposition 6.31 *Let C be a BTC and (A, m, η) a commutative algebra, and let $(X, \mu), (X', \mu') \in {}_A C$. Let $\alpha : X \otimes X' \to X''$ be a coequalizer of the pair of morphisms*

$$\mu_1 = \mu \otimes \mathrm{id}_{X'}, \quad \mu_2 = (\mathrm{id}_X \otimes \mu') \circ (c_{A,X} \otimes \mathrm{id}_{X'}) : A \otimes X \otimes X' \to X \otimes X'.$$

By the universal property of α, there is a unique $\mu'' : A \otimes X'' \to X''$ such that $\mu'' \circ (\mathrm{id}_A \otimes \alpha) = \alpha \circ \mu_1 = \alpha \circ \mu_2$. Now $(X'', \mu'') \in {}_A C$. With $(X, \mu) \otimes (X', \mu') := (X'', \mu'')$, ${}_A C$ is a tensor category. (Commutativity of A is needed for the interchange law $(s \otimes t) \circ (s' \otimes t') = (s \circ s') \otimes (t \circ t')$ in ${}_A C$, namely functoriality of $\otimes$ on morphisms.)

Lemma 6.32 *If C is braided and (A, m, η) a commutative algebra in C then ${}_A C$ is a tensor category (with the above tensor product). The free module functor $F_A : C \to {}_A C$ is a tensor functor.*

We would like to prove that ${}_A C$ is braided or symmetric. This requires additional assumptions. If the algebra (A, m, η) is étale, one can prove the following.

(i) If C is symmetric then ${}_A C$ is symmetric and the functor F_A is symmetric.
(ii) If $A \in Z_2(C)$ then ${}_A C$ is braided and the functor F_A is braided.
(iii) If $A \in Z_2(C)$ does not hold, the tensor category ${}_A C$ does not admit a braiding for which $F_A : C \to {}_A C$ is braided. The reason is that, every object of ${}_A C$ being a quotient of $F_A(X)$ for some X, the only possible candidate for a braiding on ${}_A C$ making F_A braided is the push-forward '$F_A(c)$' of the braiding c of C. However, when $A \notin Z_2(C)$, the would-be braiding $F_A(c)$ is natural only with respect to one of its arguments. Reformulating this positively, one obtains:
(iv) F_A can always be considered as a braided tensor functor $C \to Z_1({}_A C)$. More precisely, the tensor functor $F_A : C \to {}_A C$ admits a central structure $\hat{F}_A : C \to Z_1({}_A C)$ in the sense of defn 6.28 (Drinfeld et al., 2007; Davydov et al., 2012).

(v) An A-module (X, μ) is called *dyslectic* (Pareigis, 1995; see also Kirillov Jr. and Ostrik, 2002) or *local* when $\mu \circ c_{X,A} = \mu \circ \tilde{c}_{X,A}$. The full subcategory $_A\mathcal{C}^0 \subset {}_A\mathcal{C}$ of dyslectic modules is monoidal and in fact inherits a braiding from $\mathcal{C}$. (Notice that $_A\mathcal{C}^0 = {}_A\mathcal{C}$ when $A \in Z_2(\mathcal{C})$, thus in particular when $\mathcal{C}$ is symmetric.) The BTC $_A\mathcal{C}^0$ will play an important role in the sequel.

(vi) In order to define a monoidal structure on $_A\mathcal{C}$, where A is an algebra in $\mathcal{C}$, one actually does not need a braiding on all of $\mathcal{C}$. Reviewing how the tensor structure was defined above, one realizes that it suffices to be able to commute A with all objects of $\mathcal{C}$. More precisely, one should have a commutative algebra $((A, e_A), m, \eta)$ in $Z_1(\mathcal{C})$! In this situation, one has a natural monoidal structure on $_A\mathcal{C}$ (Schauenburg, 2001). (When $\mathcal{C}$ is braided and $A \in \mathcal{C} \xrightarrow{F_1} Z_1(\mathcal{C})$, this monoidal structure coincides with the one above since then $F_1(A) = (A, c_{A,\bullet})$.) Furthermore, for a commutative algebra $A \in Z_1(\mathcal{C})$, Schauenburg (2001) proved the remarkable braided equivalence $Z_1(_A\mathcal{C}) \simeq {}_AZ_1(\mathcal{C})^0$.

6.4.5 Duality in braided categories—braided fusion categories

Let $\mathcal{C}$ be a tensor category equipped with a left duality $X \mapsto ({}^\vee X, e_X, d_X)$ and a braiding c. Defining

$$e'_X = e_X \circ c_{X,{}^\vee X} : X \otimes {}^\vee X \to 1, \qquad d'_X = (c_{X,{}^\vee X})^{-1} \circ d_X : 1 \to {}^\vee X \otimes X, \quad (6.2)$$

a computation shows that $({}^\vee X, e'_X, d'_X)$ is a right dual for X. Thus in a braided category, left and right duals of each object are isomorphic and

$$\alpha_X = (\mathrm{id}_{{}^{\vee\vee} X} \otimes e_X) \circ (d'_X \otimes \mathrm{id}_X) : X \xrightarrow{\cong} {}^{\vee\vee} X \quad (6.3)$$

defines a natural isomorphism $\mathrm{id} \cong {}^{\vee\vee}-$ of functors. If $\mathcal{C}$ is symmetric a computation shows that α is a monoidal natural isomorphism, thus $\mathcal{C}$ is pivotal. Another computation shows that left and right traces coincide, thus $\mathcal{C}$ is spherical.

All this breaks down if $\mathcal{C}$ is braided but not symmetric, thus simply defining a right duality in terms of a given left duality and the braiding does not give a satisfactory result. One solution is to require in addition the existence of a ribbon structure:

Definition 6.33 *Let $\mathcal{C}$ be a (strict) tensor category with braiding c and left duality $X \mapsto ({}^\vee X, e_X, d_X)$. A ribbon structure on $\mathcal{C}$ is a natural isomorphism $\Theta : \mathrm{id}_\mathcal{C} \to \mathrm{id}_\mathcal{C}$, i.e. a natural family of isomorphisms $\Theta_X : X \to X$, satisfying*

$$\Theta_{X \otimes Y} = (\Theta_X \otimes \Theta_Y) \circ c_{Y,X} \circ c_{X,Y} \quad \forall X, Y \quad (6.4)$$

$$\Theta_{{}^\vee X} = {}^\vee(\Theta_X) \quad \forall X. \quad (6.5)$$

Using the ribbon structure, we modify the formulas of eqn (6.2) as follows:

$$e'_X = e_X \circ c_{X,{}^\vee X} \circ (\Theta_X \otimes \mathrm{id}_{{}^\vee X}) : X \otimes {}^\vee X \to 1,$$
$$d'_X = (\mathrm{id}_{{}^\vee X} \otimes \Theta_X) \circ (c_{X,{}^\vee X})^{-1} \circ d_X : 1 \to {}^\vee X \otimes X.$$

Now one finds that $\{\alpha_X\}$, defined as in eqn (6.3), but using the modified definitions of d'_X, e'_X, is a monoidal natural isomorphism. Thus $\mathcal{C}$ is pivotal, and again in fact spherical (Kassel, 1995).

Remark 6.34 Occasionally, it is preferable to reverse the above logic. Namely, if $\mathcal{C}$ is a spherical category and c a braiding (with no compatibility assumed) then defining

$$\Theta_X = (Tr_X \otimes \mathrm{id}_X)(c_{X,X}),$$

one finds that $\{\Theta_X\}$ satisfies eqns (6.4 and 6.5), thus is a ribbon structure compatible with the braiding c. Furthermore, the natural isomorphism $\alpha : \mathrm{id} \to {}^{\vee\vee}-$ given as part of the spherical structure coincides with the one defined in terms of the left duality and Θ as in eqn (6.3). Therefore, for a braided category $(\mathcal{C}, c)$ with left duality, giving a pivotal (in fact spherical) structure α is equivalent to giving a ribbon structure Θ compatible with c (Yetter, 1992).

We now briefly return to the subject of algebras in braided categories and their module categories. By a braided fusion category we simply mean a fusion category equipped with a braiding, and similarly for braided spherical categories. Now one has:

Proposition 6.35 (Kirillov Jr. and Ostrik, 2002) *If $\mathcal{C}$ is a braided fusion (resp. spherical) category and $A \in \mathcal{C}$ an étale algebra then ${}_A\mathcal{C}$ is a fusion (resp. spherical) category (not necessarily braided) and*

$$\mathrm{FP\,dim}\,{}_A\mathcal{C} = \frac{\mathrm{FP\,dim}\,\mathcal{C}}{d(A)}.$$

If $\mathcal{C}$ is spherical, both instances of FPdim in this identity can be replaced by dim *as defined in terms of the spherical structures.*

The dimension of the braided fusion category ${}_A\mathcal{C}^0$ does not just depend on (FP)dim $\mathcal{C}$ and $d(A)$, but on 'how much' of the object A lies in $Z_2(\mathcal{C})$. (For example, it is evident that ${}_A\mathcal{C}^0 = {}_A\mathcal{C}$ when $A \in Z_2(\mathcal{C})$, giving the same dimension for both categories. But as we will see, it is also possible that ${}_A\mathcal{C}^0$ is trivial, i.e. has dimension 1.) However, when $Z_2(\mathcal{C})$ is trivial, i.e. $\mathcal{C}$ is non-degenerate (resp. modular), one again has a simple formula (see eqn (6.8) below).

As discussed earlier, the tensor functor $F_A : \mathcal{C} \to {}_A\mathcal{C}$, while not braided in general, always admits a central structure. In the setting of fusion categories, central functors and module categories are closely related:

Theorem 6.36 (Davydov et al., 2012, Lemmas 3.5, 3.9)

(i) *If C is a braided fusion category, $\mathcal{D}$ is a fusion category and $F : C \to \mathcal{D}$ a central functor then there exists a connected étale algebra $A \in C$ such that the category $_A C$ is monoidally equivalent to the image of F, i.e. the smallest fusion subcategory of $\mathcal{D}$ containing $F(C)$. (The object A is determined by $\mathrm{Hom}_C(X, A) \cong \mathrm{Hom}_\mathcal{D}(F(X), \mathbf{1})$ and exists since F has a right adjoint.)*

(ii) *If C is a braided fusion category and $A \in C$ a connected étale algebra then the connected étale algebra $A' \in C$ obtained by (i) from the central functor $F_A : C \to {_A C}$ is isomorphic to A.*

6.5 Modular categories

6.5.1 Basics

The rest of this chapter will be concerned with braided fusion categories over an algebraically closed field k, most often $\mathbb{C}$. Recall that every fusion category has a minimal fusion subcategory consisting only of the multiples of the unit $\mathbf{1}$. This subcategory is equivalent to Vect_k. A fusion category C is called trivial when it is itself equivalent to Vect_k, which is the same as saying that every simple $X \in C$ is isomorphic to $\mathbf{1}$.

Definition 6.37 *A braided fusion category C is called:*
- *pre-modular if it is spherical*
- *non-degenerate if $Z_2(C)$ is trivial*
- *modular if it is pre-modular and non-degenerate; such a C is called just a 'modular category', modular tensor category, or MTC.*

Remark 6.38

1. Non-degenerate braided fusion categories are related to symmetric fusion categories in the same way as von Neumann factors, i.e. von Neumann algebras M with trivial centre $Z(M)$, to commutative von Neumann algebras, where $Z(M) = M$. Since these two extremal types of von Neumann algebras play distinguished roles in the general theory, it should not come as a surprise that the analogue also holds in the setting of braided fusion categories.

2. By an important theorem of Doplicher and Roberts (1989) and independently Deligne (1990) (and see Section 6.6), symmetric fusion categories are closely related to finite groups (and supergroups). Thus classifying symmetric fusion categories is essentially equivalent to classifying finite groups, a rather difficult task that has been achieved only partially. Given the importance of modular categories in the contexts of quantum group theory, conformal field theory, and low dimensional topology, in particular topological quantum field theories, one may argue that the study and classification of modular categories is as natural and urgent as that of finite groups.

Let $\mathcal{C}$ be a pre-modular category. For $X, Y \in \mathcal{C}$, define
$$S(X, Y) = \mathrm{tr}_{X \otimes Y}(c_{Y,X} \circ c_{X,Y}) \in k.$$
By the properties of the trace, $S(X, Y)$ depends only on the isomorphism classes $[X], [Y]$. Thus if $I(\mathcal{C})$ is the set of isomorphism classes of simple objects of $\mathcal{C}$ and we choose representers X_i, we can define an $|I(\mathcal{C})| \times |I(\mathcal{C})|$-matrix S by $S_{i,j} = S(X_i, X_j)$. S is symmetric, and it is easy to see that non-triviality of $Z_2(\mathcal{C})$ implies singularity of S: if $X_i \in Z_2(\mathcal{C})$ then $S_{i,j} = \mathrm{tr}_{X_i \otimes X_j}(\mathrm{id}) = d(X_i)d(X_j)$ for all j, and thus the i-th row (and column) are proportional to the 0-th row (column) (where $X_0 \cong \mathbf{1}$).

More interestingly, for $\mathcal{C}$ pre-modular one can show (Turaev, 1992; Turaev, 1994; Rehren, 1990):

- For simple X, Y, one has $S(X, Y) = d(X)d(Y)$ if *and only if* X and Y commute.
- Let $\mathcal{K} \subset \mathcal{C}$ be a fusion subcategory. Then
$$\sum_{i \in I(\mathcal{K})} d(X_i) S_{i,j} = \begin{cases} d(X_j) \dim \mathcal{K} & \text{if } X_j \in \mathcal{K}' \\ 0 & \text{otherwise} \end{cases}$$
- If $Z_2(\mathcal{C})$ is trivial then $S^2 = \dim \mathcal{C} \, C$, where C is the 'charge-conjugation' matrix: $C_{i,j} = \delta_{i,\bar{j}}$. (Note that $C^2 = \mathbf{1}$.) Thus if we assume $\dim \mathcal{C} \neq 0$ (which is automatic over $\mathbb{C}$ (Etingof, Nikshych and Ostrik, 2005)), then S is invertible if and only if $Z_2(\mathcal{C})$ is trivial, i.e. $\mathcal{C}$ is modular.
- If $\mathcal{C}$ is modular then the 'Gauss sums' of $\mathcal{C}$, defined by
$$\Omega^{\pm}(\mathcal{C}) = \sum_{i \in I(\mathcal{C})} \Theta(X_i)^{\pm 1} d(X_i)^2 \quad (6.6)$$
satisfy $\Omega^+(\mathcal{C})\Omega^-(\mathcal{C}) = \dim \mathcal{C}$ and the diagonal $|I(\mathcal{C})| \times |I(\mathcal{C})|$-matrix T defined by $T_{i,j} = \delta_{i,j}\Theta(X_i)$ satisfies $TSTST = \Omega^+(\mathcal{C})S$. (The modular category $\mathcal{C}$ is called 'anomaly-free' when $\Omega^+ = \Omega^-$.)
- Therefore, if $\mathcal{C}$ is modular, thus S invertible, we have $S^2 = \alpha C$ and $(ST)^3 = \beta C$ with $\alpha\beta \neq 0$. This means that S and T define a projective representation of the modular group $SL(2, \mathbb{Z})$. (Recall that the latter is generated by the elements
$$s = \begin{pmatrix} 0 & 1 \\ -1 & 0 \end{pmatrix}, \quad t = \begin{pmatrix} 1 & 1 \\ 0 & 1 \end{pmatrix},$$
which satisfy $s^2 = -\mathbf{1}, (st)^3 = \mathbf{1}$.) The existence of this representation is the rationale behind the terminology 'modular categories'.

Let X be a simple object and Y invertible. Then $X \otimes Y$ is simple, thus $c_{Y,X} \circ c_{X,Y} \in \mathrm{Aut}(X \otimes Y)$ is a scalar (=element of the ground field). Multiplying this scalar by $d(X)$ gives $S(X, Y)$, but this is not the point. The point is that the map $f : I \times I_1 \to k$ obtained by the above consideration (where $I_1 \subset I$ is the subgroup of invertible isomorphism classes) is a homomorphism with respect to the second argument. One can show that $f(Z, L) = f(X, L)f(Y, L)$ whenever $N_{X,Y}^Z > 0$ (Rehren, 1990, Section 4). This implies $f(X, Z) = f(Y, Z)$ whenever $\partial X = \partial Y$, where $\partial : I \to G(I)$

is the universal group grading discussed in Section 6.3.2. Thus f descends to a bihomomorphism $G(I) \times I_1 \to k$. Remarkably one has:

Theorem 6.39 (Gelaki and Nikshych, 2008) *When $\mathcal{C}$ is modular, the above map is a non-degenerate pairing, establishing a canonical isomorphism $I_1 \to \widehat{G(I)}$.*

Remark 6.40 Recall that for a finite group H, the grading group of Rep H is given by $G(\widehat{H}) \cong \widehat{Z(H)}$. On the other hand, $I_1(\text{Rep } H) \cong \widehat{H_{ab}}$, where $H_{ab} = H/[H, H]$ is the abelianization. The abelian groups $Z(H)$ and H_{ab} have little to do with each other. In view of this, Theorem 6.39 is one of many manifestations of the observation of Rehren (1990) that a modular category is 'a self-dual object that is more symmetric than a group'.

In the next two sections, we will encounter several constructions that give rise to modular categories. However, it seems instructive to give an example already at this point.

Example 6.41 Let A be a finite abelian group and $\widehat{A}$ its character group. Let $\mathcal{C}_0(A)$ be the strict braided tensor category defined by:

(i) $\text{Obj}\,\mathcal{C}(A) = A \times \widehat{A}$
(ii) $\text{Hom}((g, \phi), (g', \phi')) = \mathbb{C}$ if $(g, \phi) = (g', \phi')$ and $= \{0\}$ otherwise
(iii) $(g, \phi) \otimes (g', \phi') = (gg', \phi\phi')$; composition and tensor product of morphisms are defined as multiplication of complex numbers
(iv) braiding given by $c_{(g,\phi),(g',\phi')} = \phi(g')\text{id}$ (this makes sense since $(g, \phi) \otimes (g', \phi') = (g', \phi') \otimes (g, \phi)$).

Now $\mathcal{C}(A)$ is the closure of $\mathcal{C}_0(A)$ with respect to direct sums. One finds $\Theta((g, \phi)) = \phi(g)$ and $S((g, \phi), (g', \phi')) = \phi(g')\phi'(g)$, from which it follows easily that (e, ϕ_0), where $\phi_0 \equiv 1$, is the only central object, thus $\mathcal{C}(A)$ is modular.

The above construction of $\mathcal{C}(A)$ admits generalization to non-abelian finite groups, but that is better done using Hopf algebra language. This leads to the quantum double $D(G)$ of a finite group or a finite dimensional Hopf algebra $D(H)$, cf. the first paragraph of Section 6.7.

6.42 For any finite abelian group, one easily proves that $\Omega^\pm(\mathcal{C}(A)) = N$. Thus the categories $\mathcal{C}(A)$ are anomaly-free. If N is odd then the full subcategory $\mathcal{D}_N \subset \mathcal{C}(\mathbb{Z}/N\mathbb{Z})$ with objects $\{(k, k) \mid k = 0, \ldots, N - 1\}$ is itself modular, cf. point 3 of Remark 6.58, and one has

$$\Omega^\pm(\mathcal{D}_N) = \sum_{k=0}^{N-1} e^{\pm \frac{2\pi i}{N} k^2},$$

which is a classical Gauss sum. (This motivates calling the quantities $\Omega^\pm(\mathcal{C})$ 'Gauss sums' in general.) By the classical computation of these Gauss sums, we find that $\Omega^+(\mathcal{D}_N)$ equals $\sqrt{N}$ when $N \equiv 1 \pmod{4}$ and $i\sqrt{N}$ when $N \equiv 3 \pmod{4}$. In view of $\Omega^-(\mathcal{C}) = \overline{\Omega^+(\mathcal{C})}$, the categories $\mathcal{D}_N$ with $N \equiv 3 \pmod{4}$ therefore provide our first examples of modular categories that are not anomaly-free.

6.43 Modular categories first appeared explicitly in Turaev (1992), but some aspects of that paper were anticipated by two years in Rehren (1990), which also was inspired by Moore and Seiberg (1989a) but in addition drew upon the well established operator algebraic approach (Haag, 1996) to axiomatic quantum field theory and in particular on Fredenhagen, Rehren and Schroer (1989, 1992). In particular, Rehren (1990) contains the first proof of the equivalence between invertibility of the S-matrix, in terms of which modularity was defined in Turaev (1992) and Turaev (1994), and triviality of $Z_2(\mathcal{C})$. (For a more general recent proof see Beliakova and Blanchet, 2001a.)

6.5.2 Digression: modular categories in topology and mathematical physics

6.44 While some *inspiration* for Turaev 1992 came from conformal field theory and in particular Moore and Seiberg 1989a (see below), the main *motivation* came from low-dimensional topology. In 1988/9, Witten (1988, 1989) had proposed an interpretation of the new 'quantum invariants' of knots and 3-manifolds (in particular Jones' polynomial knot invariant) in terms of 'topological quantum field theories' (TQFTs), defined via a non-rigorous (not-yet-rigorous?) path-integral formalism. From this work, Atiyah (1989) immediately abstracted the mathematical axioms that should be satisfied by a TQFT, and Reshetikhin and Turaev (1991) soon used the representation theory of quantum groups to rigorously construct a TQFT that should essentially be that studied by Witten. The aim of Turaev (1992) was then to isolate the mathematical structure behind the construction in Reshetikhin and Turaev so as to enable generalizations, and indeed one has a $2+1$-dimensional TQFT $F_\mathcal{C}$ for each modular category $\mathcal{C}$, wherever $\mathcal{C}$ may come from (see Turaev (1994) for a full exposition of the early work on the subject and Balakov and Kirilov Jr. (2001) for a somewhat more recent introduction).

Since there is no space for going into this subject to any depth, we limit ourselves to mentioning that a TQFT in $2+1$ dimensions gives rise to projective representations of the mapping class groups of all closed two-manifolds. The mapping class group of the 2-torus is the modular group $SL(2,\mathbb{Z})$, and its projective representation produced by the TQFT $F_\mathcal{C}$ associated with the modular category $\mathcal{C}$ is just the one encountered above in terms of S and T.

6.45 Before we turn to a brief discussion of the role of modular categories in conformal field theory, we mention another manifestation of them in mathematical physics. It has turned out that infinite quantum systems (both field theories and spin systems) in two spatial dimensions can have 'topological excitations', whose mathematical analysis leads to BTCs that often turn out to be modular. An important example is Kitaev's 'toric code' (Kitaev, 2003), a quantum spin system in two dimensions, which gives rise to the (rather simple) modular category $D(\mathbb{Z}/2\mathbb{Z}) - \mathcal{M}od$ (the $\mathcal{C}(\mathbb{Z}/2\mathbb{Z})$ of Example 6.41; see also Naaijkens, 2011). There is a generalization of the toric code to finite groups other than $\mathbb{Z}/2\mathbb{Z}$, but not everything has been worked out yet. The toric code models play a prominent role in the subject of topological quantum computing, and have been reviewed for example in Freedman *et al.* (2003), Nayak *et al.* (2008), and Wang (2010).

6.46 While Kitaev's model lives in $2+1$ dimensions and has a mass gap, braided and modular categories also arise in conformally invariant (thus massless) quantum field theories in $1+1$ or $2+0$ dimensions, a subject that has been researched very extensively. In particular, there have been two rigorous and model-independent proofs of the statement that suitable chiral CQFTs have modular representation theories. In the operator algebraic approach to CQFTs, the basic definitions are quite easy to state, and we briefly do so now. The group $\mathcal{P} = PSU(1,1)$ acts on $S^1 = \{z \in \mathbb{Z} \mid |z| = 1\}$ by $\begin{pmatrix} a & b \\ c & d \end{pmatrix} z = \frac{az+b}{cz+d}$. Let $\mathcal{I}$ be the set of connected open subsets $I \subset S^1$ such that $\varnothing \neq I \neq S^1$. (Thus $\mathcal{I}$ consists of the non-trivial connected open intervals in S^1.) If $I \in \mathcal{I}$ and $g \in \mathcal{P}$ then $gI \in \mathcal{I}$.

Definition 6.47 A chiral CQFT $\mathcal{A}$ is a quadruple $(H_0, A(\cdot), U, \Omega)$, where H_0 is a separable Hilbert space (thus essentially independent of $\mathcal{A}$), $\Omega \in H$ a unit vector, U is a strongly continuous positive-energy representation of the group $PSU(1,1)$ on H, and for each $I \in \mathcal{I}$, $A(I) \subset B(H_0)$ is a von Neumann algebra. These data must satisfy the following axioms:

(i) isotony: $I_1 \subset I_2 \Rightarrow A(I_1) \subset A(I_2)$
(ii) iocality: $I_1 \cap I_2 = \varnothing \Rightarrow [A(I_1), A(I_2)] = \{0\}$
(iii) irreducibility: $\cap_I A(I)' = \mathbb{C}\mathbf{1}$
(iv) Möbius covariance: if $I \in \mathcal{I}, g \in \mathcal{P}$ then $U(g)A(I)U(g)^* = A(gI)$
(v) $U(g)\Omega = \Omega \;\forall g \in \mathcal{P}$.

Already from these few axioms one can prove a number of important results, among which are:

(a) *factoriality:* each $A(I)$ is a factor, i.e. has trivial centre
(b) *weak additivity:* if $I_1, I_2 \in \mathcal{I}$ such that $I_1 \cup I_2 \in \mathcal{I}$ then $A(I_1) \vee A(I_2) = A(I_1 \cup I_2)$
(c) *Haag duality:* for each $I \in \mathcal{I}$ one has $A(I)' = A(I')$, where I' is the interior of $S^1 - I$.

One has a notion of representation for chiral CQFTs:

Definition 6.48 Let $\mathcal{A} = (H, A(\cdot), U, \Omega)$ be a chiral CQFT. A representation of $\mathcal{A}$ is a pair $(H, \{\pi_I\}_{I \in \mathcal{I}})$, where H is a (separable) Hilbert space and $\{\pi_I : A(I) \to B(H)\}_{I \in \mathcal{I}}$ is a family of $*$-representations satisfying $\pi_{I_2} \upharpoonright A(I_1) = \pi_{I_1}$ whenever $I_1 \subset I_2$. A morphism of representations $(H, \{\pi_I\}), (H', \{\pi'_I\})$ is a bounded operator $V : H \to H'$ such that $V\pi_I(\cdot) = \pi'_I(\cdot)V$ for each $I \in \mathcal{I}$. This defines a $*$-category Rep $\mathcal{A}$.

If $(H, \{\pi_I\})$ is a representation, we define $\dim \pi = [\pi_{I'}(A(I')') : \pi_I(A(I))] \in [1, \infty]$, where the brackets denote the Jones index, and the right-hand side is independent of $I \in \mathcal{I}$. Rep$_f \mathcal{A}$ denotes the full subcategory of Rep $\mathcal{A}$ of representations with $\dim \pi < \infty$.

It is clear that Rep $\mathcal{A}$ and Rep$_f\mathcal{A}$ are ∗-categories. But one can prove much more: both admit canonical braided monoidal structures, and Rep$_f\mathcal{A}$ is semisimple with (two-sided) duals, but not necessarily finite (Fredenhagen, Rehren and Schroer, 1992). In order to prove finiteness or modularity of Rep$_F\mathcal{A}$ one needs to assume more:

Definition 6.49 A chiral CQFT $\mathcal{A} = (H, A(\cdot), U, \Omega)$ is called completely rational if the following holds:

- (i) split property: if $I_1, I_2 \in \mathcal{I}$ satisfy $\overline{I_1} \cap \overline{I_2} = \emptyset$ then the natural map $A(I_1) \otimes_{alg} A(I_2) \to A(I_1) \vee A(I_1)$ induces an isomorphism of von Neumann algebras
- (ii) strong additivity: if $I_1, I_2 \in \mathcal{I}$ satisfy $\#(\overline{I_1} \cap \overline{I_2}) = 1$ (i.e. the intervals are disjoint but adjacent) then $A(I_1) \vee A(I_2) = A(I)$, where I is the interior of the closure of $I_1 \cup I_2$
- (iii) finiteness: let $I_1, I_2 \in \mathcal{I}$ such that $\overline{I_1} \cap \overline{I_2} = \emptyset$; then $(I_1 \cup I_2)' = I_3 \cup I_4$ for certain $I_3, I_4 \in \mathcal{I}$ and the quantity $\mu(A) = [(A(I_3) \vee A(I_4)')' : A(I_1) \vee A(I_2)]$ (which a priori is in $[1, \infty]$ and can be shown to be independent of I_1, I_2) is finite.

(Axiom (ii) is not really essential.) The following was proven in Kawahigashi et al. (2001):

Theorem 6.50 If $\mathcal{A} = (H, A(\cdot), U, \Omega)$ is a completely rational chiral CFT then Rep$_f\mathcal{A}$ is modular with dim Rep$_f\mathcal{A} = \mu(A)$.

Many examples of completely rational chiral CQFTs are known, and Theorem 6.50 is just the beginning of a rapidly growing theory. (The fact that the dimension of the representation category can (in principle) be obtained by looking at just a few local algebras is extremely useful.) See the references in Müger (2010a).

Around 2005, Huang proved a similar result in the framework of vertex operator algebras (see Huang, 2005 and references therein), assuming the property of 'C_2-cofiniteness', which is similar to (iii) in defn 6.49.

The most important chiral conformal field theories are related to:

- the projective 'positive energy' representations of the diffeomorphism group Diff$^+(S^1)$ of the circle and/or its (centrally extended) Lie algebra, the Virasoro algebra
- the positive energy representations of the loop groups $C^\infty(S^1, G)$, where G is a compact Lie group.

These representation categories can be studied without reference to conformal field theory (see e.g. Presley and Segal, 1986, and Wakimoto, 2001), but defining the tensor structure does require conformal field theory or at least techniques of the latter (Fredenhagen, Rehren and Schroer, 1989; Fredenhagen, Rehren and Schroer, 1992; Wassermann, 1998).

6.51 For any algebraically closed field k, the group $H^2(SL(2, \mathbb{Z}), k^*)$ is trivial. This implies that by rescaling the matrices S, T one can obtain a true representation of the modular group, and there are exactly six ways of doing this. There are two situations where there is a preferred choice. When $\mathcal{C}$ is anomaly-free, as when $\mathcal{C} \simeq Z_1(\mathcal{D})$, one has $\Omega^+ = \Omega^- = \pm \dim \mathcal{C}$. Then the renormalization $S' = S/\Omega^+, T' = T$ gives a true representation of $SL(2, \mathbb{Z})$, which may be considered more canonical than the others.

On the other hand, when the modular category arises as the representation category of a conformal field theory $\mathcal{A}$ (in the setting of operator algebras or vertex algebras), for each equivalence class of simple objects of $\operatorname{Rep} \mathcal{A}$, there is an analytic function $f_i : \mathbb{H} \to \mathbb{C}$, the 'character' of that representation. Collecting these $n = |I(\operatorname{Rep} \mathcal{A})|$ functions in a vector-valued function $f : \mathbb{H} \to V = \mathbb{C}^n$, one finds that f satisfies the following definition:

Definition 6.52 Let V be a finite-dimensional complex vector space and $\pi : SL(2, \mathbb{Z}) \to \operatorname{End}(V)$ a representation. A vector-valued modular form of type π is a holomorphic map $\rho : \mathbb{H} \to V$ satisfying

$$f(g^{-1}z) = \pi(g)f(z) \quad \forall z \in \mathbb{H}, \, g \in SL(2, \mathbb{Z}). \tag{6.7}$$

Here $\mathbb{H} = \{z \in \mathbb{C} \mid \operatorname{Im} z > 0\}$ and $SL(2, \mathbb{Z})$ acts on $\mathbb{H}$ by $\begin{pmatrix} a & b \\ c & d \end{pmatrix} z = \frac{az+b}{cz+d}$.

Here, π is a uniquely determined (by the CQFT) true representation of $SL(2, \mathbb{C})$ obtained by a particular renormalization of the matrices S, T associated with the modular category $\operatorname{Rep} \mathcal{A}$.

In the CQFTs associated with the representation theories of the Virasoro and the Kac–Moody algebras, the above vector valued modular forms can be studied very explicitly (Wakimoto, 2001).

6.53 In all rational chiral CQFTs that have been studied explicitly, it turned out that there is an $N \in \mathbb{N}$ such that all conformal characters χ_i are modular functions for the congruence subgroup $\Gamma(N) = \{M \in SL(2, \mathbb{Z}) \mid M \equiv 1 \pmod{N}\}$. This means that $\chi_i(gz) = \chi_i(z)$ for all $g \in \Gamma(N)$ and all i. In view of eqn (6.7) this amounts to $\Gamma(N) \subset \ker(\pi)$. This led to the folk conjecture that this 'congruence subgroup property' holds in all rational chiral CQFTs. Such a general result was indeed obtained in Bantay (2003) (where unfortunately no rigorous formalism of CFTs was used). The 'conductor' N is closely related to the order of the diagonal matrix T (all elements of which are roots of unity). It is natural to ask whether a similar result can be proven for all modular categories irrespective of whether they arise from a chiral CQFT. (The latter question is an important open problem.) The answer is 'yes', but before we state it, we revert back to 'pure' mathematics.

6.5.3 Modular categories: structure theory and module categories

The work of Bantay (2003) inspired Sommerhäuser and Zhu (2007), in which a congruence subgroup theorem was proven for the modular categories of the form

$D(H) - \mathcal{M}od$, and the very recent Ng and Schauenburg (2010), with a result valid for all modular categories:

Theorem 6.54 (Ng and Schauenburg, 2010)

(a) If $\mathcal{D}$ is a spherical fusion category and $\mathcal{C} = Z_1(\mathcal{D})$ then the kernel of the canonical (cf. eqn 6.51) true representation of $SL(2, \mathbb{Z})$ contains $\Gamma(N)$, where N is the 'Frobenius–Schur exponent' of $\mathcal{C}$.

(b) If $\mathcal{C}$ is an arbitrary modular category, then $\Gamma(N)$ is contained in the kernel of the canonical projective representation of $SL(2, \mathbb{Z})$ generated by S, T.

The following results from Müger (2003c), generalized to non-degenerate braided fusion categories in Drinfeld et al. (2010), are the first steps towards a structure theory and perhaps classification of modular categories. Part (iv) of Theorem 6.55 shows that, in a sense, modular categories are better behaved than finite groups!

Theorem 6.55 Let $\mathcal{C}$ be a modular category and $\mathcal{D}_1 \subset \mathcal{C}$ a fusion subcategory. Let $\mathcal{D}_2 = C_\mathcal{C}(\mathcal{D}_1)$. Then

(i) $\dim \mathcal{D}_1 \cdot \dim \mathcal{D}_2 = \dim \mathcal{C}$.
(ii) $C_\mathcal{C}(\mathcal{D}_2) = \overline{\mathcal{D}_1}$, where the right-hand side denotes the closure of $\mathcal{D}_1$ under isomorphisms (i.e. the smallest replete fusion subcategory of $\mathcal{C}$ containing $\mathcal{D}_1$); thus if $\mathcal{D} \subset \mathcal{C}$ is a replete fusion subcategory then $\mathcal{D}'' = \mathcal{D}$.
(iii) if $\mathcal{D} \subset \mathcal{C}$ is a full tensor subcategory then $Z_2(\mathcal{D}) = Z_2(\mathcal{D}')$; in particular, if $\mathcal{D}$ is modular then so is $\mathcal{D}'$.
(iv) if $\mathcal{D} \subset \mathcal{C}$ is modular then $\mathcal{C} \simeq \mathcal{D} \boxtimes \mathcal{D}'$ as BTC.

Definition 6.56 A modular category $\mathcal{C}$ is called prime if every modular fusion subcategory is either trivial or equivalent to $\mathcal{C}$.

Corollary 6.57 Every modular category is equivalent to a finite direct product of prime modular categories.

Remark 6.58

1. Statement (ii) is similar to the 'double commutant theorem' in the theory of von Neumann algebras: The commutant M'' of the commutant of a von Neumann algebra equals M. Statement (iv), according to which modular subcategories always are direct factors, is analogous to the fact that an inclusion $N \subset M$ of type I factors gives rise to an isomorphism $M \cong N \otimes N'$.
2. If G is a finite non-abelian simple group then the modular category $D(G) - \mathcal{M}od$ is prime. In fact it contains only one fusion subcategory, namely Rep G. It follows that the classification of prime modular categories contains the classification of finite simple (non-abelian) groups! For general finite groups, $D(G) - \mathcal{M}od$ can have many fusion subcategories. See Naidu, Nikshych and Witherspoon (2009) for a classification.
3. In general, the prime factorization of a modular category is not unique. An example is already provided by $\mathcal{C} = D(A) - \mathcal{M}od$, where $A = \mathbb{Z}/p\mathbb{Z}$ is

cyclic of prime order $p \neq 2$. In this case, the replete prime modular subcategories of $\mathcal{C}$ are labelled by the isomorphisms $\phi : A \to \widehat{A}$, and $(\mathcal{D}_\phi)' = \mathcal{D}_{\overline{\phi}}$ where $\overline{\phi}(\cdot) = \phi(\cdot)^{-1}$ (Müger, 2003c). In this example, all objects of $\mathcal{C}$ are invertible, which is crucial for the non-uniqueness: by Davydov et al. 2012, Proposition 2.2, the prime factorization of $\mathcal{C}$ is unique if $\mathcal{C}$ has no invertible objects apart from $\mathbf{1}$. More generally, prime factors having no invertibles other than $\mathbf{1}$ appear identically in every prime factorization of $\mathcal{C}$. Thus the non-uniqueness results from the possibility of 'moving invertible objects from one direct factor to another'. The details have not yet been clarified, but it is clear that homomorphisms $\phi : G(\mathcal{C}) \to I_1(\mathcal{C}) \cong \widehat{G(\mathcal{C})}$ play a role. (Recall from eqn 6.39 that the abelian grading group $G(\mathcal{C})$ and the group $I_1(\mathcal{C})$ of (isomorphism classes of) invertible objects of $\mathcal{C}$ are canonically dual to each other.)

If $\mathcal{C}$ is modular and $\mathcal{D} \subset \mathcal{C}$ a fusion subcategory then $Z_2(\mathcal{D}) = \mathcal{D} \cap \mathcal{D}' \subset \mathcal{C} \cap \mathcal{D}' = C_\mathcal{C}(\mathcal{D})$, implying

$$\dim \mathcal{C} = \dim \mathcal{D} \cdot \dim C_\mathcal{C}(\mathcal{D}) \geq \dim \mathcal{D} \cdot \dim Z_2(\mathcal{D}).$$

This motivated the conjecture of Müger (2003c):

Conjecture 6.59 *If $\mathcal{D}$ is a pre-modular category, there is a modular category $\mathcal{C}$ containing $\mathcal{D}$ as a fusion subcategory such that $\dim \mathcal{C} = \dim \mathcal{D} \cdot Z_2(\mathcal{D})$.*

(There are indications that the conjecture in this generality may be false, but see Theorem 6.66 below.)

Above, we pointed out the analogy between non-degenerate braided fusion categories and von Neumann factors. This analogy goes a bit further: since factors are simple (as algebras), their homomorphisms have trivial kernels and therefore are embeddings. Analogously, one has:

Proposition 6.60 (Davydov et al., 2012, Corollary 3.26) *Any braided tensor functor $F : \mathcal{C} \to \mathcal{D}$ between braided fusion categories with $\mathcal{C}$ (almost) non-degenerate is fully faithful.*

Sketch of Proof Replacing $\mathcal{D}$ by its subcategory $F(\mathcal{C})$, we may assume that F is surjective. By results mentioned earlier, there is an étale algebra $A \in \mathcal{C}$ and an equivalence $H : \mathcal{D} \to {}_A\mathcal{C}$ of fusion categories such that $H \circ F \cong F_A$. That F_A is braided implies $A \in Z_2(\mathcal{C})$, and $\mathcal{C}$ being non-degenerate, we have $A = \mathbf{1}$, thus $F \cong \mathrm{id}$. □

In analogy to Proposition 6.35, one has the following result concerning dyslectic module categories for étale algebras in non-degenerate braided fusion categories:

Theorem 6.61 (Kirillov Jr. and Ostrik, 2002) *Let $\mathcal{C}$ be a non-degenerate braided fusion category and $A \in \mathcal{C}$ a connected étale algebra. Then the dyslectic module category ${}_A\mathcal{C}^0$ is a non-degenerate braided fusion category and*

$$\mathrm{FP}\dim {}_A\mathcal{C}^0 = \frac{\mathrm{FP}\dim \mathcal{C}}{d(A)^2}. \tag{6.8}$$

If $\mathcal{C}$ is spherical, so is ${}_A\mathcal{C}^0$ and FPdim can be replaced by dim.

In the next two sections we discuss two ways of constructing modular categories (or, more generally, non-degenerate braided fusion categories). The first, modularization, starts from a pre-modular category and the second from a mere (spherical) fusion category.

6.6 Modularization of pre-modular categories: generalizations

Let C be a pre-modular category, namely a braided spherical fusion category. Its symmetric centre $Z_2(C)$ then is a symmetric spherical fusion category. Since non-modularity of C is equivalent to $Z_2(C)$ being non-trivial, it is natural to try to 'quotient out' the full subcategory $Z_2(C)$ in order to obtain a modular category '$C/Z_2(C)$'. Formalizing this idea one arrives at the following:

Definition 6.62 (Bruguières, 2000) *A modularization of a pre-modular category C is a functor $F : C \to D$ of braided fusion categories, where D is modular and F is surjective (or 'dominant') in the sense that every object of D is a subobject of one of the form $F(X)$ with $X \in C$.*

The fact that F is supposed to be braided and surjective implies that it must trivialize $Z_2(C)$, i.e. $F(X)$ must be a multiple of the identity whenever $X \in Z_2(C)$. The following was shown in Bruguières (2000) and Müger (2000):

Theorem 6.63 *A pre-modular category admits a modularization if and only if the symmetric category $Z_2(C)$ is purely even, i.e. all objects have trivial twist, i.e. $\theta_X = \mathrm{id}_X \ \forall X$.*

The proof relies on the following deep result:

Theorem 6.64 (Doplicher and Roberts, 1989; Deligne, 1990) *Let C be a spherical symmetric fusion category with trivial twists. Then there is a finite group G, unique up to isomorphism, such that $C \simeq \mathrm{Rep}\, G$ as STC. This equivalence is unique up to natural monoidal isomorphism.*

(Both Doplicher and Roberts (1989) and Deligne (1990) prove much more general results: in the former, C is required to be unitary.) Now, if G is a finite group, the vector space $A = \mathrm{Fun}(G, \mathbb{C})$ underlying the left regular representation also is a commutative algebra, and one finds that A is a connected étale algebra in $\mathrm{Rep}\, G$. Furthermore, the module category $_A(\mathrm{Rep}\, G)$ is trivial. We call A the regular algebra, and we also do this for the corresponding object in C when Theorem 6.64 is invoked. Combining this observation with Theorem 6.64, it is clear how to obtain a modularization of a pre-modular category C with trivial twists: take $D = {}_A C$, where A is the regular algebra of the symmetric even category $Z_2(C) \subset C$, and $F = F_A$. It is not hard to show that D indeed is modular.

If $Z_2(\mathcal{C})$ contains odd/Fermionic objects, i.e. objects with $\Theta_X = -\mathrm{id}_X$, the above approach does not work. For a symmetric spherical category $\mathcal{C}$ with non-trivial twists, one has a generalization of Theorem 6.64, giving rise to a finite group G together with an element $k \in Z(G)$ of order two. (Such a pair (G, k) is occasionally called a supergroup.) One still has an equivalence $\mathcal{C} \simeq \mathrm{Rep}\, G$ of fusion categories, but the braidings of $\mathcal{C}$ and $\mathrm{Rep}\, G$ are related by the Koszul-type rule $c_{\mathcal{C}}(X, Y) = \pm c_{\mathrm{Rep}\, G}(X, Y)$ for simple X, Y, where the minus sign applies when X and Y are both odd (Doplicher and Roberts, 1989, Section 7). The regular representation of a supergroup (G, k) again is a connected separable algebra $A \in \mathrm{Rep}(G, k)$, but it is only graded commutative. As a consequence, when $\mathcal{C}$ is a braided fusion category containing $\mathrm{Rep}(G, k)$ as a full subcategory, the module category ${}_A\mathcal{C}$ is not a k-linear tensor category but a tensor category enriched over the category SVect_k of supervector spaces. (Notice that, by contrast to tensor categories enriched over Vect, such a category is *not* a tensor category, since the interchange law holds only up to signs: $(s \otimes t) \circ (s' \otimes t') = \pm (s \circ s') \otimes (t \circ t')$.)

However, if (G, k) is a supergroup, $\{e, k\}$ is a normal subgroup and if H denotes the quotient group, $\mathrm{Rep}(G, k)$ contains the even category $\mathrm{Rep}\, H$ as a full fusion subcategory. If $A \in \mathrm{Rep}\, H$ is the regular étale algebra of H, one finds ${}_A(\mathrm{Rep}(G, k)) \cong \mathrm{SVect}$, the category of supervector spaces, to wit the representation category of the supergroup $(\{e, k\}, k)$. This shows that every pre-modular category $\mathcal{C}$ admits a surjective braided functor $F : \mathcal{C} \to \mathcal{D}$, where $\mathcal{D}$ is 'almost-modular':

Definition 6.65 (Davydov, Nikshych and Ostrik, 2011) *A braided fusion category $\mathcal{C}$ is called almost non-degenerate if $Z_2(\mathcal{C}) \simeq \mathrm{SVect}$. (Equivalently, $Z_2(\mathcal{C})$ has precisely one non-trivial simple object X satisfying $X^2 \cong 1$ and $\theta_X = -\mathrm{id}$.)*

Almost modular categories will briefly be mentioned again at the end of this review, when the recent results of Davydov, Nikshych and Ostrik (2011) will be touched upon.

The above considerations have an important generalization: Let $\mathcal{S} \subset \mathcal{C}$ be an arbitrary full symmetric fusion subcategory of the pre-modular category $\mathcal{C}$. For simplicity, we restrict to the case where $\mathcal{S}$ is even and thus equivalent (as a BTC) to the representation category of a finite group G. Let $A \in \mathcal{S}$ be the regular algebra. Since the latter contains all simple objects of $\mathcal{S}$ as direct summands, we have $A \in Z_2(\mathcal{C})$ if and only if $\mathcal{S} \subset Z_2(\mathcal{C})$. If these equivalent conditions are satisfied, ${}_A\mathcal{C}$ is braided and $F_A : \mathcal{C} \to {}_A\mathcal{C}$ is a braided functor that trivializes the subcategory $\mathcal{S} \subset \mathcal{C}$. In that case, one finds $Z_2({}_A\mathcal{C}) \simeq {}_A(Z_2(\mathcal{C}))$, which is trivial if and only if $\mathcal{S} = Z_2(\mathcal{C})$, recovering the previous result about modularization.

However, it is interesting to drop the requirement $\mathcal{S} \subset Z_2(\mathcal{C})$. Independently of this assumption, one finds that ${}_A\mathcal{C}$ is a fusion category and F_A a surjective tensor functor. Furthermore, the group G acts on the module category ${}_A\mathcal{C}$ (by monoidal self-equivalences) and one has $({}_A\mathcal{C})^G \simeq \mathcal{C}$. When $\mathcal{S} \not\subset Z_2(\mathcal{C})$, there exists no braiding on ${}_A\mathcal{C}$ rendering F_A braided, but we have seen that there is a braided functor $\widehat{F}_A : \mathcal{C} \to Z_1({}_A\mathcal{C})$. In this specific situation, one can prove more: there is a G-grading on ${}_A\mathcal{C}$, i.e. a map ∂ from the class of simple objects to G, constant on isomorphism classes and satisfying $\partial(X \otimes Y) = \partial X \cdot \partial Y$. (As a consequence, $\partial \mathbf{1} = e$ and $\partial \overline{X} = (\partial X)^{-1}$.) If $\mathcal{D}$ is a G-graded category and $g \in G$, we denote by $\mathcal{D}_g$ the full subcategory whose objects are direct sums of simple objects X with $\partial X = g$. Now we have

$$({}_A\mathcal{C})_e = {}_A(C_{\mathcal{C}}(\mathcal{S})) = {}_A\mathcal{C}^0.$$

The action of G on ${}_A\mathcal{C}$ and the G-grading are connected by the identity $\partial(gX) = g(\partial X)g^{-1}$, which is why ${}_A\mathcal{C}$ is called a G-crossed category. While ${}_A\mathcal{C}$ does not admit a braiding (in the usual sense), it does admit a generalized braiding that takes the grading and the G-action into account: For every $Y \in {}_A\mathcal{C}$ and every $X \in ({}_A\mathcal{C})_g$, there is an isomorphism $c_{X,Y} : X \otimes Y \to (gY) \otimes X$ satisfying natural generalizations of the axioms for a braiding. Thus ${}_A\mathcal{C}$ is a braided G-crossed category. Conversely, if $\mathcal{D}$ is a braided G-crossed fusion category then $\mathcal{C} = \mathcal{D}^G$ is an ordinary braided fusion category containing Rep G as a full subcategory, and if A is the regular algebra in Rep G then ${}_A\mathcal{C} \simeq \mathcal{D}$.

Most of these results are due to Kirillov Jr. (2001, 2002, 2004) and Müger (2000, 2004a) in the case of spherical fusion categories. For a (somewhat) more extensive review than the one above, see Müger (2010b). A much longer discussion, including generalizations to not-necessarily-spherical fusion categories and proofs of precise 2-equivalences between categories of braided G-crossed fusion categories and braided fusion categories containing Rep G, see Drinfeld *et al.* (2007).

Using the above results one can prove (Müger, 2010b) the following result concerning Conjecture 6.59:

Theorem 6.66 *The following are equivalent:*

(i) *Conjecture 6.59 is true for every braided fusion category $\mathcal{C}$ whose symmetric centre $Z_2(\mathcal{C})$ is even (and therefore equivalent to Rep G for a finite group G).*
(ii) *For every modular category $\mathcal{M}$ acted upon by a finite group G there is a braided crossed G-category $\mathcal{E}$ with full G-spectrum and a G-equivariant equivalence $\mathcal{E}_e \simeq \mathcal{M}$.*

We close this section by pointing out that the above results have applications to the orbifold construction in conformal field theory (see Müger 2010b, Section 6 and references given there).

6.7 The braided centre of a fusion category

As mentioned earlier, to every finite dimensional Hopf algebra H one can associate (Drinfeld, 1987) a finite-dimensional quasi-triangular Hopf algebra $(D(H), R)$, Drinfeld's 'quantum double' of H. (The construction is not restricted to finite-dimensional algebras, but it becomes more technical if H is infinite-dimensional and less relevant for the purposes of this review.) The case where H is the group algebra of a finite group G is denoted $D(G)$ and can be described very explicitly. In particular, the braided category $D(G)$-$\mathcal{M}od$ is modular (Altschuler and Coste, 1992).

More generally, (i) semisimplicity of $D(H)$ is equivalent to (ii) semisimplicity of H and of the dual Hopf algebra $\widehat{H}$ and to (iii) $S_H^2 = \mathrm{id}$ and $\dim H \neq 0$ in the ground field k. Under these assumptions, $D(H)$-$\mathcal{M}od$ is modular (Etingof and Gelaki, 1998; see Müger, 2003b, Appendix for an alternative approach.)

Since the module category of a Hopf algebra H satisfying the above conditions is a spherical fusion category satisfying $\dim D(H)$-$\mathcal{M}od = \dim_k H \neq 0$, the following results proven in Müger (2003b) generalize those on $D(G)$ and $D(H)$:

Theorem 6.67 *Let k be an algebraically closed field and $\mathcal{C}$ a k-linear semisimple spherical category satisfying $\dim \mathcal{C} \neq 0$. Then:*

(i) $Z_1(\mathcal{C})$ is semisimple
(ii) $Z_1(\mathcal{C})$ has a natural spherical structure inherited from $\mathcal{C}$ and $\dim Z_1(\mathcal{C}) = (\dim \mathcal{C})^2$
(iii) $Z_1(\mathcal{C})$ is non-degenerate, thus modular
(iv) the Gauss sums (6.6) of $Z_1(\mathcal{C})$ are given by $\Omega^{\pm}(Z_1(\mathcal{C})) = \dim \mathcal{C}$
(v) the forgetful functor $K: Z_1(\mathcal{C}) \to \mathcal{C}$ has a two-sided adjoint
(vi) if $\mathcal{C}$ already is modular, then the braided tensor functor $H: \mathcal{C} \boxtimes \widetilde{\mathcal{C}} \to Z_1(\mathcal{C})$ is an equivalence
(vii) if $\mathcal{C}_1, \mathcal{C}_2$ satisfy the above assumptions and $\mathcal{C}_1 \approx \mathcal{C}_2$ (monoidal Morita equivalence of Müger, 2003a) then $Z_1(\mathcal{C}_1) \simeq Z_1(\mathcal{C}_2)$ (braided equivalence).

Remark 6.68

1. These results have been generalized to not necessarily spherical fusion categories (Etingof and Ostrik, 2004; see also Bruguières and Virelizier (2007, 2008, 2012) for a more conceptual approach in terms of Hopf monads).
2. (viii) is an easy consequence of the definitions and a result in Schauenburg (2001). The converse of (viii) is also true (Etingof, Nikshych and Ostrik, 2011, Theorem 3.1).
3. In view of (vii) and the considerations in Section 6.4.3, one could take the equivalence $Z_1(\mathcal{C}) \simeq \mathcal{C} \boxtimes \widetilde{\mathcal{C}}$ as alternative definition of modularity/non-degeneracy.
4. Statement (iv) means that $Z_2(Z_1(\mathcal{C}))$ is trivial for any spherical fusion category $\mathcal{C}$. This should be compared with the other results of the type 'the centre of a centre is trivial' mentioned in Remark 6.27. It is tempting to conjecture that this holds more generally in the context of centres in higher category theory.
5. Corollary (vii) implies that every modular category $\mathcal{C}$ arises as a direct factor of the braided centre of some fusion category $\mathcal{D}$: just take $\mathcal{D} = \mathcal{C}$. This is interesting since the braiding of $\mathcal{C}$ is not used in defining $Z_1(\mathcal{C})$. However, it seems pointless to reduce the classification of modular categories to the classification of fusion categories, since there are many more of the latter

and there is no hope of classification. A more promising approach to 'classifying' modular categories will be discussed in the last section.

6. By (i)–(iv), the braided centre construction gives rise to many modular categories. However, not every modular category $\mathcal{C}$ is equivalent to some $Z_1(\mathcal{D})$. This follows already from (v) and the fact that there are modular categories whose two Gauss sums are not equal. A criterion for recognizing whether a modular category is of the form $Z_1(\mathcal{D})$ be given below. Even when a modular category $\mathcal{C}$ does not satisfy this, one can often find fusion categories $\mathcal{D}$ smaller than $\mathcal{C}$ such that $Z_1(\mathcal{D})$ has $\mathcal{C}$ as a direct factor.

We now turn to the question of recognizing the modular categories that are of the form $Z_1(\mathcal{D})$ for $\mathcal{D}$ fusion, which has been solved quite recently (Drinfeld et al., 2007; Davydov et al., 2012). As mentioned in Subsection 6.4.4, a commutative (étale, connected) algebra A in a braided fusion category $\mathcal{C}$ gives rise to a braided tensor functor $F_A : \mathcal{C} \to Z_1(_A\mathcal{C})$. Under very weak conditions F_A is faithful. In general, F_A need not be full, but it is so when $\mathcal{C}$ is non-degenerate. This can be shown either by direct—and tedious—computation of $\mathrm{Hom}_{Z_1(\mathcal{C})}(F_A(X), F_A(Y))$ or by invoking Proposition 6.60. Thus if $\mathcal{C}$ is non-degenerate, $F_A : \mathcal{C} \to Z_1(_A\mathcal{C})$ is an embedding of braided fusion categories, which by Theorem 6.55 (iv) gives rise to a direct factorization. The complementary factor $C_{Z_1(\mathcal{C})}(F_A(\mathcal{C}))'$ can be identified using the result of Schauenburg mentioned in the last paragraph of Section 6.4.4. Recall that if $\mathcal{C}$ is non-degenerate braided fusion then we have the braided equivalence $\mathcal{C} \boxtimes \widetilde{\mathcal{C}} \xrightarrow{F_1 \boxtimes F_2} Z_1(\mathcal{C})$. If $A \in \mathcal{C}$ is a commutative algebra then $B = F_2(A) \in Z_1(\mathcal{C})$ is a commutative algebra and $\underline{B} = A$. The equivalence $Z_1(\mathcal{C}) \to \mathcal{C} \boxtimes \widetilde{\mathcal{C}}$ maps B to $1 \boxtimes A$. Combining with Schauenburg's result, we have

$$Z_1(_A\mathcal{C}) = Z_1(_B\mathcal{C}) \simeq {}_B Z_1(\mathcal{C})^0 \simeq {}_{(1 \boxtimes A)}(\mathcal{C} \boxtimes \widetilde{\mathcal{C}})^0 = \mathcal{C} \boxtimes {}_A \widetilde{\mathcal{C}}^0 = \mathcal{C} \boxtimes \widetilde{_A\mathcal{C}^0}.$$

Thus, if $\mathcal{C}$ is non-degenerate braided and $A \in \mathcal{C}$ a connected étale algebra, there is a braided equivalence

$$Z_1(_A\mathcal{C}) \simeq \mathcal{C} \boxtimes \widetilde{_A\mathcal{C}^0}. \tag{6.9}$$

Since one can prove (Davydov et al., 2012), that $\mathcal{C} = Z_1(\mathcal{D})$ contains a connected étale algebra such that $_A\mathcal{C}^0$ is trivial, one arrives at the following characterization of Drinfeld centres of fusion categories:

Theorem 6.69 *A non-degenerate braided fusion category $\mathcal{C}$ is equivalent to $Z_1(\mathcal{D})$, where $\mathcal{D}$ is a fusion category, if and only if there is a connected étale algebra A in $\mathcal{C}$ such that $_A\mathcal{C}^0$ is trivial. In this case, one can take $\mathcal{D} = {}_A\mathcal{C}$.*

In Section 6.6, we defined fusion categories graded by a finite group. One can ask how a G-grading on a fusion category $\mathcal{C}$ is reflected in the centre $Z_1(\mathcal{C})$. This was clarified in Gelaki, Naidu and Nikshych (2009), where the following is proven:

Theorem 6.70 (Gelaki, Naidu and Nikshych, 2009, Theorem 3.5) *Let C be G-graded with degree zero component C_e. Then the relative Drinfeld centre $Z_1(C, C_e)$ mentioned in Remark 6.27.2 (which is monoidal but not braided) has a natural structure of braided G-crossed category, and there is an equivalence*

$$Z_1(C, C_e)^G \simeq Z_1(C)$$

of braided categories.

The interest of this theorem derives from the fact that the relative centre $Z_1(C, C_e)^G$ may be easier to determine than the full $Z_1(C)$.

We close this section with an important application of the braided centre Z_1 and of Theorem 6.67 to topology. Since $Z_1(C)$ is modular when C is fusion, it gives rise to a Reshetikhin–Turaev TQFT (Reshetikhin and Turaev, 1991; Turaev, 1992; Turaev, 1994). It is natural to ask whether there is a more direct construction of this TQFT in terms of the spherical category C. In fact, shortly after the Reshetikhin–Turaev construction, Turaev and Viro (Turaev and Viro, 1992; Turaev, 1994) proposed a construction of $2 + 1$-dimensional TQFTs in terms of triangulations and 'state-sums' rather than surgery. While being fundamentally different from the RT-approach, the TV construction still required a modular category as input. It was realized by various authors that the construction of a state-sum TQFT actually does not require a braiding and that a spherical fusion category suffices as input datum (see in particular Barrett and Westbury, 1996; the same observation was also made by Ocneanu and by S. Gelfand and Kazhdan). This made it natural to conjecture that the state-sum TQFT of Barrett and Westbury (1996) associated with a spherical fusion category C is isomorphic to the surgery TQFT associated with the modular category $Z_1(C)$. This conjecture was proven in 2010, independently by Turaev and Virelizier (2010), based on extensive previous work by Bruguières and Virelizier (Bruguières and Virelizier, 2007; Bruguières, Lack and Virelizier, 2011), and Balsam and Kirillov (Balsam and Kirillov Jr., 2010; Balsam, 2010).

6.8 The Witt group of modular categories

The results in this subsection are from Davydov *et al.* (2012). They are motivated by the desire to 'classify' modular categories (or non-degenerate braided fusion categories). This is a rather hopeless project since, by Theorem 6.67, $Z_1(D)$ is modular whenever D is a spherical fusion category and since there is little hope of classifying fusion categories. (Recall that, for example, every semisimple Hopf algebra gives rise to a fusion category.) The fact that Morita equivalent fusion categories have equivalent Drinfeld centres reduces the problem only marginally.

This leads to the idea of considering categories of the form $Z_1(D)$ with D fusion as 'trivial' and of classifying modular categories (or non-degenerate braided fusion categories) 'up to centres'. The following definition provides a rigorous way of doing this.

Definition 6.71 *Two non-degenerate braided fusion categories C_1, C_2 are called Witt equivalent if there are fusion categories $\mathcal{D}_1, \mathcal{D}_2$ such that there is a braided equivalence $C_1 \boxtimes Z_1(\mathcal{D}_1) \simeq C_2 \boxtimes Z_1(\mathcal{D}_2)$.*

Witt equivalence obviously is coarser than braided equivalence, and it is not hard to show that it is an equivalence relation. In fact, the Witt classes form a set W_M that actually is countable. Denoting the Witt equivalence class of C by $[C]$, W_M becomes an abelian monoid via $[C_1] \cdot [C_2] := [C_1 \boxtimes C_2]$ with unit $\mathbf{1}_{W_M} = [\text{Vect}]$. Up to this point, analogous results hold for the set of braided equivalence classes of non-degenerate braided fusion categories, of which W_M is the quotient monoid under the identification $[Z_1(\mathcal{D})] = [\text{Vect}]$ for each fusion category $\mathcal{D}$. But W_M has one crucial additional property: it is a group. Namely, defining $[C]^{-1} := [\widetilde{C}]$, we have

$$[C] \cdot [C]^{-1} = [C] \cdot [\widetilde{C}] = [C \boxtimes \widetilde{C}] = [Z_1(C)] = \mathbf{1}_{W_M},$$

where the penultimate identity crucially depends on (vii) of Theorem 6.67. Therefore, W_M is called the Witt group. (Actually, there are three Witt groups, defined in terms of non-degenerate braided fusion categories, modular categories and unitary modular categories, respectively.)

By eqn (6.9), an étale algebra $A \in C$ gives rise to a braided equivalence $Z_1(_A C) \simeq C \boxtimes \widetilde{_A C^0}$ and therefore to the identity $[C] = [_A C^0]$ in the Witt group. Using the fact that $Z_1(\mathcal{D})$, where $\mathcal{D}$ is fusion, contains an étale algebra A such that $_A C^0$ is trivial, the following is not hard to show (Davydov et al., 2012):

Theorem 6.72 *Let C_1, C_2 be non-degenerate braided fusion categories. Then the following are equivalent.*

(i) *$[C_1] = [C_2]$, i.e. C_1 and C_2 are Witt equivalent.*
(ii) *There is a fusion category $\mathcal{D}$ such that $C_1 \boxtimes \widetilde{C_2} \simeq Z_1(\mathcal{D})$.*
(iii) *There is a connected étale algebra $A \in C_1 \boxtimes \widetilde{C_2}$ such that $_A(C_1 \boxtimes \widetilde{C_2})^0$ is trivial.*
(iv) *There exist a non-degenerate braided fusion category C, connected étale algebras $A_1, A_2 \in C$ and braided equivalences $C_1 \simeq {}_{A_1} C^0$, $C_2 \simeq {}_{A_2} C^0$.*
(v) *There exist connected étale algebras $A_1 \in C_1, A_2 \in C_2$ and a braided equivalence ${}_{A_1} C_1^0 \simeq {}_{A_2} C_2^0$.*

This shows that Witt equivalence could have been defined in terms of dyslectic module categories instead of invoking the braided centre Z_1. This latter approach has a 'physical' interpretation. Consider a rational chiral conformal field theory $\mathcal{A}$, either as a (C_2-cofinite) vertex operator algebra or in terms of von Neumann algebras indexed by intervals on S^1, as in, for example, Wassermann (1998), Xu (1998), and Kawahigashi, Longo and Müger (2001). As mentioned earlier, in both settings there is a proof of modularity of the representation category Rep$\mathcal{A}$. Furthermore, in both settings, there is a notion of 'finite extension' (or conformal extension) and one can prove that the finite extensions $\mathcal{B} \supset \mathcal{A}$ are classified by the connected étale algebras $A \in \text{Rep}\mathcal{A}$ in such way that $\text{Rep}\mathcal{B} = {}_A(\text{Rep}\mathcal{A})^0$ when the extension $\mathcal{B} \supset \mathcal{A}$ corresponds to the algebra $A \in \text{Rep}\mathcal{A}$ (see Kirillov Jr. and Ostrik, 2002, and Müger, 2010a, for proof sketches). This fact implies that we have [Rep $\mathcal{B}$] = [Rep $\mathcal{A}$] for any finite extension $\mathcal{B} \supset \mathcal{A}$ of rational chiral CFTs.

A (not very precise) folk conjecture in conformal field theory (see e.g. Moore and Seiberg, 1989b), states that every modular category, to the extent that it is realized by a CFT, can be obtained from the modular categories arising from Wess–Zumino–Witten (WZW) models combined and a certain set of 'constructions' (like orbifold and coset constructions). Now, the WZW categories coincide with the representation categories of quantum groups at root-of-unity deformation parameter. Thus if one accepts that the above constructions amount to passing to finite index subtheories and extensions, one arrives at the following mathematical formulation of the Moore–Seiberg conjecture:

Conjecture 6.73 *The Witt group W_M is generated by the classes $[\mathcal{C}(\mathfrak{g}, q)]$ of the quantum group categories $\mathcal{C}(\mathfrak{g}, q)$, where $\mathfrak{g}$ is a simple Lie algebra and q a root of unity.*

The only evidence for the conjecture so far is that there are no counterexamples! While there are fusion categories that are 'exotic' in the sense of having no (known) connection with finite group theory or Lie theory, no modular categories are known that are 'genuinely exotic' in the sense of not being (related to) Drinfeld centres of exotic fusion categories. However, the existing classification of conformal extensions provides a large and presumably complete set of relations in the subgroup of the Witt group generated by the classes $[\mathcal{C}(\mathfrak{g}, q)]$. While the full group W_M is not understood, a close relative, namely the Witt group of almost non-degenerate braided fusion categories, has been computed recently (Davydov, Nikshych and Ostrik, 2011).

The circle of ideas around Witt equivalence is also relevant for the construction of two-dimensional CFTs from a pair of chiral ('one-dimensional') CFTs. The relevant mathematical structure seems to be the following:

Definition 6.74 *Let $\mathcal{C}_1, \mathcal{C}_2$ be modular categories. A modular invariant for $(\mathcal{C}_1, \mathcal{C}_2)$ is a triple (A_1, A_2, E), where $A_1 \in \mathcal{C}_1, A_2 \in \mathcal{C}_2$ are connected étale algebras and $E: {}_{A_1}\mathcal{C}_1^0 \to {}_{A_2}\mathcal{C}_2^0$ is a braided equivalence.*

In view of Theorem 6.72, it is clear that a modular invariant for $(\mathcal{C}_1, \mathcal{C}_2)$ exists if and only if $\mathcal{C}_1$ and $\mathcal{C}_2$ are Witt equivalent. But more can be said (stated in Müger, 2010a, and proven in Davydov, Nikshych and Ostrik, 2011):

Proposition 6.75 *If $\mathcal{C}_1, \mathcal{C}_2$ are modular, there is a bijection (modulo natural equivalence relations) between modular invariants (A_1, A_2, E) for $(\mathcal{C}_1, \mathcal{C}_2)$ and* *connected étale algebras $A \in \mathcal{C}_1 \boxtimes \widetilde{\mathcal{C}}_2$ such that $_A(\mathcal{C}_1 \boxtimes \widetilde{\mathcal{C}}_2)^0$ is trivial ($\Leftrightarrow$ $d(A) = \sqrt{\dim \mathcal{C}_1 \cdot \dim \mathcal{C}_2}$).*

A related result, proven in Fröhlich *et al.* (2006), involves non-commutative algebras in a modular category. The fact that an algebra over a field has a centre, which is a commutative algebra, generalizes to braided spherical categories. But since the definition of the centre of an algebra A in a braided category $\mathcal{C}$ involves the braiding, there will actually be two centres $Z_L(A), Z_R(A)$, depending on the use of c or $\widetilde{c}$. One finds

Theorem 6.76 (Fröhlich et al., 2006) *Let A be a separable connected algebra in a modular category $\mathcal{C}$. Then the centres $Z_L(A), Z_R(A)$ are connected étale and there is an equivalence*

$$E : {}_{Z_L(A)}\mathcal{C}^0 \to {}_{Z_R(A)}\mathcal{C}^0$$

of braided categories. (Thus $(Z_L(A), Z_R(A), E)$ is a modular invariant in the sense of Definition 6.74.)

Every modular invariant arises in this way from a separable connected algebra A in $\mathcal{C}$ (Kong and Runkel, 2009).

Remark 6.77

1. The categorical constructions in Fröhlich *et al.* (2006) were inspired by analogous constructions in an operator algebraic context (Böckenhauer, Evans and Kawahigashi (1999, 2000)), and conjectures in Ostrik (2003).
2. In the series of papers by Fuchs, Runkel and Schweigert (2002, 2004a, 2004b, 2006), a construction of 'topological two-dimensional CFTs' was given taking a modular category and a separable connected algebra in it as a starting point. (The quotation marks refer to the fact that a CFT is more than a TQFT: it involves infinite-dimensional Hilbert spaces, trace class operators, analytic characters, etc.)

CHAPTER 7

Scalars, Monads, and Categories

DION COUMANS AND BART JACOBS
(Radboud University, Nijmegen)

7.1 Introduction

Scalars are the elements s used in scalar multiplication $s \cdot v$, yielding, for instance, a new vector for a given vector v. Scalars are elements in some algebraic structure, such as a field (for vector spaces), a ring (for modules), a group (for group actions), or a monoid (for monoid actions).

A categorical description of scalars can be given in a monoidal category **C**, with tensor $\otimes$ and tensor unit I, as the homset $\mathbf{C}(I, I)$ of endomaps on I. In Kelly and Laplaza (1980) it is shown that such homsets $\mathbf{C}(I, I)$ always form a commutative monoid; in Abramsky and Coecke (2009, Section 3.2) this is called the 'miracle' of scalars. Recent work in the area of quantum computation has led to renewed interest in such scalars. Ssee for instance Abramsky and Coecke (2004) and Abramsky and Coecke (2009), where it is shown that the presence of biproducts makes this homset $\mathbf{C}(I, I)$ of scalars a semiring, and that daggers † make it involutive. These are first examples where categorical structure (a category which is monoidal or has biproducts or daggers) gives rise to algebraic structure (a set with a commutative monoid, semiring, or involution structure; see also Vicary, 2011, and Heunen, 2009). Such correspondences form the focus of this chapter; not only those between categorical and algebraic structure, but also involving a third element, namely structure on endofunctors (especially monads). Combined, these correspondences will be described in terms of triangles of adjunctions.

Algebraic structures such as monoids are used in linguistics in order to analyse syntax: see the 'residuated monoids' used in the early reference by Lambek (1958) and the later move to compact structures (pregroups) in Lambek (1999).

More elaborate algebraic structures are used for instance in Moortgat (2007). More recently, categorical-structure-like compactness has been used to describe sentence structure (Coecke, Sadrzadeh and Clark, 2010). Hence, also in linguistics, the interplay between algebraic and categorical structure seems to be relevant.

To start, we describe the basic triangle of adjunctions that we shall build on. At this stage it is meant as a sketch of the setting, and not as an exhaustive explanation. Let $\aleph_0$ be the category with natural numbers $n \in \mathbb{N}$ as objects. Such a number n is identified with the n-element set $\underline{n} = \{0, 1, \ldots, n-1\}$. Morphisms $n \to m$ in $\aleph_0$ are ordinary functions $\underline{n} \to \underline{m}$ between these finite sets. Hence there is a full and faithful functor $\aleph_0 \hookrightarrow \mathbf{Sets}$. The underline notation is useful to avoid ambiguity, but we often omit it when no confusion arises and write the number n for the set $\underline{n}$.

Now consider the triangle in Figure 7.1, with functor categories at the two bottom corners. We briefly explain the arrows (functors) in this diagram. The downward arrows $\mathbf{Sets} \to \mathbf{Sets}^{\mathbf{Sets}}$ and $\mathbf{Sets} \to \mathbf{Sets}^{\aleph_0}$ describe the functors that map a set $A \in \mathbf{Sets}$ to the functor $X \mapsto A \times X$. In the other, upward direction right adjoints are given by the functors $(-)(1)$ describing 'evaluate at unit 1', that is $F \mapsto F(1)$. At the bottom the inclusion $\aleph_0 \hookrightarrow \mathbf{Sets}$ induces a functor $\mathbf{Sets}^{\mathbf{Sets}} \to \mathbf{Sets}^{\aleph_0}$ by restriction: F is mapped to the functor $n \mapsto F(n)$. In the reverse direction a left adjoint is obtained by left Kan extension (Mac Lane, 1971, Chapter X). Explicitly, this left adjoint maps a functor $F: \aleph_0 \to \mathbf{Sets}$ to the functor $\mathcal{L}(F): \mathbf{Sets} \to \mathbf{Sets}$ given by:

$$\mathcal{L}(F)(X) = \left(\coprod_{i \in \mathbb{N}} F(i) \times X^i\right)/\sim,$$

where $\sim$ is the least equivalence relation such that, for each $f: n \to m$ in $\aleph_0$,

$$\kappa_m(F(f)(a), v) \sim \kappa_n(a, v \circ f), \qquad \text{where } a \in F(n) \text{ and } v \in X^m.$$

The adjunction on the left in Figure 7.1 is then in fact the composition of the other two. The adjunctions in Figure 7.1 are not new. For instance, the one at the bottom plays an important role in the description of analytic functors and species (Joyal, 1986; see also Hasegawa, 2002; Adámek and Velebil, 2008; Curien,

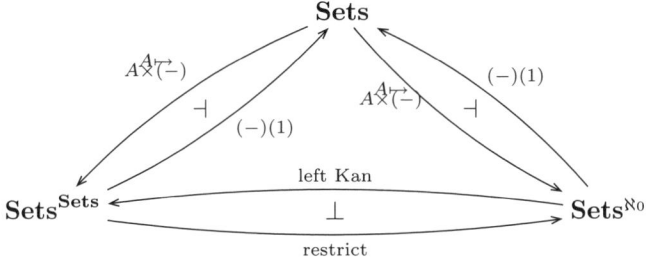

Fig 7.1 Basic triangle of adjunctions.

2012). The category of presheaves $\mathbf{Sets}^{\aleph_0}$ is used to provide a semantics for binding (see Fiore, Plotkin and Turi 1999). This chapter gives a systematic organization of correspondences in triangles, like the one in Figure 7.1, for various kinds of algebraic structures (instead of sets).

- There is a triangle of adjunctions for monoids, monads, and Lawvere theories; see Figure 7.2.
- This triangle restricts to commutative monoids, commutative monads, and symmetric monoidal Lawvere theories; see Figure 7.3.
- There is also a triangle of adjunctions for commutative semirings, commutative additive monads, and symmetric monoidal Lawvere theories with biproducts; see Figure 7.4.
- This last triangle restricts to involutive commutative semirings, involutive commutative additive monads, and dagger symmetric monoidal Lawvere theories with dagger biproducts; see Figure 7.5.

These four figures with triangles of adjunctions provide a quick way to get an overview of the chapter (the rest is just hard work). The triangles capture fundamental correspondences between basic mathematical structures.

The chapter is organized as follows. It starts with a section containing some background material on monads and Lawvere theories. The triangle of adjunctions for monoids, much of which is folklore, is developed in Section 7.3. Subsequently, Section 7.4 forms an intermezzo; it introduces the notion of additive monad and proves that a monad T is additive if and only if in its Kleisli category $\mathcal{K}\ell(T)$ coproducts form biproducts, if and only if in its category $\mathsf{Alg}(T)$ of algebras products form biproducts. These additive monads play a crucial role in Sections 7.5 and 7.6, which develop a triangle of adjunctions for commutative semirings. Finally, Section 7.7 introduces the refined triangle with involutions and daggers.

The triangles of adjunctions in this chapter are based on many detailed verifications of basic facts. We have chosen to describe all constructions explicitly but to omit most of these verifications, certainly when these are just routine. Of course, one can continue and try to elaborate deeper (categorical) structure underlying the triangles. In this chapter we have chosen not to follow that route, but rather focus on the triangles themselves.

7.2 Preliminaries

We shall assume a basic level of familiarity with category theory, especially with adjunctions and monads. This section recalls some basic facts and fixes notation. For background information we refer to Awodey (2006), Borceux (1994), and Mac Lane (1971).

In an arbitrary category $\mathbf{C}$ we write finite products as $\times, 1$, where $1 \in \mathbf{C}$ is the final object. The projections are written as π_i and tupling as $\langle f_1, f_2 \rangle$. Finite coproducts are written as $+$ with initial object 0, and with coprojections κ_i and cotupling

$[f_1, f_2]$. We write !, both for the unique map $X \to 1$ and the unique map $0 \to X$. A category is called distributive if it has both finite products and finite coproducts such that functors $X \times (-)$ preserve these coproducts: the canonical maps $0 \to X \times 0$ and $(X \times Y) + (X \times Z) \to X \times (Y + Z)$ are isomorphisms. Monoidal products are written as $\otimes, I$ where I is the tensor unit, with the familiar isomorphisms: $\alpha \colon X \otimes (Y \otimes Z) \xrightarrow{\cong} (X \otimes Y) \otimes Z$ for associativity, $\rho \colon X \otimes I \xrightarrow{\cong} X$ and $\lambda \colon I \otimes X \xrightarrow{\cong} X$ for unit, and in the symmetric case also $\gamma \colon X \otimes Y \xrightarrow{\cong} Y \otimes X$ for swap.

We write Mnd(**C**) for the category of monads on a category **C**. For convenience we write Mnd for Mnd(**Sets**). Although we shall use strength for monads mostly with respect to finite products $(\times, 1)$, we shall give the more general definition involving monoidal products $(\otimes, I)$. A monad T is called strong if it comes with a 'strength' natural transformation st with components st: $T(X) \otimes Y \to T(X \otimes Y)$, commuting with unit η and multiplication μ, in the sense that st $\circ$ $\eta \otimes \mathrm{id} = \eta$ and st $\circ$ $\mu \otimes \mathrm{id} = \mu \circ T(\mathrm{st}) \circ \mathrm{st}$. Additionally, for the familiar monoidal isomorphisms ρ and α,

$$T(Y) \otimes I \xrightarrow{\mathrm{st}} T(Y \otimes I) \qquad T(X) \otimes (Y \otimes Z) \xrightarrow{\mathrm{st}} T(X \otimes (Y \otimes Z))$$
$$\rho \searrow \quad \downarrow T(\rho) \qquad \alpha \downarrow \qquad \qquad \qquad \downarrow T(\alpha)$$
$$T(Y) \qquad (T(X) \otimes Y) \otimes Z \xrightarrow{\mathrm{st} \otimes \mathrm{id}} T(X \otimes Y) \otimes Z \xrightarrow{\mathrm{st}} T((X \otimes Y) \otimes Z)$$

Also, when the tensor $\otimes$ is a cartesian product $\times$ we sometimes write these ρ and α for the obvious maps.

The category **StMnd**(**C**) has monads with strength (T, st) as objects. Morphisms are monad maps commuting with strength. The monoidal structure on **C** is usually clear from the context.

Lemma 7.1 *Monads on* **Sets** *are always strong with respect to finite products, in a canonical way, yielding a functor* Mnd(**Sets**) $=$ Mnd $\to$ **StMnd** $=$ **StMnd**(**Sets**).

Proof For every functor $T \colon$ **Sets** $\to$ **Sets**, there exists a strength map st: $T(X) \times Y \to T(X \times Y)$, namely $\mathrm{st}(u, y) = T(\lambda x. \langle x, y \rangle)(u)$. It makes the above diagrams commute, and also commutes with unit and multiplication in the case where T is a monad. Additionally, strengths commute with natural transformations $\sigma \colon T \to S$, in the sense that $\sigma \circ \mathrm{st} = \mathrm{st} \circ (\sigma \times \mathrm{id})$. □

Given a general strength map st: $T(X) \otimes Y \to T(X \otimes Y)$ in a symmetric monoidal category one can define a swapped st$' \colon X \otimes T(Y) \to T(X \otimes Y)$ as st$' = T(\gamma) \circ \mathrm{st} \circ \gamma$, where $\gamma \colon X \otimes Y \xrightarrow{\cong} Y \otimes X$ is the swap map. There are now in principle two maps $T(X) \otimes T(Y) \rightrightarrows T(X \otimes Y)$, namely $\mu \circ T(\mathrm{st}') \circ \mathrm{st}$ and $\mu \circ T(\mathrm{st}) \circ \mathrm{st}'$. A strong monad T is called commutative if these two composites $T(X) \otimes T(Y) \rightrightarrows T(X \otimes Y)$ are the same. In that case we shall write dst for this (single) map, which is a monoidal transformation (see also Kock 1971). The powerset monad $\mathcal{P}$ is an example of a commutative monad, with dst: $\mathcal{P}(X) \times \mathcal{P}(Y) \to \mathcal{P}(X \times Y)$ given by $\mathrm{dst}(U, V) = U \times V$. Later we shall see other examples.

We write $\mathcal{K}\ell(T)$ for the Kleisli category of a monad T, with $X \in \mathbf{C}$ as objects, and maps $X \to T(Y)$ in $\mathbf{C}$ as arrows. For clarity we sometimes write a fat dot $\bullet$ for composition in Kleisli categories, so that $g \bullet f = \mu \circ T(g) \circ f$. The inclusion functor $\mathbf{C} \to \mathcal{K}\ell(T)$ is written as J, where $J(X) = X$ and $J(f) = \eta \circ f$. A map of monads $\sigma: T \to S$ yields a functor $\mathcal{K}\ell(\sigma): \mathcal{K}\ell(T) \to \mathcal{K}\ell(S)$, which is the identity on objects, and maps an arrow f to $\sigma \circ f$. This functor $\mathcal{K}\ell(\sigma)$ commutes with the Js. One obtains a functor $\mathcal{K}\ell: \mathbf{Mnd}(\mathbf{C}) \to \mathbf{Cat}$, where $\mathbf{Cat}$ is the category of (small) categories.

We will use the following standard result.

Lemma 7.2 *For $T \in \mathbf{Mnd}(\mathbf{C})$, consider the generic statement 'if $\mathbf{C}$ has $\diamond$ then so does $\mathcal{K}\ell(T)$ and $J: \mathbf{C} \to \mathcal{K}\ell(T)$ preserves $\diamond$s', where $\diamond$ is some property. This holds for:*

1. *$\diamond = $ (finite coproducts $+, 0$), or in fact any colimits*
2. *$\diamond = $ (monoidal products $\otimes, I$), in case the monad T is commutative.*

Proof Point 1 is obvious; for Point 2 one defines the tensor on morphisms in $\mathcal{K}\ell(T)$ as:

$$\left(X \xrightarrow{f} T(U)\right) \otimes \left(Y \xrightarrow{g} T(V)\right) = \left(X \otimes Y \xrightarrow{f \otimes g} T(U) \otimes T(V) \xrightarrow{\mathrm{dst}} T(U \otimes V)\right).$$

Then: $J(f) \otimes J(g) = \mathrm{dst} \circ ((\eta \circ f) \otimes (\eta \circ g)) = \eta \circ (f \otimes g) = J(f \otimes g)$. $\square$

As in this lemma we sometimes formulate results on monads in full generality, i.e. for arbitrary categories, even though our main results—see Figures 7.2, 7.3, 7.4, and 7.5—only deal with monads on **Sets**. These results involve algebraic structures such as monoids and semirings, which we interpret in the standard set-theoretic universe, and not in arbitrary categories. Such greater generality is possible, in principle, but it does not seem to add enough to justify the additional complexity.

Often we shall be interested in a 'finitary' version of the Kleisli construction, corresponding to the Lawvere theory (Lawvere, 1963; Hyland and Power, 2007) associated with a monad. For a monad $T \in \mathbf{Mnd}$ on **Sets** we shall write $\mathcal{K}\ell_{\mathbb{N}}(T)$ for the category with natural numbers $n \in \mathbb{N}$ as objects, regarded as finite sets $\underline{n} = \{0, 1, \ldots, n-1\}$. A map $f: n \to m$ in $\mathcal{K}\ell_{\mathbb{N}}(T)$ is then a function $\underline{n} \to T(\underline{m})$. This yields a full inclusion $\mathcal{K}\ell_{\mathbb{N}}(T) \hookrightarrow \mathcal{K}\ell(T)$. It is easy to see that a map $f: n \to m$ in $\mathcal{K}\ell_{\mathbb{N}}(T)$ can be identified with an n-cotuple of elements $f_i \in T(\underline{m})$, which may be seen as m-ary terms/operations.

By the previous lemma the category $\mathcal{K}\ell_{\mathbb{N}}(T)$ has coproducts given on objects simply by the additive monoid structure $(+, 0)$ on natural numbers. There are obvious coprojections $n \to n + m$, using $\underline{n + m} \cong \underline{n} + \underline{m}$. The identities $n + 0 = n = 0 + n$ and $(n + m) + k = n + (m + k)$ are in fact the familiar monoidal isomorphisms. The swap map is an isomorphism $n + m \cong m + n$ rather than an identity $n + m = m + n$.

In general, a Lawvere theory is a small category $\mathbf{L}$ with natural numbers $n \in \mathbb{N}$ as objects, and $(+, 0)$ on $\mathbb{N}$ forming finite coproducts in $\mathbf{L}$. It forms a categorical version of a term algebra, in which maps $n \to m$ are understood as n-tuples of terms t_i each with m free variables. Formally a Lawvere theory involves a functor $\aleph_0 \to \mathbf{L}$ that is the identity on objects and preserves finite coproducts 'on the nose' (up-to-identity) as opposed to up-to-isomorphism. A morphism of Lawvere theories $F: \mathbf{L} \to \mathbf{L}'$ is a functor that is the identity on objects and strictly preserves finite coproducts. This yields a category **Law**.

Corollary 7.3 *The finitary Kleisli construction $\mathcal{K}\ell_\mathbb{N}$ for monads on* **Sets**, *yields a functor* $\mathcal{K}\ell_\mathbb{N} : \mathbf{Mnd} \to \mathbf{Law}$.

In this chapter we do not discuss *models* of a Lawvere theory $\mathbf{L}$, i.e. coproduct preserving functors $\mathbf{L} \to \mathbf{C}$. It is a folklore result that for a finitary (filtered colimit preserving) monad T on **Sets**, the models $\mathcal{K}\ell_\mathbb{N}(T) \to$ **Sets** correspond to T-algebras.

7.3 Monoids

In this section we replace the category **Sets** of sets at the top of the triangle in Figure 7.1 by the category **Mon** of monoids $(M, \cdot, 1)$ and describe how to change the corners at the bottom in order to keep a triangle of adjunctions. Formally, this can be done by considering monoid objects in the three categories at the corners of the triangle in Figure 7.1 (see also Fiore, Plotkin and Turi 1999 and Curien 2012) but we prefer a more concrete description. The results in this section, which are summarized in Figure 7.2, are not new, but are included in preparation of further steps later on in this chapter.

We start by studying the interrelations between monoids and monads. In principle this part can be skipped, because the adjunction on the left in Figure 7.2 between

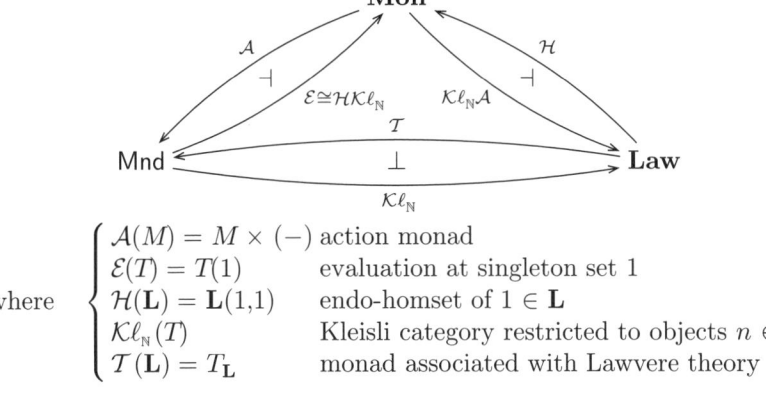

where
$$\begin{cases} \mathcal{A}(M) = M \times (-) & \text{action monad} \\ \mathcal{E}(T) = T(1) & \text{evaluation at singleton set } 1 \\ \mathcal{H}(\mathbf{L}) = \mathbf{L}(1,1) & \text{endo-homset of } 1 \in \mathbf{L} \\ \mathcal{K}\ell_\mathbb{N}(T) & \text{Kleisli category restricted to objects } n \in \mathbb{N} \\ \mathcal{T}(\mathbf{L}) = T_\mathbf{L} & \text{monad associated with Lawvere theory } \mathbf{L}. \end{cases}$$

Fig 7.2 Basic relations between monoids, monads, and Lawvere theories.

monoids and monads follows from the other two by composition. But we do make this adjunction explicit in order to completely describe the situation.

The following result is standard. We only sketch the proof.

Lemma 7.4 *Each monoid M gives rise to a monad* $\mathcal{A}(M) = M \times (-)\colon$ **Sets** $\to$ **Sets**. *The mapping* $M \mapsto \mathcal{A}(M)$ *yields a functor* **Mon** $\to$ **Mnd**.

Proof For a monoid $(M, \cdot, 1)$ the unit map $\eta \colon X \to M \times X = \mathcal{A}(M)$ is $x \mapsto (1, x)$. The multiplication $\mu \colon M \times (M \times X) \to M \times X$ is $(s, (t, x)) \mapsto (s \cdot t, x)$. The standard strength map st$\colon (M \times X) \times Y \to M \times (X \times Y)$ is given by st$((s, x), y) = (s, (x, y))$. Each monoid map $f \colon M \to N$ gives rise to a map of monads with components $f \times \mathrm{id} \colon M \times X \to N \times X$. These components commute with strength. $\square$

The monad $\mathcal{A}(M) = M \times (-)$ is called the 'action monad', as its category of Eilenberg–Moore algebras consists of M-actions $M \times X \to X$ and their morphisms. The monoid elements act as scalars in such actions.

Conversely, each monad (on **Sets**) gives rise to a monoid. In the following lemma we prove this in more generality. For a category **C** with finite products, we denote by **Mon**(**C**) the category of monoids in **C**, i.e. the category of objects M in **C** carrying a monoid structure $1 \to M \leftarrow M \times M$ with structure-preserving maps between them.

Lemma 7.5 *Each strong monad T on a category* **C** *with finite products, gives rise to a monoid* $\mathcal{E}(T) = T(1)$ *in* **C**. *The mapping* $T \mapsto T(1)$ *yields a functor* **StMnd**(**C**) $\to$ **Mon**(**C**).

Proof For a strong monad $(T, \eta, \mu, \mathrm{st})$, we define a multiplication on $T(1)$ by $\mu \circ T(\pi_2) \circ \mathrm{st} \colon T(1) \times T(1) \to T(1)$, with unit $\eta_1 \colon 1 \to T(1)$. Each monad map $\sigma \colon T \to S$ gives rise to a monoid map $T(1) \to S(1)$ by taking the component of σ at 1. $\square$

The swapped strength map st$'$ gives rise to a swapped multiplication on $T(1)$, namely $\mu \circ T(\pi_1) \circ \mathrm{st}' \colon T(1) \times T(1) \to T(1)$, again with unit η_1. It corresponds to $(a, b) \mapsto b \cdot a$ instead of $(a, b) \mapsto a \cdot b$ like in the lemma. In the case where T is a commutative monad, the two multiplications coincide, as we prove in Lemma 7.10.

The functors defined in Lemmas 7.4 and 7.5 form an adjunction. This result goes back to Wolff 1973.

Lemma 7.6 *The pair of functors* $\mathcal{A} \colon$ **Mon** $\rightleftarrows$ **Mnd** $\colon \mathcal{E}$ *forms an adjunction* $\mathcal{A} \dashv \mathcal{E}$, *as on the left in Figure 7.2.*

Proof For a monoid M and a (strong) monad T on **Sets** there are (natural) bijective correspondences:

$$\frac{\mathcal{A}(M) \xrightarrow{\sigma} T \quad \text{in Mnd}}{M \xrightarrow{f} T(1) \quad \text{in Mon}}$$

Given σ one defines a monoid map $\bar\sigma \colon M \to T(1)$ as:

$$\bar\sigma = \Big(M \xrightarrow[\cong]{\rho^{-1}} M \times 1 = \mathcal{A}(M)(1) \xrightarrow{\sigma_1} T(1)\Big),$$

where $\rho^{-1} = \langle \mathrm{id}, !\rangle$ in this cartesian case. Conversely, given f one gets a monad map $\bar f \colon \mathcal{A}(M) \to T$ with components:

$$\bar f_X = \Big(M \times X \xrightarrow{f \times \mathrm{id}} T(1) \times X \xrightarrow{\mathrm{st}} T(1 \times X) \xrightarrow[\cong]{T(\lambda)} T(X)\Big),$$

where $\lambda = \pi_2 \colon 1 \times X \xrightarrow{\cong} X$. Straightforward computations show that these assignments indeed give a natural bijective correspondence. □

Notice that, for a monoid M, the counit of the above adjunction is the projection $(\mathcal{E} \circ \mathcal{A})(M) = \mathcal{A}(M)(1) = M \times 1 \xrightarrow{\cong} M$. Hence the adjunction is a reflection.

We now move to the bottom of Figure 7.2. The finitary Kleisli construction yields a functor from the category of monads to the category of Lawvere theories (Corollary 7.3). This functor has a left adjoint, as is proven in the following two standard lemmas.

Lemma 7.7 *Each Lawvere theory* $\mathbf{L}$, *gives rise to a monad* $T_\mathbf{L}$ *on* **Sets**, *which is defined by*

$$T_\mathbf{L}(X) = \Big(\coprod_{i \in \mathbb{N}} \mathbf{L}(1, i) \times X^i\Big)/\sim, \tag{7.1}$$

where $\sim$ *is the least equivalence relation such that, for each* $f \colon i \to m$ *in* $\aleph_0 \hookrightarrow \mathbf{L}$,

$$\kappa_m(f \circ g, v) \sim \kappa_i(g, v \circ f), \qquad \text{where } g \in \mathbf{L}(1, i) \text{ and } v \in X^m.$$

Finally, the mapping $\mathbf{L} \mapsto T_\mathbf{L}$ *yields a functor* $\mathcal{T} \colon \mathbf{Law} \to \mathbf{Mnd}$.

Proof For a Lawvere theory $\mathbf{L}$, the unit map $\eta \colon X \to T_\mathbf{L}(X) = \Big(\coprod_{i \in \mathbb{N}} \mathbf{L}(1, i) \times X^i\Big)/\sim$ is given by

$$x \mapsto [\kappa_1(\mathrm{id}_1, x)].$$

The multiplication $\mu \colon T_\mathbf{L}^2(X) \to T_\mathbf{L}(X)$ is given by:

$$\mu([\kappa_i(g, v)]) = [\kappa_j((g_0 + \cdots + g_{i-1}) \circ g, [v_0, \ldots, v_{i-1}])]$$

where $g \colon 1 \to i$, and $v \colon i \to T_\mathbf{L}(X)$ is written as $v(a) = \kappa_{j_a}(g_a, v_a)$, for $a < i$, and $j = j_0 + \cdots + j_{i-1}$. It is straightforward to show that this map μ is well-defined and that η and μ indeed define a monad structure on $T_\mathbf{L}$.

For each morphism of Lawvere theories $F\colon \mathbf{L} \to \mathbf{K}$, one may define a monad morphism $\mathcal{T}(F)\colon T_{\mathbf{L}} \to T_{\mathbf{K}}$ with components $\mathcal{T}(F)_X\colon [\kappa_i(g,v)] \mapsto [\kappa_i(F(g),v)]$. This yields a functor $\mathcal{T}\colon \mathbf{Law} \to \mathbf{Mnd}$. Checking the details is left to the reader. $\square$

Lemma 7.8 *The pair of functors* $\mathcal{T}\colon \mathbf{Law} \rightleftarrows \mathbf{Mnd}\colon \mathcal{K}\ell_{\mathbb{N}}$ *forms an adjunction* $\mathcal{T} \dashv \mathcal{K}\ell_{\mathbb{N}}$, *as at the bottom in Figure 7.2.*

Proof For a Lawvere theory $\mathbf{L}$ and a monad T there are (natural) bijective correspondences:

$$\frac{\mathcal{T}(\mathbf{L}) \xrightarrow{\sigma} T \quad \text{in } \mathbf{Mnd}}{\mathbf{L} \xrightarrow{F} \mathcal{K}\ell_{\mathbb{N}}(T) \quad \text{in } \mathbf{Law}}$$

Given σ, one defines a **Law**-map $\overline{\sigma}\colon \mathbf{L} \to \mathcal{K}\ell_{\mathbb{N}}(T)$, which is the identity on objects and sends a morphism $f\colon n \to m$ in $\mathbf{L}$ to the morphism

$$n \xrightarrow{\lambda i < n.\, [\kappa_m(f \circ \kappa_i, \mathrm{id}_m)]} \mathcal{T}(\mathbf{L})(m) \xrightarrow{\sigma_m} T(m)$$

in $\mathcal{K}\ell_{\mathbb{N}}(T)$.

Conversely, given F, one defines a monad morphism $\overline{F}$ with components $\overline{F}_X\colon \mathcal{T}(\mathbf{L})(X) \to T(X)$ given, for $i \in \mathbb{N}$, $g\colon 1 \to i \in \mathbf{L}$ and $v \in X^i$, by:

$$[\kappa_i(g,v)] \mapsto (T(v) \circ F(g))(*),$$

where $*$ is the unique element of 1. $\square$

Finally, we consider the right-hand side of Figure 7.2. For each category $\mathbf{C}$ and object X in $\mathbf{C}$, the homset $\mathbf{C}(X,X)$ is a monoid, where multiplication is given by composition with the identity as unit. The mapping $\mathbf{L} \mapsto \mathcal{H}(\mathbf{L}) = \mathbf{L}(1,1)$, defines a functor $\mathbf{Law} \to \mathbf{Mon}$. This functor is right adjoint to the composite functor $\mathcal{K}\ell_{\mathbb{N}} \circ \mathcal{A}$.

Lemma 7.9 *The pair of functors* $\mathcal{K}\ell_{\mathbb{N}} \circ \mathcal{A}\colon \mathbf{Mon} \rightleftarrows \mathbf{Law}\colon \mathcal{H}$ *forms an adjunction* $\mathcal{K}\ell_{\mathbb{N}} \circ \mathcal{A} \dashv \mathcal{H}$, *as on the right in Figure 7.2.*

Proof For a monoid M and a Lawvere theory $\mathbf{L}$ there are (natural) bijective correspondences:

$$\frac{\mathcal{K}\ell_{\mathbb{N}}\mathcal{A}(M) \xrightarrow{F} \mathbf{L} \quad \text{in } \mathbf{Law}}{M \xrightarrow{f} \mathcal{H}(\mathbf{L}) \quad \text{in } \mathbf{Mon}}$$

Given F one defines a monoid map $\bar{F}\colon M \to \mathcal{H}(L) = L(1,1)$ by

$$s \mapsto F(1 \xrightarrow{\langle \lambda x.\, s,\, !\rangle} M \times 1).$$

Note that $1 \xrightarrow{\langle \lambda x.\, s,\, !\rangle} M \times 1 = \mathcal{A}(M)(1)$ is an endomap on 1 in $\mathcal{K}\ell_\mathbb{N}\mathcal{A}(M)$. Since F is the identity on objects it sends this endomap to an element of $L(1,1)$.

Conversely, given a monoid map $f\colon M \to L(1,1)$ one defines a **Law**-map $\bar{f}\colon \mathcal{K}\ell_\mathbb{N}\mathcal{A}(M) \to L$. It is the identity on objects and sends a morphism $h\colon n \to m$ in $\mathcal{K}\ell_\mathbb{N}\mathcal{A}(M)$, i.e. $h\colon n \to M \times m$ in **Sets**, to the morphism

$$\bar{f}(h) = \Big(n \xrightarrow{\big[\kappa_{h_2(i)} \circ f(h_1(i))\big]_{i<n}} m\Big).$$

Here we write $h(i) \in M \times m$ as pair $(h_1(i), h_2(i))$. We leave further details to the reader. $\square$

Given a monad T on **Sets**, $\mathcal{HK}\ell_\mathbb{N}(T) = \mathcal{K}\ell_\mathbb{N}(T)(1,1) = \mathbf{Sets}(1, T(1))$ is a monoid, where the multiplication is given by

$$(1 \xrightarrow{a} T(1)) \cdot (1 \xrightarrow{b} T(1)) = \Big(1 \xrightarrow{a} T(1) \xrightarrow{T(b)} T^2(1) \xrightarrow{\mu} T(1)\Big).$$

The functor $\mathcal{E}\colon \mathsf{Mnd}(\mathbf{C}) \to \mathsf{Mon}(\mathbf{C})$, defined in Lemma 7.5 also gives a multiplication on $\mathbf{Sets}(1, T(1)) \cong T(1)$, namely $\mu \circ T(\pi_2) \circ \mathrm{st}\colon T(1) \times T(1) \to T(1)$. These two multiplications coincide as is demonstrated in the following diagram,

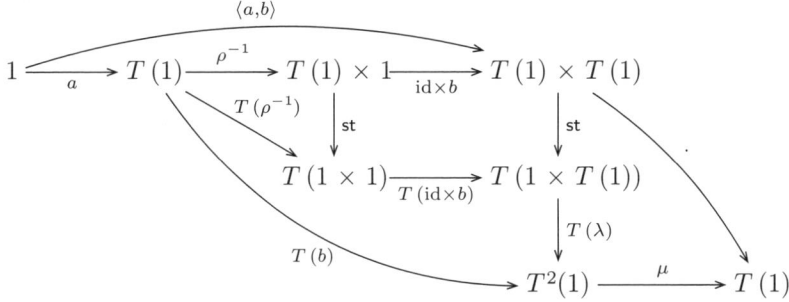

In fact, $\mathcal{E} \cong \mathcal{HK}\ell_\mathbb{N}$, which completes the picture from Figure 7.2.

7.3.1 Commutative monoids

In this subsection we briefly summarize what will change in the triangle in Figure 7.2 when we restrict ourselves to commutative monoids (at the top). This will lead to commutative monads and to tensor products. The latter are induced by Lemma 7.2.

The new situation is described in Figure 7.3. For the adjunction between commutative monoids and commutative monads we start with the following basic result.

Lemma 7.10 *Let T be a commutative monad on a category* **C** *with finite products. The monoid* $\mathcal{E}(T) = T(1)$ *in* **C** *from Lemma 7.5 is then commutative.*

Proof Recall that the multiplication on $T(1)$ is given by $\mu \circ T(\lambda) \circ \text{st}: T(1) \times T(1) \to T(1)$ and commutativity of the monad T means $\mu \circ T(\text{st}') \circ \text{st} = \mu \circ T(\text{st}) \circ \text{st}'$ where $\text{st}' = T(\gamma) \circ \text{st} \circ \gamma$, for the swap map γ; see Section 7.2. Then:

$$\mu \circ T(\lambda) \circ \text{st} \circ \gamma = \mu \circ T(T(\lambda) \circ \text{st}') \circ \text{st} \circ \gamma$$
$$= T(\lambda) \circ \mu \circ T(\text{st}') \circ \text{st} \circ \gamma$$
$$= T(\rho) \circ \mu \circ T(\text{st}) \circ \text{st}' \circ \gamma$$

(by commutativity of T, and because $\rho = \lambda\colon 1 \times 1 \to 1$)

$$= \mu \circ T(T(\rho) \circ \text{st} \circ \gamma) \circ \text{st}$$
$$= \mu \circ T(\rho \circ \gamma) \circ \text{st}$$
$$= \mu \circ T(\lambda) \circ \text{st}. \qquad \square$$

The proof of the next result is easy and left to the reader.

Lemma 7.11 *A monoid M is commutative (abelian) if and only if the associated monad* $\mathcal{A}(M) = M \times (-)\colon$ **Sets** $\to$ **Sets** *is commutative (as described in Section 7.2).*

Next, we wish to define an appropriate category **SMLaw** of Lawvere theories with symmetric monoidal structure $(\otimes, I)$. In order to do so we need to take a closer look at the category $\aleph_0$ described in the introduction. Recall that $\aleph_0$ has $n \in \mathbb{N}$ as objects whilst morphisms $n \to m$ are functions $\underline{n} \to \underline{m}$ in **Sets**, where, as described earlier, $\underline{n} = \{0, 1, \ldots, n-1\}$. This category $\aleph_0$ has a monoidal structure, given on objects by multiplication $n \times m$ of natural numbers, with $1 \in \mathbb{N}$ as tensor unit. Functoriality involves a (chosen) coordinatization, in the following way. For $f\colon n \to p$ and $g\colon m \to q$ in $\aleph_0$ one obtains $f \otimes g\colon n \times m \to p \times q$ as a function:

$$f \otimes g = \text{co}_{p,q}^{-1} \circ (f \times g) \circ \text{co}_{n,m} \colon \underline{n \times m} \longrightarrow \underline{p \times q},$$

where co is a coordinatization function

$$\underline{n \times m} = \{0, \ldots, (n \times m) - 1\} \xrightarrow[\cong]{\text{co}_{n,m}} \{0, \ldots, n-1\} \times \{0, \ldots, m-1\} = \underline{n} \times \underline{m},$$

given by

$$\text{co}(c) = (a, b) \Leftrightarrow c = a \cdot m + b. \tag{7.2}$$

We may write the inverse $\mathrm{co}^{-1}\colon \overline{n}\times \overline{m} \to \overline{n\times m}$ as a small tensor, as in $a\otimes b = \mathrm{co}^{-1}(a,b)$. Then $(f\otimes b)(a\otimes b) = f(a)\otimes g(b)$. The monoidal isomorphisms in $\aleph_0$ are then obtained from **Sets**, as in

$$\gamma^{\aleph_0} = \left(\underline{n\times m} \xrightarrow{\mathrm{co}} \underline{n}\times \underline{m} \xrightarrow{\gamma^{\mathbf{Sets}}} \underline{m}\times \underline{n} \xrightarrow{\mathrm{co}^{-1}} \underline{m\times n}\right).$$

Thus $\gamma^{\aleph_0}(a\otimes b) = b\otimes a$. Similarly, the associativity map $\alpha^{\aleph_0}\colon n\otimes (m\otimes k) \to (n\otimes m)\otimes k$ is determined as $\alpha^{\aleph_0}(a\otimes (b\otimes c)) = (a\otimes b)\otimes c$. The maps $\rho\colon n\times 1 \to n$ in $\aleph_0$ are identities.

This tensor $\otimes$ on $\aleph_0$ distributes over sum: the canonical distributivity map $(n\otimes m) + (n\otimes k) \to n\otimes (m+k)$ is an isomorphism. Its inverse maps $a\otimes b \in n\otimes (m+k)$ to $a\otimes b \in n\times m$ if $b < m$, and to $a\otimes (b-m) \in n\times k$ otherwise.

We thus define the objects of the category **SMLaw** to be symmetric monoidal Lawvere theories $\mathbf{L} \in \mathbf{Law}$ for which the map $\aleph_0 \to \mathbf{L}$ strictly preserves the monoidal structure that has just been described via multiplication $(\times, 1)$ of natural numbers; additionally the coproduct structure must be preserved, as in **Law**. Morphisms in **SMLaw** are morphisms in **Law** that strictly preserve this tensor structure. We note that for $\mathbf{L} \in \mathbf{SMLaw}$ we have a distributivity $n\otimes m + n\otimes k \xrightarrow{\cong} n\otimes (m+k)$, since this isomorphism lies in the range of the functor $\aleph_0 \to \mathbf{L}$.

By Lemma 7.2 we know that the Kleisli category $\mathcal{K}\ell(T)$ is symmetric monoidal if T is commutative. In order to see that the finitary Kleisli category $\mathcal{K}\ell_\mathbb{N}(T) \in \mathbf{Law}$ is also symmetric monoidal, we have to use the coordinatization map described in eqn (7.2). For $f\colon n\to p$ and $g\colon m\to q$ in $\mathcal{K}\ell_\mathbb{N}(T)$ we then obtain $f\otimes g\colon n\times m \to p\times q$ as

$$f\otimes g = \left(\underline{n\times m} \xrightarrow{\mathrm{co}} \underline{n}\times \underline{m} \xrightarrow{f\times g} T(\underline{p})\times T(\underline{q}) \xrightarrow{\mathrm{dst}} T(\underline{p}\times \underline{q}) \xrightarrow{T(\mathrm{co}^{-1})} T(\underline{p\times q})\right).$$

We recall from Kelly and Laplaza (1980) (see also Abramsky and Coecke, 2004, 2009) that for a monoidal category **C** the homset $\mathbf{C}(I,I)$ of endomaps on the tensor unit forms a commutative monoid. This applies in particular to Lawvere theories $\mathbf{L} \in \mathbf{SMLaw}$, and yields a functor $\mathcal{H}\colon \mathbf{SMLaw} \to \mathbf{Mon}$ given by $\mathcal{H}(\mathbf{L}) = \mathbf{L}(1,1)$, where $1 \in \mathbf{L}$ is the tensor unit. Thus we almost have a triangle of adjunctions as in Figure 7.3. We only need to check the following result.

Lemma 7.12 *The functor* $\mathcal{T}\colon \mathbf{Law} \to \mathbf{Mnd}$ *defined in* (7.1) *restricts to* $\mathbf{SMLaw} \to \mathbf{CMnd}$. *Further, this restriction is left adjoint to* $\mathcal{K}\ell_\mathbb{N}\colon \mathbf{CMnd} \to \mathbf{SMLaw}$.

Proof For $\mathbf{L} \in \mathbf{SMLaw}$ we define a map

$$\mathcal{T}(\mathbf{L})(X)\times \mathcal{T}(\mathbf{L})(Y) \xrightarrow{\mathrm{dst}} \mathcal{T}(\mathbf{L})(X\times Y)$$

$$\big([\kappa_i(g,v)], [\kappa_j(h,w)]\big) \mapsto [\kappa_{i\times j}(g\otimes h, (v\times w)\circ \mathrm{co}_{i,j})],$$

MONOIDS | 195

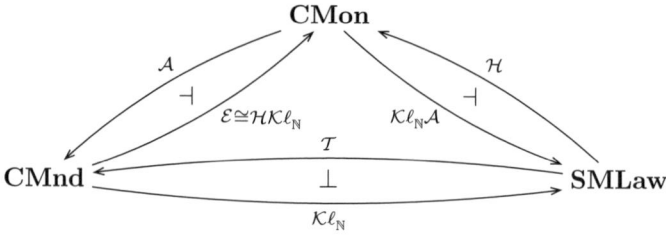

Fig 7.3 Commutative version of Figure 7.2, with commutative monoids, commutative monads and symmetric monoidal Lawvere theories.

where $g\colon 1 \to i$ and $h\colon 1 \to j$ in L yield $g \otimes h\colon 1 = 1 \otimes 1 \to i \otimes j = i \times j$, and co is the coordinatization function of eqn (7.2). Then one can show that both $\mu \circ \mathcal{T}(L)(st') \circ st$ and $\mu \circ \mathcal{T}(L)(st) \circ st'$ are equal to dst. This makes $\mathcal{T}(L)$ a commutative monad.

In order to check that the adjunction $\mathcal{T} \dashv \mathcal{K}\ell_\mathbb{N}$ restricts, we only need to verify that the unit $L \to \mathcal{K}\ell_\mathbb{N}(\mathcal{T}(L))$ strictly preserves tensors. This is easy. □

7.4 Additive monads

Having an adjunction between commutative monoids and commutative monads (Figure 7.3) raises the question of whether the category of commutative semirings is in adjunction with some specific class of monads. This is indeed the case and the monads needed are so-called additive commutative monads. Before we prove this in Section 7.5, we study additive (commutative) monads on their own and see how additivity of a monad relates to having biproducts in its Kleisli category or, equivalently, in its Eilenberg–Moore category. We get to the adjunction between commutative semirings and additive commutative monads in Section 7.5.

In this section we work with monads on a category **C** with both finite products and coproducts. If, for a monad T on **C**, the object $T(0)$ is final—i.e. satisfies $T(0) \cong 1$—then 0 is both initial and final in the Kleisli category $\mathcal{K}\ell(T)$. Such an object that is both initial and final is called a *zero object*.

Also the converse is true: if $0 \in \mathcal{K}\ell(T)$ is a zero object, then $T(0)$ is final in **C**. Although we do not use this observation in the remainder of this chapter, we also mention a related result on the category of Eilenberg–Moore algebras. The proofs are simple and are left to the reader.

Lemma 7.13 *For a monad T on a category* **C** *with finite products* $(\times, 1)$ *and coproducts* $(+, 0)$, *the following statements are equivalent:*

1. *$T(0)$ is final in* **C**
2. *$0 \in \mathcal{K}\ell(T)$ is a zero object*
3. *$1 \in \mathsf{Alg}(T)$ is a zero object.*

A zero object yields, for any pair of objects X, Y, a unique 'zero map' $0_{X,Y}: X \to 0 \to Y$ between them. In a Kleisli category $\mathcal{K}\ell(T)$ for a monad T on $\mathbf{C}$, this zero map $0_{X,Y}: X \to Y$ is the following map in $\mathbf{C}$

$$0_{X,Y} = \Big(X \xrightarrow{!_X} 1 \cong T(0) \xrightarrow{T(!_Y)} T(Y)\Big). \qquad (7.3)$$

For convenience, we make some basic properties of this zero map explicit.

Lemma 7.14 *Assume $T(0)$ is final, for a monad T on $\mathbf{C}$. The resulting zero maps $0_{X,Y}: X \to T(Y)$ from eqn (7.3) make the following diagrams in $\mathbf{C}$ commute*

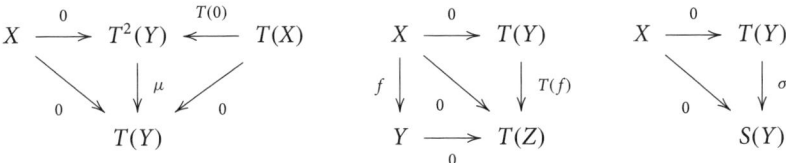

where $f: Y \to Z$ is a map in $\mathbf{C}$ and $\sigma: T \to S$ is a map of monads.

Still assuming that $T(0)$ is final, the zero map (7.3) enables us to define a canonical map

$$\mathsf{bc} \stackrel{\text{def}}{=} \Big(T(X+Y) \xrightarrow{\langle \mu \circ T(p_1), \mu \circ T(p_2)\rangle} T(X) \times T(Y)\Big), \qquad (7.4)$$

where

$$p_1 \stackrel{\text{def}}{=} \Big(X+Y \xrightarrow{[\eta, 0_{Y,X}]} T(X)\Big), \quad p_2 \stackrel{\text{def}}{=} \Big(X+Y \xrightarrow{[0_{X,Y}, \eta]} T(Y)\Big). \qquad (7.5)$$

Here we assume that the underlying category $\mathbf{C}$ has both finite products and finite coproducts. The abbreviation 'bc' stands for 'bicartesian', since this maps connects the coproducts and products. The auxiliary maps p_1, p_2 are sometimes called projections, but should not be confused with the (proper) projections π_1, π_2 associated with the product $\times$ in $\mathbf{C}$.

We continue by listing a series of properties of this map bc that will be useful in what follows.

Lemma 7.15 *In the context just described, the map* $\mathsf{bc}: T(X+Y) \to T(X) \times T(Y)$ *in (7.4) has the following properties.*

1. This bc is a natural transformation, and it commutes with any monad map $\sigma : T \to S$, as in:

$$\begin{array}{ccc} T(X+Y) & \xrightarrow{\text{bc}} & T(X) \times T(Y) \\ {\scriptstyle T(f+g)} \downarrow & & \downarrow {\scriptstyle T(f) \times T(g)} \\ T(U+V) & \xrightarrow{\text{bc}} & T(U) \times T(V) \end{array} \qquad \begin{array}{ccc} T(X+Y) & \xrightarrow{\text{bc}} & T(X) \times T(Y) \\ {\scriptstyle \sigma_{X+Y}} \downarrow & & \downarrow {\scriptstyle \sigma_X \times \sigma_Y} \\ S(X+Y) & \xrightarrow{\text{bc}} & S(X) \times S(Y) \end{array}$$

2. It also commutes with the monoidal isomorphisms (for products and coproducts in **C**):

$$\begin{array}{ccc} T(X+0) & \xrightarrow{\text{bc}} & T(X) \times T(0) \\ & {\scriptstyle T(\rho)} \searrow \cong & \cong \downarrow \rho \\ & & T(X) \end{array} \qquad \begin{array}{ccc} T(X+Y) & \xrightarrow{\text{bc}} & T(X) \times T(Y) \\ {\scriptstyle T([\kappa_2,\kappa_1])} \downarrow \cong & & \cong \downarrow {\scriptstyle \langle \pi_2, \pi_1 \rangle} \\ T(Y+X) & \xrightarrow{\text{bc}} & T(Y) \times T(X) \end{array}$$

$$\begin{array}{ccccc} T((X+Y)+Z) & \xrightarrow{\text{bc}} & T(X+Y) \times T(Z) & \xrightarrow{\text{bc} \times id} & (T(X) \times T(Y)) \times T(Z) \\ {\scriptstyle T(\alpha)} \downarrow \cong & & & & \cong \downarrow \alpha \\ T(X+(Y+Z)) & \xrightarrow{\text{bc}} & T(X) \times T(Y+Z) & \xrightarrow{id \times \text{bc}} & T(X) \times (T(Y) \times T(Z)) \end{array}$$

3. The map bc interacts with η and μ in the following manner:

$$\begin{array}{ccc} & X+Y & \\ {\scriptstyle \eta} \downarrow & \searrow {\scriptstyle \langle p_1, p_2 \rangle} & \\ T(X+Y) & \xrightarrow[\text{bc}]{} & T(X) \times T(Y) \end{array}$$

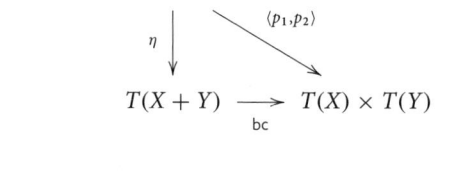

4. If **C** is a distributive category, bc commutes with strength st as follows:

$$\begin{array}{ccccc}
T(X+Y) \times Z & \xrightarrow{\text{bc} \times \text{id}} & (T(X) \times T(Y)) \times Z & \xrightarrow{\text{dbl}} & (T(X) \times Z) \times (T(Y) \times Z) \\
\text{st} \downarrow & & & & \downarrow \text{st} \times \text{st} \\
T((X+Y) \times Z) & \xrightarrow{\cong} & T((X \times Z) + (Y \times Z)) & \xrightarrow{\text{bc}} & T(X \times Z) \times T(Y \times Z)
\end{array}$$

where dbl is the 'double' map $\langle \pi_1 \times \text{id}, \pi_2 \times \text{id}\rangle \colon (A \times B) \times C \to (A \times C) \times (B \times C)$.

Proof These properties are easily verified, using Lemma 7.14 and the fact that the projections p_i are natural, both in **C** and in $\mathcal{K}\ell(T)$. □

The definition of the map bc also makes sense for arbitrary set-indexed (co)products (see Jacobs 2010), but here we only consider finite ones. Such generalized bc-maps also satisfy (suitable generalizations of) the properties in Lemma 7.15 above.

We will study monads for which the canonical map bc is an isomorphism. Such monads will be called 'additive monads'.

Definition 7.16 A monad T on a category **C** with finite products $(\times, 1)$ and finite coproducts $(+, 0)$ will be called additive if $T(0) \cong 1$ and if the canonical map bc: $T(X+Y) \to T(X) \times T(Y)$ from (7.4) is an isomorphism.

We write **AMnd(C)** for the category of additive monads on **C** with monad morphism between them, and similarly **ACMnd(C)** for the category of additive and commutative monads on **C**. In Kock 2011 the phrase 'T takes finite coproducts to products' is used for additivity of T.

Example 7.17 The powerset monad $\mathcal{P}$ on **Sets** is an example of an additive monad: the (well-known) isomorphisms $\mathcal{P}(0) \cong 1$ and $\mathcal{P}(X+Y) \cong \mathcal{P}(X) \times \mathcal{P}(Y)$ are instances of the above map bc. Similarly, the downset monad on the category of posets is an example. But our main examples of additive monads are the 'multiset' monads introduced in Section 7.5.

A basic result is that additive monads T induce a commutative monoid structure on objects $T(X)$. This result is sometimes taken as definition of additivity of monads (cf. Goncharov, Schröder and Mossakowski, 2009).

Lemma 7.18 Let T be an additive monad on a category **C** and X an object of **C**. There is an addition $+$ on $T(X)$ given by

$$+ \stackrel{\text{def}}{=} \left(T(X) \times T(X) \xrightarrow{\text{bc}^{-1}} T(X+X) \xrightarrow{T(\nabla)} T(X) \right),$$

where $\nabla = [\mathrm{id}, \mathrm{id}]$. Then:

1. this $+$ is commutative and associative,
2. and has unit $0_{1,X}: 1 \to T(X)$;
3. this monoid structure is preserved by maps $T(f)$ as well as by multiplication μ;
4. the mapping $(T,X) \mapsto (T(X), +, 0_{1,X})$ yields a functor $\mathcal{A}d\colon \mathbf{AMnd}(\mathbf{C}) \times \mathbf{C} \to \mathbf{CMon}(\mathbf{C})$.

Proof The first three statements follow by the properties of bc from Lemma 7.15. For instance, 0 is a (right) unit for $+$ as demonstrated in the following diagram.

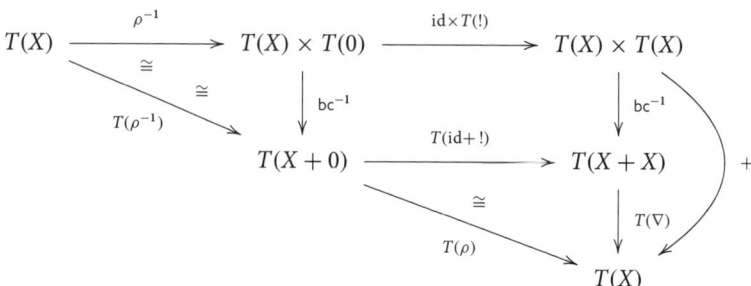

Regarding (iv) we define, for a pair of morphisms $\sigma\colon T \to S$ in $\mathbf{AMnd}(\mathbf{C})$ and $f\colon X \to Y$ in $\mathbf{C}$,

$$\mathcal{A}d((\sigma, f)) = \sigma \circ T(f)\colon T(X) \to S(Y),$$

which is equal to $S(f) \circ \sigma$ by naturality of σ. Preservation of the unit by $\mathcal{A}d((\sigma, f))$ follows from Lemma 7.14. The following diagram demonstrates that addition is preserved.

where we use point (i) of Lemma 7.15 and the naturality of σ. It is easily checked that this mapping defines a functor. □

By Lemma 7.2, for a monad T on a category $\mathbf{C}$ with finite coproducts, the Kleisli construction yields a category $\mathcal{K}\ell(T)$ with finite coproducts. Below we will prove that, under the assumption that $\mathbf{C}$ also has products, these coproducts form biproducts in $\mathcal{K}\ell(T)$ if and only if T is additive. Again, as in Lemma 7.13, a related result holds for the category $\mathsf{Alg}(T)$.

Definition 7.19 *A category with biproducts is a category $\mathbf{C}$ with a zero object $0 \in \mathbf{C}$, such that, for any pair of objects $A_1, A_2 \in \mathbf{C}$, there is an object $A_1 \oplus A_2 \in \mathbf{C}$ that is both a product with projections $\pi_i: A_1 \oplus A_2 \to A_i$ and a coproduct with coprojections $\kappa_i: A_i \to A_1 \oplus A_2$, such that*

$$\pi_j \circ \kappa_i = \begin{cases} \mathrm{id}_{A_i} & \text{if } i = j \\ 0_{A_i, A_j} & \text{if } i \neq j. \end{cases}$$

Theorem 7.20 *For a monad T on a category $\mathbf{C}$ with finite products $(\times, 1)$ and coproducts $(+, 0)$, the following are equivalent.*

1. *T is additive.*
2. *The coproducts in $\mathbf{C}$ form biproducts in the Kleisli category $\mathcal{K}\ell(T)$.*
3. *The products in $\mathbf{C}$ yield biproducts in the category of Eilenberg-Moore algebras $\mathsf{Alg}(T)$.*

Here we shall only use this result for Kleisli categories, but we include the result for algebras for completeness.

Proof First we assume that T is additive and show that $(+, 0)$ is a product in $\mathcal{K}\ell(T)$. As projections we take the maps p_i from eqn (7.5). For Kleisli maps $f: Z \to T(X)$ and $g: Z \to T(Y)$ there is a tuple via the map bc, as in

$$\langle f, g \rangle_{\mathcal{K}\ell} \stackrel{\text{def}}{=} \Big(Z \xrightarrow{\langle f, g \rangle} T(X) \times T(Y) \xrightarrow{\mathrm{bc}^{-1}} T(X+Y) \Big).$$

One obtains $p_1 \bullet \langle f, g \rangle_{\mathcal{K}\ell} = \mu \circ T(p_1) \circ \mathrm{bc}^{-1} \circ \langle f, g \rangle = \pi_1 \circ \mathrm{bc} \circ \mathrm{bc}^{-1} \circ \langle f, g \rangle = \pi_1 \circ \langle f, g \rangle = f$. The remaining details are left to the reader.

Conversely, assuming that the coproduct $(+, 0)$ in $\mathbf{C}$ forms a biproduct in $\mathcal{K}\ell(T)$, we have to show that the bicartesian map $\mathrm{bc}: T(X+Y) \to T(X) \times T(Y)$ is an isomorphism. As $+$ is a biproduct, there exist projection maps $q_i: X_1 + X_2 \to X_i$ in $\mathcal{K}\ell(T)$ satisfying

$$q_j \bullet \kappa_i = \begin{cases} \mathrm{id}_{X_i} & \text{if } i = j \\ 0_{X_i, X_j} & \text{if } i \neq j. \end{cases}$$

From these conditions it follows that $q_i = p_i$, where p_i is the map defined in eqn (7.5). The ordinary projection maps $\pi_i: T(X_1) \times T(X_2) \to T(X_i)$ are maps

$T(X_1) \times T(X_2) \to X_i$ in $\mathcal{K}\ell(T)$. Hence, as $+$ is a product, there exists a unique map $h\colon T(X_1) \times T(X_2) \to X_1 + X_2$ in $\mathcal{K}\ell(T)$, i.e. $h\colon T(X_1) \times T(X_2) \to T(X_1 + X_2)$ in **C**, such that $p_1 \bullet h = \pi_1$ and $p_2 \bullet h = \pi_2$. It is readily checked that this map h is the inverse of bc.

To prove the equivalence of (1) and (3), first assume that the monad T is additive. In the category $\mathsf{Alg}(T)$ of algebras there is the standard product

$$\left(T(X) \xrightarrow{\alpha} X\right) \times \left(T(Y) \xrightarrow{\beta} Y\right) \stackrel{\text{def}}{=} \left(T(X \times Y) \xrightarrow{\langle \alpha \circ T(\pi_1), \beta \circ T(\pi_2)\rangle} X \times Y\right).$$

In order to show that $\times$ also forms a coproduct in $\mathsf{Alg}(T)$, we first show that for an arbitrary algebra $\gamma\colon T(Z) \to Z$ the object Z carries a commutative monoid structure. We do so by adapting the structure $(+, 0)$ on $T(Z)$ from Lemma 7.18 to $(+_Z, 0_Z)$ on Z via

$$+_Z \stackrel{\text{def}}{=} \left(Z \times Z \xrightarrow{\eta \times \eta} T(Z) \times T(Z) \xrightarrow{+} T(Z) \xrightarrow{\gamma} Z\right)$$

$$0_Z \stackrel{\text{def}}{=} \left(1 \xrightarrow{0} T(Z) \xrightarrow{\gamma} Z\right)$$

This monoid structure is preserved by homomorphisms of algebras. Now, we can form coprojections $k_1 = \langle \mathsf{id}, 0_Y \circ\, !\rangle\colon X \to X \times Y$, and a cotuple of algebra homomorphisms $(TX \xrightarrow{\alpha} X) \xrightarrow{f} (TZ \xrightarrow{\gamma} Z)$ and $(TY \xrightarrow{\beta} X) \xrightarrow{g} (TZ \xrightarrow{\gamma} Z)$ given by

$$[f, g]_{\mathsf{Alg}} \stackrel{\text{def}}{=} \left(X \times Y \xrightarrow{f \times g} Z \times Z \xrightarrow{+_Z} Z\right).$$

Again, remaining details are left to the reader.

Finally, to show that (3) implies (1), consider the algebra morphisms:

$$\left(T^2(X_i) \xrightarrow{\mu} T(X_i)\right) \xrightarrow{T(\kappa_i)} \left(T^2(X_1 + X_2) \xrightarrow{\mu} T(X_1 + X_2)\right).$$

The free functor $\mathbf{C} \to \mathsf{Alg}(T)$ preserves coproducts, so these $T(\kappa_i)$ form a coproduct diagram in $\mathsf{Alg}(T)$. As $\times$ is a coproduct in $\mathsf{Alg}(T)$, by assumption, the cotuple $[T(\kappa_1), T(\kappa_2)]\colon T(X_1) \times T(X_2) \to T(X_1 + X_2)$ in $\mathsf{Alg}(T)$ is an isomorphism. The coprojections $\ell_i\colon T(X_i) \to T(X_1) \times T(X_2)$ satisfy $\ell_1 = \langle \pi_1 \circ \ell_1, \pi_2 \circ \ell_2\rangle = \langle \mathsf{id}, 0\rangle$, and similarly, $\ell_2 = \langle 0, \mathsf{id}\rangle$. Now we compute:

$$\begin{aligned}
\mathsf{bc} \circ [T(\kappa_1), T(\kappa_2)] \circ \ell_1 &= \langle \mu \circ T(p_1), \mu \circ T(p_2)\rangle \circ T(\kappa_1) \\
&= \langle \mu \circ T(p_1 \circ \kappa_1), \mu \circ T(p_2 \circ \kappa_1)\rangle \\
&= \langle \mu \circ T(\eta), \mu \circ T(0)\rangle \\
&= \langle \mathsf{id}, 0\rangle \\
&= \ell_1.
\end{aligned}$$

Similarly, bc ∘ $[T(\kappa_1), T(\kappa_2)] \circ \ell_2 = \ell_2$, so that bc ∘ $[T(\kappa_1), T(\kappa_2)] = \mathrm{id}$, making bc an isomorphism. □

It is well-known (see for instance Kelly and Laplaza, 1980, and Abramsky and Coecke, 2004) that a category with finite biproducts $(\oplus, 0)$ is enriched over commutative monoids: each homset carries a commutative monoid structure $(+, 0)$, and this structure is preserved by pre- and post-composition. The addition operation $+$ on homsets is obtained as

$$f + g \stackrel{\text{def}}{=} \left(X \xrightarrow{\langle \mathrm{id}, \mathrm{id}\rangle} X \oplus X \xrightarrow{f \oplus g} Y \oplus Y \xrightarrow{[\mathrm{id}, \mathrm{id}]} Y \right). \tag{7.6}$$

The zero map is neutral element for this addition. One can also describe a monoid structure on each object X as

$$X \oplus X \xrightarrow{[\mathrm{id}, \mathrm{id}]} X \xleftarrow{0} 0. \tag{7.7}$$

We have just seen that the Kleisli category of an additive monad has biproducts, using the addition operation from Lemma 7.18. When we apply the sum description of eqn (7.7) to the biproducts of such a Kleisli category, we obtain precisely the original addition from Lemma 7.18, since the codiagonal $\nabla = [\mathrm{id}, \mathrm{id}]$ in the Kleisli category is given by $T(\nabla) \circ \mathrm{bc}^{-1}$.

7.4.1 Additive commutative monads

In the remainder of this section we focus on the category **ACMnd**(C) of monads that are both additive and commutative on a distributive category **C**. As usual, we simply write **ACMnd** for **ACMnd**(**Sets**). For $T \in \mathbf{ACMnd}(\mathbf{C})$, the Kleisli category $\mathcal{K}\ell(T)$ is both symmetric monoidal—with $(\times, 1)$ as monoidal structure; see Lemma 7.2—and has biproducts $(+, 0)$. Moreover, it is not hard to see that this monoidal structure distributes over the biproducts via the canonical map $(Z \times X) + (Z \times Y) \to Z \times (X + Y)$ that can be lifted from **C** to $\mathcal{K}\ell(T)$.

We shall write **SMBLaw** ↪ **SMLaw** for the category of symmetric monoidal Lawvere theories in which $(+, 0)$ form not only coproducts but biproducts. Notice that a projection $\pi_1 \colon n + m \to n$ is necessarily of the form $\pi_1 = [\mathrm{id}, 0]$, where $0 \colon m \to n$ is the zero map $m \to 0 \to n$. The tensor $\otimes$ distributes over $(+, 0)$ in **SMBLaw**, as it already does so in **SMLaw**. Morphisms in **SMBLaw** are functors that strictly preserve all the structure.

The following result extends Corollary 7.3.

Lemma 7.21 *The (finitary) Kleisli construction on a monad yields a functor* $\mathcal{K}\ell_\mathbb{N} \colon \mathbf{ACMnd} \to \mathbf{SMBLaw}$.

Proof It follows from Theorem 7.20 that $(+, 0)$ form biproducts in $\mathcal{K}\ell_\mathbb{N}(T)$, for T an additive commutative monad (on **Sets**). This structure is preserved by functors $\mathcal{K}\ell_\mathbb{N}(\sigma)$, for $\sigma \colon T \to S$ in **ACMnd**. □

We have already seen in Lemma 7.12 that the functor $\mathcal{T}\colon \mathbf{Law} \to \mathbf{Mnd}$ defined in Lemma 7.7 restricts to a functor between symmetric monoidal Lawvere theories and commutative monads. We now show that it also restricts to a functor between symmetric monoidal Lawvere theories with biproducts and commutative additive monads. Again, this restriction is left adjoint to the finitary Kleisli construction.

Lemma 7.22 *The functor* $\mathcal{T}\colon \mathbf{SMLaw} \to \mathbf{CMnd}$ *from Lemma 7.12 restricts to* $\mathbf{SMBLaw} \to \mathbf{ACMnd}$. *Further, this restriction is left adjoint to the finitary Kleisli construction* $\mathcal{K}\ell_\mathbb{N}\colon \mathbf{ACMnd} \to \mathbf{SMBLaw}$.

Proof First note that $T_L(0)$ is final:

$$T_L(0) = \coprod_i L(1,i) \times 0^i \cong L(1,0) \times 0^0 \cong 1,$$

where the last isomorphism follows from the fact that $(+, 0)$ is a biproduct in L and hence 0 is terminal. The resulting zero map $0_{X,Y}\colon X \to T(Y)$ is given by

$$x \mapsto [\kappa_0(!\colon 1 \to 0, !\colon 0 \to Y)].$$

To prove that the bicartesian map $\mathrm{bc}\colon T_L(X+Y) \to T_L(X) \times T_L(Y)$ is an isomorphism, we introduce some notation. For $[\kappa_i(g,v)] \in T_L(X+Y)$, where $g\colon 1 \to i$ and $v\colon i \to X+Y$, we form the pullbacks (in **Sets**)

$$\begin{array}{ccccc} i_X & \longrightarrow & i & \longleftarrow & i_Y \\ {\scriptstyle v_X}\downarrow & & \downarrow{\scriptstyle v} & & \downarrow{\scriptstyle v_Y} \\ X & \xrightarrow{\kappa_1} & X+Y & \xleftarrow{\kappa_2} & Y \end{array}$$

By universality of coproducts we can write $i = i_X + i_Y$ and $v = v_X + v_Y\colon i_X + i_Y \to X+Y$. Then we can also write $g = \langle g_X, g_Y \rangle\colon 1 \to i_X + i_Y$. Hence, for $[\kappa_i(g,v)] \in T_L(X+Y)$,

$$\mathrm{bc}([\kappa_i(g,v)]) = \big([\kappa_{i_X}(g_X, v_X)], [\kappa_{i_Y}(g_Y, v_Y)]\big). \tag{7.8}$$

It then easily follows that the map $T_L(X) \times T_L(Y) \to T_L(X+Y)$ defined by

$$([\kappa_i(g,v)], [\kappa_j(h,w)]) \mapsto [\kappa_{i+j}(\langle g, h \rangle, v+w)]$$

is the inverse of bc.

Checking that the unit of the adjunction $\mathcal{T}\colon \mathbf{SMLaw} \leftrightarrows \mathbf{CMnd}\colon \mathcal{K}\ell_\mathbb{N}$ preserves the product structure is left to the reader. This proves also that the restricted functors form an adjunction. □

In the next two sections we will see how additive commutative monads and symmetric monoidal Lawvere theories with biproducts relate to commutative semirings.

7.5 Semirings and monads

As announced above, we prove in this section that there is an adjunction between commutative semirings and additive commutative monads. We begin with some clarifications about semirings and modules. To describe the adjunction, we show that a commutative semiring gives rise to a 'multiset' monad, which is both commutative and additive. Conversely, we will see that the 'evaluate at unit 1'-functor yields, for each additive commutative monad, a commutative semiring.

A commutative semiring in **Sets** consists of a set S together with two commutative monoid structures, one additive $(+, 0)$ and one multiplicative $(\cdot, 1)$, where the latter distributes over the former: $s \cdot 0 = 0$ and $s \cdot (t + r) = s \cdot t + s \cdot r$. For more information on semirings, see Golan (1999). Here we only consider commutative ones. Typical examples are the natural numbers $\mathbb{N}$, or the non-negative rationals $\mathbb{Q}_{\geq 0}$, or the reals $\mathbb{R}_{\geq 0}$.

One way to describe semirings categorically is by considering the additive monoid $(S, +, 0)$ as an object of the category **CMon** of commutative monoids, carrying a multiplicative monoid structure $I \xrightarrow{1} S \leftarrow S \otimes S$ in this category **CMon**. The tensor guarantees that multiplication is a bihomomorphism, and thus distributes over additions.

In the present context of categories with finite products we do not need to use these tensors and can give a direct categorical formulation of such semirings as a pair of monoids $1 \xrightarrow{0} S \xleftarrow{+} S \times S$ and $1 \xrightarrow{1} S \xleftarrow{\cdot} S \times S$ making the following distributivity diagrams commute.

$$
\begin{array}{ccc}
S \times 1 \xrightarrow{\mathrm{id} \times 0} S \times S & (S \times S) \times S \xrightarrow{dbl} (S \times S) \times (S \times S) \xrightarrow{\cdot \times \cdot} S \times S \\
\downarrow{!} \quad \quad \downarrow{\cdot} & \quad \downarrow{+ \times \mathrm{id}} \quad \quad \quad \downarrow{+} \\
1 \xrightarrow{0} S & S \times S \xrightarrow{\hspace{4cm}} S
\end{array}
$$

where $dbl = \langle \pi_1 \times \mathrm{id}, \pi_2 \times \mathrm{id} \rangle$ is the doubling map that was also used in Lemma 7.15. With the obvious notion of homomorphism between semirings this yields a category **CSRng(C)** of (commutative) semirings in a category **C** with finite products.

Associated with a semiring S there is a notion of module over S. It consists of a commutative monoid $(M, 0, +)$ together with a (multiplicative) action $\star \colon S \times M \to M$, which is an additive bihomomorphism; that is, the action preserves the additive structure in each argument separately. We recall that the properties of an action are given categorically by $\star \circ (\cdot \times \mathrm{id}) = \star \circ (\mathrm{id} \times \star) \circ \alpha^{-1} \colon (S \times S) \times M \to M$ and

$\star \circ (1 \times \mathrm{id}) = \pi_2 \colon 1 \times M \to M$. The fact that $\star$ is an additive bihomomorphism is expressed by

$$
\begin{array}{ccccc}
S \times (M \times M) & \xrightarrow{dbl'} & (S \times M) \times (S \times M) & \xleftarrow{dbl} & (S \times S) \times M \\
\downarrow{\scriptstyle \mathrm{id} \times +} & & \downarrow{\scriptstyle \star \times \star} & & \downarrow{\scriptstyle + \times \mathrm{id}} \\
 & & M \times M & & \\
 & & \downarrow{\scriptstyle +} & & \\
S \times M & \xrightarrow{\ \star\ } & M & \xleftarrow{\ \star\ } & S \times M
\end{array}
$$

where dbl' is the obvious duplicator of S. Preservation of zeros is simply $\star \circ (0 \times \mathrm{id}) = 0 \circ \pi_1 \colon 1 \times M \to M$ and $\star \circ (\mathrm{id} \times 0) = 0 \circ \pi_2 \colon S \times 1 \to M$.

We shall assemble such semirings and modules in one category $\mathcal{M}od(\mathbf{C})$ with triples $(S, M, \star)$ as objects, where $\star \colon S \times M \to M$ is an action as above. A morphism $(S_1, M_1, \star_1) \to (S_2, M_2, \star_2)$ consists of a pair of morphisms $f \colon S_1 \to S_2$ and $g \colon M_1 \to M_2$ in $\mathbf{C}$ such that f is a map of semirings, f is a map of monoids, and the actions interact appropriately: $\star_2 \circ (f \times g) = g \circ \star_1$.

7.5.1 From semirings to monads

To construct an adjunction between semirings and additive commutative monads we start by defining, for each commutative semiring S, the so-called multiset monad on S and show that this monad is both commutative and additive.

Definition 7.23 *For a semiring S, define a 'multiset' functor $M_S \colon \mathbf{Sets} \to \mathbf{Sets}$ on a set X by*

$$M_S(X) = \{\varphi \colon X \to S \mid \mathrm{supp}(\varphi) \text{ is finite}\},$$

where $\mathrm{supp}(\varphi) = \{x \in X \mid \varphi(x) \neq 0\}$ *is called the support of φ. For a function $f \colon X \to Y$ one defines $M_S(f) \colon M_S(X) \to M_S(Y)$ by:*

$$M_S(f)(\varphi)(y) = \sum_{x \in f^{-1}(y)} \varphi(x).$$

Such a multiset $\varphi \in M_S(X)$ may be written as formal sum $s_1 x_1 + \cdots + s_k x_k$, where $\mathrm{supp}(\varphi) = \{x_1, \ldots, x_k\}$ and $s_i = \varphi(x_i) \in S$ describes the 'multiplicity' of the element x_i. In this notation one can write the application of M_S on a map f as $M_S(f)(\sum_i s_i x_i) = \sum_i s_i f(x_i)$. Functoriality is then obvious.

Lemma 7.24 *For each semiring S, the multiset functor M_S forms a commutative and additive monad, with unit and multiplication:*

$$X \xrightarrow{\eta} M_S(X) \qquad\qquad M_S(M_S(X)) \xrightarrow{\mu} M_S(X)$$

$$x \longmapsto 1x \qquad\qquad \sum_i s_i \varphi_i \longmapsto \lambda x \in X.\ \sum_i s_i \varphi_i(x).$$

Proof The verification that M_S with these η and μ indeed forms a monad is left to the reader. We mention that for commutativity and additivity the relevant maps are given by:

$$M_S(X) \times M_S(Y) \xrightarrow{\text{dst}} M_S(X \times Y) \qquad\qquad M_S(X+Y) \xrightarrow{\text{bc}} M_S(X) \times M_S(Y)$$

$$(\varphi, \psi) \longmapsto \lambda(x,y).\,\varphi(x) \cdot \psi(y) \qquad\qquad \chi \longmapsto (\chi \circ \kappa_1, \chi \circ \kappa_2).$$

Clearly, bc is an isomorphism, making M_S additive. □

Lemma 7.25 *The assignment $S \mapsto M_S$ yields a functor $\mathcal{M}$: CSRng $\to$ ACMnd.*

Proof Every semiring homomorphism $f\colon S \to R$, gives rise to a monad morphism $\mathcal{M}(f)\colon M_S \to M_R$ with components defined by $\mathcal{M}(f)_X(\sum_i s_i x_i) = \sum_i f(s_i) x_i$. It is left to the reader to check that $\mathcal{M}(f)$ is indeed a monad morphism. □

For a semiring S, the category $\mathsf{Alg}(M_S)$ of algebras of the multiset monad M_S is (equivalent to) the category $\mathcal{M}od_S(\mathbf{C}) \hookrightarrow \mathcal{M}od(\mathbf{C})$ of modules over S.

7.5.2 From monads to semirings

A commutative additive monad T on a category $\mathbf{C}$ gives rise to two commutative monoid structures on $T(1)$, namely the multiplication defined in Lemma 7.10 and the addition defined in Lemma 7.18 (considered for $X = 1$). In case the category $\mathbf{C}$ is distributive these two operations turn $T(1)$ into a semiring.

Lemma 7.26 *Each commutative additive monad T on a distributive category $\mathbf{C}$ with terminal object 1 gives rise to a semiring $\mathcal{E}(T) = T(1)$ in $\mathbf{C}$. The mapping $T \mapsto \mathcal{E}(T)$ yields a functor* **ACMnd(C)** $\to$ **CSRng(C)**.

Proof For a commutative additive monad T on $\mathbf{C}$, addition on $T(1)$ is given by $T(\nabla) \circ \mathsf{bc}^{-1}\colon T(1) \times T(1) \to T(1)$ with unit $0_{1,1}\colon 1 \to T(1)$ as in Lemma 7.18 and the multiplication is given by $\mu \circ T(\lambda) \circ \mathsf{st}\colon T(1) \times T(1) \to T(1)$ with unit $\eta_1\colon 1 \to T(1)$ as in Lemma 7.10.

It was shown in the lemmas just mentioned that both addition and multiplication define a commutative monoid structure on $T(1)$. The following diagram proves distributivity of multiplication over addition.

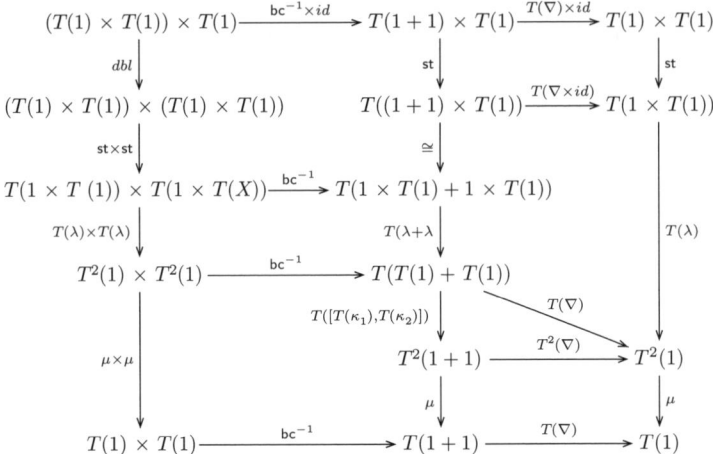

Here we rely on Lemma 7.15 for the commutativity of the upper and lower square on the left.

In a distributive category $0 \cong 0 \times X$, for every object X. In particular $T(0 \times T(1)) \cong T(0) \cong 1$ is final. This is used to obtain commutativity of the upper-left square of the following diagram proving $0 \cdot s = 0$:

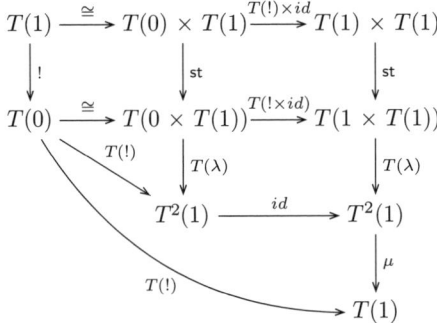

For a monad morphism $\sigma \colon T \to S$, we define $\mathcal{E}(\sigma) = \sigma_1 \colon T(1) \to S(1)$. By Lemma 7.5, σ_1 commutes with the multiplicative structure. As $\sigma_1 = T(id) \circ \sigma_1 = \mathcal{A}d((\sigma, id))$, it follows from Lemma 7.18 that σ_1 also commutes with the additive structure and is therefore a **CSRng**-homomorphism. □

7.5.3 Adjunction between monads and semirings

The functors defined in the Lemmas 7.25 and 7.26, considered on $\mathbf{C} = \mathbf{Sets}$, form an adjunction $\mathcal{M} \colon \mathbf{CSRng} \leftrightarrows \mathbf{ACMnd} \colon \mathcal{E}$. To prove this adjunction we first show that each pair (T, X), where T is a commutative additive monad on a category $\mathbf{C}$

and X an object of C, gives rise to a module on C as defined at the beginning of this section.

Lemma 7.27 *Each pair (T, X), where T is a commutative additive monad on a category $\mathbf{C}$ and X is an object of $\mathbf{C}$, gives rise to a module $\mathcal{M}od(T, X) = (T(1), T(X), \star)$. Here $T(1)$ is the commutative semiring defined in Lemma 7.26 and $T(X)$ is the commutative monoid defined in Lemma 7.18. The action map is given by $\star = T(\lambda) \circ dst\colon T(1) \times T(X) \to T(X)$. The mapping $(T, X) \mapsto \mathcal{M}od(T, X)$ yields a functor $\mathbf{ACMnd}(\mathbf{C}) \times \mathbf{C} \to \mathcal{M}od(\mathbf{C})$.*

Proof Checking that $\star$ defines an appropriate action requires some work but is essentially straightforward, using the properties from Lemma 7.15. For a pair of maps $\sigma\colon T \to S$ in $\mathbf{ACMnd}(\mathbf{C})$ and $g\colon X \to Y$ in $\mathbf{C}$, we define a map $\mathcal{M}od(\sigma, g)$ by

$$(T(1), T(X), \star^T) \xrightarrow{(\sigma_1, \sigma_Y \circ T(g))} (S(1), S(Y), \star^S).$$

Note that, by naturality of σ, one has $\sigma_Y \circ T(g) = S(g) \circ \sigma_X$. It easily follows that this defines a $\mathcal{M}od(\mathbf{C})$-map and that the assignment is functorial. $\square$

Lemma 7.28 *The pair of functors $\mathcal{M}\colon \mathbf{CSRng} \leftrightarrows \mathbf{ACMnd}\colon \mathcal{E}$ forms an adjunction, $\mathcal{M} \dashv \mathcal{E}$.*

Proof For a semiring S and a commutative additive monad T on **Sets** there are (natural) bijective correspondences:

$$\dfrac{M_S = \mathcal{M}(S) \xrightarrow{\sigma} T \quad \text{in } \mathbf{CAMnd}}{S \xrightarrow{f} \mathcal{E}(T) = T(1) \quad \text{in } \mathbf{CSRng}}$$

Given $\sigma\colon M_S \to T$, one defines a semiring map $\overline{\sigma}\colon S \to T(1)$ by

$$\overline{\sigma} = \Big(S \xrightarrow{\lambda s.(\lambda x.s)} M_S(1) \xrightarrow{\sigma_1} T(1) \Big).$$

Conversely, given a semiring map $f\colon S \to T(1)$, one gets a monad map $\overline{f}\colon M_S \to T$ with components:

$$M_S(X) \xrightarrow{\overline{f}_X} T(X) \quad \text{given by} \quad \sum_i s_i x_i \mapsto \sum_i f(s_i) \star \eta(x_i),$$

where the sum on the right-hand side is the addition in $T(X)$ defined in Lemma 7.18 and $\star$ is the action of $T(1)$ on $T(X)$ defined in Lemma 7.27.

Showing that $\bar{f}$ is indeed a monad morphism requires some work. In doing so one may rely on the properties of the action and on Lemma 7.18. The details are left to the reader. □

Notice that the counit of the above adjunction $\mathcal{E}\mathcal{M}(S) = M_S(1) \to S$ is an isomorphism. Hence this adjunction is in fact a reflection.

7.6 Semirings and Lawvere theories

In this section we extend the adjunction between commutative monoids and symmetric monoidal Lawvere theories depicted in Figure 7.3 to an adjunction between commutative semirings and symmetrical monoidal Lawvere theories with biproducts, i.e. between the categories **CSRng** and **SMBLaw**.

7.6.1 From semirings to Lawvere theories

Composing the multiset functor $\mathcal{M}\colon$ **CSRng** $\to$ **ACMnd** from the previous section with the finitary Kleisli construction $\mathcal{K}\ell_\mathbb{N}$ yields a functor from **CSRng** to **SMBLaw**. This functor may be described in an alternative (isomorphic) way by assigning to every semiring S the Lawvere theory of matrices over S, which is defined as follows.

Definition 7.29 *For a semiring S, the Lawvere theory $\mathcal{M}at(S)$ of matrices over S has, for $n, m \in \mathbb{N}$ morphisms (in **Sets**) $n \times m \to S$, i.e. $n \times m$ matrices over S, as morphisms $n \to m$. The identity $\mathrm{id}_n\colon n \to n$ is given by the identity matrix:*

$$\mathrm{id}_n(i,j) = \begin{cases} 1 & \text{if } i = j \\ 0 & \text{if } i \neq j. \end{cases}$$

The composition of $g\colon n \to m$ and $h\colon m \to p$ is given by matrix multiplication:

$$(h \circ g)(i,k) = \sum_j g(i,j) \cdot h(j,k).$$

The coprojections $\kappa_1\colon n \to n+m$ and $\kappa_2\colon m \to n+m$ are given by

$$\kappa_1(i,j) = \begin{cases} 1 & \text{if } i = j \\ 0 & \text{otherwise}. \end{cases} \qquad \kappa_2(i,j) = \begin{cases} 1 & \text{if } j \geq n \text{ and } j - n = i \\ 0 & \text{otherwise}. \end{cases}$$

Lemma 7.30 *The assignment $S \mapsto \mathcal{M}at(S)$ yields a functor $\mathbf{CSRng} \to \mathbf{Law}$. The two functors $\mathcal{M}at\mathcal{E}$ and $\mathcal{K}\ell_\mathbb{N}\colon$ **ACMnd** $\to$ **Law** are naturally isomorphic.*

Proof A map of semirings $f\colon S \to R$ gives rise to a functor $\mathcal{M}at(f)\colon \mathcal{M}at(S) \to \mathcal{M}at(R)$, which is the identity on objects and which acts on morphisms by post-composition: $h\colon n \times m \to S$ in $\mathcal{M}at(S)$ is mapped to $f \circ h\colon n \times m \to T$ in

$Mat(T)$. It is easily checked that $Mat(f)$ is a morphism of Lawvere theories and that the assignment is functorial.

To prove the second claim we define two natural transformations. First we define $\xi \colon Mat\mathcal{E} \to \mathcal{K}\ell_{\mathbb{N}}$ with components $\xi_T \colon Mat(T(1)) \to \mathcal{K}\ell_{\mathbb{N}}(T)$, which are the identity on objects and send a morphism $h \colon n \times m \to T(1)$ in $Mat(T(1))$ to the morphism $\xi_T(h)$ in $\mathcal{K}\ell_{\mathbb{N}}(T)$ given by

$$\xi_T(h) = \left(n \xrightarrow{\langle h(\_,j) \rangle_{j \in m}} T(1)^m \xrightarrow{bc_m^{-1}} T(m) \right),$$

where bc_m^{-1} is the inverse of the generalized bicartesian map

$$bc_m = \left(T(m) = T(\coprod_m 1) \longrightarrow T(1)^m \right).$$

And secondly, in the reverse direction, we define $\theta \colon \mathcal{K}\ell_{\mathbb{N}} \to Mat\mathcal{E}$ with components $\theta_T \colon \mathcal{K}\ell_{\mathbb{N}}(T) \to Mat(T(1))$, which are the identity on objects and send a morphism $g \colon n \to T(m)$ in $\mathcal{K}\ell_{\mathbb{N}}(T)$ to the morphism $\theta_T(g) \colon n \times m \to T(1)$ in $Mat(T(1))$ given by

$$\theta_T(g)(i,j) = (\pi_j \circ bc_m \circ g)(i). \tag{7.9}$$

It requires some work, but is relatively straightforward to check that the components ξ_T and θ_T are **Law**-maps. To prove preservation of the composition by ξ_T and θ_T one uses the definition of addition and multiplication in $T(1)$ and (generalizations of) the properties of the map bc listed in Lemma 7.15. A short computation shows that the functors are each other's inverses. The naturality of both ξ and θ follows from (a generalization of) point (i) of Lemma 7.15. □

The pair of functors $\mathcal{M} \colon \mathbf{CSRng} \leftrightarrows \mathbf{ACMnd} \colon \mathcal{E}$ forms a reflection, $\mathcal{EM} \cong id$ (Lemma 7.28). Combining this with the previous proposition, it follows that also the functors $Mat, \mathcal{K}\ell_{\mathbb{N}}\mathcal{M} \colon \mathbf{CSRng} \to \mathbf{Law}$ are naturally isomorphic. Hence, the functor $Mat \colon \mathbf{CSRng} \to \mathbf{Law}$ may be viewed as a functor from commutative semirings to symmetric monoidal Lawvere theories with biproducts. For a commutative semiring S the projection maps $\pi_1 \colon n + m \to n$ and $\pi_2 \colon n + m \to m$ in $Mat(S)$ are defined in a similar way as the coprojection maps from Definition 7.29. For a pair of maps $g \colon m \to p$, $h \colon n \to q$, the tensor product $g \otimes h \colon (m \times n) \to (p \times q)$ is the map $g \otimes h \colon (m \times n) \times (p \times q) \to S$ defined as

$$(g \otimes h)((i_0, i_1), (j_0, j_1)) = g(i_0, j_0) \cdot h(i_1, j_1),$$

where $\cdot$ is the multiplication from S.

7.6.2 From Lawvere theories to semirings

In Section 7.3.1, just after Lemma 7.11, we have already seen that the homset $L(1, 1)$ of a Lawvere theory $L \in \mathbf{SMLaw}$ is a commutative monoid, with multiplication given by composition of endomaps on 1. In the case where L also has biproducts we have, by eqn (7.6), an addition on this homset, which is preserved by composition. Combining those two monoid structures yields a semiring structure on $L(1, 1)$. This is standard; see e.g. Abramsky and Coecke (2004) and Kelly and Laplaza (1980). The assignment of the semiring $L(1, 1)$ to a Lawvere theory $L \in \mathbf{SMBLaw}$ is functorial and we denote this functor, as in Section 7.3.1, by $\mathcal{H} \colon \mathbf{SMBLaw} \to \mathbf{CSRng}$.

7.6.3 Adjunction between semirings and Lawvere theories

Our main, novel result is the adjunction on the right in the triangle of adjunctions for semirings; see Figure 7.4. In fact, in this case there is even an equivalence of categories.

Lemma 7.31 *The pair of functors* $\mathcal{M}at \colon \mathbf{CSRng} \rightleftarrows \mathbf{SMBLaw} \colon \mathcal{H}$, *forms an equivalence of categories* $\mathbf{CSRng} \simeq \mathbf{SMBLaw}$.

Proof We first prove that there is an adjunction $\mathcal{M}at \dashv \mathcal{H}$ and then check that the unit and counit are isomorphisms. For $S \in \mathbf{CSRng}$ and $L \in \mathbf{SMBLaw}$ there are (natural) bijective correspondences:

$$\frac{\mathcal{M}at(S) \xrightarrow{F} L \quad \text{in } \mathbf{SMBLaw}}{S \xrightarrow{f} \mathcal{H}(L) \quad \text{in } \mathbf{CSRng}}$$

Given F one defines a semiring map $\overline{F} \colon S \to \mathcal{H}(L) = L(1, 1)$ by

$$s \mapsto F(1 \times 1 \xrightarrow{\lambda x.\, s} S).$$

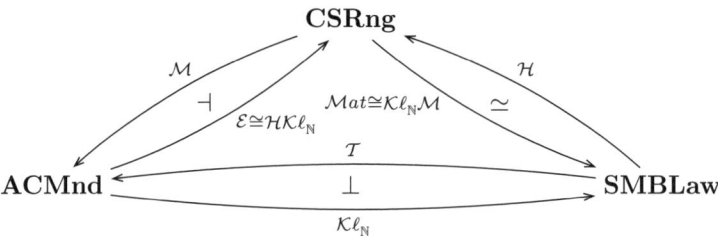

Fig 7.4 Triangle of adjunctions starting from commutative semirings, with commutative additive monads, and symmetric monoidal Lawvere theories with biproducts.

Note that $1 \times 1 \xrightarrow{\lambda x.s} S$ is an endomap on 1 in $\mathcal{M}at(S)$, which is mapped by F to an element of $L(1,1)$.

Conversely, given f one defines a **SMBLaw**-map $\bar{f}\colon \mathcal{M}at(S) \to L$ that sends a morphism $h\colon n \to m$ in $\mathcal{M}at(S)$, i.e. $h\colon n \times m \to S$ in **Sets**, to the following morphism $n \to m$ in L, forming an n-cotuple of m-tuples

$$\bar{f}(h) = \left(n \xrightarrow{\left[\langle f(h(i,j))\rangle_{j<m}\right]_{i<n}} m\right).$$

It is readily checked that $\bar{F}\colon S \to L(1,1)$ is a map of semirings. To show that $\bar{f}\colon \mathcal{M}at(S) \to L$ is a functor one has to use the definition of the semiring structure on $L(1,1)$ and the properties of the biproduct on L. One easily verifies that $\bar{f}$ preserves the biproduct. To show that it also preserves the monoidal structure one has to use that, for $s,t \in L(1,1)$, $s \otimes t = t \circ s (= s \circ t)$.

The unit $S \to \mathcal{H}(\mathcal{M}at(S))$ is an isomorphism because the endomaps $1 \to 1$ in $\mathcal{M}at(S)$ can be identified with elements of S. The counit $\mathcal{M}at(\mathcal{H}(L)) \to L$ is also an isomorphism because a map $n \to m$ in $\mathcal{M}at(\mathcal{H}(L))$ can be decomposed in maps $1 \to 1$ since the objects n and m are respectively an n- and an m-fold biproduct of 1. These maps $1 \to 1$ in $\mathcal{M}at(\mathcal{H}(L))$ are the elements of the semiring $\mathcal{H}(L)$), and thus the maps $1 \to 1$ in L. □

The results of Section 7.5 and 7.6 are summarized in Figure 7.4.

7.7 Semirings with involutions

In this final section we enrich our approach with involutions. Actually, such involutions could have been introduced for monoids already. We have not done so for practical reasons: involutions on semirings give the most powerful results, combining daggers on categories with both symmetric monoidal and biproduct structure.

An involutive semiring (in **Sets**) is a semiring $(S, +, 0, \cdot, 1)$ together with a unary operation $*$ that preserves the addition and multiplication, i.e. $(s+t)^* = s^* + t^*$ and $0^* = 0$, and $(s \cdot t)^* = s^* \cdot t^*$ and $1^* = 1$, and is involutive, i.e. $(s^*)^* = s$. The complex numbers with conjugation form an example. We denote the category of involutive semirings with homomorphisms that preserve all structure by **ICSRng**.

The adjunction $\mathcal{M}\colon \mathbf{CSRng} \leftrightarrows \mathbf{ACMnd}\colon \mathcal{E}$ considered in Lemma 7.28 may be restricted to an adjunction between involutive semirings and so-called involutive commutative additive monads (on **Sets**), which are commutative additive monads T together with a monad morphism $\zeta\colon T \to T$ satisfying $\zeta \circ \zeta = id$. We call ζ an involution on T, just as in the semiring setting. A morphism between such monads (T, ζ) and (T', ζ') is a monad morphism $\sigma\colon T \to T'$ preserving the involution, i.e. satisfying $\sigma \circ \zeta = \zeta' \circ \sigma$. We denote the category of involutive commutative additive monads by **IACMnd**.

Lemma 7.32 *The functors* $\mathcal{M}\colon \mathbf{CSRng} \leftrightarrows \mathbf{ACMnd}\colon \mathcal{E}$ *from Lemma 7.25 and Lemma 7.26 restrict to a pair of functors* $\mathcal{M}\colon \mathbf{ICSRng} \leftrightarrows \mathbf{IACMnd}\colon \mathcal{E}$. *The restricted functors form an adjunction* $\mathcal{M} \dashv \mathcal{E}$.

Proof Given a semiring S with involution $*$, we may define an involution ζ on the multiset monad $\mathcal{M}(S) = M_S$ with components

$$\zeta_X\colon M_S(X) \to M_S(X), \qquad \sum_i s_i x_i \mapsto \sum_i s_i^* x_i.$$

Conversely, for an involutive monad (T, ζ), the map ζ_1 gives an involution on the semiring $\mathcal{E}(T) = T(1)$.

A simple computation shows that the unit and the counit of the adjunction $\mathcal{M}\colon \mathbf{CSRng} \leftrightarrows \mathbf{ACMnd}\colon \mathcal{E}$ from Lemma 7.28 preserve the involution (on semirings and on monads respectively). Hence the restricted functors again form an adjunction. $\square$

The adjunction $\mathcal{M}at\colon \mathbf{CSRng} \rightleftarrows \mathbf{SMBLaw}\colon \mathcal{H}$ from Lemma 7.31 may also be restricted to involutive semirings. To do so, we have to consider dagger categories. A dagger category is a category $\mathbf{C}$ with a functor $\dagger\colon \mathbf{C}^{\mathrm{op}} \to \mathbf{C}$ that is the identity on objects and satisfies, for all morphisms $f\colon X \to Y$, $(f^\dagger)^\dagger = f$. The functor $\dagger$ is called a dagger on $\mathbf{C}$. Combining this dagger with the categorical structure we studied in Section 7.6 yields a so-called dagger symmetric monoidal category with dagger biproducts; that is, a category $\mathbf{C}$ with a symmetric monoidal structure $(\otimes, I)$, a biproduct structure $(\oplus, 0)$, and a dagger $\dagger$, such that, for all morphisms f and g, $(f \otimes g)^\dagger = f^\dagger \otimes g^\dagger$, all the coherence isomorphisms α, ρ and γ are dagger isomorphisms and, with respect to the biproduct structure, $\kappa_i = \pi_i^\dagger$, where a dagger isomorphism is an isomorphism f satisfying $f^{-1} = f^\dagger$. Further details may be found in Abramsky and Coecke (2004, 2009).

We will denote the category of dagger symmetric monoidal Lawvere theories with dagger biproducts such that the monoidal structure distributes over the biproduct structure by **DSMBLaw**. Morphisms in **DSMBLaw** are maps in **SMBLaw** that (strictly) commute with the daggers.

Lemma 7.33 *The functors* $\mathcal{M}at\colon \mathbf{CSRng} \rightleftarrows \mathbf{SMBLaw}\colon \mathcal{H}$ *forming an equivalence in Section 7.6 restrict to a pair of functors* $\mathcal{M}at\colon \mathbf{ICSRng} \rightleftarrows \mathbf{DSMBLaw}\colon \mathcal{H}$, *giving an equivalence* $\mathbf{ICSRng} \simeq \mathbf{DSMBLaw}$.

Proof For an involutive semiring S, we may define a dagger on the Lawvere theory $\mathcal{M}at(S)$ by assigning to a morphism $f\colon n \to m$ in $\mathcal{M}at(S)$ the morphism $f^\dagger\colon m \to n$ given by

$$f^\dagger(i, j) = f(j, i)^*. \tag{7.10}$$

Some short and straightforward computations show that the functor $\dagger$ is indeed a dagger on $\mathcal{M}at(S)$, which interacts appropriately with the monoidal and biproduct structure.

For a dagger symmetric monoidal Lawvere theory **L** with dagger biproduct, it easily follows from the properties of the dagger that this functor induces an involution on the semiring $\mathcal{H}(\mathbf{L}) = \mathbf{L}(1,1)$, namely via $s \mapsto s^\dagger$.

The unit and the counit of the adjunction $\mathcal{M}at\colon \mathbf{CSRng} \leftrightarrows \mathbf{SMBLaw}$ from Lemma 7.31 preserve the involution and the dagger respectively. Hence, the restricted functors also form an equivalence. □

To complete our last triangle of adjunctions, recall that, for the Lawvere theory associated with a (involutive commutative additive) monad T, $\mathcal{K}\ell_\mathbb{N}(T) \cong \mathcal{M}at(\mathcal{E}(T))$, see Proposition 7.30. Hence, using the previous two lemmas, the finitary Kleisli construction restricts to a functor $\mathcal{K}\ell_\mathbb{N}\colon \mathbf{IACMnd} \to \mathbf{DSMBLaw}$. For the other direction we use the following result.

Lemma 7.34 *The functor* $\mathcal{T}\colon \mathbf{SMBLaw} \to \mathbf{ACMnd}$ *from Lemma 7.12 restricts to* $\mathbf{DSMBLaw} \to \mathbf{IACMnd}$, *and yields a left adjoint to* $\mathcal{K}\ell_\mathbb{N}\colon \mathbf{IACMnd} \to \mathbf{DSMBLaw}$.

Proof To start, for a Lawvere theory $\mathbf{L} \in \mathbf{DSMBLaw}$ with dagger †, we have to define an involution $\zeta\colon T_\mathbf{L} \to T_\mathbf{L}$. For a set X this involves a map

$$T_\mathbf{L}(X) = \Big(\coprod_{i\in\mathbb{N}} \mathbf{L}(1,i) \times X^i\Big)/\sim \xrightarrow{\zeta_X} \Big(\coprod_{i\in\mathbb{N}} \mathbf{L}(1,i) \times X^i\Big)/\sim = T_\mathbf{L}(X)$$

$$[\kappa_i(g, v)] \mapsto [\kappa_i(\langle g_0^\dagger, \ldots, g_{i-1}^\dagger\rangle, v)],$$

where $g\colon 1 \to i$ is written as $g = \langle g_0, \ldots, g_{i-1}\rangle$ using that $i = 1 + \cdots + 1$ is not only a sum but also a product. Clearly, ζ is natural and satisfies $\zeta \circ \zeta = \mathrm{id}$. This ζ is also a map of monads; commutatution with multiplication μ requires commutativity of composition in the homset $\mathbf{L}(1,1)$.

The unit of the adjunction $\eta\colon \mathbf{L} \to \mathcal{K}\ell_\mathbb{N}(T_\mathbf{L}) \cong \mathcal{M}at(T_\mathbf{L}(1))$ commutes with daggers, since for $f\colon n \to m$ in **L** we get $\eta(f)^\dagger = \eta(f^\dagger)$ via the following argument in $\mathcal{M}at(T_\mathbf{L}(1))$. For $i < n$ and $j < m$,

$$\begin{aligned}
\eta(f)^\dagger(i,j) &= \eta(f)(i,j)^* &&\text{by eqn (7.10)} \\
&= \pi_j \mathsf{bc}_m(\kappa_m(f \circ \kappa_i, \mathrm{id}_m))^* &&\text{by eqn (7.9)} \\
&= \kappa_1(\pi_j \circ f \circ \kappa_i, \mathrm{id}_1)^* &&\text{by definition of bc, see eqn (7.8)} \\
&= \kappa_1(\pi_j \circ f \circ \kappa_i, \mathrm{id})^\dagger &&\text{since } (-)^* = (-)^\dagger \text{ on } T_\mathbf{L}(1) \\
&= \kappa_1((\pi_j \circ f \circ \kappa_i)^\dagger, \mathrm{id}) \\
&= \kappa_1(\kappa_i^\dagger \circ f^\dagger \circ \pi_j^\dagger, \mathrm{id}) \\
&= \kappa_1(\pi_i \circ f^\dagger \circ \kappa_j, \mathrm{id}) \\
&= \eta(f^\dagger)(i,j).
\end{aligned}$$
□

In the definition of the involution ζ on the monad $T_\mathbf{L}$ in this proof we have used that $+$ is a (bi)product in the Lawvere theory **L**, namely when we decompose the map $g\colon 1 \to i$ into its components $\pi_a \circ g\colon 1 \to 1$ for $a < i$. We could have avoided

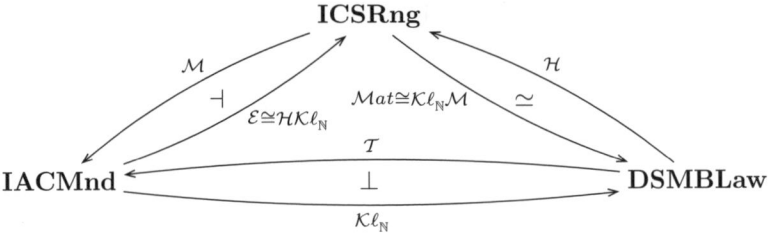

Fig 7.5 Triangle of adjunctions starting from involutive commutative semirings, with involutive commutative additive monads, and dagger symmetric monoidal Lawvere theories with dagger biproducts.

this biproduct structure by first taking the dagger $g^\dagger : i \to 1$ and then precomposing with coprojections $g^\dagger \circ \kappa_a : 1 \to 1$. Again applying daggers, cotupling, and taking the dagger one gets the same result. This is relevant if one wishes to consider involutions/daggers in the context of monoids, where products in the corresponding Lawvere theories are lacking.

By combining the previous three lemmas we obtain another triangle of adjunctions in Figure 7.5. This concludes our survey of the interrelatedness of scalars, monads, and categories.

7.8 Conclusions

In this chapter the relations between basic algebraic structures (monoids and semirings), certain monads, and certain categorical structures have been clarified, in a uniform manner, via triangles of adjunctions. This approach can be generalized. This has already been done for similar structures (effect algebras and monoids) that occur in probability theory (see Jacobs, 2011). Another extension involves inverses, both additively and multiplicatively. In ongoing work this is done via Hopf algebras and monads.

CHAPTER 8

Types and Forgetfulness in Categorical Linguistics and Quantum Mechanics

PETER HINES
(University of York)

8.1 Introduction

A recent trend in linguistics (Coecke, Sadrzadeh and Clark, 2010), also outlined in several chapters in this book, is to extend linguistic models of meaning from *words* to *sentences*. Either implicitly or explicitly, this is done via a type system—based on a categorical grammar—equipped with a notion of *evaluation*. This notion of evaluation is crucial, in that it is used to reduce all grammatically correct sentences to the same type, where they may be compared and their similarity evaluated. In this chapter, we describe this typing and evaluation process in an abstract categorical setting, based around a toy example.

We then consider the deceptively simple question of whether this evaluation process (mapping all grammatically correct sentences to entities of the same type) is reversible or not—i.e. does evaluation lose information? Our conclusion is that, in general, forgetting information is an inevitable and crucial part of this process. However, we also demonstrate that connectives are a special case, having an entirely reversible interpretation. Following this observation to its inevitable mathematical conclusion, we discover a connection between reversibility and polymorphic typing, in both the linguistic and logical sense.

The relevant structures are familiar from a wide range of settings, ranging from models of lambda calculus and the geometry of interaction, to fractals, tilings, and the Thompson groups. This chapter demonstrates a further close connection with abstract categorical models of quantum mechanics. To be precise, we derive a (lax, infinitary) form of the special sort of Frobenius algebras known as *classical structures*, around which categorical approaches to quantum information and computation are based.

8.2 Introducing typing to models of meaning

The method of comparing meaning of words known as distributional semantics is well-described elsewhere in this volume (see Chapter 6) and, as such, we restrict our description to the features that will be particularly relevant to the typing process. We then give a simple example of how typing, along with an evaluation operation, is used to allow the comparison of quantities in physics. This is followed by a formal description of what we mean by a typed system, based around the theory of monoidal closed categories, and an indication of how we expect such a categorical typing in models of meaning to allow us to compare arbitrary sentences, regardless of their grammatical structure.

As described in Chapter 6, distributional semantics provides a method of associating a vector (the *meaning vector*) with each word in a dictionary, based on its usage in some corpus. Vectors may then be compared with each other, using any of the familiar tools from linear algebra (generally, the scalar product), giving a measure of the similarity or overlap between words. The simple but ambitious aim is to extend this this process to sentences, rather than single words, using the following scheme:

1. Single words are assigned *types*, based on their role; this typing is extended to sentences, which are typed by their grammatical structure.
2. Associated with the type system is an evaluation, or reduction, process that reduces all grammatically correct sentences to elements of the same type (this is, as described elsewhere, a common approach in categorical linguistics).
3. Crucially, elements of the same type can be compared, providing a method of comparing the meaning of grammatically distinct sentences, in a similar way to distributional semantics.

It hardly needs emphasizing that this is a very ambitious program; instead of aiming to provide a complete or partial solution, this chapter describes features that such a model of meaning necessarily requires at the level of the types.

8.3 What is a type?

To a categorical logician, the answer is straightforward: *a type is an object in a (monoidal, closed) category*. To explain this, we first give a simple example of typing in basic physics, followed by the formal definition, and an illustration of why such a type system would also be useful in linguistic models of meaning.

8.3.1 Types in elementary physics

A simple, but illustrative, example of a typed system comes from basic physics, where the *units of measurement* may be thought of as the *types of quantities*. The familiar seven basic SI units (kilogram (**kg**), second (**s**), metre (**m**), lumen (**lm**), etc.) are the fundamental types, and further types may be built up recursively, using these base types and two operations known as *pairing* and *abstraction*.[1]

- *Pairing*. Given two types S, T, the pair type ST may be formed. For example *luminous energy* is measured in *lumen seconds*, and hence has type **lm s**.
- *Abstraction*. Given two types L, M, the abstraction type ML^{-1} may be formed. For example *velocity* is given in metres per second, and hence has type **m s^{-1}**.

Associated with such a type system is a notion of *evaluation* or *reduction*. A quantity of type YX^{-1} may be combined with a quantity of type X to return a quantity of type Y. For example, let us calculate how far light, with a velocity of $c = 2.997 \times 10^8$ **ms^{-1}**, travels in 1.3 s.

$$(2.998 \times 10^8) \text{ ms}^{-1} \times 1.3 \text{ s} = 3.897 \times 10^8 \text{ m} \tag{8.1}$$

Considering the typing only, we see a reduction of the form

$$\text{ms}^{-1} \times \text{s} \xrightarrow{Eval.} \text{m} \tag{8.2}$$

In this case, evaluation is simply the operation of *multiplication*.[2] Thus we observe that the type system for SI units is in fact *commutative* (i.e. the type XY is identical to the type YX). As a simple consequence of this, ordering is irrelevant, and, for example, **ms^{-1}** is equivalent to **s^{-1}m**. In general, and in categorical linguistics in particular, neither commutativity nor symmetry (i.e. commutativity up to isomorphism) may be assumed. To avoid ambiguity, we will therefore use type-theoretic notation, and write either $[S \to T]$ or $[T \leftarrow S]$ instead of TS^{-1}. Strictly, this means that we should consider two distinct evaluation operations. However, the required evaluation is often clear from the context, so for simplicity of notation we do not distinguish between the two unless absolutely essential.

8.3.2 How we wish to use types in models of meaning

By analogy with how types are used in the above simple example, we wish to consider models of meaning where words and phrases are typed according to their

---

[1] Pairing and abstraction are more commonly called *product* and *quotient*. We avoid this terminology since these are neither products nor quotients in the categorical sense.

[2] In general, evaluation in a typed system may be significantly more complex; theoretical computer scientists will be familiar with evaluation as either β-reduction in lambda calculus, or the execution of a Turing machine (Lambek and Scott, 1986; Hines, 2003).

grammatical structure, and the evaluation operation associated with the type system reduces all (grammatically correct) sentences to the same type—the *sentence type* S.

Consider the simplest possible sentence structure:

(Noun phrase, Intransitive verb)

If we assume that the *noun phrase* is of some primitive type **NP**, an intransitive verb can only have type [NP → S], where S is the sentence type. The reduction of the sentence to the type S then proceeds by direct analogy with eqn (8.2):

$$\text{NP} \times [\text{NP} \to \text{S}] \xrightarrow{Eval.} \text{S} \qquad (8.3)$$

8.3.3 Monoidal closed categories

The above notions may be formalized in the field of category theory. We refer to Chapter 1 for the basic notions of (monoidal) category theory. However, we will be forced to take a more formal approach and explicitly consider the structural isomorphisms:

Definition 8.1 Symmetric monoidal categories.
A monoidal category *is defined to be a category* C, *together with a functor* $\otimes : C \times C \to C$ *that satisfies, for all* $A, B, C \in Ob(C)$:

- unit objects: *there exists* $I \in Ob(C)$ *satisfying* $I \otimes A \cong A \cong A \otimes I$
- associativity: $A \otimes (B \otimes C) \cong (A \otimes B) \otimes C$.

If a monoidal category satisfies the additional condition:

- symmetry $A \otimes B \cong B \otimes A$

it is called a symmetric monoidal category. *The above isomorphisms exhibiting associativity or symmetry are* natural, *and satisfy various* coherence conditions *laid out in Mac Lane (1971).*

Due to Mac Lane's celebrated coherence theorem for associativity, we may treat the associativity isomorphisms as though they are strict identities, so we do not distinguish between $A \otimes (B \otimes C)$ and $(A \otimes B) \otimes C$. We follow this practice until Section 8.8, where the distinction between the two will become important.

Definition 8.2 Monoidal closed categories.
Let $(C, \otimes)$ *be a monoidal category. We say that it is* monoidal closed *when there exists a functor*

$$[\_ \to \_] : C^{op} \times C \to C$$

called the internal hom *functor, such that for fixed* $B \in Ob(C)$, *the functors given by*

$$[B \to \_] : C \to C \quad \text{and} \quad \_ \otimes B : C \to C$$

form an adjoint pair. Equivalently, for all $X, Y, Z \in Ob(\mathcal{C})$, there exists a natural isomorphism

$$\mathcal{C}(X \otimes Y, Z) \cong \mathcal{C}(X, [Y \to Z]).$$

The above definition is concise, albeit very abstract (for example, we refer to Mac Lane (1971) for the definition of an adjoint pair of functors). Instead we use the following characterization that makes the existence and role of an evaluation map central:

Theorem 8.3 *The above definition of a monoidal closed category is equivalent to the following:*

For every pair of objects $A, B \in Ob(\mathcal{C})$, there exists

- an object $[A \to B] \in Ob(\mathcal{C})$
- an arrow $Eval_{A,B} \in \mathcal{C}(A \otimes [A \to B], B)$

where, for all $f : A \otimes X \to B$, there exists unique $g \in \mathcal{C}(X, [A \to B])$ such that the following diagram commutes:

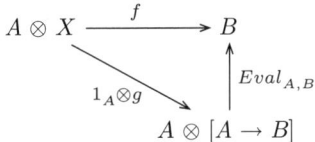

Proof Proofs may be found in any text on category theory or categorical logic (e.g. Mac Lane, 1971; Lambek and Scott, 1986). □

8.3.4 Monoidal closed categories as type systems

The connection between the theory of monoidal closed categories, and the (very elementary) type system presented in Section 8.3 should then be straightforward. More generally, we take a categorical perspective, and define a type as an object in a monoidal closed category. The operation of pairing from Section 8.3 is then simply the monoidal tensor $\_ \otimes \_$, and the operation of *abstraction* from the same section is the internal hom functor $[\_ \to \_]$. Finally, the reduction operation is simply the evaluation derived in Theorem 8.3.

The question then arises: in this setting, what is a *quantity* of a certain type, and how may such quantities be compared?

8.3.5 Elements, scalars, daggers, and duals

The objects of a monoidal closed category do not come equipped with a notion of membership, so it is not accurate to talk about 'a member x of some object N'. Instead we have the notion of *elements* of an object.

Definition 8.4 *Given a monoidal category* $(\mathcal{C}, \otimes, I)$, *an* element *of some object* $N \in Ob(\mathcal{C})$ *is a member of* $\mathcal{C}(I, N)$, *i.e. an arrow from* I *to* N. *The category* $(\mathcal{C}, \otimes, I)$ *is called* well-pointed *when, for all* $f \neq g \in \mathcal{C}(X, Y)$, *there exists some element* $a \in \mathcal{C}(I, X)$ *such that* $fa \neq ga \in \mathcal{C}(I, Y)$.

For well-pointed categories, it is easy to see how the notion of elements is a reasonable replacement for the notion of membership. Most of this chapter is based on *elements* of a category and their interaction with the monoidal structure, and the categories with which we work are generally *well-pointed*. We will point out when results depend on this assumption or when we are (unusually) referring to a non-well-pointed category.

In order to *compare* elements of an object, we need a small amount of extra structure:

Definition 8.5 *A* dagger *operation on a category* $\mathcal{C}$ *is a (contravariant) involutive endofunctor, usually written* $(\ )^\dagger : \mathcal{C}^{op} \to \mathcal{C}$, *which is the identity on objects, so* $A^\dagger = A$ *for all* $A \in Ob(\mathcal{C})$. *An arrow* $f \in \mathcal{C}(A, B)$ *satisfying* $f^\dagger f = 1_A$ *is called an* isometry, *and when this is a two-sided inverse (so* $f^\dagger$ *is also an isometry), then* f *is called* unitary.

Let $\_ \otimes \_$ *be a monoidal tensor on* $\mathcal{C}$. *When the monoidal structure has a well-behaved interaction with the dagger operation (that is, all canonical isomorphisms are unitary), then* $(\mathcal{C}, \otimes, (\ )^\dagger)$ *is called a* dagger monoidal category.

Dagger monoidal categories provide us with exactly the structure we need to compare elements:

Definition 8.6 *Following Abramsky (2005), arrows from* I *to itself in a monoidal category with daggers are called* abstract scalars. *Given two elements of the same object* $x, y \in \mathcal{C}(I, X)$, *their* generalized inner product *is the endomorphism of the unit object given by*

$$\langle x | y \rangle = x^\dagger y \in \mathcal{C}(I, I).$$

Thus generalized inner products act as *comparisons*, and give a result that is an arrow from the unit object to itself. This fits in well with our usual intuition of what it means to compare the similarity of elements, in that in various settings $\mathcal{C}(I, I)|$ is (for example) the real line $\mathbb{R}$, the complex plane $\mathbb{C}$, the natural numbers $\mathbb{N}$, the unit interval $[0, 1]$, etc. We take care to avoid using categories where the endomorphism monoid of the unit object is trivial (e.g. globally defined functions, relations on sets, vector spaces with direct sum as monoidal tensor, etc.).

Proposition 8.7 *Let* $(\mathcal{C}, \otimes, I)$ *be a monoidal category.*

1. $\mathcal{C}(I, I)$ *is an abelian monoid.*
2. *Up to canonical isomorphism,* $\alpha \otimes \beta = \alpha\beta = \beta\alpha$.
3. *When* $\mathcal{C}$ *is a dagger monoidal category, then for all* $X, Y \in Ob(\mathcal{C})$ *and elements*

$$a, b \in \mathcal{C}(I, X) \ , \ c, d \in \mathcal{C}(I, Y)$$

then

$$\langle a \otimes c | b \otimes d \rangle = \langle a | b \rangle \langle c | d \rangle.$$

Proof We refer to Abramsky (2005) for proofs. □

Much of the terminology and notation used in dagger monoidal categories comes from a canonical motivating example:

Example 8.8 *(Altschuler and Coste, 1992)* Complex finite-dimensional Hilbert spaces form a dagger monoidal category, where the monoidal tensor is the usual tensor product, and the dagger is the usual Hermitian adjoint $(\ )^H$. The unit object is then the underlying scalar field, i.e. the complex plane $\mathbb{C}$, and an endomorphism of the unit object is a linear map on a one-dimensional space—that is, multiplication by some complex scalar.

Elements of some finite-dimensional space H are then simply linear maps from $\mathbb{C}$ to H, which are, of course, in one-to-one correspondence with the points of H. Moving from points of a space to linear maps of a space is exactly the idea behind Dirac notation for states; instead of working with the point $\psi \in H$, we work with the linear map $|\psi\rangle : \mathbb{C} \to H$. The (categorical) generalized inner product is then exactly the composite $|\phi\rangle^H |\psi\rangle$, i.e. the usual inner product $\langle \phi | \psi \rangle$ of vectors in a Hilbert space, expressed in Dirac notation.

We also refer to Altschuler and Coste (1992) for the monoidal closure of this category, and the quantum-mechanical interpretation of the categorical operations such as evaluation and the dagger.

In Example 8.8, the generalized inner product is exactly the scalar product of vectors and may be used to define a *metric* on elements of a (finite-dimensional) Hilbert space. Thus, because of the first metric axiom, the generalized inner product may be used as a test of equality for elements. In other examples, (such as partial reversible functions on sets), the endomorphism monoid of the unit object is trivial, and the generalized inner product provides little or no information about elements. We axiomatize this distinction as follows:

Definition 8.9 *Let $(\mathcal{C}, \otimes, I, (\ )^\dagger)$ be a dagger monoidal category. We say that $(\ )$: $\mathcal{C}^{op} \to \mathcal{C}$ discriminates elements of $A \in Ob(\mathcal{C})$ when, for all $x, y \in \mathcal{C}(I, A)$,*

$$\langle x | y \rangle = 1_I \Leftrightarrow x = y \in \mathcal{C}(I, A).$$

When $(\ )^\dagger : \mathcal{C}^{op} \to \mathcal{C}$ discriminates elements of all objects of $\mathcal{C}$, we simply say that it discriminates elements.

8.3.6 Elements, names, and evaluation

In a monoidal *closed* category, the elements of the object $[X \to Y] \in Ob(\mathcal{C})$ have a natural interpretation as arrows from X to Y within $\mathcal{C}$, as the following result makes clear:

Proposition 8.10 *Let* $(\mathcal{C}, \otimes, [\,,\,], I)$ *be a monoidal closed category. Then, for all objects* $X, Y \in Ob(\mathcal{C})$, *there is a natural bijection between elements of* $[X \to Y]$ *and the homset* $\mathcal{C}(X, Y)$.

Proof This is a standard result from the theory of closed categories and categorical logic (Laplaza, 1977; Lambek and Scott, 1986). □

Definition 8.11 *Given a monoidal closed category* $(\mathcal{C}, \otimes, [\,,\,], I)$, *and an arrow* $f \in \mathcal{C}(X, Y)$, *then its image under the bijection of Proposition 8.10 above is called the* name *of* $f \in \mathcal{C}(X, Y)$, *written* $\ulcorner f \urcorner \in \mathcal{C}(I, [X \to Y])$.

The intuitive meaning of evaluation is that it 'promotes' an *element* (i.e. the name of an arrow $\ulcorner f \urcorner$) to actual *arrow* within the category; more formally, the following diagram is a special case of the diagram of Theorem 8.3.

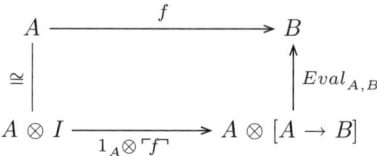

Thus, for example, we see that an element of the 'intransitive verb' object (as in Section 8.3.2) may also be considered as an arrow from the 'noun phrase' object to the 'sentence' object.

Given that *elements* of an object are themselves arrows in a category, it is natural to wonder what the name of an element is, or indeed the name of the name of an element, etc. Fortunately, such an eternal recurrence is avoided by the fact that, up to canonical isomorphism, elements 'name themselves'. Precisely, for any element $x \in \mathcal{C}(I, A)$ in a monoidal closed category, the following diagram commutes:

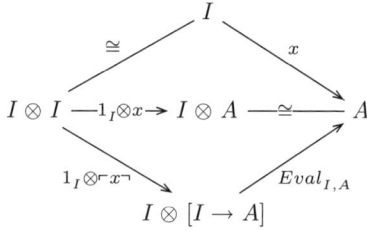

In the above diagram, lines denoting canonical coherence isomorphisms are simply labelled by '≅'. We follow this convention throughout, unless the precise coherence isomorphism is important.

8.3.7 Types for linguistics and models of meaning

In categorical linguistics, a common method of characterizing grammatically correct sentences is to assign types (i.e. objects in a monoidal closed category) to words in such a way that the evaluation map takes all grammatically correct sentences to a single distinguished type S, called the *sentence type*. In particular as shown in Coecke, Sadrzadeh and Clark (2010), this is the structure behind Lambek's *pregroup semantics* and other approaches to categorical linguistics. Thus standard categorical models of linguistics provide a type system for models of meaning applicable to arbitrary sentences, but we do not yet have actual elements or any notion of how these interact with the evaluation process, the generalized inner product, or the monoidal tensor. An obvious analogy exists with the very powerful tool of *dimensional analysis* in basic physics (Kasprzak, Lysik and Rybaczuk, 1990), which may be considered to be the underlying type system behind the SI units of Section 8.3, abstracted from consideration of actual quantities.

The remainder of this paper may be considered as an investigation of what it would mean to (re)introduce actual elements to the type system provided by categorical linguistics, and indeed what modifications must be made to the typing to account for the fact that we are interested in meaning as well as grammar.

8.3.8 Typed models of meaning—a toy example

In order to avoid becoming too abstract, we use a concrete example to illustrate how the program described above may be used to compare two sentences. We will use the following examples:

L1 *Bobby loves Marilyn Monroe.*
L2 *I like Fidel Castro and his beard.*

The classically educated reader will recognize these as lyrics from Bob Dylan songs;[3] our interest is in how a typed model of meaning could be used to compare these distinct lyrics.

The first step we take is to instantiate the variable in L2;[4] as we are familiar with these sentences as Bob Dylan lyrics, it is reasonable to replace *I* by **Bob Dylan**, and adjust the verb from the first to the third person, giving

L2' *Bob Dylan likes Fidel Castro and his beard.*

---

[3] Note that one of these lyrics is from an improvised live performance and is not part of the official Dylan canon (Dylan, 2004).

[4] Note that this is an *exophoric reference* since the pronoun *I* is not bound to any noun phrase within the text itself.

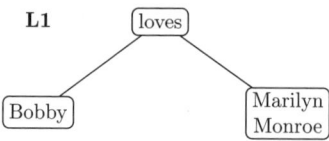

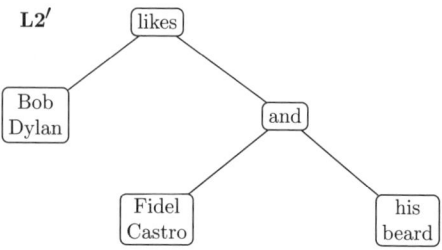

Fig 8.1 Trees for **L1** and **L2′**

We draw these two sentences L1 and L2′ in tree form as shown in Figure 8.1, and consider how both the individual constituents and the sentences as a whole may be compared.

From these trees, we see that the basic grammatical components we require are:

- *sentences* (**S**): 'Bob Dylan likes Fidel Castro and his beard', 'Bobby loves Marilyn Monroe'
- *transitive verbs* (**TV**): likes, loves
- *connectives* (**C**): and
- *noun phrases* (**NP**): Bob Dylan, Bobby, Marilyn Monroe, Fidel Castro, Fidel Castro's beard, Fidel Castro and his beard.

(We observe above that joining two noun phrases by a connective forms another noun phrase. This possibility must be reflected in the typing associated with connectives; this is explored further in Section 8.7 onwards.)

8.3.9 Categorical features for a model of meaning

We now consider the requirements for some category *Meaning* in which the meanings of L1 and L2′ may be evaluated and compared. We do not present a concrete example; rather, we use the machinery developed to lay down requirements that such a category must satisfy, and go on the consider the resulting categorical theory.

For a type system that allows us to model grammatical structure and reduce grammatical sentences to elements of the same type, we require a *monoidal closed* structure, so *Meaning* is equipped with a monoidal tensor $(\_ \otimes \_) : \mathcal{M}eaning \times \mathcal{M}eaning \to \mathcal{M}eaning$, a unit object $I \in Ob(M)$, and an internal hom,

$[\_ \to \_]: \mathcal{M}eaning^{op} \times \mathcal{M}eaning \to \mathcal{M}eaning$. In order to compare elements, we will also require a dagger operation $(\_)^\dagger : \mathcal{M}eaning^{op} \to \mathcal{M}eaning$ compatible with the monoidal tensor, giving a generalized scalar product.

As our analysis will be based on elements of objects, it is reasonable to assume that $(\mathcal{M}eaning, \otimes, [\ \to \ ], I, (\ )^\dagger)$ is well-pointed (defn 8.4). We further assume that our model of meaning is *complete*, in the sense that distinct concepts are not unnecessarily identified by the generalized scalar product;[5] categorically, this requires that the dagger operation *discriminates elements*, in the sense of defn 8.9.

The category $\mathcal{M}eaning$ must also contain objects corresponding to the grammatical components given in Section 8.3.8 above. Thus $NP, C, TV, S \in Ob(\mathcal{M}eaning)$ are the objects corresponding to the *noun phrase, connective, transitive verb, sentence* types; we take the noun phrase and sentence types $NP, S \in Ob(M)$ as primitive and build up the others in terms of their desired behaviour under evaluation.

Finally, for illustrative purposes, we take $\mathcal{M}eaning(I, I)$ to be the unit interval $[0, 1]$. The putative interpretation is that $\langle x|y \rangle = 1$ means complete equality of meaning between elements x and y, whereas $\langle x|y \rangle = 0$ means that they have nothing in common. Composition of endomorphism arrows of the unit object is, as per the requirements of Proposition 8.7, simply multiplication.

8.3.10 Comparing simple nouns

Let us start with the respective subjects of **L1** and **L2'**, the noun phrases **Bobby** and **Bob Dylan**. In a suitable typed system, these will be represented by two distinct elements of type **NP**

$$I \xrightarrow{\text{Bobby}} NP \qquad\qquad I \xrightarrow{\text{Bob Dylan}} NP$$

These elements may be compared by computing their generalized inner product, giving

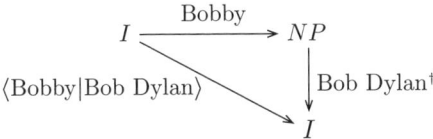

Although this chapter does not present a concrete model of meaning, observe that the above comparison would be straightforward using the distributional semantics approach described in Chapter 6. In the absence of any concrete data,

---

[5] This is not always a reasonable assumption to make, depending on the intended purpose of our model of meaning. In particular, the very successful field of *sentiment analysis* (Kim and Hovy, 2006) makes an entirely different assumption!

we make a guess for illustrative purposes, and write

$$\langle Bobby | Bob\ Dylan \rangle \;=\; 0.98 \in \mathcal{M}eaning(I, I)$$

Giving a high, if not perfect overlap[6] between **Bobby** and **Bob Dylan**.

8.3.11 Comparing transitive verbs

We now compare the central verbs of **L1** and **L2′**, i.e. we wish to assign a value to the generalized inner product $\langle \mathbf{likes}|\mathbf{loves}\rangle$. However, it is worth considering the typing that these elements must have. Recall from Section 8.3.2 that an *intransitive* verb can only have type $[NP \to S]$. Thus, we wish a *transitive* verb to have a suitable type so that, when given an object (i.e. a noun phrase) on its right-hand side, it returns something of type $[N \to S]$. Thus an intransitive verb must have type $[[NP \to S] \leftarrow NP]$.

The comparison of **likes** and **loves** elements is the following generalized inner product

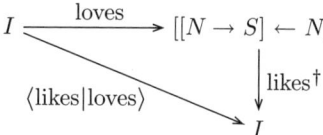

We again make an arbitrary guess,[7] and write

$$\langle likes | loves \rangle \;=\; 0.75 \in \mathcal{M}eaning(I, I)$$

8.3.12 Comparing noun phrases

In terms of comparing the primitive elements of $L1$ and $L2$, it now remains to compare the two objects of the transitive verbs: **Marilyn Monroe** and **Fidel Castro and his beard**. Leaving aside for the moment the details of how two noun phrases may be combined with a connective to produce a further noun phrase, we are happy to declare that there never has been any significant overlap between *Marilyn Monroe* and *Fidel Castro and his beard*. Thus our educated guess at this point is simply that

$$\langle Marilyn\ Monroe | Fidel\ Castro\ and\ his\ beard \rangle \;=\; 0 \in \mathcal{M}eaning(I, I).$$

---

[6] Readers familiar with 20th century US culture might assume that **Bobby** in *L1* instead refers to Robert Kennedy. Whether or not this interpretation is correct, it is neither apparent from the individual lines nor the songs as a whole. Interpreting texts in their appropriate historical and cultural context is a significant challenge for models of meaning generally.

[7] Based on a talk by Mehrnoosh Sadrzadeh (Oxford, Oct. 2010), where denotational semantics was introduced using the illustration that '*likes* is $\frac{3}{4}$ *loves* and $\frac{1}{4}$ *hates*'.

Bobby	loves	Marilyn Monroe.
Bob Dylan	likes	Fidel Castro and his beard.
$\langle Bobby|Bob\ Dylan\rangle$	$\langle likes|loves\rangle$	$\langle Fidel\ and\ his\ beard|Marilyn\ Monroe\rangle$
=	=	=
0.98	0.75	0.00

Fig 8.2 Comparisons of individual words in **L1** and **L2**

8.4 Comparing words versus comparing sentences

Bringing together the (entirely fictitious) values for the overlap between the meanings of words introduced above, we have the table shown in Figure 8.2.

The crucial question is whether these three values are enough to compare the meaning of **L1** and **L2**′? We first appeal to Part 3 of Proposition 8.7; we may compute the generalized inner product of **L1** and **L2**′, considered as elements of type $NP \otimes [[NP \to S] \leftarrow NP] \otimes NP$.

Proposition 8.12 *Using the values for the generalized scalar products of individual word proposed in Sections 8.3.10–8.3.12, the inner product of the elements*

$$\mathbf{L1} : I \to NP \otimes [[NP \to S] \leftarrow NP] \otimes NP$$
$$\mathbf{L2'} : I \to NP \otimes [[NP \to S] \leftarrow NP] \otimes NP$$

is exactly $\langle \mathbf{L1}|\mathbf{L2'}\rangle = 0 \in \mathcal{M}eaning(I, I)$.

Proof This follows from the values given in Figure 8.2, and Part 3 of Proposition 8.7, where the interaction of generalized inner products and monoidal tensors is given. □

However, we have compared these sentences *before any evaluation has taken place*, and the whole point of the typing system was that all well-formed sentences evaluate to the same sentence type **S**. The key question is then whether this matters, i.e.

Is comparison of sentences invariant under evaluation?

8.5 Inner products, evaluation, and inverses

The question at the end of Section 8.4 should properly be considered as two distinct questions.

1. Does the evaluation arrow $Eval_{A,B} =\in \mathcal{C}(A \otimes [A \to B], B)$ preserve inner products?
2. When the meaning of a word is some name $\ulcorner f \urcorner \in \mathcal{C}(I, [X \to Y])$, does the arrow $f \in \mathcal{C}(X, Y)$ preserve inner products?

Question 1 is a fundamentally category-theoretic question, whereas Question 2 is about how we expect categorical models of meaning to behave. In a dagger monoidal closed category $(\mathcal{C}, \otimes, [\_ \to \_], (\ )^\dagger)$, both isometries and unitaries preserve generalized inner products, and the canonical isomorphisms for the monoidal structure are unitary. Therefore, any dagger monoidal category contains inner product preserving arrows—Question 2 is simply asking whether any of these have a role to play in models of meaning.

As we are working within a well-pointed monoidal category with a dagger that discriminates elements, both these questions are about whether various categorical operations 'lose information', as the following result demonstrates.

Lemma 8.5.1 *Let $(\mathcal{C}, \otimes, (\ )^\dagger)$ be a well-pointed dagger monoidal category where the dagger discriminates elements, and let $F \in \mathcal{C}(A, B)$ preserve generalized scalar products, Then F is an isometry, i.e. $F^\dagger \in \mathcal{C}(B, A)$ is a left inverse of $F \in \mathcal{C}(A, B)$.*

Proof Consider arbitrary elements of $x, y \in \mathcal{C}(I, A)$. As F preserves inner products, the following diagram commutes:

$$\begin{array}{ccc} A & \xrightarrow{F} B & \xrightarrow{F^\dagger} A \\ {\scriptstyle x}\uparrow & & \downarrow{\scriptstyle y^\dagger} \\ I & \xrightarrow{x} A & \xrightarrow{y^\dagger} I \end{array}$$

Simplifying this commuting diagram, we have

$$I \xrightarrow{x} A \xrightarrow{\substack{F \\ \circlearrowleft B \\ F^\dagger}} A \xrightarrow{y^\dagger} I$$

Thus, as $\mathcal{C}$ is well-pointed with a dagger that discriminates elements, we deduce that $F^\dagger$ is a left inverse of F. □

Note that the above result does not prove that $F^\dagger$ is a two-sided inverse; indeed, in arbitrary Hilbert spaces the inner-product-preserving isomorphisms are exactly the unitary maps, whereas inner-product-preserving linear maps are simply isometries (which do indeed have a left inverse, but not necessarily a two-sided inverse).

8.6 Does evaluation preserve inner products?

We first address Question 1 of Section 8.5: is the evaluation map an isometry, i.e. is its dagger also a left inverse?

Given elements x, y of an object $A \otimes [A \to B]$ in some monoidal closed category with a discriminating dagger, we may form elements of B by composing both x and y with the canonical evaluation map $Eval_{A,B} : A \otimes [A \to B] \to B$ as shown below:

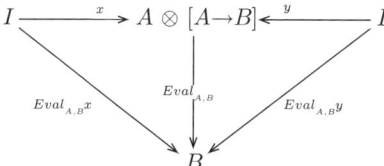

From Lemma 8.5.1, a necessary condition for the following diagram to commute

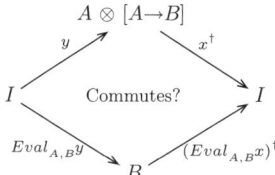

is that $Eval_{AB}$ has a left inverse. Leaving aside the irrelevant (for our purposes) case where $\mathcal{C}$ does not have a discriminating dagger, in general the above diagram does *not* commute. One of the simplest counterexamples is the motivating example of Example 8.8 and quantum-mechanical interpretations of its categorical properties, where evaluation interprets as (post-selected partial) measurement against a maximally entangled basis (Altschuler and Coste, 1992). Of course, one of the most fundamental features of the Hilbert space model of quantum mechanics is that measurement (partial or total) is certainly not a reversible operation. Other examples include models of logic or lambda calculus, where evaluation is either β-reduction or cut-elimination, neither of which are reversible operations.[8]

The question is whether this is *desirable* or *undesirable* for a model of meaning? The linguistic justification for the answer to the second question of Section 8.5 helps demonstrate that it is in fact desirable.

8.6.1 Forgetfulness: a linguistic justification

We now address Question 2 from Section 8.5. The claim that we make is that it is *vital* for the evaluation process to be *irreversible*, since we need it to forget information:

---

[8] We do not claim that, in any monoidal closed category, the evaluation arrow *cannot* be invertible. In particular, monoidal closed categories of partial reversible functions, where all objects are isomorphic, have been constructed in Hines (1998); Hines (1999) and Abramsky, Haghverdi and Scott (2002). Leaving these rather esoteric examples aside, our claim is that a (much more usual) irreversible evaluation is highly desirable and useful for typed models of meaning.

it is highly desirable that the arrows named by elements in our models of meaning do not have inverses.

As a motivating example, consider the simple noun phrase *scruffy cats*, built up from an adjective and another noun phrase:

$$I \xrightarrow{\text{Scruffy}} AD \qquad I \xrightarrow{\text{Cats}} NP$$

The noun phrase *cats* is a simple element of the object **NP**, and from its behaviour we deduce that an adjective has typing $AD = [NP \leftarrow NP]$. The term *scruffy cats*, before any reduction, is therefore is the following element:

$$I \xrightarrow{\text{Scruffy} \otimes \text{Cats}} [NP \leftarrow NP] \otimes NP$$

Momentarily forgetting about typing questions, let us assume that the 'meaning' of both *scruffy* and *cats* has been derived using some variant of the distributional semantics described in Chapter 6. The 'meaning' of *cats* will then provide information about cats generally, whether scruffy, tidy, or invisible. Similarly, the adjective *scruffy* provides information about the general concept of scruffiness, whether applied to cats, dogs, or academics.

From Proposition 8.10, an element **Scruffy** $\in \mathcal{M}eaning(I, [NP \leftarrow NP])$ is the name of some arrow $\widetilde{\textbf{Scruffy}} \in \mathcal{M}eaning(NP, NP)$. We then see that, at least in this setting, the arrow named by **Scruffy** $\in \mathcal{M}eaning(I, [NP \leftarrow NP])$ has something of the nature of a projector, or a partial identity, in that it acts to restrict a concept to a special case.

The above is not, of course, a formal justification. However, we also observe that reduction is often a multistage process, and the ability to compare sentences or sentence fragments at different levels may be a highly useful feature of a typed model of meaning. Consider sentences s_0, t_0 of some compound type $[S \leftarrow X] \otimes Y \otimes [Y \rightarrow X]$. This compound type may be reduced to $S \in Ob(\mathcal{M}eaning)$ in two stages:

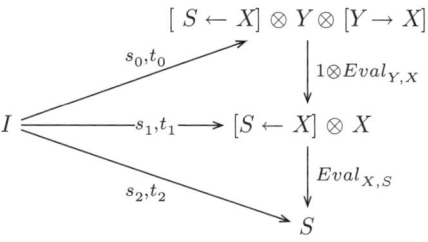

Given that $\langle s_0|t_0\rangle \neq \langle s_1|t_1\rangle \neq \langle s_2|t_2\rangle$, we observe that it is possible to compare sentences *at many different levels*, depending on how much reduction has been carried out. This unusual feature may prove useful in dealing with ambiguity, or indeed in assigning meaning to non-compositional phrases such as ***Iron Curtain***, where the meaning of this phrase is not derived by restricting the information about all possible curtains to those made of iron.

8.7 How to type connectives?

We have taken a digression in our aim of comparing the meaning of two distinct Bob Dylan lyrics; in particular, we left the question of how to deal with connectives unanswered. This was intentional, in that—as we demonstrate below—the behaviour of connectives is closely connected with questions of reversibility and evaluation.

Recall that we treated the noun phrases:

1. **Marilyn Monroe**
2. **Fidel Castro and his beard**

simply as two distinct noun phrases. However, noun phrase 2 above is clearly the *conjunction* of two distinct noun phrases; rather than being a noun phrase itself, it is a compound that should *evaluate to* a noun phrase. The question then, is simply *how should we type* **and**? As the typing will prove rather intricate, we first consider an alternative method of dealing with connectives.

8.7.1 Distributivity and conjunction

A common point of view is that, given a sentence containing the conjunction of two noun phrases, it should simply be split in two using *distributivity*, and the two sentences treated separately. For example, using distributivity,

L2 *Bob Dylan likes Fidel Castro and his beard*

would be replaced by

L2$_a$ *Bob Dylan likes Fidel Castro.*
L2$_b$ *Bob Dylan likes Fidel Castro's beard.*

This seems to be valid from a grammatical point of view, and (assuming we resolve the anaphor **his** before applying distributivity) the meaning of $L1$ is indeed the conjunction of $L1_a$ and $L1_b$. However, this is not always the case. Consider the following sentence:

T *Fidel Castro and Marilyn Monroe played tennis.*

Applying distributivity, we get the (grammatically correct)

T$_a$ *Fidel Castro played tennis.*
T$_b$ *Marilyn Monroe played tennis.*

Intuitively, we are happy to believe the conjunction of T_a and T_b, but find T rather implausible; since tennis is generally an activity indulged in by two people, we deduce that it was a joint, shared game of tennis.

Although the above example is somewhat facetious, the question of when and whether applying distributivity changes meaning has been heavily studied (Winter, 2001), including in a legal context (Vogel, 2009). See Halstead (2008) for a particular case involving arguments on whether distributivity is applicable to conjunction in the phrase 'to keep and bear arms' and whether doing so changes the meaning of this phrase. It appears that when we consider meaning as well as grammatical correctness we are forced to consider how the connectives (*and*, *or*, etc.) are typed, and behave under evaluation.

8.7.2 Typing connectives and polymorphism

The first problem is that although (based on its usage in **L1**) we might simply wish to type *and* as an element of $[[NP \to NP] \leftarrow NP]$ (or equivalently, $[NP \to [NP \leftarrow NP]]$). The word *and* is used in other settings, as the following examples demonstrate.

TABLE 8.1 Uses of the word 'and'

Type	Example
Noun phrases	*Fidel Castro and his beard*
Transitive verbs	*Bobby loves and obeys Marilyn Monroe*
Adjectives	*Fidel's big and bushy beard*
Sentences	*Bobby likes Fidel and I like Marilyn Monroe*

However, in every case, the appropriate typing appears to be

$$[[X \to X] \leftarrow X]$$

where X ranges over types, according to context. The same phenomenon appears to apply to other binary connectives.[9]

Thus it appears that the typing of binary connectives is *polymorphic*. We refer to Reynolds (1983) for the notion of parametrized types in computer science, and Girard (1971) for *System F*, the polymorphic lambda calculus. Borrowing notation from this polymorphic lambda calculus, we write the type of **and** as

---

[9] For example, **or** may be substituted for **and** in any of the above. Also, although English does not have a single connective corresponding to 'exclusive or', one could easily conceive of sentences such as *I like exactly one of Fidel Castro and his beard*, which would behave in a similar way. However, the same does *not* hold for **implies**, which is generally applied to entire sentences only.

$$\Lambda X.[[X \to X] \leftarrow X].$$

We do not give a full treatment in terms of the polymorphic lambda calculus; rather we simply treat this as shorthand for the following. Given some binary connective B, then the type of B is dependent on the context; given some element of type

$$U \otimes B \otimes V$$

together with evaluation arrows $Eval_{U,X}$ and $Eval_{V,X}$, then the 'polymorphically typed' connective B provides us with some element B_X of type

$$X \to [X \leftarrow X] \quad \text{or equiv.} \quad [X \to X] \leftarrow X.$$

8.7.3 Forgetfulness and binary connectives

In Section 8.6.1, we made an argument, based on linguistic interpretation, that the arrows of a category named by word of various types are *forgetful*: they lose information, the example given being how adjectives should, in certain cases, act as projectors or partial identities on noun phrases. However, it is clear that the connectives do not follow this general principle: when we use **and** to concatenate two sentences (or noun phrases, adverbs, etc.), we do not expect to lose any information about the constituents in this conjunction.

As a trivial example, consider taking some body of text, and replacing each full stop (period) by 'and'. Although legibility will rapidly be lost, it would be difficult to claim that any meaning or content has been erased. Thus the element

$$I \xrightarrow{and_X} [X \to [X \leftarrow X]]$$

is the name of an arrow in $\mathcal{Meaning}(X, [X \leftarrow X])$ that is *information-preserving* in the sense laid out in Sections 8.5 and 8.6.1. We consider the implications of this observation shortly, but first use some abstract category theory to simplify the types of arrows being named.

8.7.4 Revisiting types of connectives

In order to make a *considerable* simplification of the resulting theory, we now make the assumption that the left evaluation arrow and the right evaluation arrow are identical (at least, up to some canonical symmetry isomorphism). Although there is no decisive linguistic justification for this in general,[10] it is certainly satisfied by *compact closed categories* (Kelly and Laplaza, 1980), which feature heavily in

---

[10] Although, since the connectives **and** and **or** appear to be symmetric, this assumption is indeed justified for the particular examples we consider.

models of linguistics and meaning such as the vector spaces as used in distributional semantics, the more general models of meaning of Coecke, Sadrzadeh and Clark (2010), and purely grammatical models such as Lambek pregroups (Lambek, 1958).

Given this assumption, we may appeal to the defining equations of monoidal closure, from Definition 8.2, and—up to isomorphism—replace elements of type $\mathcal{M}eaning(I, [X \to [X \leftarrow X]])$ by elements of type $\mathcal{M}eaning(I, [X \otimes X \to X])$.

Thus (up to some canonical isomorphism that we elide in the following sections), a polymorphic connective such as **and** determines a family of elements

$$\mathbf{and}_{(X)} \in \mathcal{M}eaning(I, [X \otimes X \to X])$$

where X ranges over various objects, including $\{S, NP, AD, TV, \ldots\}$. Furthermore, as demonstrated in Section 8.7.3 above, in each case $\widetilde{\mathbf{and}}_X$ is the name of some arrow $\widetilde{\mathbf{and}}_X \in \mathcal{M}eaning(X \otimes X, X)$ that preserves generalized inner products and thus (from Lemma 8.5.1) has a left inverse given by its dagger.

8.7.5 Do arrows named by connectives have a right inverse?

In Section 8.7.3 above, we made the case that the object-indexed family of arrows named by the connective **and** (and, quite possibly, other binary connectives) are information-preserving, in the sense that they are *isometries*, i.e their adjoint is a left inverse, and thus they preserve the generalized inner product. In fact, it is easy to make a case that their adjoint should be a *two-sided* inverse, and they are thus unitary.[11] The justification for this is (for the connective **and**, in the case of the sentence object S) that, given some sentence $W \in \mathcal{C}(I, S)$, we can always find some pair of sentences $U, V \in \mathcal{C}(I, S)$ such that the (evaluation of the) sentence $U \otimes \mathbf{and} \otimes V$ has exactly the same intended meaning as W. A similar argument can be made for other objects in $\mathcal{M}eaning$, and for other binary connectives.

8.7.6 Frobenius algebras and self-similarity

We have seen that, for every polymorphic connective c and appropriate object $X \in Ob(\mathcal{M}eaning)$, there exists some isomorphism $\widetilde{c}_X \in \mathcal{M}eaning(X \otimes X, X)$ whose inverse is its dual $\widetilde{c}_X^{-1} = \widetilde{c}_X^\dagger \in \mathcal{M}eaning(X \to X \otimes X)$, and thus $X \cong X \otimes X$, i.e. the object X is self-similar in the sense of Hines (1998, 1999). The question we now

---

[11] We also observe that, even if this assumption should prove to be unfounded, the resulting mathematical structures will be almost identical; should we be forced to deal with some isometry $c_X \in \mathcal{M}eaning(X \otimes X \to X)$ that has a left inverse but not a right inverse, the appropriate mathematical tool will prove to be the *Karoubi envelope*, or *splitting idempotents* construction (Mac Lane, 1971), as applied to exactly this situation by Hines (1998, 2002).

address is whether, at least up to canonical isomorphism, this self-similarity gives rise to a *Frobenius algebra* structure at each of these objects in the category $\mathcal{M}$*eaning*.

Frobenius algebras in categories, definitions, diagrammatics, and various special cases and applications are well covered in other chapters, so the following exposition is brief. In particular, we refer to Chapter 7 for more detailed theory, and to Chapter 1 for a suitable string-diagram formalism.

Definition 8.13 *A Frobenius algebra in a monoidal category* $(\mathcal{C}, \otimes, I)$ *consists of a monoid structure* $(\Delta : S \to S \otimes S, \top : S \to I)$ *and a comonoid structure* $(\nabla : S \otimes S \to S, \bot : I \to S)$ *at the same object, where the monoid / comonoid pair satisfy the Frobenius condition*

$$\Delta \nabla = (1_S \otimes \Delta)(\nabla \otimes 1_S) \in \mathcal{C}(S \otimes S, S \otimes S).$$

Expanding out the definitions of a monoid and a comonoid structure, we have:

1. (associativity) $(\Delta \otimes 1_S)\Delta = (1_S \otimes \Delta)\Delta \in \mathcal{C}(S, S \otimes S \otimes S)$
2. (co-associativity) $\nabla(1_S \otimes \nabla) = \nabla(\nabla \otimes 1_S) \in \mathcal{C}(S \otimes S \otimes S, S)$
3. (unit) $\nabla(\bot \otimes 1_S) = \nabla(1_S \otimes \bot)$
4. (co-unit) $(\top \otimes_S)\Delta = 1_X \otimes \top)\Delta$.

An immediate observation is that the above axioms for the monoid/comonoid structure ignore coherence isomorphisms. In particular, they assume *strict associativity*, or at least ignore the role of associativity isomorphisms. The same also holds for the Frobenius condition, since $1_S \otimes \Delta \in \mathcal{C}(S \otimes S, S \otimes (S \otimes S))$ whereas $\nabla \otimes 1_S \in \mathcal{C}((S \otimes S) \otimes S, S \otimes S)$. With this in mind, we make the following definition.

Definition 8.14 *A lax Frobenius algebra in a monoidal category* $(\mathcal{C}, \otimes, I)$ *is defined to be an object* $S \in Ob(\mathcal{C})$, *along with arrows*

$$\Delta \in \mathcal{C}(S, S \otimes S), \ \nabla \in \mathcal{C}(S \otimes S, S), \ \top \in \mathcal{C}(S, I), \ \bot \in \mathcal{C}(I, S),$$

that satisfies the axioms of defn 8.13 above, up to canonical coherence isomorphisms.

It is also sometimes useful to consider structures that satisfy all the axioms for a Frobenius algebra—whether lax or strict—except those relating to the unit object (i.e. the existence of the arrows $\top, \bot$, and axioms 3–4 above). Such structures (S, Δ, ∇) are called unitless Frobenius algebras. These are particularly relevant when working with the unitless monoidal categories of defn 8.18 onwards.

Our claim, to be justified over the following sections, is that the arrows named by connectives do indeed provide (lax, unitless) Frobenius algebras in the category $\mathcal{M}$*eaning*. However, the details of the exact canonical coherence isomorphisms required are subtle and quite possibly controversial; we first need an in-depth investigation of the categorical structure of self-similarity.

8.8 Self-similarity, categorically

In this section, and the following sections, we do *not* appeal to Mac Lane's coherence theorem for associativity, and treat all associativity isomorphisms as though they were strict identities. For justification, we refer to Isbell's argument (quoted by Mac Lane (1971) as justification for introducing associativity up to isomorphism) and give an updating of Isbell's argument to a more general setting in Appendix 8.B.

Definition 8.15 *Let $(\mathcal{C}, \otimes, I)$ be a monoidal category. A self-similar structure $(S, \triangleleft, \triangleright)$ is defined to be an object $S \in Ob(\mathcal{C})$, together with two mutually inverse arrows*

- (code) $\triangleleft \in \mathcal{C}(S \otimes S, S)$.
- (decode) $\triangleright \in \mathcal{C}(S, S \otimes S)$.

satisfying $\triangleright \triangleleft = 1_{S \otimes S}$ and $\triangleleft \triangleright = 1_S$, so the following diagram commutes.

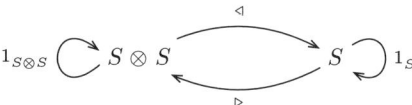

When there is a self-similar structure at some object $S \in Ob(\mathcal{C})$, we say (using the terminology of Hines (1998, 1999)) that S is a self-similar object. Note that there may be many distinct self-similar structures at the same object.

When $(\mathcal{C}, \otimes, (\ )^\dagger)$ is a dagger monoidal category and $\triangleleft = \triangleright^{-1} = \triangleright^\dagger$, we say that $(S, \triangleleft, \triangleright)$ is a dagger self-similar structure.

Motivating examples include the natural numbers $\mathbb{N}$ in various categories (relations, partial functions, partial reversible functions, etc.) with respect to various monoidal tensors (cartesian product, disjoint union). Other examples arise in the study of fractals (the Cantor set (Hines, 1998) and fractals in general (Leinster, 2011)), logical models such as Scott's celebrated domain-theoretic models of the untyped lambda calculus (see Lambek and Scott, 1986, for a categorical exposition), inverse semigroups and tilings (Lawson, 1998; Kellendonk and Lawson, 2000), the Thompson groups (Lawson, 2007), and the Cuntz C^* algebras (Cuntz, 1977).

Although there is a close connection between such self-similar structures and the canonical coherence isomorphisms of a monoidal category (Hines, 2002), we emphasize that for any given object S, there are generally many self-similar structures. Simple cardinality arguments demonstrate that the set of bijections $\{f : \mathbb{N} \to \mathbb{N} \uplus \mathbb{N}\}$ is uncountable; this is expanded on in Appendix 8.A, where an explicit correspondence between interior points of the Cantor set and order-preserving bijections from $\mathbb{N}$ to $\mathbb{N} \uplus \mathbb{N}$ is given.

Despite this, the maps between self-similar structures are particularly simple:

Definition 8.16 *Given two self-similar structures $(S, \triangleleft_1, \triangleright_1)$ and $(S, \triangleleft_2, \triangleright_2)$ at some object S of a symmetric monoidal category $(\mathcal{C}, \otimes)$, a morphism between them is an arrow $u \in \mathcal{C}(S, S)$ such that the following diagram commutes:*

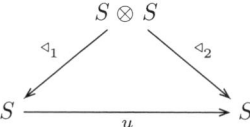

Proposition 8.17 *Let $u : (S, \triangleleft_1, \triangleright_1) \to (S, \triangleleft_2, \triangleright_2)$ be a morphism of self-similar structures. Then $u : S \to S$ is the isomorphism given by $u = \triangleleft_2 \triangleright_1$.*

Proof By definition of a self-similar structure, the following diagram commutes:

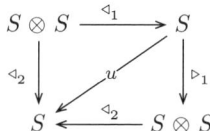

and hence $u = \triangleleft_2 \triangleright_1$. When these are dagger self-similar structures, it is also trivially unitary. □

Thus, with this definition of morphism, self-similar structures at some object $S \in Ob(\mathcal{C})$ form a *skeletal category*, where there is exactly one arrow between any two objects.

8.8.1 The generalized convolution functor

Given an arbitrary object of a monoidal category $(\mathcal{C}, \otimes)$, it generates a subcategory of $\mathcal{C}$ in the obvious way.

Definition 8.18 *Let T be an arbitrary object of a monoidal category $(\mathcal{C}, \otimes, I)$. We define the category $T^\otimes$ generated by T and $\otimes$ to be the wide subcategory of $\mathcal{C}$ with the following inductively defined objects.*

- $T \in Ob(T^\otimes)$.
- Given $A, B \in Ob(T^\otimes)$, then $A \otimes B \in Ob(T^\otimes)$.

It is immediate that $T^\otimes$ is closed under the monoidal tensor on both arrows and objects, and hence has all the structure of a monoidal category apart from the unit object I. Such categories are called unitless monoidal categories.

Unitless monoidal categories are (trivially) not well-pointed. However—as in the above example—they may arise as subcategories of well-pointed categories.

Proposition 8.19 *Given a self-similar structure $(S, \triangleleft, \triangleright)$ in a monoidal category $(\mathcal{C}, \otimes)$, then for every $X \in Ob(S^\otimes)$, there exist isomorphisms*

$$\triangleright_X : S \to X, \quad \triangleleft_X = \triangleright_X^{-1} : X \to S.$$

Proof We give these isomorphisms inductively:

- $\triangleright_S = 1_S = \triangleleft_S$
- $\triangleright_{X \otimes Y} = (\triangleright_X \otimes \triangleright_Y) \triangleright$
- $\triangleleft_{X \otimes Y} = \triangleleft (\triangleleft_X \otimes \triangleleft_Y)$.

It is straightforward to verify that $\triangleright_A \in \mathcal{C}(S, A)$ and $\triangleleft_A \in \mathcal{C}(A, S)$ are isomorphisms, and each other's inverse. Similarly, when $(S, \triangleleft, \triangleright)$ is a dagger self-similar structure then $\triangleleft_A = \triangleright_A^\dagger \in \mathcal{C}(A, S)$, for all $A \in Ob(\mathcal{C})$. □

For every self-similar structure $(S, \triangleleft, \triangleright)$ in some monoidal category $(\mathcal{C}, \otimes)$, there is an obvious functor from $S^\otimes$ to $\mathcal{C}(S, S)$, considered as a one-object category.

Definition 8.20 *Given a self-similar structure $(S, \triangleleft, \triangleright)$ in a monoidal category $(\mathcal{C}, \otimes)$, we define the* generalized convolution *functor*

$$\Phi_\triangleleft : S^\otimes \to \mathcal{C}(S, S)$$

as follows.

- (Objects) $\Phi_\triangleleft(X) = S$ for all $X \in Ob(S^\otimes)$
- (Arrows) Given $f \in S^\otimes(X, Y)$, then $\Phi_\triangleleft(f) = \triangleleft_Y f \triangleright_X$, as shown below:

$$\begin{array}{ccc} X & \xrightarrow{f} & Y \\ \triangleleft_X \uparrow & & \downarrow \triangleleft_Y \\ S & \xrightarrow{\Phi_\triangleleft(f)} & S \end{array}$$

where $\triangleleft_Y : Y \to S$ and $\triangleright_X : S \to X$ are as in Proposition 8.19.

Proposition 8.21 *Given a self-similar structure $(S, \triangleleft, \triangleright)$ in a monoidal category $(\mathcal{C}, \otimes)$, the generalized convolution $\Phi_\triangleleft : S^\otimes \to \mathcal{C}(S, S)$ defined above is indeed a functor. Further, when $(\mathcal{C}, \otimes)$ is a dagger monoidal category, and $(S, \triangleleft, \triangleright)$ is a dagger self-similar object, then the functor $\Phi_\triangleleft$ preserves the dagger, so*

$$\Phi_\triangleleft(f^\dagger) = (\Phi_\triangleleft(f))^\dagger.$$

Proof We refer to Hines (1998) for proof that $\Phi_\triangleleft : S^\otimes \to \mathcal{C}(S, S)$ is indeed a functor. Now assume that $(S, \triangleleft, \triangleright)$ is a dagger self-similar structure. By definition, for arbitrary $f \in S^\otimes(X, Y)$

$$(\Phi_\triangleleft(f))^\dagger = (\triangleleft_Y f \triangleright_X)^\dagger = (\Phi_\triangleleft(f))^\dagger = \triangleright_X^\dagger f^\dagger \triangleleft_Y^\dagger.$$

By Proposition 8.19 above, $\triangleleft_Y^\dagger = \triangleright_Y$ and $\triangleright_X^\dagger = \triangleleft_X$, so $\left(\Phi_\diamond(f)\right)^\dagger = \triangleleft_X f^\dagger \triangleright_Y = \Phi_\diamond(f^\dagger)$ as required. □

8.8.2 Untyped monoidal categories

As well as being functorial $\Phi_\diamond : S^\otimes \to \mathcal{C}(S, S)$ preserves many categorical properties, such as monoidal structures, categorical closure, categorical traces, etc. (Hines, 1999). We briefly outline how this gives $\mathcal{C}(S, S)$ the structure of a (one-object) monoidal category.

Definition 8.22 *Given a self-similar structure $(S, \triangleleft, \triangleright)$ in a monoidal category $(\mathcal{C}, \otimes)$, we define the* internal tensor *of S determined by this self-similar structure to be the monoid homomorphism*

$$\_ \otimes_\diamond \_ : \mathcal{C}(S, S) \times \mathcal{C}(S, S) \to \mathcal{C}(S, S)$$

given by the following generalized convolution:

$$\begin{array}{ccc} S \otimes S & \xrightarrow{f \otimes g} & S \otimes S \\ \triangleright \uparrow & & \downarrow \triangleleft \\ S & \xrightarrow{f \otimes_\diamond g} & S \end{array}$$

We refer to Hines (1998, 1999) for proof that this is a monoid homomorphism; this also follows from the fact that $f \otimes_\diamond g : S \to S$ is, by definition, the image of $f \otimes g : S \otimes S \to S \otimes S$ under the generalized convolution functor $\Phi_\diamond : S^\otimes \to \mathcal{C}(S, S)$.

We now demonstrate that $(\mathcal{C}(S, S), \otimes_\diamond)$ is a one-object unitless monoidal category.

Theorem 8.23 *Given $S \in Ob(\mathcal{C})$ as above, then there exists some $\tau_\diamond \in \mathcal{C}(S, S)$ satisfying, for all $f, g, h \in \mathcal{C}(S, S)$:*

1. *(naturality)* $\tau_\diamond.(f \otimes_\diamond (g \otimes_\diamond h)) = ((f \otimes_\diamond g) \otimes_\diamond h).\tau_\diamond$
2. *(pentagon condition)* $(\tau_\diamond \otimes_\diamond 1_S)\tau_\diamond(1_S \otimes \tau_\diamond) = \tau_\diamond^2$.

Proof Let $\tau_\diamond \in \mathcal{C}(S, S)$ be defined as follows:

$$\begin{array}{ccccc} S & \xrightarrow{\triangleright} & S \otimes S & \xrightarrow{1_S \otimes \triangleright} & S \otimes (S \otimes S) \\ \tau_\diamond \downarrow & & & & \downarrow \tau_{S,S,S} \\ S & \xrightarrow{\triangleleft} & S \otimes S & \xleftarrow{\triangleleft \otimes 1_S} & (S \otimes S) \otimes S \end{array}$$

Either direct calculation, or referring to Hines (1998, 1999), will demonstrate that conditions 1 and 2 of Theorem 8.23 are satisfied. Thus $(\mathcal{C}(S, S), \otimes_\diamond)$ satisfies all the axioms for a monoidal category, apart from the unit object. Note that in this category associativity can only be up to isomorphism, and is never strict; forcing the identity $\tau_\triangleleft = 1_S$ will make $C(S,S)$ collapse to an abelian monoid (see Appendix 8.B). □

A simple corollary of this is the following.

Corollary 8.24 *Given a self-similar structure $(S, \triangleleft, \triangleright)$ in a monoidal category $(C, \otimes, \tau\_\_)$, let $\Phi_\triangleleft : S^\otimes \to C(S,S)$ be the functor of defn 8.20. Then for all objects $X, Y, Z, A, B, C \in Ob(S^\otimes)$,*

$$\Phi_\triangleleft(\tau_{X,Y,Z}) = \Phi_\triangleleft(\tau_{A,B,C})$$

i.e. $\Phi_\triangleleft$ maps all associativity isomorphisms of $S^\otimes$ to $\tau_{\triangleright\triangleleft} \in C(S,S)$.

Proof This follows by the uniqueness of canonical isomorphisms in monoidal categories. □

Corollary 8.25 *Let $(S, \triangleleft, \triangleright)$ be a dagger self-similar structure of some dagger-monoidal category $(C, \otimes, (\ )^\dagger)$. Then the one-object (unitless) monoidal category $(C(S,S), \otimes_\triangleleft)$ is a (unitless) dagger-monoidal category.*

Proof We have seen that the functor $\Phi_\triangleleft$ preserves the dagger operation. Now choose arbitrary $A, B, C \in Ob(S^\otimes)$. By Corollary 8.24 above,

$$\tau_\triangleleft^\dagger = \Phi_\triangleleft(\tau_{X,Y,Z}^\dagger) = \Phi_\triangleleft(\tau_{A,B,C}^{-1}) = \tau_\triangleleft^{-1}.$$

Thus $\tau_\triangleleft$ is unitary, as required. □

As well as preserving the monoidal structure, the functor $\Phi_\triangleleft$ preserves any symmetric monoidal structure. We outline the proof, and refer to Hines (1998, 1999) for details.

Theorem 8.26 *Let $(S, \triangleleft, \triangleright)$ be a self-similar structure of some symmetric monoidal category $(C, \otimes, t, s)$. Then the functor $\otimes_\triangleleft$ of defn 8.22 is symmetric, up to a natural isomorphism satisfying Mac Lane's hexagon condition. Further, when $(S, \triangleleft, \triangleright)$ is a dagger self-similar structure, then this canonical isomorphism is also unitary.*

Proof We define the arrow $\sigma_\triangleleft \in C(S,S)$ by the following convolution:

$$\begin{array}{ccc} S \otimes S & \xrightarrow{\sigma_{S,S}} & S \otimes S \\ {\scriptstyle \triangleright} \uparrow & & \downarrow {\scriptstyle \triangleleft} \\ S & \xrightarrow{\sigma_\triangleleft} & S \end{array}$$

Equivalently, $\sigma_\lozenge = \Phi_\lozenge(\sigma_{S,S})$. The functoriality of $\Phi_\lozenge$ implies

- (*naturality*) $\sigma_\lozenge(f \otimes_\lozenge g) = (g \otimes_\lozenge f)\sigma_\lozenge$ for all $f, g \in \mathcal{C}(S, S)$.

Either direct calculation or reference to Hines (1998, 1999) will also demonstrate the following:

- (*Mac Lane's hexagon*) $\tau_\lozenge \sigma_\lozenge \tau_\lozenge = (\sigma_\lozenge \otimes_\lozenge 1_S)\tau_\lozenge(1_S \otimes_\lozenge \sigma_\lozenge)$.

Uniqueness of canonical isomorphisms will (in the same manner as Corollary 8.24) demonstrate that $\Phi_\lozenge(\sigma_{A,B}) = \sigma_\lozenge$ for arbitrary $A, B \in Ob(S^\otimes)$ and therefore, using almost identical reasoning to Corollary 8.25, when $(S, \triangleleft, \triangleright)$ is a dagger self-similar structure the arrow $\sigma_\lozenge \in \mathcal{C}(S, S)$ is also unitary. □

8.8.3 Untyped and polymorphically typed systems: a discussion

The functor $\Phi_\lozenge$ of defn 8.20 may be seen to be a general *type-erasing* construction; it maps various categorical structures to one-object (i.e. untyped) analogues of the same structures. Examples include, but are certainly not limited to, symmetric monoidal structures, categorical closure (including compact closure and cartesian closure), categorical traces, projections and injections, etc. (Hines, 1998; Hines, 1999).

The immediate question must then be: *why are we spending so much time developing a type-erasing procedure, when the whole point of the linguistics project is to introduce types into models of meaning?*

Recall that the starting point for our investigation of self-similarity was the (polymorphically typed) connectives; linguistic arguments were used to demonstrate that many of the objects corresponding to types in our models of meaning must be self-similar, with the self-similarity exhibited by the arrows named by connectives. Partly, therefore, the existence of such structures is forced upon us by the typing of connectives and their intended interpretation in some categorical model of meaning. However, there is a more fundamental justification; so far we have simply treated the polymorphically typed connectives as arrows parametrized by some class of objects. If we were to take a more foundational approach and look for models based on models of (for example) System F, we would discover a close connection between polymorphic typing and such a type-erasing procedure.

In models of polymorphic lambda calculus and related systems, the underlying categories commonly have a single object; the types of the logical system are built up from certain families of arrows, satisfying a 'biorthogonality' relation. Although this is far beyond the scope of this paper, we refer to Hyland and Schalk (2003) for an interesting point of view and details and references for this kind of approach. From a linguistic point of view, we simply remark that it is perhaps not so surprising that the polymorphically typed terms are exactly those that do not lose information, inevitably and counterintuitively leading to such a type-erasing procedure.

8.9 Self-similarity and lax Frobenius algebras

We have now developed the categorical machinery that enables us to justify the claim made at the end of Section 8.7.6 that self-similar structures form (lax unitless) Frobenius algebras (defn 8.14), i.e. they satisfy the axioms of defn 8.13 (excluding those based on the unit object) up to canonical coherence isomorphisms.

Theorem 8.27 *Let $(S, \triangleleft, \triangleright)$ be a self-similar structure in some monoidal category $(\mathcal{C}, \otimes, \tau)$, and let $(\mathcal{C}(S,S), \otimes_{\triangleleft\triangleright}, \tau_{\triangleleft\triangleright})$ be the corresponding one-object unitless monoidal category described in Theorem 8.23. Then the following conditions are satisfied.*

1. *(Unitless monoid) The arrow $\triangleleft \in \mathcal{C}(S \otimes S, S)$ is associative, up to the canonical associativity isomorphisms $\tau_{S,S,S}, \tau_{\triangleleft\triangleright}$.*
2. *(Unitless comonoid) The arrow $\triangleright \in \mathcal{C}(S, S \otimes S)$ is co-associative, up to the canonical associativity isomorphisms $\tau_{S,S,S}^{-1}, \tau_{\triangleleft\triangleright}$.*
3. *(Unitless Frobenius condition) The pair of arrows $\triangleleft \in \mathcal{C}(S \otimes S, S)$ and $\triangleright \in \mathcal{C}(S, S \otimes S)$ satisfy the Frobenius condition of defn 8.13, up to the canonical associativity isomorphisms $\tau_{S,S,S}^{-1}, \tau_{\triangleleft\triangleright}$.*

Hence the self-similar structure $(S, \triangleleft, \triangleright)$ is a (unitless, lax) Frobenius algebra.

Proof Almost by definition, the following diagrams may be seen to commute:

1. *(Lax associativity)*

$$\begin{array}{ccccc}
S & \xrightarrow{\triangleright} & S \otimes S & \xrightarrow{1_S \otimes \triangleright} & S \otimes (S \otimes S) \\
{\scriptstyle \tau_{\triangleleft\triangleright}} \downarrow & & & & \downarrow {\scriptstyle \tau_{S,S,S}} \\
S & \xrightarrow{\triangleright} & S \otimes S & \xrightarrow{\triangleright \otimes 1_S} & (S \otimes S) \otimes S
\end{array}$$

2. *(Lax co-associativity)*

$$\begin{array}{ccccc}
S \otimes (S \otimes S) & \xrightarrow{1_S \otimes \triangleleft} & S \otimes S & \xrightarrow{\triangleleft} & S \\
{\scriptstyle \tau_{S,S,S}} \downarrow & & & & \downarrow {\scriptstyle \tau_{\triangleleft\triangleright}} \\
(S \otimes S) \otimes S & \xrightarrow{\triangleleft \otimes 1_S} & S \otimes S & \xrightarrow{\triangleleft} & S
\end{array}$$

3. *(Lax Frobenius condition)*

$$\begin{array}{ccc} S \otimes S & \xrightarrow{\triangleright \otimes 1_S} & (S \otimes S) \otimes S \\ {\scriptstyle \triangleright \circ \tau_\triangleleft \circ \triangleleft} \downarrow & & \downarrow {\scriptstyle \tau^{-1}_{S,S,S}} \\ S \otimes S & \xleftarrow{1_S \otimes \triangleright} & S \otimes (S \otimes S) \end{array}$$

□

Corollary 8.28 *Let* $(S, \triangleleft, \triangleright)$ *be a dagger self-similar structure in some dagger monoidal category* $(\mathcal{C}, \otimes, \tau(\ )^\dagger)$. *Then the dagger self-similar structure* $(S, \triangleleft, \triangleright)$ *is— up to the same canonical coherence isomorphisms listed above—a (unitless, lax) dagger Frobenius algebra.*

Proof This is a simple corollary of Theorem 8.27, Proposition 8.21, and Corollary 8.25. □

Remark 8.29 *The role of coherence isomorphisms in Theorem 8.27*

From a certain point of view, the proof of Theorem 8.27 seems to be cheating, in that the diagrams used to *prove* associativity etc. up to isomorphism are minor variants of those used to *define* the associativity isomorphism $\tau_\triangleleft$. In particular, it seems highly unsurprising that associativity and co-associativity hold up to isomorphism since, after all, the code and decode arrows are both themselves isomorphisms. What rescues this from being a triviality is that $\tau_\triangleleft$ is natural and satisfies Mac Lane's pentagon condition (Theorem 8.23). Therefore, in a very strong sense, it is exactly a *canonical coherence isomorphism*.

Even so, the fact that two, rather than one, canonical isomorphisms are required may be considered to be pushing the definition of 'lax' too far, especially since they are, technically, coherence isomorphisms for two distinct monoidal categories. However, these categories and monoidal tensors are not arbitrary; instead, one is a wide subcategory of the other, and the distinct monoidal tensors and coherence arrows are mutually definable by a generalized form of convolution. What is really required is a coherence theorem for this very special situation!

In the absence of a full coherence theorem covering such a situation, we are forced to rely on the familiar coherence theorems of Mac Lane for monoidal categories. Ultimately, however, the acid test must be whether such structures behave in a similar manner to more familiar Frobenius algebras. We are thus led into particular examples of Frobenius algebras and their applications to demonstrate that this is indeed the case.

8.10 Classical structures

A particular form of Frobenius algebra, used heavily in categorical quantum mechanics, is the *classical structure* (Coecke and Pavlovic, 2007). When modelling quantum phenomena in abstract categories (e.g. as in Coecke and Pavlovic, 2007, and Pavlovic, 2009) the notion of a classical structure is fundamental in ways beyond the scope of this chapter, although fundamental to other chapters in this volume. Instead, we consider their behaviour in a particular concrete category.

In the symmetric dagger monoidal category ($\mathbf{Hilb_{FD}}, \otimes, ()^\dagger$) of finite dimensional Hilbert spaces with tensor product and Hermitian adjoint, a classical structure at an object $H \in Ob(\mathbf{Hilb_{FD}})$ is exactly an orthonormal basis for the Hilbert space H. Thus each classical structure at on object determines *matrix representations* for linear maps on this object. The connection between orthonormal bases and measurements is clear, and the (physically reasonable) unitary maps arise as isomorphisms of classical structures, or equivalently as changes of basis.

Definition 8.30 *A classical structure* $(S, \Delta, \nabla, \top, \bot)$ *in a dagger monoidal category* $(\mathcal{C}, \otimes, I, ()^\dagger)$ *is defined to be a dagger Frobenius algebra satisfying, for all* $f, g \in \mathcal{C}(S, S)$,

1. (commutativity) $(f \otimes g)\Delta = (g \otimes f)\Delta$
2. (co-commutativity) $\nabla(f \otimes g) = \nabla(g \otimes f)$
3. (the classical structure condition) $\nabla \Delta = 1_S$.

Note that defn 8.30 is somewhat over-axiomatized. In particular, as shown in Pavlovic (2009), the classical structure condition on any Frobenius algebra $(S, \Delta, \nabla, \top, \bot)$ in a dagger monoidal category will imply that $(S, \Delta, \nabla, \top, \bot)$ is a dagger Frobenius algebra. Moreover, once the identity $\Delta = \nabla^\dagger$ is satisfied, then commutativity and co-commutativity are equivalent. We have deliberately taken this over-axiomatized route, since it is not clear how many of these implications will survive the passage to the lax unitless version of the above structures.

Definition 8.31 *We define a lax classical structure in a dagger monoidal category* $(\mathcal{C}, \otimes, I, ()^\dagger)$ *to be a lax unitless dagger Frobenius algebra satisfying conditions 1–3 of defn 8.30 above, up to canonical coherence isomorphisms.*

Theorem 8.32 *Let* $(S, \triangleleft, \triangleright)$ *be a dagger self-similar structure in a symmetric dagger monoidal category* $(\mathcal{C}, \otimes, \tau, \sigma)$, *and let* $(\mathcal{C}(S, S), \otimes_{\triangleleft\triangleright}, \tau_{\triangleleft\triangleright}, \sigma_{\triangleleft\triangleright})$ *be the corresponding one-object unitless dagger symmetric monoidal category Then:*

1. $(S, \triangleleft, \triangleright)$ *is a lax unitless dagger Frobenius algebra*
2. *the arrow* $\triangleright \in \mathcal{C}(S, S \otimes S)$ *is co-commutative, up to the canonical coherence isomorphisms* $\sigma_{S,S}, \sigma_{\triangleleft\triangleright}$
3. *the arrow* $\triangleleft \in \mathcal{C}(S \otimes S, S)$ *is commutative, up to the canonical coherence isomorphisms* $\sigma_{S,S}, \sigma_{\triangleleft\triangleright}$
4. *the classical structure condition* $\triangleleft\triangleright = 1_S$ *holds, strictly,*

and hence $(S, \triangleleft, \triangleright)$ *is a lax classical structure.*

Proof As in Theorem 8.27, the following proof is almost by definition. We refer to Remark 8.29 for a discussion of the issues around this, and Section 8.11 for justification by example of why this is reasonable.

1. We refer to Theorem 8.27 and Corollary 8.28 for a proof that $(S, \triangleleft, \triangleright)$ is a lax unitless Frobenius algebra.
2. By definition of $\sigma_\triangleleft$ and naturality of canonical coherence isomorphisms, the following diagram commutes, for all $f, g \in \mathcal{C}(S, S)$:

$$\begin{array}{ccccc} S & \xrightarrow{\triangleright} & S \otimes S & \xrightarrow{f \otimes g} & S \otimes S \\ \sigma_\triangleleft \downarrow & & & & \uparrow \sigma_{S,S} \\ S & \xrightarrow{\triangleright} & S \otimes S & \xrightarrow{g \otimes f} & S \otimes S \end{array}$$

and hence by naturality of both $\sigma_{S,S}$ and $\sigma_\triangleleft$,

$$(f \otimes g) \triangleright \cong (g \otimes f) \triangleright$$

up to canonical coherence isomorphisms.

3. Similarly to point 2, since $\sigma_{S,S} = \sigma_{S,S}^{-1} = \sigma_{S,S}^\dagger$, $\sigma^{-1} = \sigma^\dagger$, and $\triangleleft^\dagger = \triangleright$, the commutativity of the above diagram for all $f, g \in \mathcal{C}(S, S)$ implies the commutativity of the following diagram:

$$\begin{array}{ccccc} S \otimes S & \xrightarrow{f \otimes g} & S \otimes S & \xrightarrow{\triangleleft} & S \\ \sigma_{S,S} \downarrow & & & & \uparrow \sigma_\triangleleft \\ S \otimes S & \xrightarrow{g \otimes f} & S \otimes S & \xrightarrow{\triangleleft} & S \end{array}$$

and thus $\triangleleft \in \mathcal{C}(S \otimes S, S)$ is commutative up to canonical coherence isomorphisms.

4. The defining equation of a self-similar structure is:

$$\triangleleft \triangleright = 1_S \, , \quad \triangleright \triangleleft = 1_{S \otimes S},$$

which is a stronger (i.e. two-sided) version of the classical structure condition. $\square$

It may be objected again (as in Remark 8.29) that using canonical isomorphisms from two distinct settings is pushing the definition of a lax structure too far. In the absence of a full coherence theorem relating the monoidal tensor of a category (and its canonical isomorphisms) with the internal tensor at an object (and its canonical isomorphisms), this is worth considering.

However, we now present an intriguing example in which distinct self-similar structures at some object S within a symmetric monoidal category determine distinct matrix representations for arrows on this object.

8.11 A self-similar structure familiar in logic (and linguistics)

One of the best-studied self-similar structures, at least in certain logical communities, is the *dynamical algebra*. There are many wide-ranging applications, including the pure untyped lambda calculus (Danos and Regnier, 1993), linear logic and the geometry of interaction (Girard, 1988), and combinatory logic (Abramsky, Haghverdi and Scott, 2002).

In Danos and Regnier 1993, the dynamical algebra is introduced as the monoid semiring $P[\mathbb{N}]$, where the monoid P may be defined in terms of generators and relations, as follows:

$$P = \langle p, q, p', q' : pp' = 1 = qq', \, pq' = 0 = qp' \rangle$$

This is of course, the inverse *polycyclic monoid* of Nivat and Perrot (1970).

Other definitions vary in the precise notion of summation used, often restricting or generalizing the summation via reference to some representation (e.g. analytic convergence in infinite-dimensional Hilbert space (Girard, 1988), suprema in the natural partial order of an inverse category (Hines, 1998), the Σ-monoid axiomatization of Haghverdi (2000), Abramsky, Haghverdi and Scott (2002), etc.). However, in every case, the representations given are self-similar structures in a monoidal category (a purely categorical explanation of how the polycyclic monoid arises from self-similarity is given in Hines, 1999).

A particularly well-studied example is the monoidal category of partial isomorphisms on sets, with disjoint union (**pInj**, $\uplus$). Here, any bijection $\mathbb{N} \cong \mathbb{N} \uplus \mathbb{N}$ gives rise to an embedding of the dynamical algebra in **pInj**$(\mathbb{N}, \mathbb{N})$ (Hines, 1998; Hines, 1999).

The key feature at this point is that, even though it does *not* have a coproduct, the monoidal category (**pInj**, $\uplus$) admits matrix representations, as observed by many authors (Hines, 1998; Lawson, 1998; Abramsky, Haghverdi and Scott, 2002). Thus, any isomorphism $\mathbb{N} \cong \mathbb{N} \uplus \mathbb{N}$ allows us to give (2 × 2) matrix representations to partial bijections on $\mathbb{N}$. However, matrix representations for these arrows are not unique; instead, each matrix representation uniquely determines, and is determined by, such an isomorphism. The details of how these dagger self-similar structures act like bases for such matrix representations, allowing for many properties more familiar from linear algebra such as matrix representations, changes of basis, diagonalizations, mutual diagonalization, etc. are the subject of a forthcoming paper.

Finally, for readers who fear that we have by now strayed way too far from any possible linguistic interpretation, we refer to the recent rediscovery of structures isomorphic to the dynamical algebra in linguistic models of meaning (Clarke, 2007; Clarke, Weir and Lutz, 2011).

Acknowledgements

The author wishes to thank: Samson Abramsky, for discussions and assistance in most of the topics covered in this chapter; Daoud Clarke, for a linguistic point of view, and particularly the linguistic interpretation of the models touched on in Section 8.11; Steve Clark, for putting up with what are probably—from a linguists point of view—very simple questions; Bob Coecke, for discussions on category theory and coherence, and their connections with both linguistics and quantum mechanics; Mark Lawson, for self-similarity from an algebraic and semigroup-theoretic viewpoint; Phil Scott, for many discussions and assistance on the logical interpretations of closed categories, and interpretations of the geometry of interaction; and Prakash Panangaden, for both a physicist's point of view, and the intuitive code/decode terminology for the arrows in a self-similar structure.

Finally, thanks are due to Mehrnoosh Sadrzadeh, from both a linguistic/categorical perspective, and for the considerable organization that made this book possible.

Appendix 8.A Order-preserving bijections $\mathbb{N} \cong \mathbb{N} \uplus \mathbb{N}$ as interior points of the Cantor set

We now give an explicit illustration of how the points of the Cantor set (excluding a distinguished subset of measure zero) may be interpreted as order-preserving bijections exhibiting the self-similarity of the natural numbers (with respect to disjoint union. However, a very similar construction applies to the natural numbers with cartesian product—this is left as an exercise).

Definition 8.33 *The Cantor set is defined to be the set of all one-sided infinite binary strings, $\mathscr{C} = \{0, 1\}^\omega$ or equivalently, the set $\mathbf{Fun}(\mathbb{N}, \{0, 1\})$ of all functions from $\mathbb{N}$ to $\{0, 1\}$. A point $a = a_0 a_1 a_2 a_3 \ldots$ of the Cantor set is called a* boundary point *if there exists some $K \in \mathbb{N}$ such that, for all $L \geq K$, $a_L = a_K$. We denote the set of all boundary points by $\mathscr{C}_\mathscr{B}$. The complement of the boundary is called the* interior, *denoted $\mathscr{C}_\mathscr{O} = \mathscr{C} \setminus \mathscr{C}_\mathscr{B}$. Members of $\mathscr{C}_\mathscr{O}$ are called* interior points *or* balanced functions. *Note that the set of boundary points is a countable subset of $\mathscr{C}$, whereas its complement, the set of interior points, has the same cardinality as the Cantor set itself*

Theorem 8.34 *Each interior point of the Cantor set $\eta \in \mathscr{C}_\mathscr{O}$ uniquely determines and is determined by an order-preserving bijection $\widetilde{\eta} : \mathbb{N} \to \mathbb{N} \uplus \mathbb{N}$.*

Proof By construction, the balanced function $\eta : \mathbb{N} \to \{0, 1\}$ divides $\mathbb{N}$ into two disjoint countably infinite subsets $\mathbb{N} = \eta^{-1}(0) \cup \eta^{-1}(1)$. Each of these subsets is totally ordered, with order inherited from $\mathbb{N}$ in the obvious way, so $\mathbb{N}$ is divided into two disjoint countably infinite chains. This is illustrated by example in Figure 8.3.

$n =$	0	1	2	3	4	5	6	7	8	9	$\cdots$
$p(n) =$	0	0	1	1	0	1	0	1	0	0	$\cdots$
$p^{-1}(0)$:	0	1			4		6		8	9	$\cdots$
$p^{-1}(1)$:			2	3		5		7			$\cdots$
$\widetilde{p}(n) =$	(0,0)	(1,0)	(0,1)	(1,1)	(2,0)	(2,1)	(3,0)	(3,1)	(4,0)	(5,0)	$\cdots$

Fig 8.3 A balanced indicator function, two chains, and a bijection.

Each interior point $\eta : \mathbb{N} \to \{0, 1\}$ uniquely determines and is determined by such a split into disjoint chains; η is simply the indicator function for chain membership. As both chains are countably infinite, this therefore determines a bijection $\widetilde{\eta} : \mathbb{N} \to \mathbb{N} \uplus \mathbb{N}$ in the obvious way. Formally, taking $\mathbb{N} \uplus \mathbb{N} \stackrel{def.}{=} \mathbb{N} \otimes \{0\} \cup \mathbb{N} \otimes \{1\}$, we have

$$\widetilde{\eta}(n) = (x, \eta(n)) \text{ where } x = \begin{cases} n - \sum_{j<n} \eta(j) & \eta(n) = 0 \\ \sum_{j<n} \eta(j) & \eta(n) = 1 \end{cases}$$

This is again illustrated in Figure 8.3.

By construction, $\widetilde{\eta} : \mathbb{N} \to \mathbb{N} \uplus \mathbb{N}$ is order-preserving, and since η is simply the indicator function for the division of $\mathbb{N}$ into two disjoint countably infinite chains, all such bijections arise in this way. Finally, it is immediate that

$$\widetilde{\eta} = \widetilde{\mu} \Leftrightarrow \eta = \mu$$

and so $\widetilde{(\ )}$ is bijective. □

The following special case is then immediate:

Corollary 8.35 *Under the correspondence of Theorem 8.34, the familiar Cantor pairing*

$$n \mapsto \begin{cases} \left(\frac{n}{2}, 0\right) & n \text{ even} \\ \left(\frac{n-1}{2}, 1\right) & n \text{ odd} \end{cases}$$

corresponds to the alternating point $0101010101\ldots$ *of the Cantor set, or equivalently the function* $n \mapsto n \pmod 2$.

Although the above example is simple, we emphasize that Theorem 8.34 is about *all* interior points of the Cantor set, not simply the computable ones. For example, we could consider an enumeration of Turing machines $\mathfrak{T}_0, \mathfrak{T}_1, \mathfrak{T}_2, \mathfrak{T}_3, \ldots$ and define the interior point $u \in \mathscr{C}_\mathcal{O}$ by $u(n) = 0$ if $\mathfrak{T}_n$ halts, and $u(n) = 1$ otherwise. By Theorem 8.34, this also corresponds to an (uncomputable) order-preserving bijection $\mathbb{N} \to \mathbb{N} \uplus \mathbb{N}$.

It is entirely possible that, should we restrict ourselves to *computable* interior points of the Cantor set, we would be able to develop a consistent theory of *computable self-similar structures*. However, this is work that remains to be carried out.

Appendix 8.B Isbell's argument in a general setting

Mac Lane (1971) introduces associativity up to isomorphism, by reference to a result of Isbell, on denumerable objects in the category (**Fun**, ×) of sets and functions with cartesian product. Isbell demonstrated that, for any object $D \in Ob(\mathbf{Fun})$, the existence of a bijection between D and $D \times D$ would, in the presence of *strict* associativity, force all endomorphism arrows of D to be identified (i.e. D would necessarily be isomorphic to the unit object of (**Fun**, ×)).

This gave a very strong argument for the necessity of considering associativity up to isomorphism, rather than strict associativity; requiring strict associativity forces an identification of the natural numbers $\mathbb{N}$ with the one-object set $\{*\}$, in the category (**Fun**, ×). A slight variation of Isbell's argument, in more modern language, demonstrates that this is not unique to the cartesian closed category **Fun**.

Let S be a self-similar object of some symmetric monoidal category $(\mathcal{C}, \otimes)$, and let the subcategory $S^\otimes$ be the full subcategory of S given in defn 8.18. It is stated in this definition that $S^\otimes$ is a *unitless* monoidal category. This is correct. However, it is worth considering the objects of $S^\otimes$ to see why none of these act as the unit object for $S^\otimes$. In particular, for every object $X \in Ob(S^\otimes)$, it is a simple corollary of Proposition 8.19 that

$$X \otimes S \cong X \cong S \otimes X.$$

The natural question then, is: 'why is $S \in Ob(S^\otimes)$ *not* the unit object?' The simple answer is that, although $X \otimes S \cong X \cong S \otimes X$, the arrows exhibiting these isomorphisms do not satisfy the coherence conditions given in Mac Lane 1971. In particular, their interaction with the other canonical isomorphisms fails.

However, when we ignore canonical isomorphisms and coherence conditions this is no longer an obstacle, and we are forced to conclude that the category $(S^\otimes, \otimes)$ does indeed have a unit object: S itself, with all this implies about the endomorphism monoid of S (see Proposition 8.7).

CHAPTER 9

From Sentence to Concept

ANNE PRELLER

(Laboratoire d'Informatique, Robotique et Microélectronique de Montpellier)

9.1 Introduction

The present work attempts to relate two semantic representations of natural language: the functional logical models and the distributional vector models. The former deals with individuals and their properties, the latter with concepts and how they can be approximated.

Montague semantics and similar functional logical models for natural language are extensional and compositional. Meaningful expressions designate individuals, sets of individuals, functions from and to (sets of) individuals, truth–value functions and so on. The meaning of a grammatical string of words is computed from the meanings of the constituents using functional application or composition. This semantics requires prior grammatical analysis, where every word contributes to the meaning, including 'noise' like negation, determiners, quantifiers, relative pronouns, etc.

The semantic vector models are based on the principle that the content of a word is measured in relation to the content of other words. They handle probabilistic estimations of concepts. Words, with the exclusion of 'noise', are represented by vectors in a finite-dimensional space over the field of real numbers. Frequency counts of co-occurrences with other words determine the coordinates of a word. Semantic vector models excel in detecting similarity of words. They confound opposites.

Compositionality of vector semantics remains an open question and is subject of intensive research. One approach to compositionality is quantum logic on the lattice of projectors of Hilbert spaces; see Rijsbergen (2004) for an overview oriented towards information retrieval or Widdows (2004) for a discussion of geometric properties of meaning. There is, however, no general algorithm that transforms a string of words into a vector respecting the logic. Another approach is composition of vectors by the tensor product, Smolensky (1988) invoking computational principles of cognitive science, or Clark and Pulman (2007) and Clark, Coecke and Sadrzadeh (2008) using syntactical analysis. Again, 'noise' is not included in composition.

The present work outlines a method that takes into account the logical content of 'noise' and transforms the compositional extensional representation into a conceptual representation. Both representations are based on biproduct dagger compact closed categories.

On one hand, a *concept space*, that is to say a tensor product of two-dimensional spaces, hosts both the words (concepts) and their probabilistic approximations. Concept spaces are the linguistic analogue to compound systems in quantum mechanics. On the other hand, the logical functional representations of (strings of) words are also recast as vectors. These vectors are, roughly speaking, the names of the functions representing the words. Their construction involves syntactical analysis.

The examples presented here are analysed via pregroup grammars (Lambek, 1999), based on compact bilinear logic (Lambek, 1993). The pregroup calculus is a simplification of the syntactic calculus by the same author (Lambek, 1958). Compact bilinear logic 'compacts' the higher order of categorial grammars into second-order logic with general models, or equivalently, into two-sorted first-order logic (van Benthem and Doets, 1983). Moreover, the category of types and proofs of compact bilinear logic is the free compact 2-category (Preller and Lambek, 2007).

Categorical semantics in compact 2-categories for pregroup grammars was first proposed in Preller (2005), and reformulated in Preller (2007) in terms of functions in two-sorted first-order logic. This reformulation rests on the fact that sets and two-sorted functions form a compact closed category. The embedding of the category of two-sorted functions in the category of semimodules over a real interval (Preller and Sadrzadeh, 2011c), establishes the connection to semantic vector models.

The formulation of functional logic in a biproduct dagger compact closed category has been chosen to facilitate a comparison with quantum logic. It is based on Abramsky and Coecke (2004), casting quantum mechanics in the abstract setting of a biproduct dagger compact closed category. The result is an embedding of functional two-sorted first-order logic into the lattice of projectors of concept spaces.

Section 9.2 introduces the semantical and syntactical categories. The category of two-sorted functions follows in Section 9.3 with its two-sorted first-order logic. An embedding transfers them to an arbitrary bicategory dagger compact closed category.

The algorithm in Section 9.4, constructing meanings of strings from meanings of words, is based on syntactical analysis. Examples from natural language provide the graphs depicting the computation of the meaning by 'information flow'.

Concept spaces and the logical properties of their intrinsic projectors are investigated in Section 9.5. Subsection 9.5.1 deals with propositional logic and Subsection 9.5.2 with predicate logic. The truth-preserving one-to-one correspondence between predicates on and intrinsic projectors of concept spaces is the subject of Section 9.5.3. This correspondence is used in Section 9.5.4 to compute the meaning of strings directly in concept spaces and to view arbitrary word vectors as a probabilistic approximation of concepts.

9.2 Notations, basic properties

Natural language processing involves both syntactical analysis and logical representation. Both can be formulated in the language of compact bicategories, also known as non-symmetric star autonomous categories.

Throughout this paper, the *syntactical category* is the compact bicategory freely generated by a 'basic' category. It is not symmetric. The *semantic category* C is any biproduct dagger compact closed category in which all objects have a chosen finite basis, for example the category $\mathcal{RI}$ of free semimodules generated by finite sets over the lattice of the real interval $[0, 1]$.

9.2.1 The syntactical category

The syntactical category $C2(\mathcal{B})$ is the free compact bicategory generated by a category $\mathcal{B}$. It is notationally convenient to replace the canonical associativity and unit isomorphisms by identities, for example $A \otimes (B \otimes C) = (A \otimes B) \otimes C$, $A \otimes I = A = I \otimes A$. Strictly speaking, the bicategory is treated like a 2-category.

Saying that a bicategory is compact means that every 1-cell A has a left adjoint A^ℓ and a right adjoint A^r. Let $\eta_A : I \longrightarrow A^r \otimes A$ be the unit and $\epsilon_A : A \otimes A^r \longrightarrow I$ the counit for the right adjoint. Then $A \simeq A^{\ell r}$ is a right adjoint to A^ℓ so that $\eta_{A^\ell} : I \longrightarrow A \otimes A^\ell$ and $\epsilon_{A^\ell} : A^\ell \otimes A \longrightarrow I$ act as unit and counit for the left adjunction of A^ℓ to A.

Starting with any 1-cell A that is an object of $\mathcal{B}$, one obtains the *iterated right adjoints* $A^\ell, A^{\ell\ell}, A^{\ell\ell\ell}, \ldots$ and the *iterated left adjoints* $A^r, A^{rr}, A^{rrr}, \ldots$ of A, but no mixed adjoints, because $A^{\ell r}$ and $A^{r\ell}$ are both isomorphic to A.

The *morphisms*, i.e. the 2-cells of $C2(\mathcal{B})$, are represented by graphs where the vertices are objects of $\mathcal{B}$ and the oriented links are labelled by morphisms of $\mathcal{B}$. Examples are:

$$\eta_A = \begin{matrix} I \\ \downarrow {\scriptstyle 1_A} \\ A^r \otimes A \end{matrix} \quad , \quad \eta_{A^\ell} = \begin{matrix} I \\ \downarrow {\scriptstyle 1_A} \\ A \otimes A^\ell \end{matrix} \quad , \quad \epsilon_A = \begin{matrix} A \otimes A^r \\ \downarrow {\scriptstyle 1_A} \\ I \end{matrix} \quad , \quad \epsilon_{A^\ell} = \begin{matrix} A^\ell \otimes A \\ \downarrow {\scriptstyle 1_A} \\ I \end{matrix}$$

Note that the graphs display the domain of the morphism above, the codomain below.

In the case where the label is an identity, it is in general omitted. An arbitrary $f : A \longrightarrow B \in \mathcal{B}$ also creates labels for links, for example:

$$\ulcorner f \urcorner = \begin{array}{c} I \\ \uparrow {\scriptstyle f} \\ A^r \otimes B \end{array} \quad , \quad (f \otimes 1_{A^\ell}) \circ \eta_{A^\ell} = \begin{array}{c} I \\ \overset{f}{\frown} \\ B \otimes A^\ell \end{array} \quad , \quad \llcorner f \lrcorner = \begin{array}{c} I \\ \underset{f}{\frown} \\ A \otimes B^r \end{array} \quad , \text{etc.}$$

Note that the labels of the links are morphisms of $\mathcal{B}$. Stripping the tail of the link of its adjoints, we obtain the domain of the label in $\mathcal{B}$. Similarly, the head without the adjoints is the codomain of the label.

Composition of morphism is computed by connecting the graphs at the joint interface and walking paths, picking up and composing the labels in the order in which they are encountered.

Here are few examples involving $f : A \longrightarrow B$, $g : B \longrightarrow C$

$$f^\ell = \begin{array}{c} B^\ell \\ \uparrow {\scriptstyle f} \\ A^\ell \end{array} = \begin{array}{c} B^\ell \\ B^\ell \otimes \overset{f}{\frown} \otimes A^\ell \\ A^\ell \end{array} = (\epsilon_{B^\ell} \otimes 1_{A^\ell}) \circ (1_{B^\ell} \otimes ((f \otimes 1_{A^\ell}) \circ \eta_{A^\ell}))$$

Recall that the domain of the morphism f^ℓ is the top line in the graph, the codomain is the bottom line, i.e. $f^\ell : B^\ell \longrightarrow A^\ell$.

$$\epsilon_{B^\ell} \circ (1_{B^\ell} \otimes f) = \begin{array}{c} B^\ell \otimes A \\ \uparrow \quad \downarrow {\scriptstyle f} \\ B^\ell \otimes B \\ \smile \\ I \end{array} = \begin{array}{c} B^\ell \otimes A \\ \overset{f}{\frown} \\ I \end{array} = \begin{array}{c} B^\ell \otimes A \\ {\scriptstyle f} \uparrow \quad \downarrow \\ A^\ell \otimes A \\ \smile \\ I \end{array} = \epsilon_{A^\ell} \circ (f^\ell \otimes 1_A)$$

An equality of graphs is far easier to compute than the equality of the corresponding algebraic expressions. For example, the equality $(\llcorner f \lrcorner \otimes 1_C) \circ (1_A \otimes \ulcorner g \urcorner) = g \circ f = (1_C \otimes (\epsilon_{B^\ell} \circ (1_{B^\ell} \otimes f))) \circ (((g \otimes 1_{B^\ell}) \circ \eta_{B^\ell}) \otimes 1_A) : A \longrightarrow C$ is proved thus:

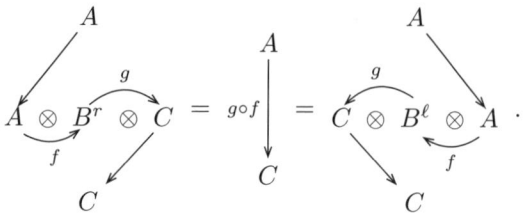

Note that links do not cross in the graphs of the syntactical category. The benefit of orienting and labelling links will become evident through the examples of natural language processing in Section 9.4.1.

9.2.2 The semantic category

The general definitions and properties of biproduct dagger compact closed categories can be found in Selinger (2007) or Abramsky and Coecke (2004). Two semantic categories are specially tailored to natural language semantics, namely the category $2\mathcal{SF}$ of two-sorted functions, Subsection 9.3.1, and the category $\mathcal{RI}$ of free semimodules over the lattice of the real interval $[0, 1]$ generated by finite sets.

Its importance to natural language processing resides in the fact that semantic vector models interpret words as vectors the coordinates of which are obtained by frequency counts of co-occurrences in context-windows. Without loss of generality, one may assume that the coordinates belong to the real interval $[0, 1]$.

Recall that the linear order on the real numbers in $[0, 1]$ induces a distributive and implication-complemented lattice structure on $[0, 1]$, namely

$$\alpha \vee \beta = \max\{\alpha, \beta\} \text{ and } \alpha \wedge \beta = \min\{\alpha, \beta\}$$
$$\alpha \rightarrow \beta = \max\{\gamma \in I : \alpha \wedge \gamma \leq \beta\}$$
$$\neg \alpha = \alpha \rightarrow 0.$$

This lattice is not Boolean, because $\neg\neg\alpha = 1 \neq \alpha$ for $0 < \alpha < 1$. The subset $\{0, 1\}$, however, forms a Boolean algebra.

The lattice operations define a semiring structure on $[0, 1]$ with neutral element 0 and unit 1 by

$$\alpha + \beta = \alpha \vee \beta \qquad \alpha \cdot \beta = \alpha \wedge \beta.$$

9.2.2.1 BASIC PROPERTIES OF BIPRODUCT DAGGER COMPACT CLOSED CATEGORIES

Objects of an arbitrary biproduct dagger compact closed category are called *spaces*, morphisms *linear maps*. A morphism $v : I \longrightarrow V$, where I is the unit of the tensor product, is called a *vector of V*. Write $v \in V$ for vectors $v : I \longrightarrow V$ and $f(v) \in W$ for $f \circ v : I \longrightarrow W$, where $f : V \longrightarrow W$.

Vectors $b_1, \ldots, b_n$ of V form a *basis* of V if every vector of V can be written in a unique way as a linear combination of the vectors $b_1, \ldots, b_n$. A space is n-dimensional if it has a basis of cardinality n. The dimension is unique. A space with chosen basis $B = b_1, \ldots, b_n$ is denoted V_B. All spaces are assumed to be finite-dimensional from now on.

Linear maps identify with matrices such that multiplication of matrices corresponds to composition of maps. Indeed, let $A = a_1, \ldots, a_m$ and $B = b_1, \ldots, b_n$. A linear map $f : V_A \longrightarrow V_B$ is determined by its values on the basis vectors $a_1, \ldots, a_m \in A$ and is characterized by the matrix (ϕ_{ij}) where $\phi_{ij} \in I$ is the i-th coordinate of $f(a_j) = \sum_{i=1}^{n} \phi_{ij} b_i$. Such matrices can be identified with vectors in $V_A \otimes W_B$. The basis vectors of the latter are $a_i \otimes b_j$, for $i = 1, \ldots, m$ and $j = 1, \ldots, n$.

The *inner product* $\langle v|w \rangle$ of vectors $v = \sum_{j=1}^{m} \alpha_j a_j$ and $u = \sum_{j=1}^{m} \beta_j a_j$ in $\mathcal{RI}$ is given by

$$\langle v|u \rangle = \langle \sum_{j=1}^{m} \alpha_j a_j | \sum_{j=1}^{m} \beta_j a_j \rangle = \sum_{j=1}^{m} \alpha_j^\dagger \beta_j .$$

Vectors are *orthogonal* if $\langle v|u \rangle = 0$. In the case of $\mathcal{RI}$, we have $\alpha = \alpha^\dagger$ for all scalars $\alpha \in [0, 1]$. Hence vectors with coordinates in $[0, 1]$ are orthogonal in $\mathcal{RI}$ exactly when they are orthogonal in the category of Hilbert spaces.

The category of semimodules $\mathcal{RI}$ is a biproduct dagger compact closed category with monoidal unit $I = [0, 1]$. Every object V of $\mathcal{RI}$ has a unique finite basis A, which we express by $V = V_A$. It is its own adjoint, $V_A = V_A^*$. The unit $\eta_{V_A} : I \longrightarrow V_A \otimes V_A$ and counit $\epsilon_{V_A} : V_A \otimes V_A \longrightarrow I$ of the adjunction are given by

$$\eta_{V_A}(1) = \sum_{a \in A} a \otimes a \qquad \epsilon_{V_A}(a \otimes b) = \begin{cases} 1 & \text{if } a = b \\ 0 & \text{else} \end{cases} = \langle a|b \rangle.$$

The name and coname of $f : V_A \longrightarrow V_B$ are defined by

$$\ulcorner f \urcorner(1) = \sum_{a \in A} a \otimes f(a)$$
$$\llcorner f \lrcorner(a \otimes b) = \langle f(a)|b \rangle \text{ for } a \in A, b \in B.$$

By definition, $V_A = V_A^\dagger$. The adjoint of $f : V_A \longrightarrow V_B$ is the morphism $f^* = f^\dagger$ induced by the transpose of the matrix of f.

9.2.2.2 THE LOGIC OF VECTORS

Definition 9.1 (Boolean vector) Let $\mathcal{C}$ be any semantic category, $0 = 0_{0I} : 0 \longrightarrow I$, $1 = 1_I : I \longrightarrow I$ and V_B any space in $\mathcal{C}$. A vector $v = \sum_{i=1}^{n} \alpha_i b_i \in V_B$ is Boolean if $\alpha_i \in \{0, 1\}$, for $i = 1, \ldots, n$.

The connectives $\neg, \wedge$ etc. are operators on the set $\{0, 1\}$ satisfying

$$\neg 0 = 1, \neg 1 = 0 \text{ and } 1 \wedge 1 = 1, 0 \wedge 0 = 1 \wedge 0 = 0 \wedge 1 = 0 \text{ etc.}$$

They lift to the Boolean vectors, where they are defined coordinate by coordinate

$$\neg \sum_{i=1}^{n} \alpha_i b_i = \sum_{i=1}^{n} (\neg \alpha_i) b_i, \quad (\sum_i \alpha_i b_i) \wedge (\sum_j \beta_j b_j) = \sum_i (\alpha_i \wedge \beta_i) b_i \text{ etc.} \quad (9.1)$$

and induce a partial order by the postulate

$$v \leq w \text{ if and only if } v \wedge w = v.$$

We have:

1. The Boolean vectors together with the logical vector connectives form a Boolean algebra.
2. Every Boolean v defines a unique subset $K \subseteq B$ such that $v = \sum_{b \in K} b$ and vice versa.
3. The *null vector* $\vec{0}$ (with coordinates all equal to 0) is the smallest and the *full vector* $\vec{1}$ (with coordinates all equal to 1) the largest vector.

Lemma 9.2 *If V_B is a space of $\mathcal{RI}$ the equalities (9.1) define a distributive, implication complemented lattice structure on V_B such that the following equivalences hold:*

$$\neg \neg v = v \iff v \vee \neg v = \vec{1} \iff \text{the coordinates of } v \text{ are 0 or 1.} \quad (9.2)$$

Moreover, vector conjunction is the linear map $\wedge : V_B \otimes V_B \longrightarrow V_B$ defined on the basis vectors by

$$\wedge (b \otimes b) = b, \quad \wedge (b \otimes b') = \vec{0}, \text{ for } b \neq b' \in B.$$

9.2.2.3 THE LOGIC OF PROJECTORS

Let $\mathcal{C}$ be any semantic category and E an n-dimensional space with chosen basis $B = b_1, \ldots, b_n$.

Recall that a morphism $p : E \longrightarrow E$ is a *projector* if it is idempotent and self-adjoint:

$$p \circ p = p, \quad p^\dagger = p.$$

In $\mathcal{RI}$, the latter equality means that the matrix of p is symmetric.

A projector p of a Hilbert space determines a subspace, namely the set of vectors invariant under p

$$E_p = \{w : w = p(w)\} = p(E).$$

These subspaces are in one-to-one correspondence with the projectors. Hence the quantum connectives are defined on the set of projectors/subspaces (see Rijsbergen, 2004) by

$$\neg E_p = E_p^\perp, \ E_p \vee E_q = E_p + E_q, \ E_p \wedge E_q = E_{p \circ q},$$
$$E_p \rightarrow E_q = \{u : q(p(u)) = p(u)\}.$$

The quantum connectives induce a not necessarily distributive lattice structure on the set of projectors such that $p \leq q$ is equivalent to $p \wedge q = q$. Thinking of projectors as propositions and of 1_E as the true proposition, the equality $p \longrightarrow q = 1_E$ is read as 'p implies q'.

This approach is not possible in an arbitrary semantic category. Any two-dimensional space $V_{\{a_1, a_2\}}$ of $\mathcal{RI}$ has subspaces that are not images of any projector. An example is the span of the vectors $u = \alpha a_1$, $v = \beta a_2$, $w = \gamma a_1 + \beta a_2$, where $0 < \beta < \alpha < \gamma \leq 1$. We need a property that connects subspaces and projectors in an arbitrary semantic category.

Definition 9.3 (Intrinsic morphism) *A linear map of $\mathcal{C}$ is intrinsic if it sends every basis vector to a basis vector or to the null vector.*

Intrinsic linear maps are closed under composition.

Lemma 9.4 *A projector $p : V_B \longrightarrow V_B$ is intrinsic if and only if*

$$p(b_i) = b_i \text{ or } p(b_i) = \vec{0}, \text{ for } i = 1, \ldots, n. \quad (9.3)$$

Proof Let p be an intrinsic projector and (π_{ij}) its matrix. This matrix is symmetric, because p is self-adjoint and the entries π_{ij} are 0 or 1.

We must show that $p(b_k) = b_l$ implies $k = l$. From $p(b_k) = b_l$ follows $p(b_l) = b_l$, because p is idempotent. The latter equality implies $\pi_{ll} = 1$ and $\pi_{il} = 0$ for $i \neq l$. Moreover, $p(b_k) = b_l$ implies $\pi_{lk} = 1$ and $\pi_{ik} = 0$ for $i \neq l$. By symmetry, $\pi_{kl} = 1$. Hence $k = l$. □

Hence every intrinsic projector has the form $\sum_{k \in K} |b_k\rangle\langle b_k|$ where $K \subseteq \{1, \ldots, n\}$, where $|b_i\rangle\langle b_i|$ maps b_i to itself and every other basis vector to the null vector. The composition $q \circ p$ of intrinsic projectors $p : V_B \longrightarrow V_B$ and $q : V_B \longrightarrow V_B$ is again an intrinsic projector satisfying

$$(q \circ p)(x) = q(x) \wedge p(x), \text{ for all } x \in B.$$

Intrinsic projectors are in one-to-one correspondence with Boolean vectors. Indeed, let $v = \alpha_1 b_1 + \cdots + \alpha_n b_n$ be a vector of V_B. Define the linear map $p_v : V_B \longrightarrow V_B$ by its values on the basis vectors

$$p_v(b_i) = \alpha_i b_i, \text{ for } i = 1, \ldots, n. \quad (9.4)$$

If v is Boolean some obvious properties are:

1. p_v is an intrinsic projector
2. $p_v(w) = v \wedge w$ for every Boolean vector w; in particular $p_v(\vec{1}) = v$
3. the map $v \mapsto p_v$ is one-to-one
4. $p_{\vec{1}} = 1_{V_B}$.

Lemma 9.5 *For every intrinsic projector p there is a Boolean vector v such that $p = p_v$. For every Boolean vector v, the subspace E_{p_v} of vectors invariant under p_v coincides with the subspace generated by the basis vectors b_i satisfying $b_i \leq v$. Every subspace generated by a subset of basis vectors is the invariance space of an intrinsic projector.*

Proof Let p be an intrinsic projector. Define

$$K = \{k : p(b_k) = b_k \ \& \ 1 \leq k \leq n\} \text{ and } v = \sum_{k \in K} b_k.$$

Then $p(b_i) = p_v(b_i)$, for $i = 1, \ldots, n$. Hence the map $v \mapsto p_v$ is onto the set of intrinsic projectors. Moreover, E_p is generated by the set of basis vectors $\{b_k : k \in K\}$ and $b_i \leq v$ if and only if $i \in K$. □

Theorem 9.6 *The map $v \mapsto p_v$ is a negation preserving lattice isomorphism $\mathcal{H}$ from the Boolean vectors onto the intrinsic projectors of V_B.*

Proof Writing $E_v = E_{p_v}$, we must prove that

$$E_v^\perp = E_{\neg v}, \ E_v \vee E_w = E_{v \vee w}, \ p_v \circ p_w = p_{v \wedge w}, \ E_v \to E_w = E_{v \to w}. \quad (9.5)$$

The first three equalities follow immediately from Property 2 and Property 4 above.

To prove the last equality, let $v = \sum_{i=1}^n \alpha_i b_i$, $w = \sum_{i=1}^n \beta_i b_i$ be Boolean vectors, and $u = \sum_{i=1}^n \gamma_i b_i$ be any vector. By definition, $v \longrightarrow w = \sum_{i=1}^n (\neg \alpha_i \vee \beta_i) b_i$. On one hand,

$$u \in E_{v \to w} \Leftrightarrow u = p_{v \to w}(u)$$
$$\Leftrightarrow \gamma_i = (\neg \alpha_i \vee \beta_i) \gamma_i, \text{ for } i = 1, \ldots, n.$$

On the other hand, recall that $E_v \to E_w = \{u : p_w(p_v(u)) = p_v(u)\}$. Hence

$$u \in E_v \to E_w \Leftrightarrow \alpha_i \gamma_i = \alpha_i \beta_i \gamma_i, \text{ for } i = 1, \ldots, n.$$

The equalities $\gamma_i = (\neg \alpha_i \vee \beta_i) \gamma_i$ and $\alpha_i \gamma_i = \alpha_i \beta_i \gamma_i$ both hold trivially if $\alpha_i = 0$ or $\beta_i = 1$. In the case where $\alpha_i = 1$ and $\beta_i = 0$ they are both equivalent to the equality $\gamma_i = 0$. □

It follows that the sublattice of intrinsic projectors is Boolean. Moreover, intrinsic projectors are monotone increasing on Boolean vectors.

The linear map defined by Equalities (9.4) is a projector in $\mathcal{RI}$ for an arbitrary vector and Equalities (9.5) hold for all vectors.

9.3 Two-sorted first-order logic in compact closed categories

The relevance of two-sorted first-order logic for natural language resides in the fact that it is equivalent to second-order logic with general models (see van Benthem and Doets, 1983), and a widespread belief that second-order logic suffices for natural language semantics.

9.3.1 The category of two-sorted functions

Two-sorted functions are tailored to natural language, because they accept elements (sort 1) as well as sets (sort 2) as arguments. In a similar way, verbs accept both singulars and plurals. Functions in two-sorted first-order logic were first used in Preller (2007). The presentation below follows Preller and Sadrzadeh (2011c).

Definition 9.7 (Two-sorted function) *A function $f : A \longrightarrow B$ is two-sorted if it maps elements and subsets of A to elements or subsets of B and satisfies*

$$\begin{aligned} f(\{a\}) &= f(a) \text{ for } a \in A \\ f(\emptyset) &= \emptyset \\ f(X \cup Y) &= f(X) \cup f(Y) \text{ for } X, Y \subseteq A. \end{aligned} \tag{9.6}$$

Obviously, a two-sorted function defined on a finite set is determined by its values on the elements. Examples are the *two-sorted identity* and the *two-sorted diagonal*.

$$1_A(a) = a, \text{ for } a \in A \qquad 2_A(a) = \langle a, a \rangle, \text{ for } a \in A.$$

Lemma 9.8 *The category $2\mathcal{SF}$ of finite sets and two-sorted functions is a dagger biproduct compact closed category.*

Proof The biproduct is the disjoint union of sets. Hence $V_A = A$ for every finite set A.

The monoidal structure is given by the cartesian product of sets. A two-sorted notation for the cartesian product brings the same notational advantages as the tensor product

$$\begin{aligned} a \times_2 b &= \langle a, b \rangle \\ a \times_2 B &= \{a\} \times B \\ A \times_2 b &= A \times \{b\} \\ A \times_2 B &= A \times B. \end{aligned}$$

The two-sorted product $f \times_2 g : A \times_2 C \longrightarrow B \times_2 D$ for $f : A \longrightarrow B$ and $g : C \longrightarrow D$ is determined by it values on the elements of $A \times_2 C$, namely

$$(f \times_2 g)(a \times_2 b) = f(a) \times_2 g(b), \text{ for } a \in A, b \in C.$$

The monoidal unit is a distinguished singleton set $I = \{a_0\}$.

Every object is its own adjoint, $A = A^* = A^\dagger$, the unit $\eta_A : I \longrightarrow A \times A$ and counit $\epsilon_A : A \times A \longrightarrow I$ of the adjunction are given by

$$\eta_A(a_0) = \{a \times_2 a : a \in A\} \qquad \epsilon_A(a \times_2 b) = \begin{cases} a_0 & \text{if } a = b \\ \varnothing & \text{else} \end{cases}$$

The name, coname, dual and dagger of $f : A \longrightarrow B$ are

$$\ulcorner f \urcorner(a_0) = \{a \times_2 b : f(a) = b \text{ or } b \in f(a), a \in A, b \in B\}$$
$$\llcorner f \lrcorner(a \times_2 b) = \begin{cases} a_0 & \text{if } f(a) = b \text{ or } b \in f(a) \\ \varnothing & \text{else} \end{cases}$$
$$f^*(b) = \{a \in A : f(a) = b \text{ or } b \in f(a)\} = f^\dagger(b). \qquad \square$$

The category $2\mathcal{SF}$ has an abundance of projectors. Here is one, which is not a projector in the category of Hilbert spaces.

Example 9.9 A two-sorted projector that is not intrinsic.
The two-sorted function $p : \{a, b\} \longrightarrow \{a, b\}$ defined by

$$p(a) = \{a, b\} \text{ and } p(b) = \{a, b\}$$

is a projector in $2\mathcal{SF}$. The corresponding matrix is

$$(\pi_{ij}) = \begin{pmatrix} 1 & 1 \\ 1 & 1 \end{pmatrix}.$$

The same matrix induces an endomorphism in any semantic category. This endomorphism is again a projector in $\mathcal{RI}$, but not, in general, when addition is not idempotent.

Luckily, the semantics of natural language only involves projectors that live in every semantic category, namely the intrinsic projectors. The characterization in eqn (9.3) of intrinsic projectors is equivalent in $2\mathcal{SF}$ to

$$p(Y) = \{x \in B : p(x) = x\} \cap Y, \text{ for every } Y \subseteq B. \tag{9.7}$$

9.3.2 The embedding

Given a set $A = \{a_1, \ldots, a_m\}$, the map $\mathcal{J}_A$ is defined for elements $a \in A$ and subsets $X \subseteq A$. Thus

$$\mathcal{J}_A(a) = a \in V_A, \qquad \mathcal{J}_A(X) = \sum_{a \in X} a \in V_A. \tag{9.8}$$

The map $\mathcal{J}_A$ has an inverse $\mathcal{J}_A^{-1}$ that sends a sum of distinct basis vectors in V_A to the corresponding subset of A, i.e. for $\{i_1, \ldots, i_k\} \subseteq \{1, \ldots, n\}$

$$\mathcal{J}_A^{-1}\left(\sum_{j=1,\ldots,k} a_{i_j}\right) = \{a_{i_1}, \ldots, a_{i_k}\}. \tag{9.9}$$

The following holds for every Boolean vector $v \in V_A$ and every $X \subseteq A$

$$\mathcal{J}_A^{-1} \circ \mathcal{J}_A(X) = X$$
$$\mathcal{J}_A \circ \mathcal{J}_A^{-1}(v) = v.$$

In fact, $\mathcal{J}_A$ is an isomorphism of Boolean algebras that maps subsets of A onto Boolean vectors of V_A. Indeed, $\mathcal{J}_A$ commutes with the logical connectives

$$\mathcal{J}_A(X \cup Y) = \mathcal{J}_A(X) \vee \mathcal{J}_A(Y)$$
$$\mathcal{J}_A(X \cap Y) = \mathcal{J}_A(X) \wedge \mathcal{J}_A(Y), \text{ for } X, Y \subseteq A$$
$$\mathcal{J}_A(A \setminus X) = \neg \mathcal{J}_A(X)$$

Finally, for any finite set A let

$$\mathcal{J}(A) = V_A$$

and for any two-sorted function $f : A \longrightarrow B$ let $\mathcal{J}(f) : V_A \longrightarrow V_B$ be the linear map satisfying

$$\mathcal{J}(f)(a) = \mathcal{J}_B(f(a)) \text{ for } a \in A. \tag{9.10}$$

The restriction of $\mathcal{J}(f) : V_A \longrightarrow V_B$ to Boolean vectors is a Boolean homomorphism in $\mathcal{RI}$, because vector disjunction coincides with vector addition in this category.

Definitions 9.8–9.10 make sense in the category of finite-dimensional Hilbert spaces as well. The maps $\mathcal{J}_A$ are Boolean isomorphisms, but the linear map $\mathcal{J}(f)$ is not a Boolean homomorphism when restricted to Boolean vectors.

The following *embedding lemma* is one of the reasons why $\mathcal{RI}$ is especially appropriate for natural language semantics.

Lemma 9.10 (Embedding Lemma) *The map $\mathcal{J} : 2\mathcal{SF} \longrightarrow \mathcal{RI}$ is a one-to-one functor that preserves the biproduct dagger compact closed structures and the logical connectives.*

The restriction of $\mathcal{J}$ to the subcategory of intrinsic maps is a functor in an arbitrary semantic category.

Proof The proof is straightforward and essentially that given in Preller and Sadrzadeh (2011c). □

Only the equality $\mathcal{J}(g \circ f) = \mathcal{J}(g) \circ \mathcal{J}(f)$, which holds in $\mathcal{RI}$, does not hold in an arbitrary semantic category unless both f and g are intrinsic.

Indeed, let $f'(a) = f(a)$ if $f(a)$ is a set and $f'(a) = \{f(a)\}$ if $f(a)$ is an element. Define $g'(b)$ similarly and let $K = \bigcup_{b \in f'(a)} g'(b)$. Then on one hand,

$$(g \circ f)(a) = \bigcup_{b \in f'(a)} \bigcup_{c \in g'(b)} \{c\} = \bigcup_{c \in K} \{c\} \text{ and therefore } \mathcal{J}(g \circ f)(a) = \sum_{c \in K} c.$$

On the other hand, $\mathcal{J}(f)(a) = \mathcal{J}_B(f(a)) = \sum_{b \in f'(a)} b$. Hence

$$\mathcal{J}(g) \circ \mathcal{J}(f)(a) = \sum_{b \in f'(a)} \mathcal{J}(g)(b) = \sum_{b \in f'(a)} \sum_{c \in g'(b)} c = \sum_{c \in K} c. \tag{9.11}$$

The rightmost equality of (9.11) holds in $\mathcal{RI}$, because vector addition is idempotent. This is not the case in an arbitrary semantic category (Example 9.9), unless f and g are intrinsic. If they are intrinsic, however, the sets $f'(a)$ and $g'(b)$ are either empty or singleton sets and the equalities (9.11) above hold. Hence again $\mathcal{J}(g) \circ \mathcal{J}(f) = \mathcal{J}(g \circ f)$.

9.3.3 Two-sorted truth

Let S be a fixed two-dimensional space of the semantic category $\mathcal{C}$ with basis vectors $\top$ and $\bot$. Think of $\top$ as 'true' and of $\bot$ as 'false'. The Boolean vectors of S are called two-sorted *truth-values*. The basis vectors are of sort 1, the null-vector $\vec{0}$ and the full vector $\vec{1} = \top + \bot$ of sort 2. The null-vector stands for 'neither true nor false' and $\top + \bot$ for 'partly true and partly false'. Two-sorted truth–values reflect second-order properties of natural language; Subsection 9.4.3 provides some examples.

The two-sorted connectives on S are linear maps, which are determined by their values on basis vectors. They are intrinsic and thus live in every compact closed category with biproducts.

The *two-sorted conjunction* $\text{and}_S : S \otimes S \longrightarrow S$, the *two-sorted disjunction* $\text{or}_S : S \otimes S \longrightarrow S$ and the *two-sorted negation* $\text{not}_S : S \longrightarrow S$ are given by

$$\text{and}_S(\top \otimes \top) = \top, \ \text{and}_S(\bot \otimes \top) = \text{and}_S(\top \otimes \bot) = \text{and}_S(\bot \otimes \bot) = \bot$$
$$\text{or}_S(\bot \otimes \bot) = \bot, \ \text{or}_S(\bot \otimes \top) = \text{or}_S(\top \otimes \bot) = \text{or}_S(\top \otimes \top) = \top$$
$$\text{not}_S(\top) = \bot, \ \text{not}_S(\bot) = \top.$$

The two-sorted connectives are Boolean operators on S, because

$$\begin{aligned} \text{not}_S \circ \text{not}_S &= 1_S \\ \text{or}_S \circ (\text{not}_S \otimes \text{not}_S) &= \text{not}_S \circ \text{and}_S. \end{aligned} \tag{9.12}$$

Two-sorted negation coincides with vector negation on basis vectors, but they are not identical. In fact, not_S is the symmetry isomorphism that exchanges the two basis vectors of S, whereas vector negation is not even an endomorphism of S:

$$\mathrm{not}_S(\top) = \neg\top, \ \mathrm{not}_S(\bot) = \neg\bot,$$
$$\text{but } \mathrm{not}_S(\vec{0}) = \vec{0} \text{ and } \neg\vec{0} = \vec{1}.$$

In general, the two-sorted connectives differ from the vector connectives, even on basis vectors. For example,

$$\mathrm{and}_S(\bot \otimes \top) = \bot, \text{ whereas } \bot \wedge \top = \vec{0}.$$

Natural language chooses the two-sorted connectives on S, a fact acknowledged by the notation.

9.3.4 Two-sorted predicates

Let E be any object of $\mathcal{C}$; think of the basis vectors of E as 'individuals'. Abbreviate the n-fold tensor product of E by $E^n = E \otimes \ldots \otimes E$.

Definition 9.11 (Two-sorted predicate) *A two-sorted predicate is an intrinsic morphism with codomain S.*

A two-sorted predicate on E is a two-sorted predicate that maps basis vectors of E to basis vectors of S.

An n-ary two-sorted predicate on E is a two-sorted predicate on E^n.

Lemma 9.12 *The n-ary predicates on E together with the two-sorted connectives form a Boolean algebra.*

More precisely, assume that $p : E^n \longrightarrow S$ and $r : E^n \longrightarrow S$ are n-ary two-sorted predicates on E. Then the linear maps

$$\mathrm{not}_S \circ p, \ \mathrm{and}_S \circ (p \otimes r), \ \mathrm{or}_S \circ (p \otimes r), \ \mathrm{and}_S \circ (p \otimes r) \circ 2_{E^n}, \ \mathrm{or}_S \circ (p \otimes r) \circ 2_{E^n}$$

are again two-sorted predicates on E such that

$$\begin{aligned} \mathrm{not}_S \circ \mathrm{not}_S \circ p &= p \\ \mathrm{not}_S \circ \mathrm{and}_S \circ (p \otimes r) &= \mathrm{or}_S \circ ((\mathrm{not}_S \circ p) \otimes (\mathrm{not}_S \circ r)). \end{aligned} \quad (9.13)$$

Proof It suffices to check the equalities (9.13) for basis vectors, which follows from (9.12). □

Two-sorted predicates take two possible values on individuals. For sets, however, they have a wider range of values. The following lemma tells us when.

Lemma 9.13 (Fundamental Property) *Let $p : V_B \longrightarrow S$ be a two-sorted predicate on V_B and $A = \{b_{i_1}, \ldots, b_{i_k}\}$ a subset of k basis vectors. Then there is a non-negative integer $k_1 \leq k$ such that*

$$p(b_{i_1} + \cdots + b_{i_k}) = k_1 \cdot \top + (k - k_1) \cdot \bot$$

holds. In particular, the following equivalences hold in the categories $2\mathcal{SF}$ and $\mathcal{RI}$

$$p(\textstyle\sum_{x\in A} x) = \vec{0} \Leftrightarrow A = \emptyset$$
$$p(\textstyle\sum_{x\in A} x) = \top \Leftrightarrow p(x) = \top \text{ for all } x \in A \text{ and } A \neq \emptyset$$
$$p(\textstyle\sum_{x\in A} x) = \bot \Leftrightarrow p(x) = \bot \text{ for all } x \in A \text{ and } A \neq \emptyset \quad (9.14)$$
$$p(\textstyle\sum_{x\in A} x) = \vec{1} \Leftrightarrow p(x) = \top \text{ and } p(y) = \bot \text{ for some } x, y \in A.$$

Proof Let k_1 be the number of elements of the set $A_1 = \{x \in A : p(x) = \top\}$. Then $k - k_1$ is the number of elements of $A_2 = \{x \in A : p(x) = \bot\}$. By linearity, $p(\sum_{x\in A} x) = \sum_{x\in A_1} p(x) + \sum_{x\in A_2} p(x) = k_1 \cdot \top + (k - k_1) \cdot \bot$. The particular case follows because addition is idempotent. □

In Hilbert spaces, the value a two-sorted predicate assigns to the vector $\sum_{x\in A} x$ is the result of two 'counts'. One counts the elements of A for which the predicate is true, the other one those for which the predicate is false. It suffices to divide by the number of elements k in the set A, to obtain a frequency distribution. The step from two-sorted predicates to probabilities is minimal.

Corollary 9.14 *The following equivalences hold in the categories $2\mathcal{SF}$ and $\mathcal{RI}$, for any element x and any subset Y of B*

$$p(x) = \bot \Leftrightarrow p(x) \neq \top \Leftrightarrow \text{not}_S(p(x)) = \top$$

$$p(Y) = \bot \Leftrightarrow \text{not}_S(p(Y)) = \top. \quad (9.15)$$

In general, however, $p(Y) \neq \top$ does not imply $p(Y) = \bot$.

Proof The equivalence $p(Y) = \bot \Leftrightarrow \text{not}_S(p(Y)) = \top$ follows from the definition of the two-sorted negation.

By the fundamental property, $p(Y) = \bot$ is equivalent to $\forall_x (x \in Y \Rightarrow p(x) = \bot)$ for a non-empty set Y. Now assume that Y has two distinct elements x and y and that $p(x) = \top$ and $p(y) = \bot$. Then $p(Y) = \{\top, \bot\} = \vec{1} \neq \bot$. □

In the context of natural language, different font shapes help to distinguish the set $X \subseteq B$ from the vector $\mathcal{J}_B(X)$, namely italic font for the former and typewriter font for the latter. For example, if $Bank \subseteq B$

$$\text{bank} = \sum_{x \in Bank} x = \mathcal{J}_B(Bank).$$

The same applies when distinguishing a one-sorted predicate on B and the corresponding two-sorted predicate on V_B. For example, the one-sorted predicate *Rich* corresponding to rich : $V_B \longrightarrow S$ satisfies

$$x \in Rich \Leftrightarrow Rich(x) \Leftrightarrow \text{rich}(x) = \top, \text{ for all } x \in B.$$

Here are a few examples of how two-sorted predicates work in an arbitrary semantic category.

Under the assumption that a noun designates a non-empty set of individuals, the following equivalences hold in $2\mathcal{SF}$ and $\mathcal{RI}$. The proofs are straightforward via the Fundamental Property, eqn (9.14).

Example 9.15 The following are equivalent

$$\text{rich}(\text{bank}) = \top$$
$$\forall x (x \in \text{Bank} \Rightarrow x \in \text{Rich}).$$

Example 9.16 The following are equivalent

$$\text{not}_S(\text{rich}(\text{bank})) = \top$$
$$\forall x (x \in \text{Bank} \Rightarrow x \notin \text{Rich}).$$

Important: the first-order formula in Example 9.16 is not the negation of the first-order formula in Example 9.15. But then the equality $\text{rich}(\text{bank}) = \bot$ is not the negation of the equality $\text{rich}(\text{bank}) = \top$, because there are more than two truth values.

9.4 Semantics via pregroup grammars

Let E be a finite-dimensional space with basis B. Call its basis vectors *individuals*. Single basis vectors of E correspond to singulars and sums to plurals of natural language. The examples below concern unary predicates only. The generalization to ordered pairs, triples etc. of individuals is straightforward. The space S is the two-dimensional space of two-sorted truth introduced in Section 9.3.3.

9.4.1 The computation of meanings

Like every other categorial grammar, a pregroup grammar is given by a lexicon and a calculus. The pregroup calculus is *compact bilinear logic* where proofs identify with morphisms in the free compact bicategory $\mathcal{C}2(\mathcal{B})$ generated by a partially ordered set $\mathcal{B}$.

The 1-cells are called *types* and the tensor product is written as concatenation. Hence a *type* is a string of simple types, where a *simple type* is either an element of $x, y, \cdots \in \mathcal{B}$ or an iterated adjoint $x^\ell, y^\ell, \ldots, x^{\ell\ell}, y^{\ell\ell}, \ldots, x^r, y^r, \ldots, x^{rr}, y^{rr}, \ldots$ etc. The types $x, y, \cdots \in \mathcal{B}$ are called *basic types*. They stand for grammatical notions.

As a consequence, every functor from $\mathcal{B}$ into a semantic category $\mathcal{C}$ extends into a functor from $\mathcal{C}2(\mathcal{B})$ to $\mathcal{C}$ preserving the structure of compact bicategories.

A *lexicon* is a finite list of entries. An *entry* is a triple $w : T :: m$, where w is a word, T a type and m a meaning expression.

This description differs from the original one in Lambek (1999). There, only pregroup 'dictionaries' are considered where the entries are pairs $w : T$, without meaning expressions. The latter must be added explicitly, because the functional semantics of categorial grammars with higher-order types has been lost by the pregroup types.

The *meaning* in the entry is a formal expression $m : I \longrightarrow V$ in the language of compact closed categories or, equivalently, a string of two-sorted functions. It depends functionally on the pair $w : T$.

Consider the following entries

$$no : ss^\ell n_2 c_2{}^\ell :: I \xrightarrow{\overline{no}} S \otimes S^* \otimes E \otimes E^* \quad some : n_2 c_2{}^\ell :: I \xrightarrow{\overline{some}} E \otimes E^*$$

$$are : n_2{}^r ss^\ell \bar{n} :: I \xrightarrow{\overline{are}} E^* \otimes S \otimes S^* \otimes E \quad big : c_2 c_2{}^\ell :: I \xrightarrow{\overline{big}} E \otimes E^*$$

$$banks: c_2 \quad :: I \xrightarrow{\overline{bank}} E$$

$$and: s^r ss^\ell \quad :: I \xrightarrow{\overline{ands}} S^* \otimes S \otimes S^* \quad rich : \bar{n}^r s :: I \xrightarrow{\overline{rich}} E^* \otimes S$$

The basic types $c_2, n_2, \bar{n}, s$, stand for 'plural count noun', 'plural noun-phrase', 'dummy noun-phrase', and 'sentence', in that order. Moreover, $c_2 < n_2$. The properties of the meaning vector m in the entry $w : T :: m$ depend on the logical content of the word, given in due course, and on the type.

The lexicon defines a *canonical* functor from $\mathcal{B}$ into the compact closed category $\mathcal{C}$. For example, in the list of entries above, the basic type s is interpreted by S and each of the basic types $c_2, n_2, \bar{n}$ by E. The canonical functor maps the inequality $c_2 < n_2$ to the identity 1_E and left and right adjoints of a type to 'the' adjoint space, because right and left adjoints may be identified in a symmetric monoidal category.

All meanings are presented in the form of names, up to a symmetry isomorphism that arranges the factors of the tensor product in the order given by the simple types. For example, $\overline{big} = \sigma_{E^*,E} \circ \ulcorner big \urcorner : 1 \longrightarrow E \otimes E^*$ where $big : E \longrightarrow E$.

The meaning of a grammatical string involves both the meanings of the words and the syntactical analysis of the string. An ambiguous string with several parsings has several meanings. A non-grammatical string has no meaning.

A string of words $w_1 \ldots w_n$ is *grammatical* if there are entries $w_1 : T_1 :: m_1, \ldots, w_n : T_n :: m_n$ and a basic type b such that

$$T_1 \ldots T_n \vdash b$$

is provable in compact bilinear logic (or stated otherwise, if there is a morphism $f : T_1 \ldots T_n \longrightarrow b$ in the syntactic category). Due to a theorem in Lambek (1999) the graph of the proof involves only underlinks and is called a *reduction*.

For example, the reduction corresponding to the graph on the left below analyses *big banks* as a plural noun-phrase. The graph on the right is the corresponding morphism in the semantic category.

$$1_{c_2} \otimes c_2 = \quad \begin{array}{c} \text{big} \quad \text{banks} \\ (c_2\ c_2^\ell)\ (c_2) \\ \downarrow \\ c_2 \end{array} \qquad r_1 = 1_E \otimes\ _E = \quad \begin{array}{c} (E \otimes E^*) \otimes E \\ \downarrow \\ E \end{array}$$

The reason why the syntactic category must not be symmetric is obvious in this example. The type $c_2 c_2 c_2^\ell$ of the non-grammatical string *banks big* has no reduction to a basic type. If we had symmetry all order variants of a grammatical string would be grammatical.

The meaning vector of a lexical entry also identifies with a graph, for example

$$I \xrightarrow{\overline{big}} E \otimes E^* = \quad \begin{array}{c} I \\ \swarrow \\ \text{big} \\ E \otimes E^* \end{array} \qquad \overline{\text{bank}} : I \longrightarrow E = \quad \begin{array}{c} I \\ \downarrow \\ \text{bank} \\ \downarrow \\ E \end{array}.$$

Again, the domain of the morphism is at the top of the graph, the codomain at the bottom.

For a grammatical string $w_1 \ldots w_n$, let $w_1 : T_1 :: m_1, \ldots, w_n : T_n :: m_n$ be a choice of entries and r a reduction of $T_1 \ldots T_n$ to a basic type. The corresponding meaning is

$$r \circ (m_1 \otimes \ldots \otimes m_n),$$

where now r denotes the linear map obtained by applying the canonical functor to the reduction.

The meaning of a string can be computed graphically. Connect the graphs at their joint interface and follow the paths from top to bottom picking up the labels along the way. For example, the graphs

$$\overline{\text{big}} \otimes \overline{\text{bank}} = \quad \begin{array}{c} I \otimes I \\ \swarrow \quad \searrow \\ \text{big} \quad \text{bank} \\ \downarrow \\ (E \otimes E^*) \otimes (E) \end{array} \qquad r_1 = \quad \begin{array}{c} (E \otimes E^*) \otimes (E) \\ \downarrow \\ E \end{array}$$

when connected at their joint interface compute to

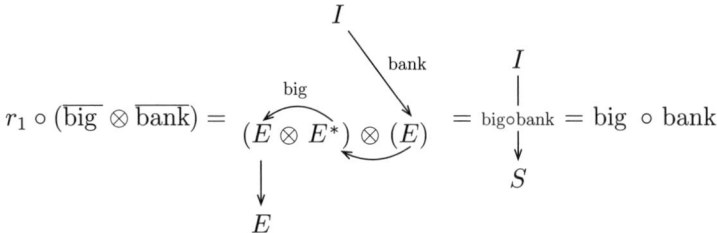

The meaning vector $\overline{are} : I \longrightarrow E^* \otimes S \otimes S^* \otimes E$ is up to a symmetry isomorphism the name of the linear map are : $E \otimes S \longrightarrow E \otimes S$. The following postulate renders the *logical property* of the word *are*

$$are = 1_E \otimes 1_S.$$

Hence the graph of $\overline{are}$ is

$$\overline{are} = \quad \underset{E^* \otimes S \otimes S^* \otimes E}{\overset{I}{\frown}} \triangleright$$

The lexicon assigns to the pair *rich* : $\bar{n}^r s$ a meaning vector $\overline{rich} : I \longrightarrow E^* \otimes S$. Its graph has the form

$$\overline{rich} = \quad \underset{E^* \otimes S}{\overset{I}{\underset{rich}{\frown}}} \quad , \text{ where } rich : E \longrightarrow S \triangleright$$

The reduction of the sentence *big banks are rich* is the graph

$$r = \quad \underset{(c_2\, c_2^\ell)\,(c_2)\,(n_2^r\, s\, s^\ell\, \bar{n})\,(\bar{n}^r\, s)}{\overset{big \quad banks \quad are \quad rich}{\cdots}}$$

Compute the meaning by composing the tensor product of the word vectors with the reduction

$r \circ (\overline{\text{big}} \otimes \overline{\text{bank}} \otimes \overline{\text{are}} \otimes \overline{\text{rich}})$

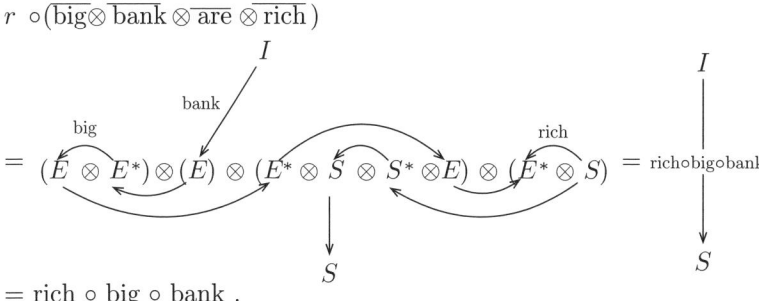

$= \text{rich} \circ \text{big} \circ \text{bank}.$

The sentence *All banks are rich* is computed by the same graph except that the label big is replaced by the label all. A similar remark applies to the sentence *Some banks are rich*.

The last example concerns the computation of the meaning vector of the sentence *no banks are rich*. The reduction of the sentence is

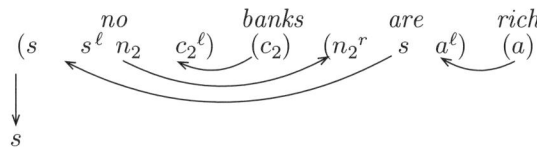

The meaning vector $\overline{\text{no}} : I \longrightarrow S \otimes S^* \otimes E \otimes E^*$ is defined as the name of $\text{not}_S \otimes 1_E : S \otimes E \longrightarrow S \otimes E$ with the corresponding graph

$$\overline{\text{no}} = \begin{array}{c} I \\ \text{not}_S \\ S \otimes S^* \otimes E \otimes E^* \end{array}$$

The meaning of the sentence *no banks are rich* is

$r \circ (\overline{\text{no}} \otimes \overline{\text{bank}} \otimes \overline{\text{are}} \otimes \overline{\text{rich}})$

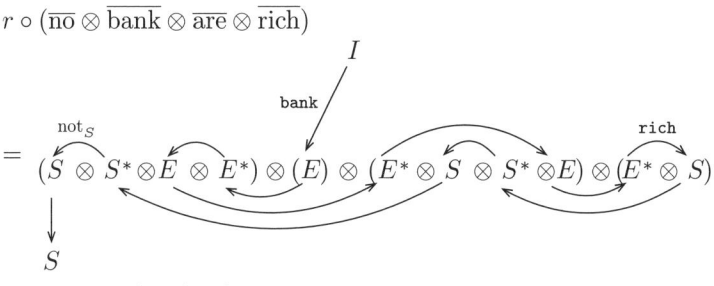

$= \text{not}_S \circ \text{rich} \circ \text{bank}.$

SEMANTICS VIA PREGROUP GRAMMARS | 271

'Walking graphs' makes the computation linear in the number of links. The latter is proportional to the length of the string of words, because the lexicon is finite and thus the length of its types is bounded.

9.4.2 The logical content of words

The preceding examples mention the logical content of some words. More generally, the description of the logical content can be organized according to the type of the words. We postulate the following.

1. Any $f : E^n \longrightarrow S$ occurring in a lexical entry is an n-ary two-sorted predicate.
2. Any $f : E \longrightarrow E$ occurring in a lexical entry is an intrinsic projector.
3. Any $f : I \longrightarrow E$ occurring in a lexical entry is a Boolean vector.

Under these postulates, adjectives in predicative position and intransitive verbs are unary two-sorted predicates. The meaning map $\overline{\text{word}} : E \longrightarrow E$ in $\text{word} : n_i c_i^\ell ::$ $\overline{\text{word}} : I \longrightarrow E \otimes E^*$ interpreting an adjective in attributive position or a determiner is an intrinsic projector. The universal determiner satisfies an even stronger property

$$\overline{\text{all}} = 1_E.$$

Recall that B is the canonical basis of E and that any subset of B is identified with a Boolean vector, by equality (9.8). Assume $\overline{\text{word}} : E \longrightarrow E$ is an intrinsic projector occurring in the lexicon. Use the abbreviation

$$\text{Word} = \overline{\text{word}}(B).$$

This abbreviation, combined with Equality (9.7), implies

$$\overline{\text{word}}(Y) = \text{Word} \cap Y. \qquad (9.16)$$

For example, $\text{Bank} = \overline{\text{bank}}(B)$ and $\overline{\text{big}}(\text{Bank}) = \text{Big} \cap \text{Bank}$.

9.4.3 Examples

The examples below concern sentences and their meaning vectors in the categories $2\mathcal{SF}$ and $\mathcal{RI}$. Represent the truth of a sentence by the fact that the meaning vector computes to $\top$. Then show that this representation coincides with the 'usual' translation of the sentence in logic.

Example 9.17 *All banks are rich* / rich(all(bank))

The following are equivalent

$$\forall x (x \in \text{Bank} \Rightarrow \text{Rich}(x))$$
$$\overline{\text{rich}}(\overline{\text{all}}(\overline{\text{bank}})) = \top.$$

Proof Recall that all $= 1_E$. Hence rich(all(bank)) = rich(bank). For the proof of the equivalence of rich(bank) $= \top$ with $\forall x(x \in Bank \Rightarrow Rich(x))$ refer to Example 9.15. □

Example 9.18 *Big banks are rich* / rich(big(bank))

The following are equivalent:

$$\forall x(x \in Big \cap Bank \Rightarrow Rich(x))$$
$$\text{rich(big(bank))} = \top.$$

Proof Recall that bank $= \sum_{x \in Bank} x$ and therefore the vector big(bank) is identified with the set big(Bank) $= Big \cap Bank$ by eqn (9.16). The equivalence now follows from the Fundamental Property of eqn (9.14). □

Example 9.19 *No banks are rich* / $\text{not}_S(\text{rich(bank)})$

The following are equivalent

$$\text{not}_S(\text{rich(bank)}) = \top$$
$$\forall x(x \in Bank \Rightarrow x \notin Rich)$$

Proof See Example 9.16. □

Example 9.20 *Some banks are rich*/rich(some(bank))

The implication

$$\text{rich(some(bank))} = \top \Rightarrow \exists_x(x \in Bank \,\&\, Rich(x))$$

holds, but its converse does not hold in general.

Proof The equality rich(some(bank)) $= \top$ implies some(bank) $\neq \emptyset$ and $\forall_x(x \in$ some(bank) $\Rightarrow \text{rich}(x) = \top$), by the Fundamental Property. The first-order formula follows, because some(bank) $\subseteq$ bank.

If the first-order formula holds, take a witness $b \in Bank$ for which $Rich(b)$ holds. Let some$_b$ be the intrinsic projector that maps b to itself and every other individual to the null vector. Clearly some$_b$(bank) $= b$ and rich(b) $= \top$. Hence,

$$\text{rich(some}_b\text{(bank))} = \top,$$

but this does not imply that the particular intrinsic projector some$_b$ coincides with the original some. □

Natural language confronts us with a problem. On one hand, the interpretation of '*some Y*' changes from occurrence to occurrence, for example in *Some banks are rich and some banks are not rich*. On the other hand, '*some Y*' acts like a well-determined reference set. In *Some banks are rich. They scare me*, the personal pronoun *they* refers to the set *some banks* of the preceding sentence.

The interpretation of some given above may vary from occurrence to occurrence and at the same time it defines a set at each occurrence, which is available for later reference, e.g. they = some(bank).

The interpretation by a generalized quantifier (see Barwise and Cooper, 2002), takes into account the change of meaning in different occurrences, but it does not construct the set to which the noun-phrase refers.

9.5 Compositional semantics in concept spaces

Quantum logic stands for 'logic of projectors in a concept space' and *concept* for 'Boolean vector in a concept space'.

9.5.1 Classical propositional calculus in concept spaces

Let $P = \{p_1, \ldots, p_d\}$ be a non-empty set. Call *compound system* or *concept space* the tensor product

$$C(P) = C(p_1) \otimes \ldots \otimes C(p_d),$$

where $C(p_i)$ is a two-dimensional space with basis vectors $p_{i\top}, p_{i\bot}$, for $i = 1, \ldots, d$.

The space $C(p_i)$ is a 'basic variable' in quantum protocols and a 'basic concept' in semantics for natural language. For example, key-words of Roget's (or the speaker's mental) thesaurus provide sets of basic concepts.

Any basis vector b_f of $C(P)$ is a tensor product of basis vectors of the factors

$$b_f = f(1) \otimes \ldots \otimes f(d), \text{ where } f \in \prod_{i=1}^{d} \{p_{i\top}, p_{i\bot}\}.$$

Due to the fine-grained structure of the basis vectors, the Boolean algebra of intrinsic projectors of a concept space is isomorphic to the Boolean algebra freely generated by the set P. The rest of this subsection is devoted to the proof of this fact.

For every $i = 1, \ldots, d$, define the two so-called *primitive vectors*

$$\vec{p_i} = \vec{1} \otimes \ldots \otimes \vec{1} \otimes p_{i\top} \otimes \vec{1} \otimes \ldots \otimes \vec{1} = \sum_{f, f(i)=p_{i\top}} b_f$$
$$\vec{\neg p_i} = \vec{1} \otimes \ldots \otimes \vec{1} \otimes p_{i\bot} \otimes \vec{1} \otimes \ldots \otimes \vec{1} = \sum_{f, f(i)=p_{i\bot}} b_f.$$

The two primitive vectors defined by $p_i \in P$ are orthogonal to each other. In fact, one is the negation of the other one

$$\neg \vec{p_i} = \vec{\neg p_i} \text{ and } \neg(\vec{\neg p_i}) = \vec{p_i}.$$

Every Boolean vector can be written as a disjunction of conjunctions of primitive vectors. Indeed, let $\{j_1, \ldots, j_k\}$ be a subset of $\{1, \ldots, d\}$. Assume that $g \in \prod_{i=1}^{d} \{p_{i\top}, p_{i\bot}, \vec{1}\}$ satisfies for $i = 1, \ldots, d$

$$g(i) \in \{p_{i\top}, p_{i\bot}\} \text{ if and only if } i \in \{j_1, \ldots, j_k\}. \tag{9.17}$$

The *partial choice vector* associated to g is

$$v_g = g(1) \otimes \ldots \otimes g(d).$$

Lemma 9.21 *Every partial choice vector v_g is a conjunction of primitive vectors. In particular, every basis vector is a conjunction of primitive vectors.*

Proof Assume that g satisfies eqn (9.17). Let $q_{j_l} = g(j_l) \in \{p_{j_l\top}, p_{j_l\bot}\}, l = 1, \ldots, k$, and $G = \left\{ f \in \prod_{i=1}^d \{p_{i\top}, p_{i\bot}\} : f(j_l) = q_{j_l}, \text{ for } l = 1, \ldots, k \right\}$. Then

$$v_g = \sum_{f \in G} b_f = \vec{q_{j_1}} \wedge \cdots \wedge \vec{q_{j_k}}. \tag{9.18}$$

□

Theorem 9.22 *The free Boolean algebra $B(P)$ generated by P is isomorphic to the lattice of intrinsic projectors of $C(P)$. The map $p \mapsto \vec{p}$ extends to an isomorphism $\mathcal{K}$ from $B(P)$ onto the lattice of Boolean vectors of $C(P)$.*

Proof Partial choice vectors are Boolean vectors by eqn (9.18). Hence, the map $p \mapsto \vec{p}$ extends to a unique Boolean homomorphism $\mathcal{K}$ from $B(P)$ into the Boolean algebra of Boolean vectors of $C(P)$. A classical theorem (Halmos, 1974) states that the free Boolean algebra $B(P)$ is isomorphic to the set algebra generated by the following subsets of $\prod_{j=1}^d \{0, 1\}$:

$$p_i \simeq \left\{ h \in \prod_{j=1}^d \{0, 1\} : h(i) = 1 \right\}, \quad \neg p_i \simeq \left\{ h \in \prod_{j=1}^d \{0, 1\} : h(i) = 0 \right\},$$

where i varies from 1 to d. Every singleton set $\{h\}$ can be written as a conjunction of subsets of the form p_i or $\neg p_i$. Therefore the homomorphism $\mathcal{K}$ maps $\{h\}$ to the corresponding basis vector in $C(P)$. It follows that the homomorphism $\mathcal{K}$ maps $B(P)$ onto the set of Boolean vectors, because every Boolean vector can be written uniquely as a sum of basis vectors.

Compose $\mathcal{K}$ with the isomorphism $\mathcal{H}$ of Theorem 9.6 to obtain the isomorphism $\mathcal{H} \circ \mathcal{K}$ onto the lattice of intrinsic projectors. □

By Theorem 9.22, the *primitive projectors* $p_{\vec{p_i}}$ and $p_{\neg \vec{p_i}}$ play an important role in the lattice of projectors of $C(P)$:

1. Every intrinsic projector is a finite disjunction of finite conjunctions of primitive projectors
2. The lattice of intrinsic projectors in a compound system is the classical propositional calculus modulo equiderivability.

3. One can use *induction on the complexity of propositions* for defining and proving properties of Boolean vectors/intrinsic projectors.

Propositional complexity creates a somewhat unusual hierarchy on subspaces. Recall that $com(p_i) = 0$, $com(\neg p) = 1 + com(p)$, and $com(p \wedge q) = 1 + max(com(p), com(q))$. The primitive subspace $E_{\vec{p_i}}$ has complexity 0 and $E_{\neg \vec{p_i}}$ has complexity 1, but both have dimension 2^{d-1}. The one-dimensional subspace generated by a single basis vector b_f has complexity $d - 1$ if $f(i) = p_i$ for $i = 1, \ldots, d$, and complexity d otherwise.

9.5.2 Concept spaces and two-sorted truth

A *classification system* consists of

1. a set B (of individuals, pairs of individuals etc.)
2. a set $P = \{p_1, \ldots, p_d\}$ (of properties)
3. a relation $\models \;\subseteq B \times P$.

Read $x \models p$ as 'x satisfies p'.

Extend the relation $\models$ to arbitrary concepts in $C(P) = C(P_1) \otimes \ldots \otimes C(P_d)$ for every individual $x \in B$ using induction on the complexity of concepts:

$$\begin{aligned} x &\models \vec{p_i} & &\text{if and only if } x \models p_i \\ x &\models \neg v & &\text{if and only if } x \not\models v \\ x &\models v \wedge w & &\text{if and only if } x \models v \,\&\, x \models w \\ x &\models v \vee w & &\text{if and only if } \models v \text{ or } x \models w. \end{aligned}$$

Clearly, either $x \models v$ or $x \models \neg v$ holds for every individual $x \in B$ and every concept $v \in C(P)$.

Extend satisfaction to every non-empty subset Y of B and every concept v:

$$Y \models v \text{ if and only if } x \models v \text{ for all } x \in Y. \tag{9.19}$$

Read $Y \models v$ as 'Y has property v in general'.

Note that $Y \not\models v$ and $Y \not\models \neg v$ may hold simultaneously. It suffices that Y has an element satisfying v and another one that does not satisfy v.

The satisfaction system induces a representation of (sets of) individuals by concepts in $C(P)$. For any $x \in B$ let

$$\begin{aligned} q(x)_i &= p_{i\top} \text{ if } x \models p_i \\ q(x)_i &= p_{i\bot} \text{ if } x \not\models p_i, \end{aligned}$$

for $i = 1, \ldots, d$. This choice determines a basis vector, the *concept v_x internalizing x*,

$$v_x = q(x)_1 \otimes \ldots \otimes q(x)_d.$$

For any non-empty subset $Y \subseteq B$ define the *concept v_Y internalizing Y*

$$v_Y = \sum_{x \in Y} v_x.$$

Different individuals may be internalized by the same basis vector. This means that they are indiscernible by the properties listed in P.

Lemma 9.23 *For any concept $c \in C(P)$ and any individual $x \in B$*

$$x \models c \text{ if and only if } v_x \leq c. \tag{9.20}$$

In particular, for any basis vector $b_f \in C(P)$

$$x \models b_f \text{ if and only if } v_x = b_f. \tag{9.21}$$

For $Y \neq \emptyset$

$$Y \models c \text{ if and only if } v_Y \leq c. \tag{9.22}$$

Proof Show eqn (9.20), the equivalence concerning individuals, by induction on the propositional complexity of c. Equivalence (9.21) is a particular case of (9.20). The equivalence concerning sets, eqn (9.22), now follows from the equivalence for individuals. □

One consequence of the lemma above is that satisfaction in a classification system coincides with the conditional logic for Boolean vectors/projectors. Indeed, the inequality $v \leq w$ is equivalent to $v \to w = \vec{1}$, where the full vector $\vec{1}$ stands for 'true'.

Another consequence is that the concept v_x is the best possible description of the individual x in the classification system and the same holds for v_Y and the set Y.

9.5.3 Intrinsic projectors and two-sorted predicates

Let E be a space with basis B. One can think of any satisfaction system $(B, P = \{p_1, \ldots, p_d\}, \models)$ as a model of the language generated by P. It suffices to think of $p \in P$ as the two-sorted predicate $p(.) : E \longrightarrow S$ satisfying

$$p(x) = \begin{cases} \top & \text{if } x \models p \\ \bot & \text{if } x \not\models p \end{cases}, \text{ for all } x \in B.$$

Recall that $x \models v$ is equivalent to $v_x \leq v$. The expressiveness of the logic remains unchanged if the individuals in B are replaced by the basis vectors internalizing them. Indeed, any individuals x, y for which $v_x = v_y$ are indiscernible in the logic.

The compound system $C(P)$ is endowed with a *canonical satisfaction system*, namely

$$P = \{p_1, \ldots, p_d\}$$
$$B = \left\{b_f : f \in \prod_{i=1}^{d} \{p_{i\top}, p_{i\bot}\}\right\}$$
$$x \models p \Leftrightarrow x \leq \overrightarrow{p}, \text{ for all } x \in B, p \in P.$$

In the canonical satisfaction system, a basis vector can be both an individual and a concept. More generally, every Boolean vector is both a set of individuals and a concept.

Given $p \in P$, define a two-sorted predicate $\mathcal{L}(\overrightarrow{p})$ on $C(P)$ by stipulating

$$\begin{aligned} \mathcal{L}(\overrightarrow{p})(x) = \top &\Leftrightarrow x \leq \overrightarrow{p} \\ \mathcal{L}(\overrightarrow{p})p(x) = \bot &\Leftrightarrow x \not\leq \overrightarrow{p} \end{aligned}, \text{ for all } x \in B. \quad (9.23)$$

Theorem 9.24 (Definability of predicates) *The map $\overrightarrow{p} \mapsto \mathcal{L}(\overrightarrow{p})$ extends to a Boolean isomorphism from the lattice of concepts of $C(P)$ onto the Boolean algebra of two-sorted predicates on $C(P)$ satisfying*

$$\begin{aligned} \mathcal{L}(\neg v) &= \text{not}_S \circ \mathcal{L}(v) \\ \mathcal{L}(v \wedge w) &= \text{and}_S \circ (\mathcal{L}(v) \otimes \mathcal{L}(w)) \circ 2_E. \end{aligned} \quad (9.24)$$

Moreover, if K is a non-empty subset of k basis vectors and $w = \sum_{x \in K}$ the following equivalences hold

$$\begin{aligned} \mathcal{L}(v)(w) = k \cdot \top &\Leftrightarrow w \leq v \\ \mathcal{L}(v)(w) = k \cdot \bot &\Leftrightarrow w \leq \neg v. \end{aligned} \quad (9.25)$$

Proof The extension of $\mathcal{L}$ to all Boolean vectors such that eqn (9.24) holds is guaranteed by Theorem 9.22. Next, prove eqn (9.25) in the particular case where K consists of a single basis vector, i.e. prove that

$$\begin{aligned} \mathcal{L}(v)(x) = \top &\Leftrightarrow x \leq v \\ \mathcal{L}(v)(x) = \bot &\Leftrightarrow x \leq \neg v \end{aligned}, \text{ for all } x \in B. \quad (9.26)$$

Use induction on the propositional complexity of v. If the complexity is 0 then $v = \overrightarrow{p}$, for some $p \in P$. The two equivalences of eqn (9.26) hold for $\overrightarrow{p}$ by eqn (9.23) and the fact that $x \leq \neg \overrightarrow{p}$ if and only if $x \not\leq \overrightarrow{p}$.

For the induction step, assume that eqn (9.26) holds for v. Recall that not_S is the symmetry isomorphism that exchanges the two basis vectors $\top$ and $\bot$. Thus

$$\mathcal{L}(\neg v)(x) = \text{not}_S \circ \mathcal{L}(v)(x) = \top \Leftrightarrow \mathcal{L}(v)(x) = \bot.$$

The right-hand equality above is equivalent to $x \leq \neg v$ by induction hypothesis. The equivalence $\mathcal{L}(\neg v)(x) = \top \Leftrightarrow x \leq \neg v$ follows.

Next, assume that eqn (9.26) holds for the concepts v and w. Then

$$\mathcal{L}(v \wedge w)(x) = \text{and}_S \circ (\mathcal{L}(v)(x) \otimes \mathcal{L}(w)(x)).$$

Therefore, $\mathcal{L}(v \wedge w)(x) = \top$ holds exactly if both $\mathcal{L}(v)(x) = \top$ and $\mathcal{L}(w)(x) = \top$ hold, by definition of and_S. The latter two equalities are equivalent to $x \leq v$ and $x \leq w$ by induction hypothesis, and to $x \leq v \wedge w$ by the definition of vector conjunction. This terminates the proof that the first equivalence of eqn (9.26) holds for $v \wedge w$. The proof of the second equivalence is similar. Hence the particular case of eqn (9.26) holds for all Boolean vectors.

Next, eqn (9.26) implies that $\mathcal{L}$ is one-to-one. Indeed, if v and w are different Boolean vectors there is a basis vector x such that $x \leq v$ and $x \leq \neg w$.

Next, for showing that $\mathcal{L}$ maps the set of Boolean vectors onto the set of two-sorted predicates, assume that $r : C(P) \longrightarrow S$ is an arbitrary two-sorted predicate on $C(P)$. Let $K = \{x \in B : r(x) = \top\}$ and $v = \sum_{x \in K} x$. Then $r(x) = \bot$ for all basis vectors $x \notin K$, because $\bot$ is the only other possible value of r for a basis vector. Thus $r(x) = \top$ if and only if $x \leq v$ and $r(x) = \bot$ if and only if $x \leq \sum_{x \in B \setminus K} x = \neg v$. The equality $r = \mathcal{L}(v)$ follows by eqn (9.26).

To show eqn (9.25) in the general case, let K be a non-empty subset of k basis vectors and $w = \sum_{x \in K} x$. Then $\mathcal{L}(v)(w) = k \cdot \top$ if and only if $v(x) = \top$ for all $x \in K$. The latter is equivalent to $w \leq v$, because of eqn (9.26). The proof of the second equivalence of eqn (9.25) is similar. □

A succinct summary of the theorem above says that every predicate on $C(P)$ can be expressed as a Boolean combination of the predicates $\mathcal{L}(\vec{p})$ for $p \in P$. The fact that the space is the concept space $C(P)$ is essential here. Assume that a and b are distinct individuals of some space E that are indiscernible by the properties $p_i \in P$. Then the two-sorted predicate r on E that maps x to $\top$ if and only if $x = b$ is not definable by a concept of $C(P)$.

Note that the homomorphism $\mathcal{L}$ maps the full vector $\vec{1} \in C(P)$ to the predicate that is 'everywhere' true, where 'everywhere' means 'for all basis vectors'. Switching from Boolean vectors to intrinsic projectors, identify the intrinsic projector p_v with the two-sorted predicate $\mathcal{L}(v)$. In particular, $1_{C(P)} = p_{\vec{1}}$ identifies with the predicate that is 'everywhere' true. The equalities (9.24), recast in terms of projectors, connect predicate logic and projector logic, thus

$$\text{not}_S \circ \mathcal{L}(v) = p_v^\perp, \quad \text{and}_S \circ (\mathcal{L}(v) \otimes \mathcal{L}(w)) \circ 2_{C(P)} = p_v \circ p_w.$$

Theorems 9.6 and 9.24 bring a new understanding to projectors in a compound system $C(P)$. The slogans 'negation is orthogonality' and 'conjunction is composition of projectors' can be extended to 'every grammatical string corresponds to a projector such that predicate logic becomes quantum logic'. It suffices to make $E = C(P)$ in Subsection 9.4.1 to compute the projector.

Thus, the sample sentences of Section 9.4.3 have two interpretations in $C(P)$. One is a two-sorted predicate and the other one a projector. Assuming that we evaluate the former in $2\mathcal{SF}$ or $\mathcal{RI}$, we have the following equivalences concerning the two interpretations:

All banks are rich / rich(all(bank)) / $p_{\text{bank}} \to p_{\text{rich}}$

$$\text{rich(all(bank))} = \top \Leftrightarrow (p_{\text{bank}} \to p_{\text{rich}}) = 1_{C(P)}.$$

Big banks are rich / rich(big(bank)) / $p_{\text{big}} \circ p_{\text{bank}} \to p_{\text{rich}}$

$$\text{rich(big(bank))} = \top \Leftrightarrow (p_{\text{big}} \circ p_{\text{bank}} \to p_{\text{rich}}) = 1_{C(P)}.$$

No banks are rich / $\text{not}_S(\text{rich(bank)})$ / $p_{\text{bank}} \to p_{\text{rich}}^{\perp}$

$$\text{not}_S(\text{rich(bank)}) = \top \Leftrightarrow (p_{\text{bank}} \to p_{\text{rich}}^{\perp}) = 1_{C(P)}.$$

Some banks are rich / rich(some(bank)) / $p_{\text{some(bank)}} \to p_{\text{rich}}$

$$\text{rich(some(bank))} = \top \Leftrightarrow (p_{\text{some(bank)}} \to p_{\text{rich}}) = 1_{C(P)}.$$

This means that projector equalities translate to quantified formulas of two-sorted first-order logic; in other words, quantum logic includes two-sorted first-order logic. For example,

$$(p_{\text{bank}} \to p_{\text{rich}}^{\perp}) = 1_{C(P)} \Leftrightarrow \forall x (x \in \text{Bank} \Rightarrow x \notin \text{Rich}).$$

These equivalences concern the categories $2\mathcal{SF}$ or $\mathcal{RI}$, where predicates cannot count but only assert. The next subsection deals with Hilbert spaces, where predicates count the elements that satisfy them.

9.5.4 States in a concept space

In this subsection, $\mathcal{C}$ is the category of finite-dimensional real Hilbert spaces.

A satisfaction relation requires a yes or no answer for every individual and every basic property p_i. For practical reasons such a precise information may not be available. Instead, real numbers $\alpha_{iY} \in [0, 1]$ are available representing the probability that an arbitrary individual in Y has property p_i, $i = 1, \ldots, d$.

Let $0 \le \alpha_{iY} \le 1$ and $\beta_{iY} = 1 - \alpha_{iY}$ for $i = 1, \ldots, d$. The set Y is represented in $C(p_1) \otimes \ldots \otimes C(p_d)$ by its state vector

$$\mu_Y = (\alpha_{1Y} p_{1\top} + \beta_{1Y} p_{1\perp}) \otimes \ldots \otimes (\alpha_{dY} p_{d\top} + \beta_{dY} p_{d\perp}).$$

Lemma 9.25 *The coordinates of μ_Y define a probability on the event space $B(P)$ generated by the $\vec{p_i}$ s. Moreover, α_i is equal to the sum of the coordinates of $\vec{p_i} \wedge \mu_Y$ and β_i to the sum of the coordinates of $\neg \vec{p_i} \wedge \mu_Y$.*

Proof Let γ_f be the coordinate of μ_Y for $f \in \prod_{i=1}^d \{p_{i\top}, p_{i\bot}\}$. Then $\vec{p_i} \wedge \mu_Y = \sum_{f, f(i)=p_{i\top}} \gamma_f b_f$. Hence the assertions follow from the equalities

$$\sum_f \gamma_f = 1, \quad \alpha_i = \sum_{f(i)=p_{i\top}} \gamma_f, \quad \beta_i = \sum_{f(i)=p_{i\bot}} \gamma_f.$$

Prove the first equality by induction on d. The case $d = 1$ is trivial. For the induction step, let $d' = d - 1$, $P' = \{1, \ldots, d'\}$ and

$$\mu'_Y = (\alpha_{1Y} p_{1\top} + \beta_{1Y} p_{1\bot}) \otimes \ldots \otimes (\alpha_{d'Y} p_{d'\top} + \beta_{d'Y} p_{d'\bot}).$$

Let δ_g be the coordinate of μ'_Y in $C(P')$, i.e. $\mu'_Y = \sum_{g \in \prod_{i=1}^{d'} \{p_{i\top}, p_{i\bot}\}} \delta_g b_g$. Then $\sum_g \delta_g = 1$ by induction hypothesis. We have

$$\mu_Y = \mu'_Y \otimes (\alpha_{dY} p_{d\top} + \beta_{dY} p_{d\bot}) = \sum_g \delta_g \alpha_{dY} (b_g \otimes p_{d\top}) + \sum_g \delta_g \beta_{dY} (b_g \otimes p_{d\bot}).$$

This finishes the proof, because for every basis vector $b_f \in C(P)$ there is a unique $g \in \prod_{i=1}^{d-1} \{p_{i\top}, p_{i\bot}\}$ such that either $b_f = b_g \otimes p_{d\top}$ or $b_f = b_g \otimes p_{d\bot}$. □

The projector $p_{\vec{p_i}}$ of $C(P)$ maps the state vector μ_Y to a vector $p_{\vec{p_i}}(\mu_Y) = \vec{p_i} \wedge \mu_Y$, the coordinates of which sum up to α_i. Hence the projector $p_{\vec{p_i}}$, equivalently, the two-sorted predicate $\mathcal{L}(\vec{p_i})$, returns for μ_Y the probability that an arbitrary individual in Y satisfies p_i.

Return to vector semantics in information retrieval systems. Choose a set $P = p_1, \ldots, p_d$ of basic properties, for example the most frequent words in a (set of) document(s). Represent words by vectors in the d-dimensional space V_P, where the coordinate γ_i of word w is the frequency of co-occurrence with p_i. The projection onto the one-dimensional subspace of V_P generated by p_i is the vector $\gamma_i p_i$.

The scalar γ_i may be interpreted as the similarity of the word with p_i, but not as the probability that an arbitrary individual designated by w has property p_i, because positive and negative occurrences like *some banks are safe, some banks are not safe* contribute both to γ_i. 'Reasoning by probability' based on frequency counts requires a distinction between positive and negative occurrences.

9.6 Conclusion

New in this approach is that two separate notions of truth, one for concepts and one for sentences, are handled formally inside a single mathematical frame with a resulting equivalence of the two representations. The geometrical properties of quantum logic and the functional application of logic are preserved. On a technical level, both the tensor product and syntactical analysis intervene when composing meanings.

Many interesting questions have not been addressed. For example, biproducts of concept spaces are necessary to handle predicates of an arbitrary arity simultaneously. Ambiguous words as well live in a biproduct of different concept spaces. Disambiguation by context uses the probability that the meaning factors through one branch rather than the other of the biproduct.

The most challenging questions belong to the probabilistic approach to natural language semantics and its relation to compositionality. How to distinguish between opposites? (The usual probabilistic approach confounds them.) How to capture the intuitive interaction of statistical learning of concepts and their logical use?

CHAPTER 10

Proof Nets for the Lambek–Grishin Calculus

MICHAEL MOORTGAT AND
RICHARD MOOT
(Utrecht University and LaBRI(CNRS), Université de Bordeaux)

10.1 Background, motivation

In his two seminal papers (Lambek, 1958; Lambek, 1961), Jim Lambek introduced the 'parsing as deduction' method in linguistics: the traditional parts of speech (noun, verb, adverb, determiner, etc.) are replaced by logical formulas, or types if one takes the computational view; the judgement whether an expression is well-formed is the outcome of a process of logical deduction, or, reading formulas as types, a computation in the type calculus.

$$np \otimes (np\backslash s) \otimes (((np\backslash s)\backslash(np\backslash s))/np) \otimes (np/n) \otimes n \to s \quad (10.1)$$
$$\text{time} \quad \text{flies} \quad \text{like} \quad \text{an} \quad \text{arrow}$$

What is the precise nature of grammatical composition, the $\otimes$ operation in the example above? The 1958 and 1961 papers present two views on this: in the 1958 paper, types are assigned to *strings* of words; in the 1961 paper, they are assigned to *phrases*, bracketed strings, with a grouping into constituents. The Syntactic Calculus, under the latter view, is extremely simple. The derivability relation between types is given by the preorder laws of eqn (10.2) and the residuation principles of eqn (10.3).

$$A \to A \quad ; \quad \text{from } A \to B \text{ and } B \to C \text{ infer } A \to C \quad (10.2)$$
$$A \to C/B \quad \text{iff} \quad A \otimes B \to C \quad \text{iff} \quad B \to A\backslash C \quad (10.3)$$

To obtain the 1958 view, one adds the *non-logical axioms* of eqn (10.4), attributing associativity properties to the $\otimes$ operation.

$$(A \otimes B) \otimes C \to A \otimes (B \otimes C) \quad ; \quad A \otimes (B \otimes C) \to (A \otimes B) \otimes C \quad (10.4)$$

The syntactic calculus in its two incarnations—the basic system **NL** given by eqns (10.2) and (10.3) and the associative variant **L**, which adds the postulates of eqn (10.4)—recognizes only context-free languages. It is well known that to capture the dependencies that occur in natural languages, one needs expressivity beyond the context-free languages. Here are some characteristic patterns from formal language theory that can be seen as suitable idealizations of phenomena that occur in the wild.

$$\begin{aligned} \text{copying: } & \{w^2 \mid w \in \{a,b\}^+\} \\ \text{counting dependencies: } & \{a^n b^n c^n \mid n > 0\} \\ \text{crossed dependencies: } & \{a^n b^m c^n d^m \mid n, m > 0\} \end{aligned} \quad (10.5)$$

In the tradition of extended rewriting systems, there is a large group of grammar formalisms that handle these and related patterns gracefully: tree adjoining grammars, linear indexed grammars, combinatory categorial grammars, minimalist grammars, multiple context-free grammars, ... (Kallmeyer, 2010). Also in the Lambek tradition, extended type-logical systems have been proposed with expressive power beyond context-free languages: multimodal grammars (Morrill, 1994; Moortgat, 1996), discontinuous calculi (Morrill, Fadda and Valentin, 2007), etc. These extensions, as well as the original Lambek systems, respect an 'intuitionistic' restriction: in a sequent presentation, derivability is seen as a relation between (a structured configuration of) hypotheses $A_1, \ldots, A_n$ and a *single* conclusion B. In a paper antedating linear logic by a couple of years, Grishin (1983) proposes a generalization of the Lambek calculus that removes this intuitionistic restriction. Linguistic application of Grishin's ideas is fairly recent. In the present paper, we study the system presented in Moortgat (2009), which we will refer to as **LG**.

10.1.1 Dual residuation principles, linear distributivities

In **LG** the inventory of type-forming operations is doubled: in addition to the familiar operators $\otimes, \backslash, /$ (product, left and right division), we find a dual family $\oplus, \oslash, \ominus$: coproduct, right and left difference.

$$\begin{aligned} A, B ::= p \mid & \quad \text{atoms: } s, np, \ldots \\ & A \otimes B \mid B \backslash A \mid A/B \mid \quad \text{product, left vs right division} \\ & A \oplus B \mid A \oslash B \mid B \ominus A \quad \text{coproduct, right vs left difference} \end{aligned} \quad (10.6)$$

Some clarification about the notation: we follow Lambek (1993) in writing $\oplus$ for the coproduct, which is a multiplicative operation, like $\otimes$. We read $B\backslash A$ as 'B under A', A/B as 'A over B', $B \ominus A$ as 'B from A' and $A \oslash B$ as 'A less B'. For the difference operations then, the quantity that is subtracted is under the circled (back)slash, just as we have the denominator under the (back)slash in the case of left and right division types. In a formulas-as-types spirit, we will feel free to refer to the division operations as implications, and to the difference operations as co-implications.

10.1.1.1 DUAL RESIDUATION PRINCIPLES

The most basic version of **LG** is the symmetric generalization of **NL**, which means that to eqns (10.2) and (10.3) we add the dual residuation principles of eqn (10.7).

$$B \oslash C \to A \quad \text{iff} \quad C \to B \oplus A \quad \text{iff} \quad C \oslash A \to B \tag{10.7}$$

To get a feeling for the consequences of the preorder laws of eqn (10.2) and the (dual) residuation principles of eqns (10.3) and (10.7), here are some characteristic theorems and derived rules of inference. First, the *compositions* of the product and division operations, and of the co-product and difference operation give rise to the expanding and contracting patterns of eqn (10.8). The rows here are related by a left–right symmetry, the columns by arrow reversal.

$$
\begin{array}{ll}
A \otimes (A \backslash B) \to B \to A \backslash (A \otimes B) & (B/A) \otimes A \to B \to (B \otimes A)/A \\
(B \oplus A) \oslash A \to B \to (B \oslash A) \oplus A & A \oslash (A \oplus B) \to B \to A \oplus (A \oslash B)
\end{array}
\tag{10.8}
$$

Secondly, one can show that the type-forming operations have the monotonicity properties summarized in the following schema, where $\uparrow$ ($\downarrow$) is an isotone (antitone) position:

$$(\uparrow \otimes \uparrow), (\uparrow / \downarrow), (\downarrow \backslash \uparrow), (\uparrow \oplus \uparrow), (\uparrow \oslash \downarrow), (\downarrow \oslash \uparrow)$$

In other words, the following inference rules are valid:

$$
\frac{A' \to A \quad B' \to B}{A' \otimes B' \to A \otimes B} \qquad \frac{A \to A' \quad B \to B'}{A \oplus B \to A' \oplus B'}
\tag{10.9}
$$

$$
\frac{A' \to A \quad B \to B'}{A \backslash B \to A' \backslash B'} \quad \frac{A' \to A \quad B \to B'}{A' \oslash B' \to A \oslash B} \quad \frac{A' \to A \quad B \to B'}{B/A \to B'/A'} \quad \frac{A' \to A \quad B \to B'}{B' \oslash A' \to B \oslash A}
\tag{10.10}
$$

10.1.1.2 INTERACTION: DISTRIBUTIVITY PRINCIPLES

As we saw above, one could extend the inferential capabilities of this minimal system by adding postulates of associativity and/or commutativity for $\otimes$ and $\oplus$. From a substructural perspective, each of these options destroys structure-sensitivity for a particular dimension of grammatical organization: word order in the case of commutativity, constituent structure in the case of associativity. In **LG** there is an alternative that leaves the sensitivity for linear order and phrasal structure intact: instead of considering structural options for the individual $\otimes$ and $\oplus$ families, one can consider *interaction* principles for the communication between them. We will consider the following group:

$$(A \oslash B) \otimes C \to A \oslash (B \otimes C) \qquad C \otimes (B \oslash A) \to (C \otimes B) \oslash A$$
$$C \otimes (A \oslash B) \to A \oslash (C \otimes B) \qquad (B \oslash A) \otimes C \to (B \otimes C) \oslash A \qquad (10.11)$$

These postulates have come to be called *linear distributivity principles* (e.g. Cockett and Seely, 1996): linear, because they respect resources (no material gets copied). Moot (2007) models the adjunction operation of tree adjoining grammars using the interaction principles of eqn (10.11) and shows how through this modelling the mildly context-sensitive patterns of eqn (10.5) can be obtained within **LG**.

10.1.2 Arrows: LG as a deductive system

Lambek (1988) studies the syntactic calculus from a categorical perspective. Types are seen as the objects of a category and one studies morphisms between these objects, arrows $f : A \longrightarrow B$. For each A, there is an identity arrow 1_A. Then there are inference rules to produce new arrows from arrows already obtained. Among these is the composition $g \circ f$, defined when $dom(g) = cod(f)$. Composition is associative, i.e. one has the equation $f \circ (g \circ h) = (f \circ g) \circ h$. Also, $f \circ 1_A = f = 1_B \circ f$, where $f : A \longrightarrow B$.

$$1_A : A \longrightarrow A \qquad \frac{f : A \longrightarrow B \quad g : B \longrightarrow C}{g \circ f : A \longrightarrow C} \qquad (10.12)$$

In this paper, we will not pursue the categorical interpretation of **LG**: our emphasis in the following sections is on the *sequent* calculus for this logic, the term language coding sequent proofs, and the correspondence between these proofs and proof nets. Our aim in this section is simply to have a handy language for naming proofs in the deductive presentation, and to use this in Section 10.2.1 to establish the equivalence between the deductive and the sequent presentations.

To obtain a**LG**, one adds to eqn (10.12) further rules of inference for the residuation principles and their duals. (For legibility, type subscripts $\triangleright_{A,B,C} f$ are omitted.)

$$\frac{f : A \otimes B \longrightarrow C}{\triangleright f : A \longrightarrow C/B} \qquad \frac{f : A \otimes B \longrightarrow C}{\triangleleft f : B \longrightarrow A\backslash C} \qquad (10.13)$$

$$\frac{g : A \longrightarrow C/B}{\triangleright^{-1} g : A \otimes B \longrightarrow C} \qquad \frac{g : B \longrightarrow A\backslash C}{\triangleleft^{-1} g : A \otimes B \longrightarrow C} \qquad (10.14)$$

$$\frac{f : C \longrightarrow B \oplus A}{\triangleleft f : B \oslash C \longrightarrow A} \qquad \frac{f : C \longrightarrow B \oplus A}{\triangleright f : C \oslash A \longrightarrow B} \qquad (10.15)$$

$$\frac{g : B \oslash C \longrightarrow A}{\triangleleft^{-1} g : C \longrightarrow B \oplus A} \qquad \frac{g : C \oslash A \longrightarrow B}{\triangleright^{-1} g : C \longrightarrow B \oplus A} \qquad (10.16)$$

As remarked above, the Lambek–Grishin calculus exhibits two involutive symmetries at the level of types and proofs: a left–right symmetry $\cdot^\natural$ and an arrow-reversing symmetry $\cdot^\dagger$ such that

$$f^\natural : A^\natural \longrightarrow B^\natural \quad \text{iff} \quad f : A \longrightarrow B \quad \text{iff} \quad f^\dagger : B^\dagger \longrightarrow A^\dagger \qquad (10.17)$$

with, on the type level, the translation tables below (abbreviating a long list of defining equations $(A \otimes B)^\natural \doteq B^\natural \otimes A^\natural, (B \otimes A)^\natural \doteq A^\natural \otimes B^\natural, \ldots$)

$$\frac{A \otimes B \quad A/B \quad A \oplus B \quad A \oslash B}{B \otimes A \quad B\backslash A \quad B \oplus A \quad B \oslash A} \natural \qquad \frac{A/B \quad A \otimes B \quad B\backslash A}{B \oslash A \quad B \oplus A \quad A \oslash B} \dagger$$

and on the level of proofs $(1_A)^\natural = 1_{A^\natural}, (g \circ f)^\natural = g^\natural \circ f^\natural, (1_A)^\dagger = 1_{A^\dagger}, (g \circ f)^\dagger = f^\dagger \circ g^\dagger$, and the list of defining equations $(\triangleleft f)^\natural \doteq \triangleright f^\natural, (\triangleleft f)^\dagger \doteq \blacktriangleright f^\dagger, \ldots$ corresponding to the translation tables above.

The distributivity principles, in **aLG**, take the form of extra axioms (primitive arrows). Below we show arrows **d**, **b** for the interaction between $\oslash$ and $\otimes$. For the left–right symmetric pair $\mathbf{d}^\natural, \mathbf{q}^\natural$ we write **b**, **p**

$$\begin{aligned}\mathbf{d}_{A,B,C} &: (A \oslash B) \otimes C \longrightarrow A \oslash (B \otimes C) \\ \mathbf{q}_{A,B,C} &: C \otimes (A \oslash B) \longrightarrow A \oslash (C \otimes B)\end{aligned} \qquad (10.18)$$

To establish the equivalence between **aLG** and the sequent calculus **sLG**, which we discuss in the next section, we will use the fact that the monotonicity rules are derived rules of inference of **aLG**. For example, f/g can be defined as in eqn (10.19).

$$\frac{f : A \longrightarrow A' \quad g : B \longrightarrow B'}{f/g : A/B' \longrightarrow A'/B} \qquad (10.19)$$

$$f/g \doteq (\triangleright(f \circ (\triangleright^{-1} 1_{A/B}))) \circ (\triangleright \triangleleft^{-1}((\triangleleft \triangleright^{-1} 1_{A/B'}) \circ g))$$

Similarly, for the distributivity postulates, we will rely on a rule form, which for **d** would be

$$\frac{B \otimes C \rightarrow A \oplus D}{A \oslash B \rightarrow D/C} \qquad (10.20)$$

The inference rule of eqn (10.20) is derived as shown in eqn (10.21).

$$\frac{\mathbf{d}_{A,B,C} : (A \oslash B) \otimes C \longrightarrow A \oslash (B \otimes C) \quad \dfrac{f : B \otimes C \longrightarrow A \oplus D}{\blacktriangleleft f : A \oslash (B \otimes C) \longrightarrow D}}{\dfrac{(\blacktriangleleft f) \circ \mathbf{d}_{A,B,C} : (A \oslash B) \otimes C \longrightarrow D}{\triangleright((\blacktriangleleft f) \circ \mathbf{d}_{A,B,C}) : A \oslash B \longrightarrow D/C}} \qquad (10.21)$$

10.2 Display sequent calculus and proof nets

Is there a decision procedure to determine whether $A \to B$ holds? In the presence of *expanding* patterns, as we saw in eqn (10.8), this is not immediately clear. For the language with $/, \otimes, \backslash$, the key result of Lambek's original papers was to establish decidability by applying Gentzen's method: the syntactic calculus is recast as a sequent calculus; for the sequent presentation one then shows that the cut rule (the sequent form of transitivity) is admissible; backward-chaining, cut-free proof search then yields the desired decision procedure.

In Section 10.2.1 below, we work through a similar agenda for **LG**. We introduce **sLG**, a sequent system for the Lambek–Grishin calculus in the style of display logic (Goré, 1997), and show that it is equivalent to **aLG**. The sequent presentation enjoys cut elimination; decidability follows. Sequent proof search, though decidable, remains suboptimal in that it allows a great many derivations for what in effect one would like to consider as 'the same' proof. In Section 10.2.2, we introduce proof nets for **LG** and show how these nets remove the spurious forms of non-determinism of sequent proof search.

10.2.1 sLG: display sequent calculus

The arrows of **aLG** are morphisms between *types*. In the sequent calculus, derivability is a relation between *structures* built from types. We will present the sequent calculus for **LG** in the format of a display logic (see Goré, 1997, for a comprehensive display logical view on the substructural landscape). The characteristic feature of display logic is that for every logical connective, there is a corresponding structural connective. We use the same symbols for the logical operations and their structural counterparts; structural operations are marked off by centre dots. The grammar for input (sequent left-hand side), and output structures (sequent right-hand side) is shown below.

$$\begin{aligned} \mathcal{I} &::= \mathcal{F} \mid \mathcal{I} \cdot \otimes \cdot \mathcal{I} \mid \mathcal{I} \cdot \oslash \cdot \mathcal{O} \mid \mathcal{O} \cdot \oslash \cdot \mathcal{I} \\ \mathcal{O} &::= \mathcal{F} \mid \mathcal{O} \cdot \oplus \cdot \mathcal{O} \mid \mathcal{I} \cdot \backslash \cdot \mathcal{O} \mid \mathcal{O} \cdot / \cdot \mathcal{I} \end{aligned}$$

The rules of **sLG** come in three groups: the identity group (Axiom, Cut), the structural group (display postulates, distributivity postulates), and the logical group (left and right introduction rules for the logical connectives). Variables X, Y, Z in these rules range over structures, input or output, depending on whether they appear left or right of the sequent arrow.

1. *Axiom, Cut.*

$$\frac{}{A \Rightarrow A}\, \text{Ax} \qquad \frac{X \Rightarrow A \quad A \Rightarrow Y}{X \Rightarrow Y}\, \text{Cut} \qquad (10.22)$$

2. *Display postulates.* The (dual) residuation principles are formulated at the structural level. These rules ensure that any formula constituent of a sequent can be displayed as the single occupant of the sequent left-hand side or right-hand side—hence the name.

$$\frac{\overline{X \Rightarrow Z \cdot / \cdot Y}}{\overline{\overline{X \cdot \otimes \cdot Y \Rightarrow Z}}} \, rp \qquad \frac{\overline{Y \cdot \oslash \cdot Z \Rightarrow X}}{\overline{\overline{Z \Rightarrow Y \cdot \oplus \cdot X}}} \, drp \tag{10.23}$$
$$\overline{Y \Rightarrow X \cdot \backslash \cdot Z} \, rp \qquad \overline{Z \cdot \oslash \cdot X \Rightarrow Y} \, drp$$

3. *Distributivity postulates.* The linear distributivities motivate the choice for a *display* sequent calculus. The distributivity postulates, in their rule form in eqn (10.20), in the sequent format become *structural* rules. In a Gentzen-style sequent calculus, formulating such structural rules would be impossible: one only has structural punctuation marks for $\otimes$ and $\oplus$ (the antecedent and succedent comma). However, one could not formulate eqn (10.20) as a *logical* rule either: it introduces two operations simultaneously.

$$\frac{X \cdot \otimes \cdot Y \Rightarrow Z \cdot \oplus \cdot W}{Z \cdot \oslash \cdot X \Rightarrow W \cdot / \cdot Y} \, G1 \qquad \frac{X \cdot \otimes \cdot Y \Rightarrow Z \cdot \oplus \cdot W}{Y \cdot \oslash \cdot W \Rightarrow X \cdot \backslash \cdot Z} \, G3$$

$$\frac{X \cdot \otimes \cdot Y \Rightarrow Z \cdot \oplus \cdot W}{Z \cdot \oslash \cdot Y \Rightarrow X \cdot \backslash \cdot W} \, G2 \qquad \frac{X \cdot \otimes \cdot Y \Rightarrow Z \cdot \oplus \cdot W}{X \cdot \oslash \cdot W \Rightarrow Z \cdot / \cdot Y} \, G4 \tag{10.24}$$

4. *Logical rules.* For each connective there is a left and a right introduction rule. One of these is a one-premise rewrite rule, exchanging the logical connective for its structural counterpart; the other rule puts together a complex formula alongside the matching complex structure.

   (a) *Rewrite rules.*
   $\$ \in \{\otimes, \oslash, \oslash\}, \# \in \{\oplus, \backslash, /\}$.

$$\frac{A \cdot \$ \cdot B \Rightarrow Y}{A \, \$ \, B \Rightarrow Y} \, \$L \qquad \frac{X \Rightarrow A \cdot \# \cdot B}{X \Rightarrow A \, \# \, B} \, \#R \tag{10.25}$$

The rewrite rules are invertible. As an example, compare ($\otimes L$) and ($\otimes L$)$^{-1}$.

$$\frac{A \cdot \otimes \cdot B \Rightarrow Y}{A \otimes B \Rightarrow Y} \otimes L \qquad \frac{\dfrac{A \Rightarrow A \quad B \Rightarrow B}{A \cdot \otimes \cdot B \Rightarrow A \otimes B} \otimes R \quad A \otimes B \Rightarrow Y}{A \cdot \otimes \cdot B \Rightarrow Y} \, \text{Cut} \tag{10.26}$$

(b) *Two-premise rules.* The ($/L$), ($\oslash R$) rules are left–right symmetric.

$$\frac{X \Rightarrow A \quad Y \Rightarrow B}{X \cdot \otimes \cdot Y \Rightarrow A \otimes B} \otimes R \qquad \frac{A \Rightarrow X \quad B \Rightarrow Y}{A \oplus B \Rightarrow X \cdot \oplus \cdot Y} \oplus L \tag{10.27}$$

$$\frac{X \Rightarrow A \quad B \Rightarrow Y}{A\backslash B \Rightarrow X \cdot \backslash \cdot Y} \backslash L \qquad \frac{X \Rightarrow A \quad B \Rightarrow Y}{X \cdot \oslash \cdot Y \Rightarrow A \oslash B} \oslash R \qquad (10.28)$$

10.2.1.1 EQUIVALENCE

For every arrow $f : A \longrightarrow B$, there is a sequent proof $A \Rightarrow B$. For every sequent proof $X \Rightarrow Y$, there is an arrow $f : X^\circ \longrightarrow Y^\circ$, where X°, Y° are the formulas obtained from X, Y by replacing the structural connectives by their logical counterparts.

From arrows to sequent proofs

1_A and composition $g \circ f$ are immediate. We use the invertibility of the rewrite rules to prove the residuation/adjoints laws in the sequent calculus. Below, as an example, is a sequent proof for $\triangleright f$.

$$\frac{f : A \otimes B \longrightarrow C}{\triangleright f : A \longrightarrow C/B} \quad \rightsquigarrow \quad \cfrac{\cfrac{\cfrac{A \otimes B \Rightarrow C}{A \cdot \otimes \cdot B \Rightarrow C} (\otimes L)^{-1}}{A \Rightarrow C \cdot / \cdot B} rp}{A \Rightarrow C/B} /R \qquad (10.29)$$

From sequent proofs to arrows

Under the mapping $\cdot^\circ$, Cut turns into composition of arrows, the (dual) display postulates into the (dual) residuation rules, and the distributivity postulates into the rule form of the arrows **d, q, b, p**, which in eqn (10.21) we have shown to be derivable in **aLG**. For the logical group, the premise and conclusion of the rewrite rules are identified. The two-premise logical rules become the monotonicity rules, which are derivable rules of inference in **aLG**, as we saw.

10.2.1.2 CUT ELIMINATION, DECIDABILITY (MOORTGAT, 2007)

In **sLG**, Cut is an admissible rule: every theorem has a cut-free derivation.

Decidability is a nice property to have. However, the astute reader at this point may feel disappointed: the goal-driven, backward-chaining, cut-free proof search of the decision procedure presupposes that the *structure* of the goal sequent is given. Parsing, as it is standardly understood, means deciding whether a string is well-formed, and assigning it a proper structure. Here, to start backward-chaining sequent proof search, we have to assume that the correct structure is already given. A generate-and-test approach is obviously not feasible here, the number of binary bracketings over a string of length n being the Catalan number C_n. We have not addressed the parsing problem, in other words. Turning to *proof nets* in Section 10.2.2, this situation will change: the construction algorithm for **LG** nets will work in a *data-driven* mode, effectively *computing* the structure of the goal sequent.

10.2.2 Proof nets

Proof nets are a graphical way of representing proofs, introduced first for linear logic (Girard, 1987). Proof nets can either be seen as a sort of 'parallellized' sequent proofs or as a sort of multi-conclusion natural deduction. Proof nets are defined as a subclass of a larger class of graphs called proof structures. Where proof nets correspond to sequent proofs, proof structures in general may not, but we can distinguish proof nets from other proof structures based only on properties of the graph.

The proof nets for the Lambek–Grishin calculus we present in this section are a simple extension of the proof nets for the multimodal Lambek calculus of Moot and Puite (2002). A proof structure is a (hyper)graph where the vertices are labeled by formulas and the edges connect these formulas. In what follows we will often speak of formula occurrences (or simply *formulas* if there is no possibility of confusion) instead of vertices labelled by formulas. The hyperedges correspond to the logical rules, linking the active formulas and the main formula of the rule and keeping track of whether one is dealing with a non-invertible two-premise rule or with an invertible one-premise rule. We will call these *tensor* and *cotensor* links respectively.

10.2.2.1 PROOF STRUCTURES AND ABSTRACT PROOF STRUCTURES

Definition 10.1 *A link is a tuple* $\langle t, p, c, m \rangle$ *where:*

- *t is the type of the link—tensor or cotensor*
- *p is the list of premises of the link*
- *c is the list of conclusions of the link*
- *m, the main vertex/formula of the link, is either a member of p, a member of c, or the constant 'nil'.*

In the case where m is a member of p we speak of a left *link (corresponding to the left rules of the sequent calculus, where the main formula of the link occurs in the antecedent) and in case m is a member of c we speak of a* right *link.*

Graphically, links are displayed as shown below. A central node links together the premises and conclusions of the link; when we need to refer to the connections between the central node and the vertices, we will call them its *tentacles*. The interior of this central node is white for a tensor link and black for a cotensor link. The premises are drawn, in left-to-right order, above the central node; the conclusions, also in left-to-right order, are drawn below it. The main formula of cotensor links is drawn as an arrow to the member of the premises or the conclusions that is the main formula of the link. The main formula of tensor links are not distinguished visually, but can be determined by inspection of the formula labels.

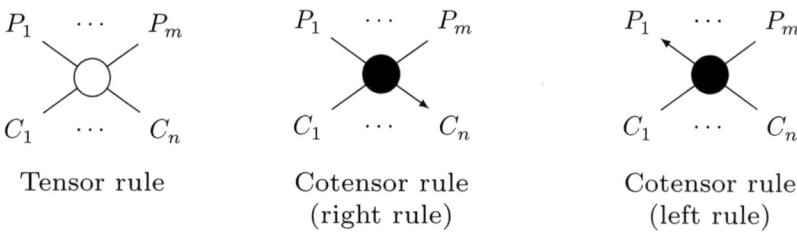

Tensor rule   Cotensor rule (right rule)   Cotensor rule (left rule)

Figure 10.1 shows the links for the Lambek–Grishin calculus. There are two links for each connective: one link where the main formula is a premise of the link (a left link) and one link where the main formula is a conclusion of the link (a right link). The symmetry between the Lambek connectives and the Grishin connectives is immediately clear: the links for the Grishin connectives are up-down symmetric versions of the links for the Lambek connectives.

Definition 10.2 *A proof structure $\langle S, \mathcal{L} \rangle$ is a finite set of formula occurrences S and a set of links $\mathcal{L}$ from those shown in Figure 10.1 such that:*

- *each formula is at most once the premise of a link*
- *each formula is at most once the conclusion of a link.*

Formulas that are not the conclusion of any link are called the hypotheses *of the proof structure. Formulas that are not the premise of any link are called the* conclusions *of the proof structure. We will say that a proof structure with hypotheses $H_1, \ldots, H_m$ and conclusions $C_1, \ldots, C_n$ is a proof structure of $H_1, \ldots, H_m \Rightarrow C_1, \ldots C_n$.*

Example 10.3 Figure 10.2 shows the hypothesis unfolding of $(s \oslash s) \obslash np$ and the conclusion unfolding of $s / (np \backslash s)$. Both are obtained by simple application of the rules of Figure 10.1 until we reach the atomic subformulas.

Although the figure satisfies the conditions for being a proof structure (note, for example, that connectedness is not a requirement, so a proof structure is allowed to have one connected component for each of the unfolded formulas), it is a proof structure of $(s \oslash s) \obslash np, s, s, np \Rightarrow s / (np \backslash s), s, s, np$. We can obtain a proof structure of $(s \oslash s) \obslash np \Rightarrow s / (np \backslash s)$ by identifying atomic formulas (this node identification corresponds to the 'axiom links' of linear logic proof nets). In this case, we choose to identify the top s of the left subgraph with the bottom s of the right subgraph and perform the unique choice for the remaining atomic formulas. The result is the proof structure shown in Figure 10.3 on the left.

Let us take a closer look at this new proof structure. We have connected the minor premise of the implication and co-implication links by a curve. This is due to the graphical constraints of writing these proof nets on the plane: we want to draw the $np \backslash s$ node *below* the cotensor link at the bottom of the figure, since it is a conclusion of this link. However, we would have to draw the figure on a cylinder to make this work; in other words, following down a path premise–link–conclusion

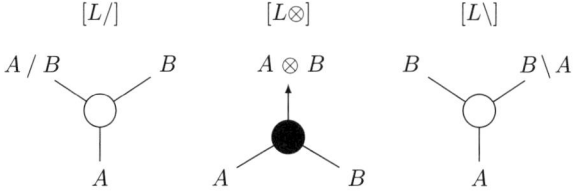

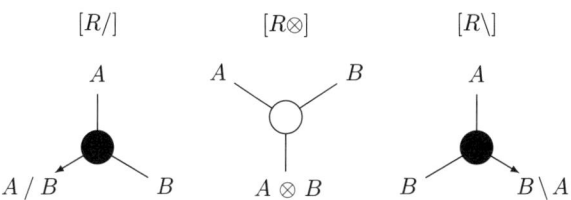

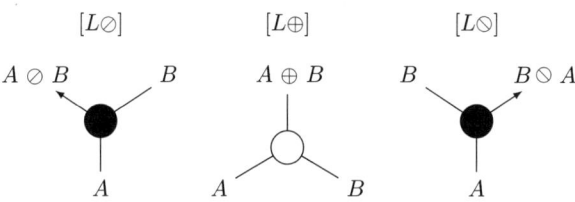

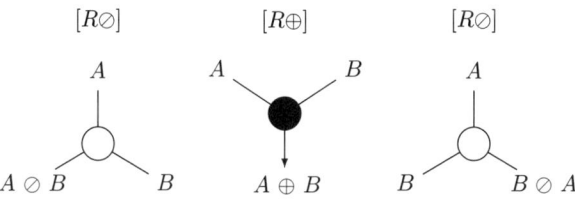

Fig 10.1 Links for proof structures of the Lambek–Grishin calculus.

does not necessarily give a total order but can give a cyclic order on the formulas in the proof structure; for proof *nets*, these cyclic paths can only pass through the minor premise of a cotensor (co-)implication link. As indicated by the drawing, the connection to the $np \setminus s$ node from the cotensor link arrives from above, indicating it is a conclusion of this link. Similarly, we go down from the $s \oslash s$ node to arrive at the other cotensor link.

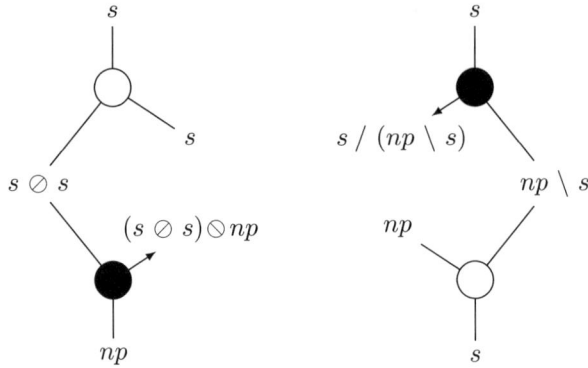

Fig 10.2 Lexical unfolding.

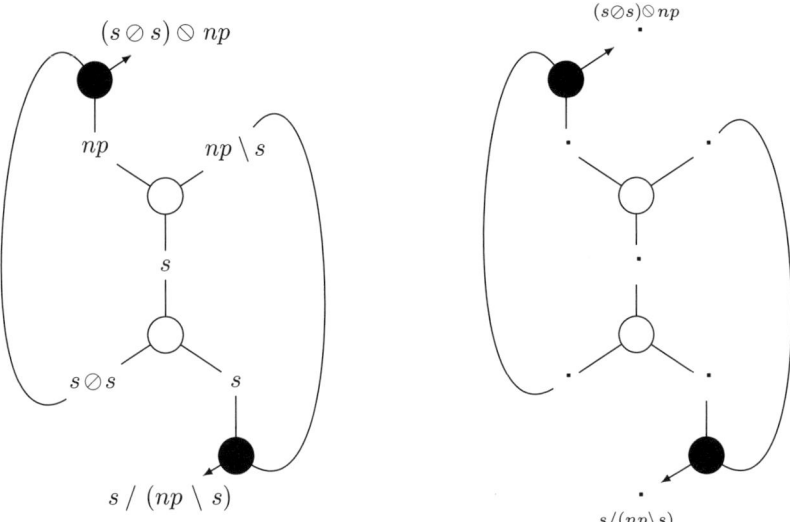

Fig 10.3 Proof structure of $(s \oslash s) \obslash np \Rightarrow s / (np \setminus s)$ corresponding to the lexical unfolding of Figure 10.2 and its corresponding abstract proof structure.

A comparison with the introduction rule for the implication in natural deduction is another way to make this clear. For the introduction rule, we hypothesize a formula $np \setminus s$ (here, a conclusion of the cotensor rule), then derive s (here a premise of the cotensor rule). The introduction rule then indicates we can withdraw this hypothesis and conclude $s / (np \setminus s)$, with some indexing indicating which hypotheses are withdrawn at which rule. In the proof structure above, the connection between the cotensor link and the $np \setminus s$ rule plays exactly the role of this indexing (although, since a proof structure is not necessarily a proof, we have no guarantee yet that the introduction rule is correctly applied; the contractions introduced later will remedy this).

With this in mind, we can verify that the proof structure in Figure 10.3 corresponds exactly to the one in Figure 10.2 with the stated node identifications: we have the same formula occurrences and the links have the same premises as well as the same conclusions.

So while the logical rules of a sequent proof correspond directly to the links of a proof net, the axioms and cut rules of a sequent proof correspond to *formulas*. An axiomatic formula is a formula that is not the main formula of any link. A cut formula is a formula that is the main formula of two links. So on the left of Figure 10.3, the np formula and both s formulas are axiomatic.

Definition 10.4 *An abstract proof structure $\langle V, \mathcal{L}, h, c \rangle$ is a set of vertices V, a set of (unlabeled) links $\mathcal{L}$, and two functions h and c, such that:*

- *each formula is at most once the premise of a link*
- *each formula is at most once the conclusion of a link*
- *h is a function from the hypotheses of the abstract proof structure to formulas*
- *c is a function from the conclusions of the abstract proof structure to formulas.*

Note that the abstract proof structure corresponding to a two-formula sequent $A \Rightarrow B$ has only a single vertex v, with $h(v) = A$ and $c(v) = B$.

The transformation from proof structure to abstract proof structure is a forgetful mapping: we transform a proof structure into an abstract proof structure by erasing all formula information from the internal vertices, keeping only the formula labels of the hypotheses and the conclusions. Visually, we remove the formula labels of the graph and replace them by simple vertices (.) and we indicate the results of the functions h and c above (resp. below) the vertices (those which are hypotheses and conclusions of the abstract proof structure respectively). As a result, we have the following four types of vertices in an abstract proof structure:

Internal   Hypothesis   Conclusion   Both

Example 10.5 Figure 10.3 shows (on the right) the transformation of the proof structure on its left into an abstract proof structure. In the abstract proof structure, we can no longer distinguish which vertices are axioms: only the cotensor links still allow us to distinguish between the main and active vertices of the link by means of the arrow.

Definition 10.6 *A tree is an acyclic, connected abstract proof structure that does not contain any cotensor links.*

The trees of defn 10.6 correspond to sequents in a rather direct way. In fact, they have the pleasant property of 'compiling away' the display rules of the sequent calculus. Or, in other words, trees represent a class of sequents that is equivalent up to the display postulates.

Definition 10.7 *Given an abstract proof structure A, we say that A contracts in one step to A', written A → A', iff A' is obtained from A by replacing one of the subgraphs of the form shown in Figures 10.4 and 10.5 by a single vertex*

$$H$$
$$\bullet$$
$$C$$

H represents the result of the function h for the indicated node (relevant only in the case this node is a hypothesis of the abstract proof structure). Similarly, C represents the formula assigned by the function c to the indicated node.

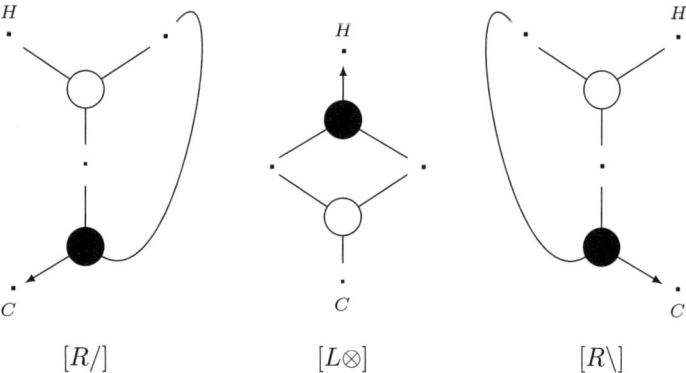

Fig 10.4 Contractions—Lambek connectives.

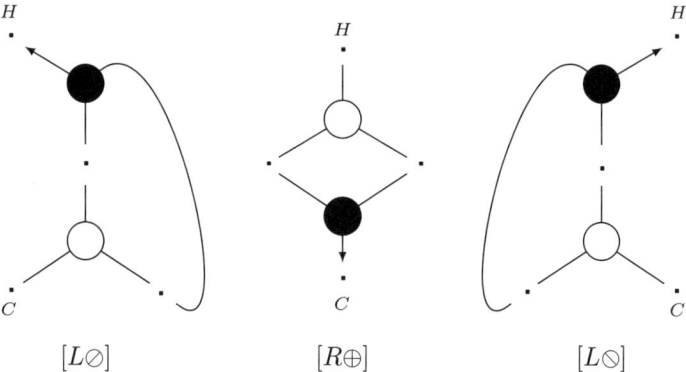

Fig 10.5 Contractions—Grishin connectives.

Given an abstract proof structure A we say that A contracts to *an abstract proof structure A'* if there is a sequence of zero or more one-step contractions from A to A'.

When we say that a *proof structure P contracts to an abstract proof structure A'* we will mean that the underlying abstract proof structure A of P contracts to A'.

As we saw in Section 10.1.2, to obtain expressivity beyond context-free languages, we are interested in **LG** with added interaction principles. The (rule forms of the) postulates of eqn (10.11) correspond to additional rewrite rules on the abstract proof structures. Figures 10.6 and 10.7 give the rewrite rules corresponding to the postulates **d, b** and **q, p** respectively;[1] a total of four rewrite rules, (G1) to (G4). All four rewrite rules start from the same initial configuration and replace it by one of the four possible configurations indicated in the figures.

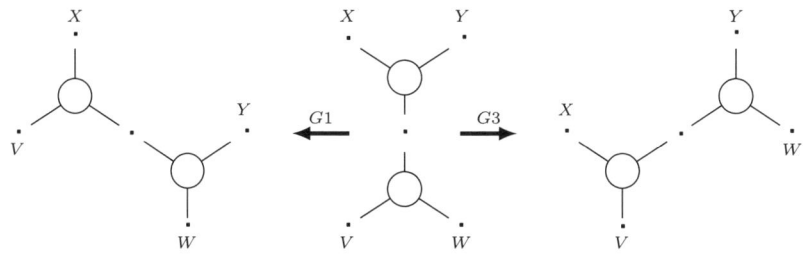

Fig 10.6 Grishin interactions I—'mixed associativity'.

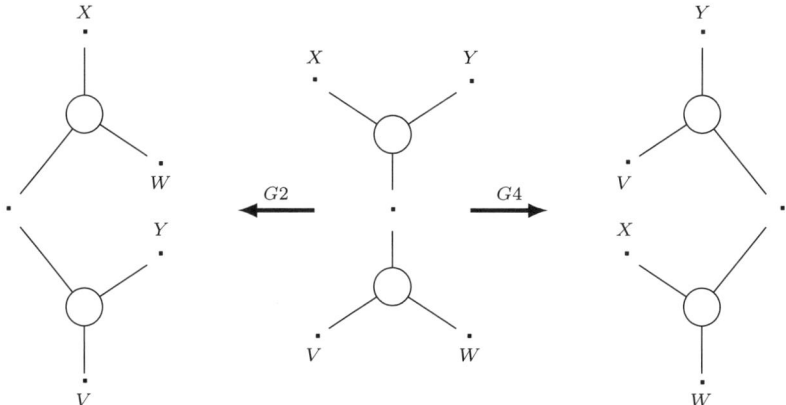

Fig 10.7 Grishin interactions II—'mixed commutativity'.

---

[1] These are Grishin's Class IV interactions. His Class I (not discussed in this paper) can be obtained by reversing all four arrows in the two figures.

10.2.2.2 PROOF NETS

Definition 10.8 *A proof structure P is a proof net iff its underlying abstract proof structure A converts to a tree using the contractions of Figures 10.4 and 10.5 and the structural rules of Figures 10.6 and 10.7.*

Theorem 10.9 *A proof structure P is a proof net—that is, P converts to a tree T—iff there is a sequent proof of T.*

The proof is an easy adaptation of the proof of Moot and Puite (2002). A detailed proof can be found in Moot (2007).

In the following section, we will inspect the structure of the conversion sequence in more detail. The notion of a proof net *component* will be useful in this respect:

Definition 10.10 *Given a proof net P, a component C of P is a maximal subnet of P containing only tensor links.*

From a proof net, we can obtain its components by simply erasing all cotensor links. The components will be the connected components (in the graph-theoretic sense) of the resulting graph. In what follows we will implicitly use the word 'component' to refer only to components containing at least one tensor link. Though there is no problem in allowing a component to be a single vertex, the correspondence between focused sequent proofs and proof nets is more clear when components are non-trivial.

Let us look now at some examples of proof structures, and the conversions that identify them as bona fide proof nets.

Example 10.11 To show that the proof structure of Figure 10.3 is a proof net, we need to show it can be contracted to a tree. Inspection of the contractions shows that none of them apply, but the interaction rules do: the two tensor links in the center of the figure match the input configuration for rule ($G1$). Applying this rule produces the abstract proof structure shown in Figure 10.8 on the right. At this point, the two cotensor links can be contracted in any order. Figure 10.9 shows the result of first applying the ($L\oslash$), then the ($R/$) contraction.

Example 10.12 For a second example (to be taken up again when we discuss focused proof search in Section 10.3) we turn to Figure 10.10, which shows the lexical proof structures for a generalized quantifier noun phrase, a transitive verb, a determiner, and a lexical noun.

Consider the sentence 'everyone likes the teacher'. In the unfocused sequent calculus **sLG**, the sequent $(np \mathbin{/} n) \otimes n, (np \backslash s) \mathbin{/} np, np \mathbin{/} n, n \Rightarrow s$ has at least seven proofs, depending on the order of application of the introduction rules for the five occurrences of the logical connectives involved: $\otimes$ (once), $/$ (three times), $\backslash$ (once). Figure 10.11 gives, on the left, the *single* possible identification of n and np formulas that gives rise to a proof net with the lexical entries in the desired order. The corresponding abstract proof structure is given on the right. This abstract proof structure allows us to apply a contraction directly, as shown in Figure 10.12.

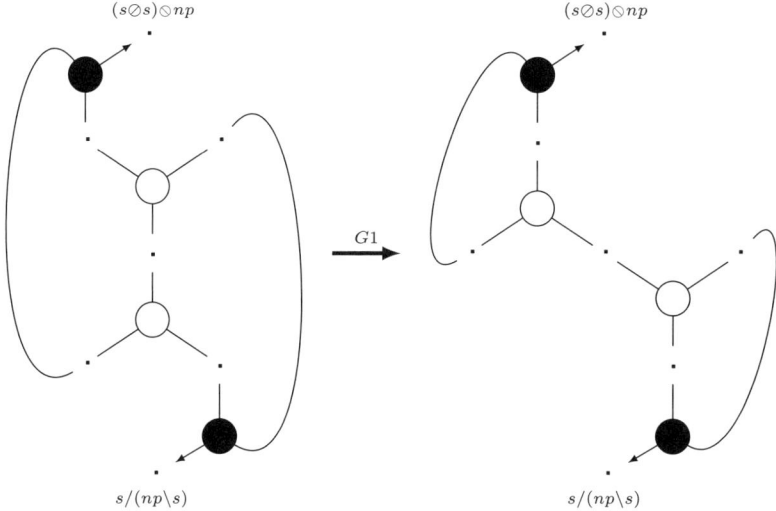

Fig 10.8 Applying rule (G1) to the abstract proof structure of Figure 10.3.

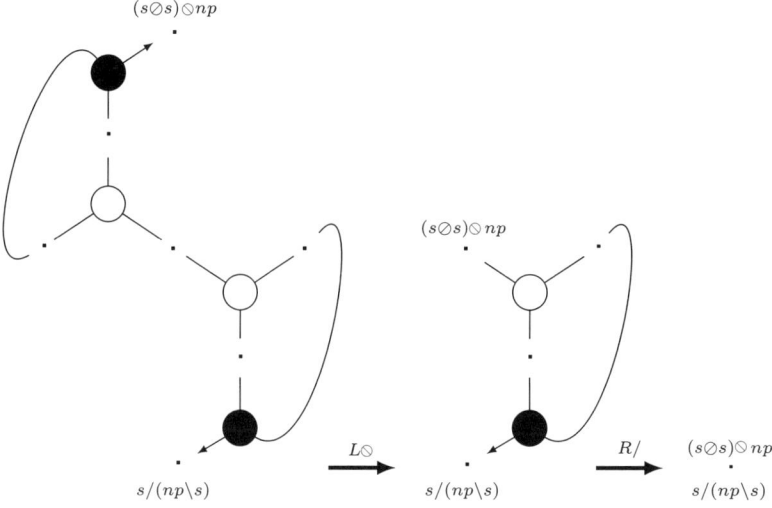

Fig 10.9 Applying rule (L⊘) and (R/) contractions to the abstract proof structure of Figure 10.8.

Discussion: proof nets and sequent proofs

Table 10.1 summarizes the correspondence between proof nets and sequent proofs.

The invertible one-premise rules correspond to both a link and a contraction and the interaction rules are invisible in the proof structure, appearing only in the conversion sequence.

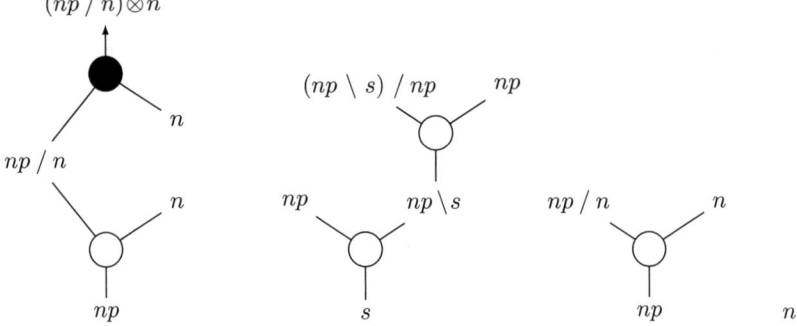

Fig 10.10 Lexical proof structures for a generalized quantifier noun phrase, a transitive verb, a determiner, and a noun.

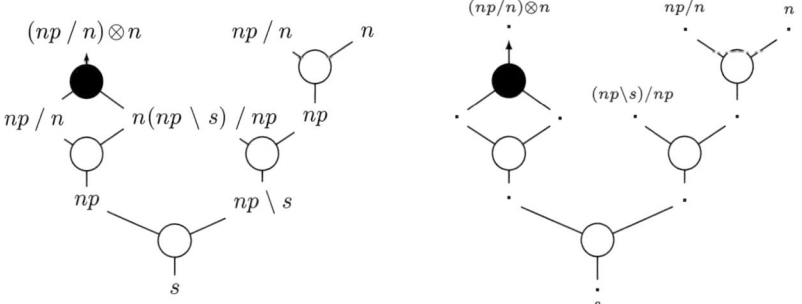

Fig 10.11 Judgement $(np\,/\,n) \otimes n, (np\,\backslash\,s)\,/\,np, np\,/\,n, n \Rightarrow s$: proof structure and abstract proof structure.

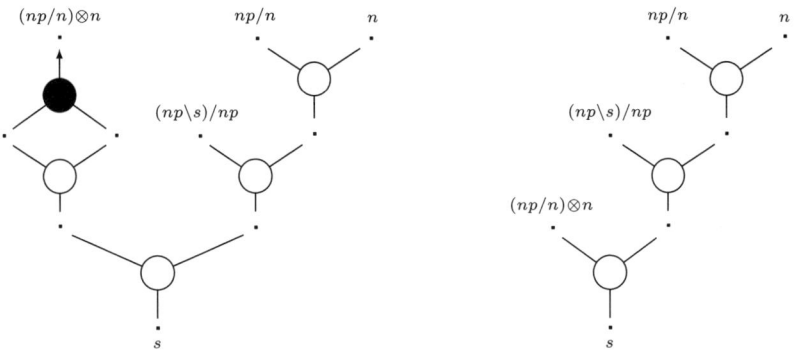

Fig 10.12 Abstract proof structure and contraction.

TABLE 10.1 Proof nets and sequent proofs

Sequent calculus	Proof structure	Conversion
axiom	axiomatic formula	—
cut	cut formula	—
two-premise rule	tensor link	—
one-premise rule	cotensor link	contraction
interaction rule	—	rewrite

With a bit of extra effort in the sequentialization proof we can show that these correspondences are one-to-one; that is, each axiomatic formula in a proof net corresponds to exactly one axiom rule in the sequent proof, each non-invertible two-premise rule corresponds to exactly one link in the proof net, and each invertible one-premise rule to exactly one link in the proof net and exactly one contraction in its conversion sequence.

Summing up, the proof net approach offers the following benefits in comparison to sequent proof search:

- *Parsing.* Whereas for sequent proof search the structure of the sequent has to be given, the contraction sequence that identifies a proof structure as a proof net actually *computes* this structure.
- *Removal of spurious ambiguity.* Proof nets, like (product-free) natural deduction, have different proof objects only for proofs of a judgement that differ essentially. The combinatorial possibilities for such readings, which are obtained by finding a complete matching of the atomic formulas, can easily be enumerated for a given sequence of formulas.
- *Display rules compiled away.* The tensor trees associated with well-formed proof nets represent a class of sequents that is equivalent up to the display postulates.

10.3 Proof nets and focused display calculus

The spurious non-determinism of naive backward-chaining proof search can also be addressed within the sequent calculus itself, by introducing an appropriate notion of 'normal' derivations. In Section 10.3.1, we introduce **fLG**, a focused version of the sequent calculus for **LG**. In Section 10.3.2, we then study how to interpret focused derivations from a proof net perspective.

10.3.1 fLG: focused display calculus

The strategy of focusing has been well-studied in the context of linear logic, starting with the work of Andreoli (2001). It is based on the distinction between *asyn-*

chronous and *synchronous* non-atomic formulas. The introduction rule for the main connective of an asynchronous formula is *invertible*; it is non-invertible for the synchronous formulas. Backward-chaining focused proof search starts with an asynchronous phase where invertible rules are applied deterministically until no more candidate formulas remain. At that point, a non-deterministic choice for a synchronous formula must be made: this formula is put 'in focus', and decomposed to its subformulae by means of non-invertible rules until no more non-invertible rules are applicable, at which point one reenters an asynchronous phase. The main result of Andreoli (2001) is that focused proofs are complete for linear logic.

Focused proof search for the Lambek–Grishin calculus has been studied by Bastenhof (2011), who uses a one-sided presentation of the calculus. In this section, we implement his focusing regime in the context of the two-sided sequent format of Bernardi and Moortgat (2010). We proceed in two steps. First we introduce **fLG**, the focused version of the sequent calculus of Section 10.2.1. **fLG** makes a distinction between focused and unfocused judgements, and has a set of inference rules to switch between these two. **fLG** comes with a term language that is in Curry–Howard correspondence with its derivations. This term language is a directional refinement of the $\overline{\lambda}\mu\widetilde{\mu}$ language of Curien and Herbelin (2000).

The second step is to give a constructive interpretation for **LG** derivations by means of a continuation-passing-style (CPS) translation: a mapping $\lceil \cdot \rceil$ that sends derivations of the multiple-conclusion source logic to (natural deduction) proofs in a fragment of single-conclusion multiplicative intuitionistic linear logic (MILL) (in the categorial terminology: **LP**). For the translation of Bastenhof (2011) that we follow here, the target fragment has linear products and negation $A^\perp$, i.e. a restricted form of linear implication $A \multimap \perp$, where $\perp$ is a distinguished atomic type: the response type. Focused source derivations then can be shown to correspond to distinct *normal* natural deduction proofs in the target calculus.

$$\mathbf{fLG}^{\mathcal{A}}_{/,\otimes,\backslash,\oslash,\oplus,\obar} \xrightarrow{\lceil \cdot \rceil} \mathbf{LP}^{\mathcal{A}\cup\{\perp\}}_{\otimes,\cdot^\perp} \left(\xrightarrow{\cdot^\ell} \mathbf{IL}^{\{e,t\}}_{\times,\to} \right)$$

For the linguistic illustrations in Section 10.3.2.3, we compose the CPS translation $\lceil \cdot \rceil$ with a second mapping $\cdot^\ell$, which establishes the connection with Montague-style semantic representations. This mapping sends the linear constructs to their intuitionistic counterparts, and allows *non-linear* meaning recipes for the translation of the lexical constants.

10.3.1.1 FLG: PROOFS AND TERMS

We set up **fLG** in the Curry–Howard proofs-as-programs fashion, starting from a term language for which the sequent logic then provides the type system. The term language encodes the *logical* steps of a derivation (left and right introduction rules, and the new set of left and right (de)focusing rules, to be introduced below); structural rules (residuation, distributivity) leave no trace in the proof terms.

Sequent structures, as in Section 10.2.1, are built out of formulae. Input formulas now are labeled with variables $x, y, z, \ldots$, output formulas with covariables $\alpha, \beta, \gamma, \ldots$. To implement the focusing regime, we allow sequents to have one displayed formula *in focus*. Writing the focused formula in a box, **fLG** will have to deal with three types of judgements: sequents with no formula in focus (we'll call these *structural*), and sequents with a succedent or antecedent formula in focus.

$$X \vdash Y \qquad X \vdash \boxed{A} \qquad \boxed{A} \vdash Y$$

Corresponding to the types of sequents, the term language has three types of expression: commands, values, and contexts. For commands, we use the metavariables c, C; for values v, V; and for contexts e, E. The typing rules below provide the motivation for the subclassification.

$$v ::= \mu\alpha.C \mid V \quad ; \quad V ::= x \mid v_1 \otimes v_2 \mid v \oslash e \mid e \oslash v$$
$$e ::= \tilde{\mu}x.C \mid E \quad ; \quad E ::= \alpha \mid e_1 \oplus e_2 \mid v\backslash e \mid e/v \tag{10.30}$$
$$c ::= \langle x \upharpoonright E \rangle \mid \langle V \upharpoonright \alpha \rangle$$
$$C ::= c \mid \tfrac{x\,y}{z}.C \mid \tfrac{x\,\beta}{z}.C \mid \tfrac{\beta\,x}{z}.C \mid \tfrac{\alpha\,\beta}{\gamma}.C \mid \tfrac{x\,\beta}{\gamma}.C \mid \tfrac{\beta\,x}{\gamma}.C$$

10.3.1.2 TYPING RULES

To enforce the alternation between asynchronous and synchronous phases of focused proof search, formulas are associated with a polarity:

- *positive* for non-atomic formulas with invertible left introduction rule: $A \otimes B$, $A \oslash B$, $B \oslash A$
- *negative* for non-atomic formulas with invertible right introduction rule: $A \oplus B$, $A\backslash B$, B/A.

For atomic formulas, one can fix an arbitrary polarity. Different choices lead to different proof-theoretic behaviour (and to different interpretations, once we turn to the CPS translation). We will assume that atoms are assigned a bias (positive or negative) in the lexicon. Below the typing rules for **fLG** (restricting attention to the cut-free system).

(Co-)Axiom, (de)focusing

$$\frac{}{x : A \vdash \boxed{x : A}} \text{Ax} \qquad \frac{}{\boxed{\alpha : A} \vdash \alpha : A} \text{CoAx}$$

$$\frac{X \vdash \boxed{V : A}}{\langle V \upharpoonright \alpha \rangle : (X \vdash \alpha : A)} \mu^* \qquad \frac{\boxed{E : A} \vdash X}{\langle x \upharpoonright E \rangle : (x : A \vdash X)} \tilde{\mu}^*$$

$$\frac{C : (x : A \vdash X)}{\boxed{\tilde{\mu}x.C : A} \vdash X} \tilde{\mu} \qquad \frac{C : (X \vdash \alpha : A)}{X \vdash \boxed{\mu\alpha.C : A}} \mu$$

First we have the focused version of the axiomatic sequents, and rules for focusing and defocusing that are new with respect to the unfocused presentation of Section 10.2.1. There is a polarity restriction on the formula A in these rules: the boxed formula has to be negative for CoAx, μ, $\tilde{\mu}^*$; for Ax, $\tilde{\mu}$, μ^* it has to be positive. In the (Co-)Axiom cases, A can be required to be atomic.

From a backward-chaining perspective, the $\mu, \tilde{\mu}$ rules *remove* the focus from a focused succedent or antecedent formula. The result is an unfocused premise sequent, the domain of applicability of the invertible rules, i.e. one enters the asynchronous phase. From the same perspective, the rules $\mu^*, \tilde{\mu}^*$ place a succedent or antecedent formula in focus, shifting control to the non-invertible rules of the synchronous phase. The $\mu^*, \tilde{\mu}^*$ rules are in fact instances of Cut where one of the premises is (co-)axiomatic.

Invertible rules

The term language makes a distinction between simple commands c (the image of the focusing rules $\tilde{\mu}^*, \mu^*$: $\langle x \upharpoonright E \rangle, \langle V \upharpoonright \alpha \rangle$) and extended commands C. The latter start with a sequence of invertible rewrite rules replacing a logical connective by its structural counterpart. We impose the requirement that in the asynchronous phase all formulas to which an invertible rule is applicable are indeed decomposed.

$$\frac{C : (x : A \cdot \otimes \cdot y : B \vdash X)}{\frac{xy}{z}.C : (z : A \otimes B \vdash X)} \otimes L \qquad \frac{C : (X \vdash \alpha : A \cdot \oplus \cdot \beta : B)}{\frac{\alpha\beta}{\gamma}.C : (X \vdash \gamma : A \oplus B)} \oplus R$$

$$\frac{C : (x : A \cdot \oslash \cdot \beta : B \vdash X)}{\frac{x\beta}{z}.C : (z : A \oslash B \vdash X)} \oslash L \qquad \frac{C : (X \vdash x : A \cdot \backslash \cdot \beta : B)}{\frac{x\beta}{\gamma}.C : (X \vdash \gamma : A\backslash B)} \backslash R$$

$$\frac{C : (\beta : B \cdot \obslash \cdot x : A \vdash X)}{\frac{\beta x}{z}.C : (z : B \obslash A \vdash X)} \obslash L \qquad \frac{C : (X \vdash \beta : B \cdot / \cdot x : A)}{\frac{\beta x}{\gamma}.C : (X \vdash \gamma : B/A)} /R$$

Non-invertible rules

When a positive (negative) formula has been brought into focus in the succedent (antecedent), one is committed to transfer the focus to its subformulae.

$$\frac{\boxed{e_1 : B} \vdash Y \quad \boxed{e_2 : A} \vdash X}{\boxed{e_1 \oplus e_2 : B \oplus A} \vdash Y \cdot \oplus \cdot X} \oplus L \qquad \frac{X \vdash \boxed{v_1 : A} \quad Y \vdash \boxed{v_2 : B}}{X \cdot \otimes \cdot Y \vdash \boxed{v_1 \otimes v_2 : A \otimes B}} \otimes R$$

$$\cfrac{\boxed{X \vdash \boxed{v : A}} \quad \boxed{\boxed{e : B} \vdash Y}}{\boxed{v \backslash e : A \backslash B} \vdash X \cdot \backslash \cdot Y} \backslash L \qquad \cfrac{\boxed{X \vdash \boxed{v : A}} \quad \boxed{\boxed{e : B} \vdash Y}}{X \cdot \oslash \cdot Y \vdash \boxed{v \oslash e : A \oslash B}} \oslash R$$

$$\cfrac{\boxed{\boxed{e : B} \vdash Y} \quad \boxed{X \vdash \boxed{v : A}}}{\boxed{e/v : B/A} \vdash Y \cdot / \cdot X} /L \qquad \cfrac{\boxed{\boxed{e : B} \vdash Y} \quad \boxed{X \vdash \boxed{v : A}}}{Y \cdot \oslash \cdot X \vdash \boxed{e \oslash v : B \oslash A}} \oslash R$$

Derived inference rules: focus shifting

To highlight the correspondence with the algorithm for proof net construction to be discussed in Section 10.2.2, we will use a derived rule format for shifting between a conclusion- and premise-focused formula. A branch from ($\tilde{\mu}^*$) via a sequence (possibly empty) of structural rules and rewrite rules to (μ) is compiled in a derived inference rule with the $\tilde{\mu}^*$ restrictions on A and the μ restrictions on B.

$$\cfrac{\boxed{E : A} \vdash Y}{\langle x \uparrow E \rangle : (x : A \vdash Y)} \tilde{\mu}^*$$

$$\vdots$$

$$(res, distr, rewrite)$$

$$\vdots$$

$$\cfrac{(\div)\langle x \uparrow E \rangle : (X \vdash \beta : B)}{X \vdash \boxed{\mu \beta.(\div)\langle x \uparrow E \rangle : B}} \mu \qquad \rightsquigarrow \qquad \cfrac{\boxed{E : A} \vdash Y}{X \vdash \boxed{\mu \beta.(\div)\langle x \uparrow E \rangle : B}} \rightleftharpoons$$

For the combinations of $\mu^*, \tilde{\mu}^*$ and $\mu, \tilde{\mu}$, this results in the focus-shifting rules below. We leave it to the reader to add the terms.

$$\cfrac{\boxed{A} \vdash Y}{X \vdash \boxed{B}} \rightleftharpoons \qquad \cfrac{X' \vdash \boxed{A}}{X \vdash \boxed{B}} \Rightarrow \qquad \cfrac{X \vdash \boxed{A}}{\boxed{B} \vdash Y} \rightleftharpoons \qquad \cfrac{\boxed{A} \vdash Y'}{\boxed{B} \vdash Y} \Leftarrow \qquad (10.31)$$

10.3.1.3 ILLUSTRATIONS

We illustrate the effect of the focusing regime with some alternative ways of assigning a polarity bias to atomic formulas with a simple Subject–Transitive Verb–Object sentence. Examples with lexical material filled in would be 'everyone seeks/finds a unicorn'.

$$(np/n \otimes n) \cdot \otimes \cdot ((np\backslash s)/np \cdot \otimes \cdot (np/n \cdot \otimes \cdot n)) \vdash s \quad (10.32)$$

For the Object we have a Determiner–Noun combination. For the Subject, we take a product type $(np/n) \otimes n$, so that we have a chance to illustrate the working of the asynchronous phase of the derivation. In the discussion of Figure 10.10, we saw that the sequent of eqn (10.32) has multiple proofs in the unfocused sequent calculus, but only one proof net, i.e. one way of matching the premise and conclusion atomic formulas.

What about the focused calculus **fLG**? Before answering this question we have to decide on the polarization of the atomic types. Suppose we give them uniform negative bias. There is only one focused proof then: 'goal driven' (or top-down, to use parsing terminology). In the proof terms, we write tv for the transitive verb; det for the object determiner; noun for the object common noun; and subj for the subject noun phrase.

$$\mu\beta.(\frac{y\,z}{\text{subj}}.\langle\,\text{tv}\,|\,((Q \backslash \beta) / Q')\,\rangle) \quad \text{with}$$

$$Q : \mu\gamma.\langle\,y\,|\,(\gamma / \mu\gamma'.\langle\,z\,|\,\gamma'\rangle)\rangle\,,\ Q' : \mu\alpha.\langle\,\text{det}\,|\,(\alpha / \mu\alpha'.\langle\,\text{noun}\,|\,\alpha'\rangle)\rangle$$

(10.33)

As an alternative, suppose basic type s keeps its negative bias, resetting the sentence continuation for each clausal domain, but the other basic types are assigned positive bias. We now have *two* focused derivations: 'data driven' and bottom-up. To make sense of this difference, we will have to look at the CPS translation of these proofs, to be introduced below.

$$
\begin{array}{c}
np \stackrel{x_1}{\vdash} \boxed{np} \qquad \boxed{s^-} \stackrel{\alpha_0}{\vdash} s^- \\
\hline
\boxed{np\backslash s^-} \vdash np \cdot \backslash \cdot s^- \qquad np \stackrel{y_1}{\vdash} \boxed{np} \\
\hline
\boxed{(np\backslash s^-)/np} \vdash (np \cdot \backslash \cdot s^-) \cdot / \cdot np \\
\hline
\boxed{np} \vdash (np\backslash s^-)/np \cdot \backslash \cdot (np \cdot \backslash \cdot s^-) \qquad n \stackrel{\text{noun}}{\vdash} \boxed{n} \\
\hline
\boxed{np/n} \vdash ((np\backslash s^-)/np \cdot \backslash \cdot (np \cdot \backslash \cdot s^-)) \cdot / \cdot n \\
\hline
\boxed{np} \vdash s^- \cdot / \cdot ((np\backslash s^-)/np \otimes \cdot (np/n \otimes \cdot n)) \qquad n \stackrel{z_0}{\vdash} \boxed{n} \\
\hline
\boxed{np/n} \vdash (s^- \cdot / \cdot ((np\backslash s^-)/np \otimes \cdot (np/n \otimes \cdot n))) \cdot / \cdot n \\
\hline
(np/n) \otimes \cdot n \otimes \cdot ((np\backslash s^-)/np \otimes \cdot (np/n \otimes \cdot n)) \vdash \boxed{s^-}
\end{array}
$$

$$\mu\alpha.(\frac{x'\ z}{\text{subj}}.\langle x' \mid (\widetilde{\mu}x.\langle \text{det} \mid (\widetilde{\mu}y.\langle \text{tv} \mid ((x \backslash \alpha)/y)\rangle /\text{noun})\rangle /z)\rangle) \qquad (10.34)$$

$$
\begin{array}{c}
np \stackrel{x}{\vdash} \boxed{np} \qquad \boxed{s^-} \stackrel{\alpha}{\vdash} s^- \\
\hline
\boxed{np\backslash s^-} \vdash np \cdot \backslash \cdot s^- \qquad np \stackrel{y}{\vdash} \boxed{np} \\
\hline
\boxed{(np\backslash s^-)/np} \vdash (np \cdot \backslash \cdot s^-) \cdot / \cdot np \\
\hline
\boxed{np} \vdash s^- \cdot / \cdot ((np\backslash s^-)/np \otimes \cdot np) \qquad n \stackrel{z}{\vdash} \boxed{n} \\
\hline
\boxed{np/n} \vdash (s^- \cdot / \cdot ((np\backslash s^-)/np \otimes \cdot np)) \cdot / \cdot n \\
\hline
\boxed{np} \vdash (np\backslash s^-)/np \cdot \backslash \cdot ((np/n \otimes \cdot n) \cdot \backslash \cdot s^-) \qquad n \stackrel{\text{noun}}{\vdash} \boxed{n} \\
\hline
\boxed{np/n} \vdash ((np\backslash s^-)/np \cdot \backslash \cdot ((np/n \otimes \cdot n) \cdot \backslash \cdot s^-)) \cdot / \cdot n \\
\hline
(np/n) \otimes \cdot n \otimes \cdot ((np\backslash s^-)/np \otimes \cdot (np/n \otimes \cdot n)) \vdash \boxed{s^-}
\end{array}
$$

$$\mu\alpha.(\frac{x'\ z}{\text{subj}}.\langle \text{det} \mid (\widetilde{\mu}y.\langle x' \mid (\widetilde{\mu}x.\langle \text{tv} \mid ((x \backslash \alpha)/y)\rangle /z)\rangle /\text{noun})\rangle) \qquad (10.35)$$

10.3.1.4 CPS TRANSLATION

Let us turn then to the translation that associates the proofs of the multiple-conclusion source logic **fLG** with a constructive interpretation, i.e. a linear lambda term of the target logic MILL/LP. CPS translations for **LG** were introduced by Bernardi and Moortgat (2007, 2010), who adapt the call-by-value and call-by-name regimes of Curien and Herbelin (2000) to a directional environment. The translation of Bastenhof (2011) (following Girard, 1991) is an improvement in that it avoids the 'administrative redexes' of the earlier approaches: the image of **LG** source derivations, under the mapping from Bastenhof (2011) that we present below, are *normal* LP terms.

The target language, on the type level, has the same atoms as the source language, and in addition a distinguished atom $\perp$, the response type. Complex types are linear products $- \otimes -$ and a defined negation $A^\perp \doteq A \multimap \perp$. The CPS translation $\lceil \cdot \rceil$ maps fLG source types, sequents and their proof terms to the target types and terms in Curry–Howard correspondence with normal natural deduction proofs.

Types

For positive atoms, $\lceil p \rceil = p$, for negative atoms $\lceil p \rceil = p^\perp$. For complex types, the value of $\lceil \cdot \rceil$ depends on the polarities of the subtypes as shown in Table 10.2.

Terms

The action of $\lceil \cdot \rceil$ on terms is given in eqn (10.36). We write $\tilde{x}, \tilde{\alpha}$ for the target variables corresponding to source x, α. The (de)focusing rules correspond to application/abstraction in the target language. Non-invertible (two-premise) rules are mapped to linear pair terms; invertible rewrite rules to the matching deconstructor, the **case** construct (ϕ, ψ, ξ metavariables for the (co)variables involved).

TABLE 10.2 CPS translation: non-atomic types

pol(A)	pol(B)	$\lceil A \otimes B \rceil$	$\lceil A/B \rceil$	$\lceil B \backslash A \rceil$
−	−	$\lceil A \rceil^\perp \otimes \lceil B \rceil^\perp$	$\lceil A \rceil \otimes \lceil B \rceil^\perp$	$\lceil B \rceil^\perp \otimes \lceil A \rceil$
−	+	$\lceil A \rceil^\perp \otimes \lceil B \rceil$	$\lceil A \rceil \otimes \lceil B \rceil$	$\lceil B \rceil \otimes \lceil A \rceil$
+	−	$\lceil A \rceil \otimes \lceil B \rceil^\perp$	$\lceil A \rceil^\perp \otimes \lceil B \rceil^\perp$	$\lceil B \rceil^\perp \otimes \lceil A \rceil^\perp$
+	+	$\lceil A \rceil \otimes \lceil B \rceil$	$\lceil A \rceil^\perp \otimes \lceil B \rceil$	$\lceil B \rceil \otimes \lceil A \rceil^\perp$
pol(A)	pol(B)	$\lceil A \oplus B \rceil$	$\lceil A \oslash B \rceil$	$\lceil B \obar A \rceil$
−	−	$\lceil A \rceil \otimes \lceil B \rceil$	$\lceil A \rceil^\perp \otimes \lceil B \rceil$	$\lceil B \rceil \otimes \lceil A \rceil^\perp$
−	+	$\lceil A \rceil \otimes \lceil B \rceil^\perp$	$\lceil A \rceil^\perp \otimes \lceil B \rceil^\perp$	$\lceil B \rceil^\perp \otimes \lceil A \rceil^\perp$
+	−	$\lceil A \rceil^\perp \otimes \lceil B \rceil$	$\lceil A \rceil \otimes \lceil B \rceil$	$\lceil B \rceil \otimes \lceil A \rceil$
+	+	$\lceil A \rceil^\perp \otimes \lceil B \rceil^\perp$	$\lceil A \rceil \otimes \lceil B \rceil^\perp$	$\lceil B \rceil^\perp \otimes \lceil A \rceil$

(co)var	$\lceil x \rceil = \tilde{x}$;	$\lceil \alpha \rceil = \tilde{\alpha}$
linear application	$\lceil \langle x \uparrow E \rangle \rceil = (\tilde{x} \lceil E \rceil)$;	$\lceil \langle V \restriction \alpha \rangle \rceil = (\tilde{\alpha} \lceil V \rceil)$
linear abstraction	$\lceil \tilde{\mu} x.C \rceil = \lambda \tilde{x}.\lceil C \rceil$;	$\lceil \mu \alpha.C \rceil = \lambda \tilde{\alpha}.\lceil C \rceil$ (10.36)
linear pair	$\lceil \phi \# \psi \rceil = \langle \lceil \phi \rceil, \lceil \psi \rceil \rangle$	$(\# \in \{\otimes, /, \backslash, \oplus, \oslash, \odot\})$
case	$\lceil \frac{\phi \, \psi}{\xi}.C \rceil = \textbf{case}\ \tilde{\xi}\ \textbf{of}\ \langle \tilde{\phi}, \tilde{\psi} \rangle.\lceil C \rceil$	

Sequents

For sequent hypotheses/conclusions, we have

$$\begin{array}{c|cc} \text{pol}(A) & \lceil x : A \rceil & \lceil \alpha : A \rceil \\ \hline + & \tilde{x} : \lceil A \rceil & \tilde{\alpha} : \lceil A \rceil^\perp \\ - & \tilde{x} : \lceil A \rceil^\perp & \tilde{\alpha} : \lceil A \rceil \end{array} \quad (10.37)$$

Table 10.2 then specifies how the translation extends to sequents (replace logical connectives by their structural counterparts, and target $\otimes$ by the comma for multiset union).

$$\begin{aligned} \lceil C : (X \vdash Y) \rceil &= \lceil X \rceil, \lceil Y \rceil \vdash_{\text{LP}} \lceil C \rceil : \perp \\ \left\lceil X \vdash \boxed{v : A} \right\rceil &= \lceil X \rceil \vdash_{\text{LP}} \lceil v \rceil : \lceil A \rceil \\ \left\lceil \boxed{e : A} \vdash Y \right\rceil &= \lceil Y \rceil \vdash_{\text{LP}} \lceil e \rceil : \lceil A \rceil^\perp \end{aligned} \quad (10.38)$$

Illustrations

We return to our sample derivations. In eqn (10.39) one finds the CPS image of the source types for transitive verb and determiner under the different assignments of bias to the atomic subformulas, and the composition with $\cdot^\ell$, assuming $np^\ell = e$ (entities) and $s^\ell = \perp^\ell = t$ (truth values). For the lexical constants of the illustration, Table 10.3 gives $\cdot^\ell$ translations compatible with the typing. In Table 10.4, these lexical recipes are substituted for the parameters of the CPS translation.

	LG	$\lceil \cdot \rceil^\perp$	$(\lceil \cdot \rceil^\perp)^\ell$
a.	$(np^+ \backslash s^-)/np^+$	$((np \otimes s^\perp) \otimes np)^\perp$	$((e \times (tt)) \times e) \to t$
b.	np^+/n^+	$(np^\perp \otimes n)^\perp$	$((et) \times (et)) \to t$
c.	$(np^- \backslash s^-)/np^-$	$((np^{\perp\perp} \otimes s^\perp) \otimes np^{\perp\perp})^\perp$	$((((et)t) \times (tt)) \times ((et)t)) \to t$
d.	np^-/n^-	$(np^\perp \otimes n^{\perp\perp})^\perp$	$((et) \times (((et)t)t)) \to t$

(10.39)

TABLE 10.3 Constants: lexical translations

$(np^+\backslash s^-)/np^+$	finds	$\lambda\langle\langle x,c\rangle,y\rangle.(c\ (\text{FIND}^{eet}\ y\ x))$
$(np^+/n^+)\otimes n^+$	everyone	$\langle\lambda\langle x,y\rangle.(\forall\ \lambda z.(\Rightarrow (y\ z)\ (x\ z))), \text{PERSON}^{et}\rangle$
np^+/n^+	some	$\lambda\langle x,y\rangle.(\exists\ \lambda z.(\wedge\ (y\ z)\ (x\ z)))$
n^+	unicorn	UNICORN^{et}
$(np^-\backslash s^-)/np^-$	needs	$\lambda\langle\langle q,c\rangle,q'\rangle.(q\ \lambda x.(\text{NEED}^{((et)t)et}\ q'\ x))$
$(np^-/n^-)\otimes n^-$	everyone	$\langle\lambda\langle x,w\rangle.(\forall\ \lambda z.(\Rightarrow (w\ \lambda y.(y\ z))\ (x\ z))), \lambda k.(k\ \text{PERSON}^{et})\rangle$
np^-/n^-	some	$\lambda\langle x,w\rangle.(\exists\ \lambda z.(\wedge\ (w\ \lambda y.(y\ z))\ (x\ z)))$
n^-	unicorn	$\lambda k.(k\ \text{UNICORN}^{et})$

TABLE 10.4 Compositional translations

$\lceil(10.33)\rceil = \lambda\widetilde{\beta}.(\text{case subj}^\ell \text{ of } \langle\widetilde{\gamma},\widetilde{z}\rangle.(\text{tv}^\ell\ \langle\langle\lambda\widetilde{\gamma}.(\widetilde{y}\ \langle\widetilde{\gamma},\lambda\widetilde{\gamma}'.(\widetilde{z}\ \widetilde{\gamma}')\rangle),\widetilde{\beta}\rangle, \lambda\widetilde{\alpha}.(\text{det}^\ell\ \langle\widetilde{\alpha},\lambda\widetilde{\alpha}'.(\text{noun}^\ell\ \widetilde{\alpha}')\rangle))))$

$\lceil(10.33)\rceil^\ell = \lambda c.(\forall\ \lambda x.((\Rightarrow\ (\text{person } x))\ (c\ ((\text{NEEDS } \lambda w.(\exists\ \lambda y.((\wedge\ (\text{unicorn } y))\ (w\ y))))\ x))))$

$\lceil(10.34)\rceil = \lambda\widetilde{\alpha}.(\text{case subj}^\ell \text{ of } \langle\widetilde{x}',\widetilde{z}\rangle.(\widetilde{x}'\ \langle\lambda\widetilde{x}.(\text{det}^\ell\ \langle\lambda\widetilde{y}.(\text{tv}^\ell\ \langle\langle\widetilde{x},\widetilde{\alpha}\rangle,\widetilde{y}\rangle), \text{noun}^\ell\rangle),\widetilde{z}\rangle))$

$\lceil(10.34)\rceil^\ell = \lambda c.(\forall\ \lambda x.((\Rightarrow\ (\text{PERSON } x))\ (\exists\ \lambda y.((\wedge\ (\text{UNICORN } y))\ (c\ ((\text{LIKES } y)\ x))))))$

$\lceil(10.35)\rceil = \lambda\widetilde{\alpha}.(\text{case subj}^\ell \text{ of } \langle\widetilde{x}',\widetilde{z}\rangle.(\text{det}^\ell\ \langle\lambda\widetilde{y}.(\widetilde{x}'\ \langle\lambda\widetilde{x}.(\text{tv}^\ell\ \langle\langle\widetilde{x},\widetilde{\alpha}\rangle,\widetilde{y}\rangle),\widetilde{z}\rangle), \text{noun}^\ell\rangle))$

$\lceil(10.35)\rceil^\ell = \lambda c.(\exists\ \lambda y.((\wedge\ (\text{UNICORN } y))\ (\forall\ \lambda x.((\Rightarrow\ (\text{PERSON } x))\ (c\ ((\text{LIKES } y)\ x))))))$

10.3.2 Proof nets and focusing

In this section, we introduce term-labeled proof nets, and show how a proof term can be read off from the *composition graph* associated with a net. Our approach is comparable to that of de Groote and Retoré (1996), who present an algorithm to compute a linear lambda term from a traversal of the dynamic graph associated with a proof net for a derivation in the Lambek calculus. Whereas in the case of the single-conclusion Lambek calculus the term associated with a given proof net is unique, in the case of multiple-conclusion LG there will be the possibility that the term computation algorithm associates more than one term with a proof net. These multiple results will then be shown to correspond to the derivational ambiguity of focused proof search.

10.3.2.1 REDUCTION TREE

When P is a proof net (and therefore converts to a tensor tree using a sequence ρ of conversions and contractions) the components of P can be seen as a parallel representation of the synchronous phases in sequent proof search. Taking a closer look at the conversion sequence ρ, we see that all interaction rules operate in one component C, the cotensor rules and the corresponding contractions operate on a component to which it is attached by both of its active tentacles (i.e. the tentacles without the arrow) and the contraction removes a tensor link from this component. If the main tentacle points to a vertex attached to a non-trivial component C' then a new component is formed by merging C (minus the contracted tensor link) and C' into a new component. When multiple cotensor links have both active tentacles attached to a single component (Figure 10.8 shows an example), we can apply all contractions simultaneously: since a contraction connects a tensor and a cotensor link at two out of three tentacles, there cannot be a conflict (multiple cotensor links connected to a tensor link with *both* contractions being impossible without violating the definition of proof structures). In addition, when the main vertex of a cotensor link is the active vertex of another cotensor link, then, if the other active vertex of this link is connected to the current component as well, we can apply this contraction immediately.

So instead of seeing ρ as a *sequence* of reductions, we can see it as a rooted *tree* of reductions: the initial components are its leaves (synchronous phases) and the contractions connecting multiple components to form new components are its branches (the branches from the active components to their parents correspond to asynchronous phases) and the final tree—a single component—is its root.

Example 10.13 Figure 10.13 shows an example of how the view of components given above allows us to see a proof net as a tree of components. The shaded subnet boxes are components and contain only tensor links. For clarity, the cotensor links are shown in the figure as well.

Each interaction rule takes place completely in one of the C_i. In the figure, the components that do not contain the main vertex of a cotensor link are shown in a darker shade; we will call these components *active*. In Figure 10.13, C_2 and C_4 are active. Now, it is easy to show that whenever there exists a conversion sequence ρ, we can transform it into a conversion sequence ρ' where conversions take place only in the active components: any conversions in C_1 can be delayed until after the contraction connecting C_1 and C_2, since only C_2 is relevant for this contraction (it contains both active vertices of the cotensor link and therefore also the tensor link it contracts with), and any conversions in C_5 can be delayed until the final component C_{1-5}.

In addition, the two active components C_2 and C_4 are independent: we can apply conversions to these two components in parallel.

10.3.2.2 NETS AND TERM LABELING

When assigning a term label to a proof net, we will be interested in assigning labels to larger and larger subnets of a given proof net, until we have computed a term for the complete proof net. Like in the sequent calculus, we distinguish between

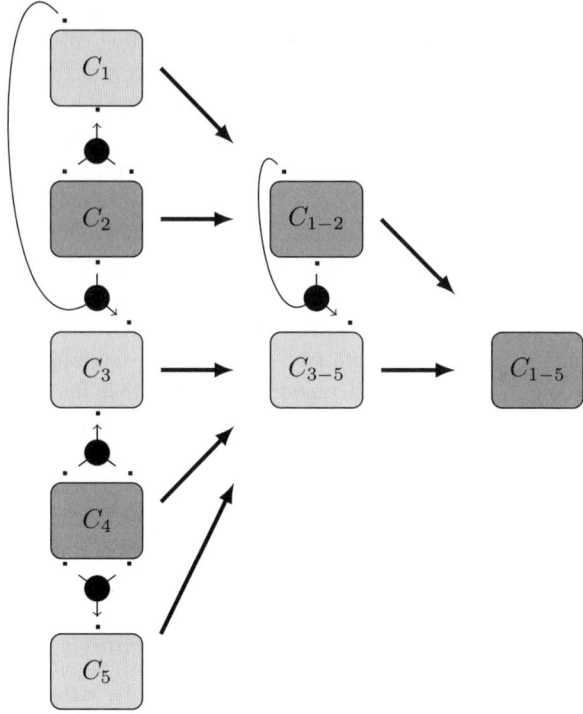

Fig 10.13 A reduction sequence seen as a rooted tree.

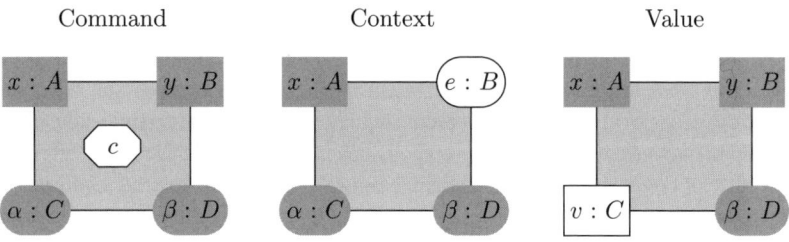

Fig 10.14 Proof nets with term labels: commands, context, and values.

subnets which are commands, contexts and values. Figure 10.14 shows how we will distinguish these visually: the main formula of a subnet is drawn in white, other formulas are drawn in light grey; values are drawn inside a rectangle, contexts inside an oval.

Figure 10.15 gives the term-labeled version of the proof net links corresponding to the logical rules of the sequent calculus. The flow of information is shown by the

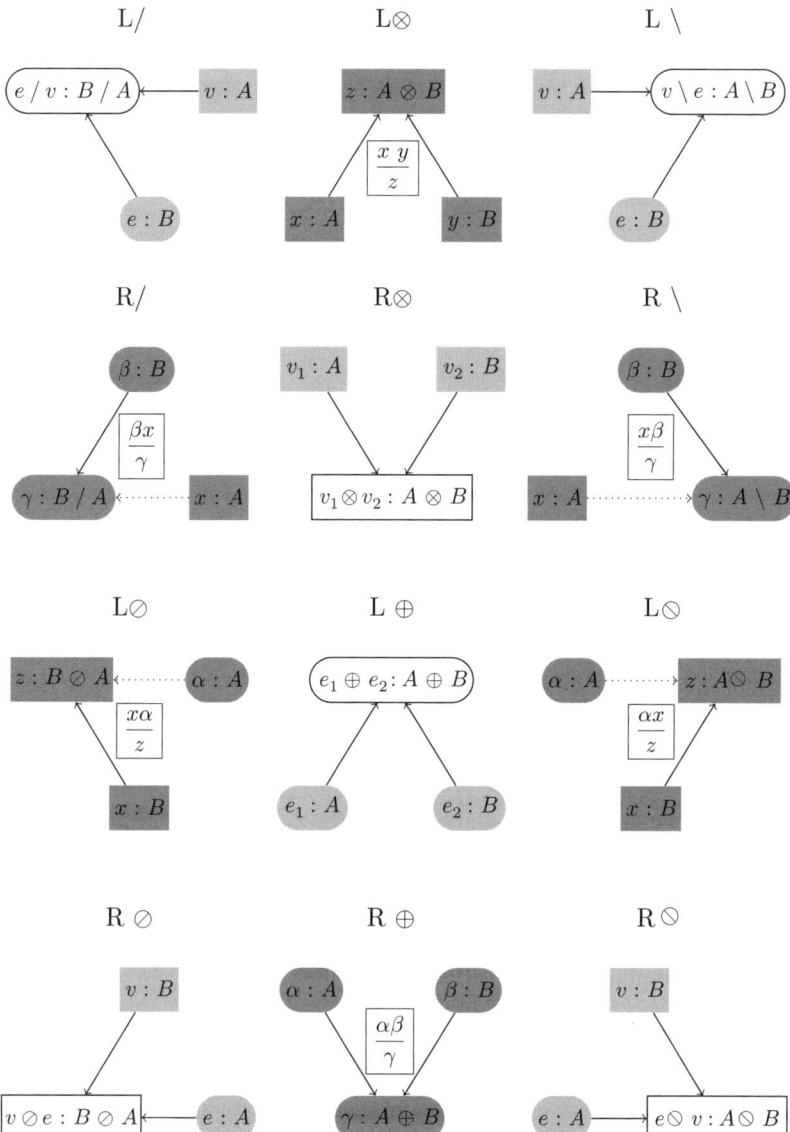

Fig 10.15 LG links with term labeling.

arrows: information flow is always from the active formulas to the main formula of a link, and as a consequence the complex term can be assigned either to a conclusion or to a premise of the link. This is the crucial difference with term labelling for the single-conclusion Lambek calculus, where the complex term is always assigned to a conclusion. The cotensor rules, operating on commands, indicate the prefix for the command corresponding to the term assignment for the rule (we will see later how commands are formed).

The proof term of an **LG** derivation is computed on the basis of the *composition graph* associated with its proof net.

Definition 10.14 *Given a proof net P, the associated* composition graph $cg(P)$ *is obtained as follows.*

1. *All vertices of P with formula label A are expanded into* axiom links: *edges connecting two vertices with formula label A; all links are replaced by the corresponding links of Figure 10.15.*
2. *All vertices in this new structure are assigned atomic terms of the correct type (variable or covariable) and the terms for the tensor rules are propagated from the active formulas to the main formula.*
3. *All axiom links connecting terms of the same type (value or context) are collapsed.*

Figure 10.16 gives an example of the composition graph associated with a net. In all, the expansion stage gives rise to four types of axiom links, depending on the type of the term assigned to the A premise and the A conclusion. These cases are summarized in Figure 10.17. The substitution links are collapsed in the final stage of the construction of the composition graph; the command and $\mu/\tilde{\mu}$ cases are the ones that remain.

Given the composition graph $cg(P)$ associated with a proof net P, we compute terms for it as follows:

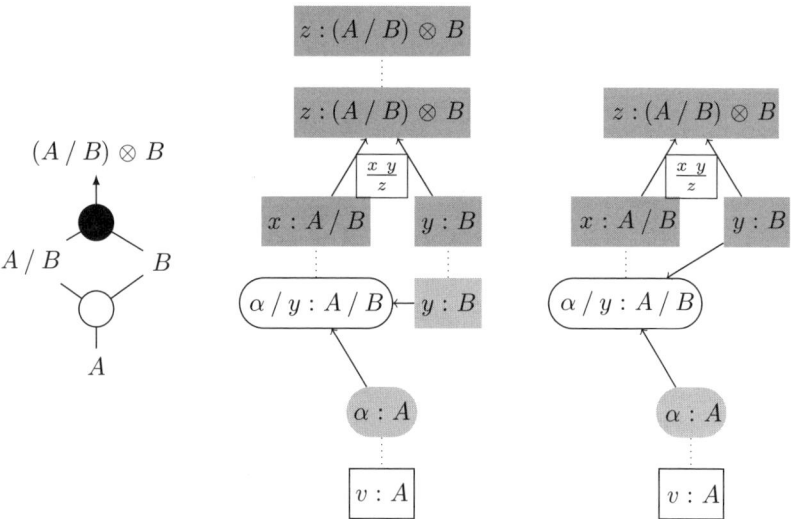

Fig 10.16 Proof net (left) and its associated composition graph; in the middle the expanded net with term annotations; on the right the result of contracting the substitution links.

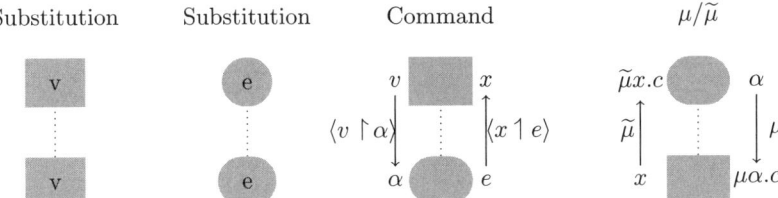

Fig 10.17 Types of axiom links.

1. We compute all maximal subnets of $cg(P)$ consisting of a set of tensor links with a single main formula, marking all these links as visited.
2. While $cg(P)$ contains unvisited links, do the following:
   (a) follow an unvisited command link attached to a previously calculated maximal subnet, forming a correct command subnet; like before, we restrict to *active* subnets which do not contain (or allow us to reach through an axiom) the main formula of a negative link;
   (b) for each negative link with both active formulas attached to the current command subnet, pass to the main formula of the negative link, forming a new command, repeat this step until no such negative links remain attached;
   (c) follow a μ or $\widetilde{\mu}$ link to a new vertex, forming a larger value or context subnet and replacing the variable previously assigned to the newly visited vertex by the μ value or $\widetilde{\mu}$ context.

The algorithm stays quite close to the focused derivations of the previous section: the maximal subnets of step 1 are *rooted* versions of the components we have used before, with the directions of the arrows potentially splitting components into multiple rooted components (Figure 10.19 will give an example). The asynchronous phases, which joined these rooted components by means of contractions, will now consist of a passage through a command link, followed by zero or more cotensor links, followed by either a μ or a $\widetilde{\mu}$ link, the result being a new, larger subnet. The term-assignment algorithm is a way to enumerate the non-equivalent proof terms of a net. Given that these terms are isomorphic to focused sequent proofs, it is no coincidence that the computation of the proof terms looks a lot like the sequentialization algorithm.[2] The following lemma is easy to prove:

Lemma 10.15 *If P is a proof net (with a pairing of command and $\mu/\widetilde{\mu}$ links) and v is a term calculated for P using this pairing then there is a sequent proof π which is assigned v as well.*

---

[2] The connection between proof net sequentialization and focusing for linear logic is explored in Andreoli and Maieli 1999.

This lemma is easily proved by induction on the depth of the tree: it holds trivially for the leaves (which are rooted components), and, inductively, each command, contensor, $\mu/\tilde{\mu}$ sequence will produce a sequent proof of the same term: in fact each such step corresponds exactly to the derived inference rules for focus shifting discussed in Section 10.3.1.2.

To summarize, the difference between computing terms for proof nets in the Lambek calculus and in **LG** can be characterized by the following statements:

- Lambek calculus: the (potential) terms are given through a bijection between premise and conclusion atomic formulas (ie. a complete matching of the axioms).
- **LG**: the (potential) terms are given through a bijection between premise and conclusion atomic formulas *plus* a bijection between command and $\mu/\tilde{\mu}$ axioms.

We speak of *potential* terms, since in the case of the Lambek calculus only proof nets can be assigned a term, whereas in the **LG** case we need proof nets plus a coherent bijection between command and $\mu/\tilde{\mu}$ axioms, where the μ or $\tilde{\mu}$ rule is applied to one of the free variables of the command c.

10.3.2.3 ILLUSTRATIONS

Figure 10.18 shows how to compute the term for the example proof net of Figure 10.16, starting from the composition graph (on the right). We first look for the components (step 1). Since there is only a single tensor link, this is simple. Figure 10.18 shows, on the left, the context subnet corresponding to this link.

Now, there is only one command to follow from here (step 2a), which produces the command shown in the middle of Figure 10.18. Applying the cotensor link

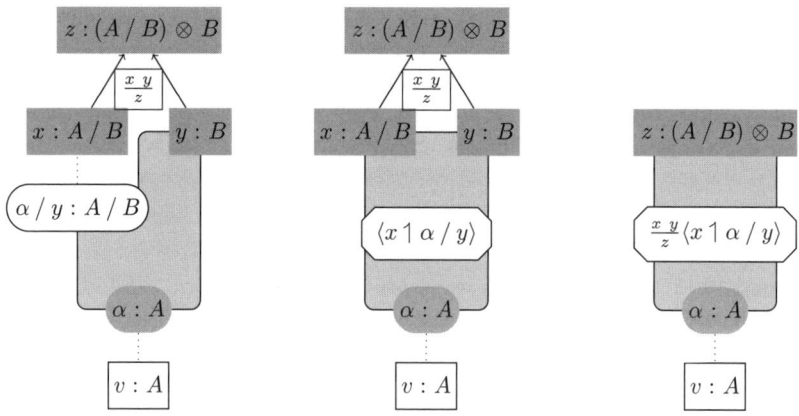

Fig 10.18 Computing the proof term from a composition graph.

(step 2b) produces the figure shown on the right. The final μ link (step 2c, not shown) produces the completed term for this proof net. a

$$v = \mu\alpha.\frac{x\,y}{z}\langle x \mid \alpha \mathbin{/} y\rangle$$

Some remarks about this example. First, some of the axioms can be traversed in only one of the two possible directions: in cut-free proof nets, command links move either towards the active formulas of cotensor links or towards 'dead ends': hypotheses, or conclusions of the proof net. And since we want to compute the value of v for the example proof net, it only makes sense to apply a μ rule to compute this value: we always 'exit' the proof net from a designated conclusion. With a slight modification to the algorithm that reads off terms from a composition graph, we could also compute *commands* for proof nets, or compute the *context* for a designated *premise* of the net.

Figure 10.19 returns to our 'subj tv det noun' example. On the left we see the composition graph for the example of Figure 10.11.

The only cotensor link in the figure has the node subj : $(np \mathbin{/} n) \otimes n$ as its main formula. When we compute the rooted components, we see that there are three, shown on the right of the figure. There are three command axioms, one for the root node of each of the three components C_1 to C_3 on the right-hand side of the figure; these commands are numbered $c1$ to $c3$ next to the corresponding links with the same number as the corresponding component. There are also three $\mu/\tilde{\mu}$ links (numbered μ_1 to μ_3). Figure 10.20 gives a schematic representation of the proof net of Figure 10.19.

Since we are interested in calculating the value of x', $(\mu 1)$ will be the last link we pass in the proof net and therefore we will pass it downwards, producing a term of the form $\mu\beta.c$. Figure 10.20 gives a schematic representation of the proof net of Figure 10.19. The arrows next to the $\mu/\tilde{\mu}$ links indicate the different possibilities for traversing the link and whether this traversal corresponds to a μ or a $\tilde{\mu}$ link.

If both np arguments of the transitive verbs are lexically assigned a positive bias, then we can only pass the two axioms $\mu_2/\tilde{\mu}_2$ and $\mu_3/\tilde{\mu}_3$ in the $\tilde{\mu}_2$ and $\tilde{\mu}_3$ directions, following the arrows away from component C_2. This will necessarily mean that the first command is c_2 and that we can follow this command either with $\tilde{\mu}_2$ (going to the component of the subject and producing the narrow scope reading for the subject quantifier) or with $\tilde{\mu}_3$ (going to the component of the object determiner and producing the narrow scope reading for the object quantifier). The figure shows (in the middle) the result of choosing $c_2 - \tilde{\mu}_3$.

The term computed for component C_2 by following command c_2 is $\langle \text{tv} \mid (x\backslash\beta)/y\rangle$ and the $\tilde{\mu}_3$ link joins components C_2 and C_3, replacing covariable α by the complex context $\tilde{\mu}y.\langle \text{tv} \mid (x\backslash\beta)/y\rangle$, producing the configuration shown schematically in the middle of Figure 10.20 (refer back to Figure 10.19 to see the initial labels).

From this middle configuration and given the restriction to $\tilde{\mu}_2$ for the link connecting the two remaining components, only the command c_3 is a possible in

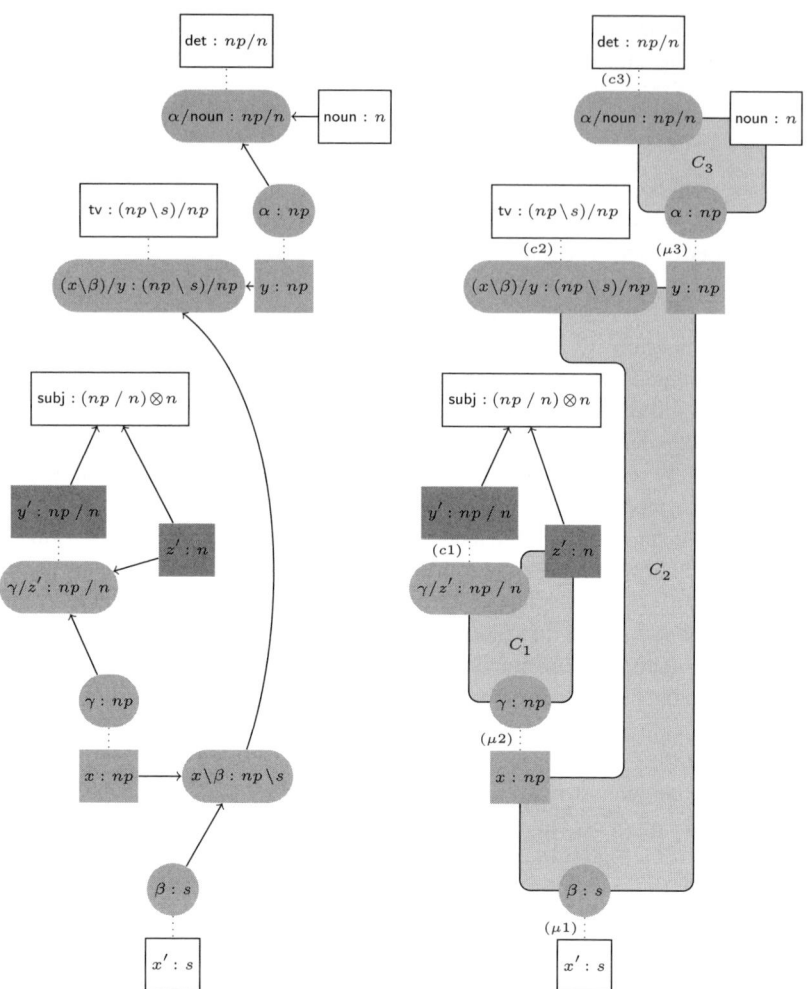

Fig 10.19 Composition graph (left) and initial components (right) for the 'subj tv det noun' example.

combination with the $\widetilde{\mu}_2$ link. Command c_3 would produce ⟨det ↑ α/noun⟩ from the configuration shown in Figure 10.19 but given the previous substitution for α it will now produce ⟨det ↑ ($\widetilde{\mu}y$.⟨tv ↑ $(x\backslash\beta)/y$⟩)/noun⟩ and the $\widetilde{\mu}_2$ link will replace covariable γ by the context $\widetilde{\mu}x$.⟨det ↑ ($\widetilde{\mu}y$.⟨tv ↑ $(x\backslash\beta)/y$⟩)/noun⟩ and produce the configuration shown schematically on the right of Figure 10.20.

The final command c_1 produces ⟨y' ↑ ($\widetilde{\mu}x$.⟨det ↑ ($\widetilde{\mu}y$.⟨tv ↑ $(x\backslash\beta)/y$⟩)/noun⟩)/z'⟩, the cotensor link gives rise to the extended command

$$\frac{y'z'}{\text{subj}}.\langle y' \uparrow (\widetilde{\mu}x.\langle \text{det} \uparrow (\widetilde{\mu}y.\langle \text{tv} \uparrow (x\backslash\beta)/y\rangle)/\text{noun}\rangle)/z'\rangle \qquad (10.40)$$

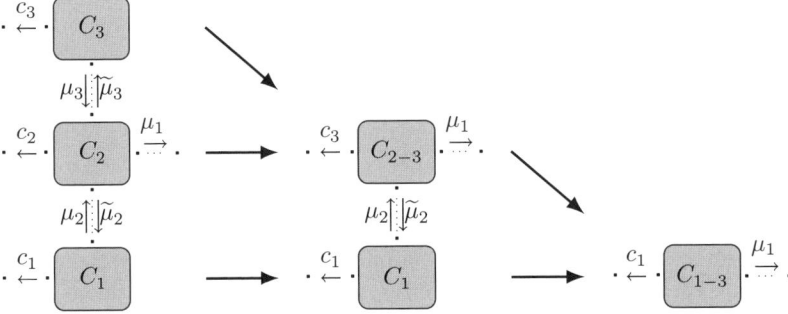

Fig 10.20 Matching: $c_2 - \tilde{\mu}_3, c_3 - \tilde{\mu}_2, c_2 - \mu_1$. Reading: subj < det < tv.

and, finally, by μ_1 the term for the complete proof net (and its command-$\mu/\tilde{\mu}$ pairing):

$$\mu\gamma.\frac{y'z'}{\text{subj}}.\langle y' \mid (\tilde{\mu}x.\langle \det \mid (\tilde{\mu}y.\langle \text{tv} \mid (x\backslash\beta)/y\rangle)/\text{noun}\rangle)/z'\rangle \tag{10.41}$$

Similarly, starting with the $c_2 - \tilde{\mu}_2$ pairing will produce.

$$\mu\beta.\langle \det \mid (\tilde{\mu}y.\frac{y'z'}{\text{subj}}.\langle y' \mid (\tilde{\mu}x.\langle \text{tv} \mid (x\backslash\beta)/y\rangle)/z'\rangle)/\text{noun}\rangle. \tag{10.42}$$

These are the only two readings available with positive bias for the two atomic *np* arguments of the transitive verb, and, as we have seen before, this gives the right quantifier scope possibilities for an extensional transitive verb such as 'likes' that we have seen in eqns (10.34) and (10.35) (apart from the variable names, eqn (10.42) differs from (10.35) in that the extended command fraction in the latter term is at the innermost position, but the terms are equivalent up to commutative conversions).

When we use a negative bias for the two *np* arguments of the transitive verb, we obtain the following term, corresponding to equation (10.33):

$$\mu\beta.\frac{y'z'}{\text{subj}}.\langle y' \mid (\tilde{\mu}x.\langle \text{tv} \mid (x\backslash\beta)/(\mu\alpha.\langle \det \mid \alpha/\text{noun}\rangle)\rangle)/z'\rangle \tag{10.43}$$

10.4 Conclusions

The Lambek–Grishin calculus is a symmetric version of the Lambek calculus. Together with the interaction principles, it allows for the treatment of patterns beyond context-free languages that cannot be satisfactorily handled in the Lambek calculus. We have compared two proof systems for **LG**: focused sequent proofs and proof nets. Focused proofs avoid the spurious non-determinism of backward-chaining search in the sequent calculus; they provide a natural interface to semantic interpretation via their continuation-passing-style translation. Proof nets present the essence of a derivation in a visually appealing form; they do away with the syntactic clutter of sequent proofs and compute the structure of the end-sequent in a data-driven manner, where this structure has to be given before one can a start backward-chaining sequent derivation. Proof terms are read off from the composition graph associated with a net. The computation of these terms depends both on a bijection between premise and conclusion atomic formulas and between command and $\mu/\widetilde{\mu}$ axioms. As a result, one net can be associated with multiple construction recipes (proof terms), corresponding to multiple derivations in the focused sequent calculus.

Acknowledgement

We thank Arno Bastenhof for helpful comments on an earlier version.

CHAPTER 11

Algebras over a Field and Semantics for Context-Based Reasoning

DAOUD CLARKE
(University of Sussex)

11.1 Introduction

This chapter introduces context algebras and demonstrates their application to combining logical and vector-based representations of meaning. Other chapters in this volume consider approaches that attempt to reproduce aspects of logical semantics within new frameworks. The approach we present here is different: we show how logical semantics can be embedded within a vector space framework, and use this to combine distributional semantics, in which the meanings of words are represented as vectors, with logical semantics, in which the meaning of a sentence is represented as a logical form.

The ideas discussed here are present (at least implicitly) in earlier work, but we have introduced some notions that allow the mathematics to be tidied considerably:

- When context algebras were introduced (Clarke, 2007) they were applied only to functions from a free monoid A^* to $\mathbb{R}$. In fact, this construction generalizes to functions from A^* to an arbitrary vector space V. The proof of the general case is identical to the specific one, and is reproduced here unchanged.
- This more general construction gives us an elegant way of embedding logical semantics within an algebraic framework. The embedding presented here follows similar lines to the thinking in Clarke (2007), but uses the new, more general, context algebras.

- The method of combining logical semantics with vector-based lexical semantics is new, but follows similar lines to an approach suggested in Clarke, Weir and Lutz (2011).

11.1.1 Motivation

Like other work in this book, we are concerned with the question of how to compose vector-based representations of meaning so that phrases and sentences are also represented as vectors. We wish to preserve the wonderful flexibility and fine-grained distinctions of meaning that vector spaces allow, and which have been so successful in lexical semantics, to build a complete framework for natural language semantics encompassing words, phrases, sentences, and beyond.

Unlike other work, in the approach presented here we do not attempt to reconstruct logical semantics from scratch, instead embedding logical representations within a vector space. This has some benefits:

- Doing natural language semantics well is difficult, and a lot of work has gone into getting logical semantics for natural language right. It includes worrying about things like anaphora resolution, generalized quantifiers, and negation; reproducing this work from scratch in a vector-based framework is a mammoth task. Our approach allows us to reuse existing work while incorporating vector-based lexical semantics.
- There is the potential to reuse existing tools for natural language semantics, although computation in general is a problem with our approach.

The downside to our approach is that we do not yet have an efficient way of computing with it, although we have ideas for how this may be achieved. Another potential criticism of this approach is that the flexibility in how vector representations are combined with logic may be hindered by requiring the wholesale adoption of existing formalisms, rather than the more tailored approaches of other work.

11.2 Theory of meaning

We first recall some basic definitions:

Definition 11.1 (Algebra over a field) *An algebra over a field is a vector space $\mathcal{A}$ over a field K together with a binary operation $(a, b) \mapsto ab$ on $\mathcal{A}$ that is bilinear,*

$$a(\alpha b + \beta c) = \alpha ab + \beta ac \tag{11.1}$$

$$(\alpha a + \beta b)c = \alpha ac + \beta bc \tag{11.2}$$

*for all $a, b, c \in \mathcal{A}$ and all $\alpha, \beta \in K$. If we additionally have the property $(ab)c = a(bc)$ then $\mathcal{A}$ is called **associative**. An algebra is called **unital** if it has a distinguished **unity*** element 1 satisfying $1x = x1 = x$ for all $x \in A$. We are generally only interested in *real* associative algebras, where K is the field of real numbers, $\mathbb{R}$.

Examples of associative algebras are given by square matrices of order n under normal matrix multiplication and entrywise vector operations. The field of the algebra is the field of the elements of the matrices, so real-valued matrices form a real associative algebra.

11.2.1 Meaning as context

The distributional hypothesis of Harris (1968) states that words will have similar meanings if and only if they occur in similar contexts. We formalize this idea, and examine the resultant mathematical properties.

Let A be some set, which we imagine to be the set of words of a natural language. Let A^* be a free monoid on A, with all the strings constructed from elements of *a* as elements, the empty string as unit, and concatenation as the monoidal operation. If V is a vector space, we define a *general language for V* (or simply a *language* when there is no ambiguity) as a function from the free monoid A^* to V. For each string $x \in A^*$, we have associated with it a vector in V that may have several interpretations:

- V may simply be the real numbers $\mathbb{R}$, and the language may describe a probability distribution over strings in A^*, in which case we can view the language as a generative model of a natural language, describing the probability of observing each possible string as a sentence or a document.
- V may be a vector space describing the meaning of strings, for example a representation of model-theoretic semantics. In this case, the language attaches a meaning to each possible string in A^*.

Given a general language L we define the context vector $\hat{x}$ of a string x as a function from $A^* \times A^*$ to V:

$$\hat{x}(y, z) = L(yxz).$$

Thus, as in the study of formal languages, we consider the context of a string to be the pair of strings surrounding it. We think of $\hat{x}$ as an element of the vector space $V^{A^* \times A^*}$, the space of functions from $A^* \times A^*$ to V. This is a vector space with operations defined pointwise, i.e. if $f, g \in V^{A^* \times A^*}$ and $\alpha \in K$ where K is the field of V then $(\alpha f)(x, y) = \alpha f(x, y)$ and $(f + g)(x, y) = f(x, y) + g(x, y)$ for all $x, y \in A^*$.

Definition 11.2 (Generated subspace $\mathcal{A}$) *The subspace $\mathcal{A}$ of $V^{A^* \times A^*}$ is the set defined by*

$$\mathcal{A} = \{a : a = \sum_{x \in A^*} \alpha_x \hat{x} \text{ for some } \alpha_x \in \mathbb{R}\} \tag{11.3}$$

In other words, it is the space of all vectors formed from linear combinations of context vectors.

Given this definition, we can define multiplication on $\mathcal{A}$, by assuming linearity and making the multiplication compatible with the underlying multiplication of A^*; that is, we want to define a product $\cdot$ on $\mathcal{A}$ such that $\hat{x} \cdot \hat{y} = \widehat{xy}$ for all $x, y \in A^*$. However, in general there is more than one basis for $\mathcal{A}$ formed from elements $\hat{x}$, for $x \in A^*$. We need to confirm that multiplication will be the same, regardless of which basis we choose.

Proposition 11.3 (Context algebra) *Multiplication on $\mathcal{A}$ is the same irrespective of the choice of basis B.*

Proof We say $B \subseteq A^*$ defines a basis $\mathcal{B}$ for $\mathcal{A}$ when $\mathcal{B}$ is a basis such that $\mathcal{B} = \{\hat{x} : x \in B\}$. Assume there are two sets $B_1, B_2 \subseteq A^*$ that define corresponding bases $\mathcal{B}_1$ and $\mathcal{B}_2$ for $\mathcal{A}$. We will show that multiplication in basis $\mathcal{B}_1$ is the same as in the basis $\mathcal{B}_2$.

We represent two basis elements $\hat{u}_1$ and $\hat{u}_2$ of $\mathcal{B}_1$ in terms of basis elements of $\mathcal{B}_2$:

$$\hat{u}_1 = \sum_i \alpha_i \hat{v}_i \quad \text{and} \quad \hat{u}_2 = \sum_j \beta_j \hat{v}_j \tag{11.4}$$

for some $u_i \in B_1$, $v_j \in B_2$ and $\alpha_i, \beta_j \in \mathbb{R}$. First consider multiplication in the basis $\mathcal{B}_1$. Note that $\hat{u}_1 = \sum_i \alpha_i \hat{v}_i$ means that $L(xu_1 y) = \sum_i \alpha_i L(xv_i y)$ for all $x, y \in A^*$. This includes the special case where $y = u_2 y'$ so

$$L(xu_1 u_2 y') = \sum_i \alpha_i L(xv_i u_2 y') \tag{11.5}$$

for all $x, y' \in A^*$. Similarly, we have $L(xu_2 y) = \sum_j \beta_j L(xv_j y)$ for all $x, y \in A^*$, which includes the special case $x = x' v_i$, so $L(x' v_i u_2 y) = \sum_j \beta_j L(x' v_i v_j y)$ for all $x', y \in A^*$. Inserting this into the above expression yields

$$L(xu_1 u_2 y) = \sum_{i,j} \alpha_i \beta_j L(xv_i v_j y) \tag{11.6}$$

for all $x, y \in A^*$, which we can rewrite as

$$\hat{u}_1 \cdot \hat{u}_2 = \widehat{u_1 u_2} = \sum_{i,j} \alpha_i \beta_j (\hat{v}_i \cdot \hat{v}_j) = \sum_{i,j} \alpha_i \beta_j \widehat{v_i v_j} \tag{11.7}$$

Conversely, the product of u_1 and u_2 using the basis $\mathcal{B}_2$ is

$$\hat{u}_1 \cdot \hat{u}_2 = \sum_i \alpha_i \hat{v}_i \cdot \sum_j \beta_j \hat{v}_j = \sum_{i,j} \alpha_i \beta_j (\hat{v}_i \cdot \hat{v}_j) \tag{11.8}$$

thus showing that multiplication is defined independently of what we choose as the basis. □

Multiplication as defined above makes $\mathcal{A}$ an algebra. Moreover, it is easy to see that it is associative since the multiplication on A^* is associative. It has a unity, which is given by $\hat{\epsilon}$, where ϵ is the empty string.

11.2.2 Entailment

Our notion of entailment is founded on the idea of *distributional generality* (Weeds, Weir and McCarthy, 2004). This is the idea that the distribution over contexts not only has implications for similarity of meaning but can also describe how general a meaning is. A term t_1 is considered distributionally more general than another term t_2 if t_1 occurs in a wider range of contexts than t_2. It is proposed that distributional generality may be connected to semantic generality. For example, we may expect the term *animal* to occur in a wider range of contexts than the term *cat* since the first is semantically more general.

We translate this to a mathematical definition by making use of an implicit partial ordering on the vector space:

Definition 11.4 (Partially ordered vector space) *A partially ordered vector space V is a real vector space together with a partial ordering $\leq$ such that:*

if $x \leq y$ then $x + z \leq y + z$
if $x \leq y$ then $\alpha x \leq \alpha y$

for all $x, y, z \in V$ and for all $\alpha \geq 0$. Such a partial ordering is called a vector space order *on V. An element u of V satisfying $u \geq 0$ is called a* positive element; *the set of all positive elements of V is denoted V^+. If $\leq$ defines a lattice on V then the space is called a* vector lattice *or* Riesz space.

If V is a vector lattice, then the vector space of contexts, $V^{A^* \times A^*}$, is a vector lattice, where the lattice operations are defined component-wise: $(u \wedge v)(x, y) = u(x, y) \wedge v(x, y)$, and $(u \vee v)(x, y) = u(x, y) \vee v(x, y)$. For example, $\mathbb{R}$ is a vector lattice with meet as the min operation and join as max, so $\mathbb{R}^{A^* \times A^*}$ is also a vector lattice. In this case, where the value attached to a context is an indication of its frequency of occurrence, $\hat{x} \leq \hat{y}$ means that y occurs at least as frequently as x in every context.

Note that, unlike the vector operations, the lattice operations are dependent on the basis: a different basis gives different operations. This makes sense in the linguistic setting, since there is nearly always a distinguished basis, originating in the contexts from which the vector space is formed.

11.3 From logical forms to algebra

Model-theoretic approaches generally deal with a subset of all possible strings—the language under consideration—translating sequences in the language to a logical form, expressed in another, logical language. Relationships between logical forms are expressed by an entailment relation on this logical language.

This section is about the algebraic representation of logical languages. Representing logical languages in terms of an algebra will allow us to incorporate statistical information about language into the representations. For example, if we have multiple parses for a sentence, each with a certain probability, we will be able to represent the meaning of the sentence as a probabilistic sum of the representations of its individual parses.

By a logical language we mean a language $\Lambda \subseteq A'^*$ for some alphabet A', together with a relation $\vdash$ on Λ that is reflexive and transitive; this relation is interpreted as entailment on the logical language. We will show how each element $u \in \Lambda$ can be associated with a projection on a vector space; it is these projections that define the algebra. Later we will show how this can be related to strings in the natural language λ that we are interested in.

For a subset T of a set S, we define the projection P_T on $L^\infty(S)$ (the set of all bounded real-valued functions on S) by

$$P_T e_s = \begin{cases} e_s & \text{if } s \in T \\ 0 & \text{otherwise} \end{cases}$$

where e_s is the basis element of $L^\infty(S)$ corresponding to the element $s \in S$. Given $u \in \Lambda$, define $\downarrow_\vdash(u) = \{v : v \vdash u\}$.

As a shorthand we write P_u for the projection $P_{\downarrow_\vdash(u)}$ on the space $L^\infty(\Lambda)$. The projection P_u can be thought of as projecting onto the space of logical statements that entail u. This is made formal in the following proposition:

Proposition 11.5 $P_u \leq P_v$ if and only if $u \vdash v$.

Proof Recall that the partial ordering on projections is defined by $P_u \leq P_v$ if and only if $P_u P_v = P_v P_u = P_u$ (Aliprantis and Burkinshaw, 1985). Clearly

$$P_u P_v e_w = \begin{cases} e_w & \text{if } w \vdash u \text{ and } w \vdash v \\ 0 & \text{otherwise} \end{cases}$$

so if $u \vdash v$ then since $\vdash$ is transitive, if $w \vdash u$ then $w \vdash v$ so we must have $P_u P_v = P_v P_u = P_u$.

Conversely, if $P_u P_v = P_u$ then it must be the case that $w \vdash u$ implies $w \vdash v$ for all $w \in \Lambda$, including $w = u$. Since $\vdash$ is reflexive, we have $u \vdash u$, so $u \vdash v$, which completes the proof. □

To help us understand this representation better, we will show that it is closely connected to the ideal completion of partial orders. Define a relation $\equiv$ on Λ by $u \equiv v$ if and only if $u \vdash v$ and $v \vdash u$. Clearly $\equiv$ is an equivalence relation; we denote the equivalence class of u by $[u]$. Equivalence classes are then partially ordered by $[u] \leq [v]$ if and only if $u \vdash v$. Then note that $\bigcup \downarrow_\vdash([u]) = \downarrow_\vdash(u)$, and thus P_u projects onto the space generated by the basis vectors corresponding to the elements $\bigcup \downarrow_\vdash ([u])$, the ideal completion representation of the partially ordered equivalence classes.

What we have shown here is that logical forms can be viewed as projections on a vector space. Since projections are operators on a vector space, they are themselves vectors; viewing logical representations in this way allows us to treat them as vectors, and we have all the flexibility that comes with vector spaces: we can add them, subtract them, and multiply them by scalars; since the vector space is also a vector lattice, we also have the lattice operations of meet and join. As we will see in the next section, in some special cases such as that of the propositional calculus, the lattice meet and join coincide with logical conjunction and disjunction.

11.3.1 Example: propositional calculus

In this section we apply the ideas of the previous section to an important special case: that of the propositional calculus. We choose as our logical language Λ the language of a propositional calculus with the usual connectives $\vee$, $\wedge$ and $\neg$, the logical constants $\top$ and $\bot$ representing 'true' and 'false' respectively, with $u \vdash v$ meaning 'infer v from u', behaving in the usual way. Then:

$$P_{u \wedge v} = P_u P_v$$

$$P_{\neg u} = 1 - P_u + P_\bot$$

$$P_{u \vee v} = P_u + P_v - P_u P_v$$

$$P_\top = 1.$$

To see this, note that the equivalence classes of $\vdash$ form a Boolean algebra under the partial ordering induced by $\vdash$, with

$$[u \wedge v] = [u] \wedge [v]$$

$$[u \vee v] = [u] \vee [v]$$

$$[\neg u] = \neg [u].$$

Note that while the symbols $\wedge$, $\vee$ and $\neg$ refer to logical operations on the left-hand side; on the right-hand side they are the operations of the Boolean algebra of equivalence classes; they are completely determined by the partial ordering associated with $\vdash$.[1]

Since the partial ordering carries over to the ideal completion we must have

$$\downarrow [u \wedge v] = \downarrow [u] \cap \downarrow [v]$$

$$\downarrow [u \vee v] = \downarrow [u] \cup \downarrow [v]$$

---

[1] In the context of model theory, the Boolean algebra of equivalence classes of sentences of some theory T is called the Lindenbaum–Tarski algebra of T (Hinman, 2005).

where $\downarrow(x) = \{y : y \leq x\}$, with $\leq$ as the partial ordering induced by $\vdash$. Since $u \vdash \top$ for all $u \in \Lambda$, it must be the case that $\downarrow[\top]$ contains all sets in the ideal completion. However, the Boolean algebra of subsets in the ideal completion is larger than the Boolean algebra of equivalence classes; the latter is embedded as a Boolean subalgebra of the former. Specifically, the least element in the completion is the empty set, whereas the least element in the equivalence class is represented as $\downarrow[\bot]$. Thus negation carries over with respect to this least element:

$$[\neg u] = ([\top] - [u]) \cup [\bot].$$

We are now in a position to prove the original statements:

- Since $\downarrow[\top]$ contains all sets in the completion, $\bigcup \downarrow[\top] = \downarrow_\vdash(\top) = \Lambda$, and $P_\top$ must project onto the whole space, that is $P_\top = 1$.
- Using the above expression for $\downarrow[u \wedge v]$, taking unions of the disjoint sets in the equivalence classes we have $\downarrow_\vdash(u \wedge v) = \downarrow_\vdash(u) \cap \downarrow_\vdash(v)$. Making use of the equation in the proof to Proposition 11.5, we have $P_{u \wedge v} = P_u P_v$.
- In the above expression for $\downarrow[\neg u]$, note that $\downarrow[\top] \subseteq \downarrow[u] \subseteq \downarrow[\bot]$. This allows us to write, after taking unions and converting to projections, $P_{\neg u} = 1 - P_u + P_\bot$, since $P_\top = 1$.
- Finally, we know that $u \vee v \equiv \neg(\neg u \wedge \neg v)$, and since equivalent elements in Λ have the same projections we have:

$$P_{u \vee v} = 1 - P_{\neg u \wedge \neg v} + P_\bot$$
$$= 1 - P_{\neg u} P_{\neg v} + P_\bot$$
$$= 1 - (1 - P_u + P_\bot)(1 - P_v + P_\bot) + P_\bot$$
$$= P_u + P_v - P_u P_v - 2P_\bot + P_\bot P_u + P_\bot P_v$$
$$= P_u + P_v - P_u P_v$$

It is also worth noting that in terms of the vector lattice operations $\vee$ and $\wedge$ on the space of operators on $L^\infty(\Lambda)$, we have $P_{u \vee v} = P_u \vee P_v$ and $P_{u \wedge v} = P_u \wedge P_v$.

11.3.2 From logic to context algebras

In the simplest case, we may be able to assign to each natural language sentence a single sentence in the logical language (its interpretation). Let $\Gamma \subseteq A^*$ be a formal language consisting of natural language sentences x with a corresponding interpretation in the logical language Λ, which we denote $\rho(x)$. The function ρ maps natural language sentences to their interpretations, and may incorporate tasks such as word-sense disambiguation, anaphora resolution, and semantic disambiguation.

We can now define a general language to represent this situation. We take as our vector space V the space generated by projections $\{P_u : u \in \Lambda\}$ on $L^\infty(\Lambda)$. For $x \in A^*$ we define

$$L(x) = \begin{cases} P_{\rho(x)} & \text{if } x \in \Gamma \\ 0 & \text{otherwise} \end{cases}$$

Given the discussions in the preceding sections, it is clear that for $x, y \in \Gamma$, $L(x) \leq L(y)$ if and only if $\rho(x) \vdash \rho(y)$, so the partial ordering of the vector space encodes the entailment relation of the logical language.

The context algebra constructed from L gives meaning to any substring of elements of Γ, so any natural language expression that is a substring of a sentence with a logical interpretation has a corresponding non-zero element in the algebra. If Γ has the property that no element of Γ is a substring of any other element of Γ (for example, Γ consists of natural language sentences starting and ending with a unique symbol), then we will also have $\hat{x} \leq \hat{y}$ if and only if $\rho(x) \vdash \rho(y)$ for all $x, y \in \Gamma$. In this case, the only context which maps to a non-zero vector is the pair of empty strings, (ϵ, ϵ).

11.3.3 Incorporating word vectors

The construction of the preceding section is not very useful on its own, as it merely encodes the logical reasoning within a vector space framework. However, we will show how this construction can be used to incorporate vector-based representations of lexical semantics.

In the general case, we assume that associated with each word $a \in A$ there is a vector $\psi(a)$ in some finite-dimensional vector space $L^\infty(S)$ that represents its lexical semantics, perhaps obtained using latent semantic analysis or some other technique. S is a finite set indexing the lexical semantic vector space, which we interpret as containing *aspects* of meaning. This set may be partitioned into subsets containing aspects for different parts of speech; for example, we may wish that the vector for a verb is always disjoint to the vector for a noun. Similarly, there may be a single aspect for each closed class or function word in the natural language, to allow these words not to have a vector nature.

Instead of mapping directly from strings of the natural language to the logical language, we map from strings of aspects of meaning. Let $\Delta \subset S^*$ be a formal language consisting of all meaningful strings of aspects, i.e. those with a logical interpretation. As before, we assume a function ρ from Δ to a logical language Λ. The corresponding context algebra $\mathcal{A}$ describes composition of aspects of meaning.

We can now describe the representation of the meaning $\tilde{a}$ of a word $a \in A$ in terms of elements of $\mathcal{A}$:

$$\tilde{a} = \sum_{s \in S} \psi(a)_s \hat{s}$$

thus a term is represented as a weighted sum of the context vectors for its aspects. Composition of $\mathcal{A}$ together with the distributivity of the algebra is then enough to define vectors for any string in A^*. For $x \in A^*$, we define $\tilde{x}$ by

$$\tilde{x} = \tilde{x}_1 \cdot \tilde{x}_2 \cdots \tilde{x}_n$$

where $x = x_1 x_2 \cdots x_n$ for $x_n \in A$.

This construction achieves the goal of combining vector space representations of lexical semantics with existing logical formalisms. It has the following properties:

- The entailment relation between the logical expressions associated with sentences is encoded in the partial ordering of the vector space.
- The vector space representation of a sentence includes a sum over all possible logical sentences, where each word has been represented by one of its aspects.

Discussion

This second property is actually inconsistent with the idea that distributional generality determines semantic generality. To see why, consider the sentences

1. No animal likes cheese
2. No cat likes cheese

Although the first sentence entails the second, the term *animal* is more general than *cat*; the quantifier *no* has reversed the direction of entailment. The distributivity of the algebra means that if the term *animal* occurs in a wider range of contexts than *cat* (i.e. it is distributionally more general), then this generality must persist through to the sentence level as terms are multiplied.

There are two ways of viewing this inconsistency:

1. The construction is incorrect because the distributional hypothesis does not hold universally at the lexical level. In fact, this idea is justified by our example above, in which the more general term, *animal*, would be expected to occur in a smaller range of contexts than *cat* when preceded by the quantifier *no*. Under this view, the construction needs to be altered so that quantifiers such as *no* can reverse the direction of entailment. In fact, for this to be the case, we would need to dispense with using an algebra altogether, since distributivity is a fundamental property of the algebra.
2. The construction is correct as long as the 'aspects' of words are really their senses, so the vector space nature only represents semantic ambiguity and not semantic generality. In reality, aspects may be more subtle than what is normally considered a word sense, and the vector space representation would capture this subtlety. In this view, distributional generality is correlated with semantic generality at the sentence level, but not necessarily the word level. Vector representations describe only semantic ambiguity and not generality, which should be incorporated into the associated logical representation. This would mean that the algorithms used for obtaining vectors for terms would have to have more in common with automatic word

sense induction than the more general semantic induction associated with the distributional hypothesis.

An alternative solution for this type of quantifier (another example is *all*) and negation in general, is to use a construction similar to that proposed in Preller and Sadrzadeh (2011b), which uses Bell states to swap dimensions in a similar manner to qubit operators.

11.3.4 Partial entailment

In general, it is unlikely that any two strings $x, y \in A^*$ that humans would judge as entailing would have vectors such that $\tilde{x} \leq \tilde{y}$ because of the nature of the automatically obtained vectors. Instead, it makes sense to consider a *degree* of entailment between strings. In this case, we assume we have a linear functional ϕ on $V^{A^* \times A^*}$. The degree to which x entails y is then given defined as

$$\frac{\phi(\tilde{x} \wedge \tilde{y})}{\phi(\tilde{x})}$$

This has many of the properties we would expect from a degree of entailment. In certain cases ϕ can be used to make the space an abstract Lebesgue space (Abramovich and Aliprantis, 2002), in which case we can interpret the above definition as a conditional probability. This idea, and methods of defining ϕ are discussed in detail in Clarke (2007).

For the vector space V generated by projections on $L^\infty(\Lambda)$, one possible definition of ϕ would be given by

$$\phi(u) = \sum_{l \in \Lambda} P(l) \| u(\epsilon, \epsilon) e_l \|$$

where $P(l)$ is some probability distribution over elements of Λ. The interpretation of this is: all contexts except those consisting of a pair of empty strings are ignored (so only strings that have logical interpretations make a contribution to the value of the linear functional); the value is then given by summing over all strings of the logical language and multiplying the probability of each string l by the size of the vector resulting from the action of the operator on the basis element corresponding to l. The probability distribution over Λ needs to be estimated; this could perhaps be done using machine learning:

- Given a corpus, build a model for each sentence in the corpus, and its negation, using a model builder.
- Train a support vector machine using the models for the sentence and its negation as the two classes. This requires defining a kernel on models.
- Given a new string $l \in \Lambda$, build a model, then use the probability estimate of the support vector machine for the model belonging to the positive class, together with a normalization function.

11.3.5 Computational issues

In general it will be very hard to compute the degree of entailment between two strings using the preceding definitions. The number of logical interpretations that need to be considered increases exponentially with the length of the string. One possible way of tackling this would be to use Monte Carlo techniques, for example by sampling dimensions of the vector space and computing degrees of entailment for the sample.

A more principled approach would be to exploit symmetries in the mapping ρ from strings of aspects to the logical language. A further possibility is that a deeper analysis of the algebraic properties of context algebras leads to a simpler method of computation. Further work is undoubtedly necessary to tackle this problem.

11.4 Conclusion

We have introduced a more general definition of context algebras, and shown how they can be used to combine vector-based lexical semantics with logic-based semantics at the sentence level. Whilst computational issues remain to be resolved, our approach allows the reuse of the abundance of work in logic-based natural language semantics.

CHAPTER 12

Distributional Semantic Models

STEPHEN PULMAN
(University of Oxford)

12.1 Introduction

In theoretical linguistics, specifically in formal semantics, compositional logical models of the semantics of natural languages have been an object of study for the past 30 years or more, originally inspired by the pioneering work of Richard Montague. Distributional models of semantics, computationally implemented, although prefigured in early work in linguistics and computational linguistics, have only come to prominence within the last 10 or 15 years, with the availability of large-scale machine-readable corpora. A vigorous recent strand of research in computational linguistics has been the attempt to combine logical compositional models of semantics (which typically take word meanings as unexplained primitives) with empirically derived distributional models of word meaning, with the envisaged benefit being a combination of the power of compositional mechanisms with the robustness of distributional models. In this chapter I describe what is meant by compositional and distributional models in semantics and describe some of the problems that must be faced in any attempt at combining the two. I also relate the notion of distribution to some perennial—and some more recent—philosophical issues which are theoretically interesting, if perhaps of little practical importance for the aims of computational linguistics. I survey some recent work in compositional distributional semantics and argue that many recent attempts claiming to display composition of distributionally derived word meanings are in fact doing no such thing, but demonstrating the ability of certain kinds of combination of distributional models to effect word-sense disambiguation. I go on to suggest some tests that would more convincingly require a genuine compositional distributional semantics to be developed.

12.2 Compositional syntax and semantics

The principle of compositionality—that the meaning of a complex expression is determined by its structure and by the meanings of its constituents, or that the meaning of a complex expression is some function of the meanings of its components—is attributed originally to Frege (1884) and has been one of the main tenets of work in formal semantics of natural language. When coupled with the recursive nature of the rules governing the syntax of language it has been one of the principles helping to explain the infinite productivity of languages or the 'infinite use of finite means' property claimed (e.g. Chomsky, 1965; the phrase is originally from Wilhelm von Humboldt) to be a unique characteristic of human languages. It has also been held to be a necessary, though not sufficient, condition for languages to be learnable (Davidson, 1984).

There are many subtleties in the precise formulation of this principle (see Szabó, 2008). There is debate over whether Frege actually held the principle in its commonly understood form (Pelletier, 2001). The situation is complicated by the fact that Frege also appears to have held some version of the view that word senses were determined distributionally. This view has been called his 'context principle', namely that only within a complete sentence do words have meaning. He says 'it is enough if the sentence as whole has meaning; thereby also its parts obtain their meanings' (Szabó, 2008). Again, there are weaker and stronger versions of this principle, and it is not clear whether he regarded this as an alternative way of thinking about compositionality or whether it was a different principle altogether. If the latter, it is a remarkable anticipation of more recent distributional claims. See Pelletier (2001) and Janssen (2001) for some discussion of these issues.

Here we will stick to the version of compositionality, which is the working underlying principle of formal semantics. The meaning of a word is identified with its denotation, and the denotation of a phrase is some function of the denotation of its components. The denotation of a sentence is a truth value, or in more sophisticated theories, a function from possible worlds and/or contexts to truth values.

That is not to say that the principle of compositionality applies to every aspect of language. There are some obvious phenomena in language that seem not to follow the compositionality principle: there are many apparently fixed phrases like 'in spite of', where the component word 'spite' is unconnected with the usual meaning of 'spite' as a noun (they are etymologically distinct). Likewise, you cannot predict that 'put up with' means 'suffer' or 'endure' on the basis of the meanings of 'put', 'up', and 'with' (if they do have independent meanings at all, that is). Idioms like 'skate on thin ice' (meaning 'take a risk') or 'put the cat among the pigeons' (meaning 'cause a disturbance') have meanings not fully predictable from their literal meanings, although there are traces of compositionality of some mysterious sort still present. For example, 'skating on very thin ice' means 'taking a big risk'.

Another exception to full compositionality comes from context-dependent items such as pronouns, ellipsis, or indexicals, which require information often from outside the domain of the sentence, which is usually regarded as the maximal unit for which compositionality holds. And certain kinds of complex nominals such as 'stone lion', or 'plastic gun', also represent units for which the denotation of the entire phrase cannot be easily predicted from the denotation of its components.

Some scholars have questioned whether compositionality—even while restricted to the appropriate domain—is empirically supported by the facts (Higginbotham, 2003). Others (Zadrozny, 1992) have questioned whether the principle has any empirical content at all, offering methods that purport to make any putatively non-compositional method compositional, if somewhat trivially. However, I shall take it as a given that it is possible to build a compositional, syntax driven, semantic description for at least the core grammatical descriptions of English—and other languages—which produces what has become known as a 'quasi logical form' (Alshawi and van Eijck, 1989), capturing the main predicate argument structure of sentences, along with grammatically governed modifications of different aspects of that structure, but abstracting away from all types of context dependency like anaphora, tense, and resolution of different types of operator or quantifier scope.

Compositional formal semantics derives from the work of Richard Montague (1973). He was the first to show how a fragment of a natural language could be treated in a way essentially parallel to that of an artificial logical language like first-order logic, where an interpretation with respect to a model is supplied for each of an infinite number of well-formed formulae. The main problem to solve in moving from an artificial to a natural language is how to make sure that all syntactically motivated constituents receive an independent interpretation. In fact, a typical semantics for first-order logic does not quite achieve this, even for such a relatively simple language: natural languages are of course much more syntactically complex. Montague showed how using a typed higher-order logic, one could associate with constituents meanings that were capable of combining with each other in such a way as to build up a logical form representing the meaning of the whole sentence, and in a way which paralleled the syntactic structure of the sentence, thus capitalizing on the ability of a syntax to generate and infinite number of distinct sentences from a finite set of words and rules.

Montague himself used as his syntactic framework a form of categorial grammar that has not subsequently met with wide acceptance. But the general technique is not restricted to any particular syntactic theory. Within a more usual context-free grammatical framework, analyses within the Montague tradition usually play out along lines roughly as illustrated in the following tiny example, showing syntax-driven semantic assembly, with each constituent having its own denotation. These rules are to be paraphrased as follows: 'A sentence can consist of a noun phrase (NP) followed by a verb phrase (VP), and the meaning of the sentence is obtained by applying the meaning of the VP (which will be a function) to the meaning of the NP', and so on.

1. S → NP VP : VP(NP)
2. NP → Jack, John etc : jack, john etc.
3. VP → V_{trans} NP : V_{trans}(NP)
4. V_{trans} → hits : $\lambda y.\lambda x.hit(x, y)$

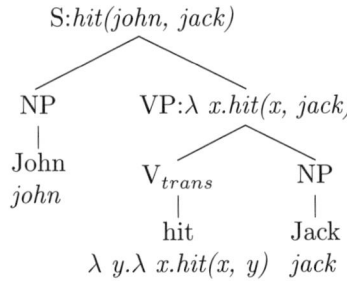

One thing to note—an observation that we will return to—is that in approaches like these word-sense disambiguation plays no interesting part. It is generally assumed that such disambiguation has been performed already and the relevant sense of a word is just translated to the appropriate logical constant. In fact, as Davidson (1967) remarked in his discussion of truth conditional theories of meaning, 'a theory of the kind proposed leaves the whole business of what individual words mean exactly where it was'. The main contribution of formal semantics is to show how complex meaning *structures* can arise out of simpler ones: with the exception of a few function words that serve as 'glue' to build the final logical form, the enterprise has nothing to say about individual word meanings or their disambiguation. At best, this version of semantics is sensitive only to the semantic 'class' of a word, which often will correspond to a distinct 'type', in a logical sense of type. For example, although attributive adjectives are a fairly homogeneous category syntactically, they fall into several distinct semantic subclasses: the intersective ('Welsh', 'pregnant', 'wooden'), the gradeable ('tall', 'old', 'clever'), and the privative ('former', 'alleged', 'plastic N', where N is not stereotypically made of plastic).

12.3 Harris's notion of distribution

There are actually several different notions of 'distribution' in the linguistics literature. One, deriving from Zellig Harris and the American structuralist linguists (Matthews, 2001), refers to a principle by which it is supposed to be possible to identify linguistic units of all sizes—phonemes upwards—by observing the contexts such units occur in, by a relatively simple process of formal comparison and contrast. To illustrate, consider the following 'corpus' from a Mexican language (Nida, 1946) (it is worth recalling that the distributional approach was in part developed in the context of linguistic fieldworkers recording disappearing North American native languages):

ikalwewe	his big house
ikalsosol	his old house
ikalci-n	his little house
komitwewe	his big cooking-pot
komitsosol	his old cooking-pot
komitci-n	his little cooking-pot

It is not too difficult (in this case) to isolate the various morphemes concerned. We can go on to choose particular contexts to be used as diagnostic frames: for example, in English, anything occurring after some form of the auxiliary verb 'be' and suffixed with '-ing' we will decide to recognize as a verb. Anything that will appear in the frame 'the ___ is' we will call a noun, and so on.

How might these notions of distribution work out for identifying syntactic constituents above the word level? For English, at least, a workable strategy for identifying different constituent types is to fix on a set of representative function or other low content words ('targets') and deem that any sequences of words that can replace a target word while still retaining grammaticality is a phrase of the appropriate type (see Harris, 1957). Thus we might choose 'it' or 'they' as nominal targets, defining NP, a simple intransitive verb like 'exists' as VP target, something like 'outside' as a PP target etc. Of course, this presupposes some kind of distinction between function and content words—even if only of a degree: cf. the use of 'outside' as a representative intransitive preposition—but this may be defensible on objective grounds such as corpus frequency counts. It also presupposes that the notion of 'grammatical' as opposed to 'meaningful' makes sense, since some concatenations will sound semantically odd. These are familiar issues. If we assume they can be satisfactorily resolved then we can use Harrisian techniques to arrive at a set of statements defining a 'grammar' by the language L represented by a corpus C:

1. Define a 'verb' in C as {X | X appears in 'is/are X in'}.
2. Define a ''noun' as {X | the X is/are…}.
3. A 'pronoun' is one of the following…
4. A 'noun phrase' is any sequence of words that when substituted for a pronoun in a sentence of C results in a sentence that is either already in C or could be.
5. A 'verb phrase' is any sequence of words that when substituted for 'exists' in 'NP exists' for some NP, results in a sentence in C, or a sentence which could be in C.
6. A 'declarative sentence' is any sequence of words analysable as 'NP VP'.

Notions of hierarchical structure will emerge from these distributional statements, such as that 'if A occurs in environment X and AB does too, but B does not, then A is the head of AB' (Harris, 1954), i.e. the X in a phrase of the type XP—its 'head'—can always (or mostly) appear wherever the XP itself can appear. Similarly we may discover that a conjunction 'XP and/or/but XP' can occur wherever it is possible for a single XP to occur.

It should be intuitively clear that this notion of distribution is wholly compatible with the notion of semantic compositionality, in that from it issues a set of units capable of occurring in different environments and of being combined and recombined in many different ways. While compositionality may not follow logically from Harrisian distributional methods, the independent syntactic units provided by the latter seem to be a prerequisite for the former. In concrete terms, we can in practice go on from such observations to abstract a set of phrase structure rules, and given the right kind of semantic assembly language, construct compositional translations from syntactically analysed sequences of words to expressions of some meaning representation as per the Montagovian programme.

12.4 Firth's notion of collocation

The other linguistic sense of 'distribution' used, particularly in the computational linguistics literature, is that deriving from J. R. Firth (1957, Chapter 15) who referred to it as 'collocation'. The idea here was that the behaviour of a word could be linked to or partially explained by the contexts in which the word was observed to occur. Firth only seemed to have individual words or small compounds in mind, not larger units like phrases or sentences. One of his interests was to get a handle on the notion of different word senses. He argued that different senses of an ambiguous word could be revealed by looking at the different contexts and environments that the target word can occur in. And vice-versa: observing that a word occurs in systematically different contexts suggests that there is a distinction of meaning to be found. The interesting thing about Firth's notion of distribution is that it was explicitly not tied to any notion of compositionality; in fact it was a way of explaining the meaning of non-compositional, quasi-idiomatic uses of words. One example might be a word like 'wax', as a verb. Part of what you need to know about 'wax' as an English speaker is that it can be transitive but with a limited range of objects: 'wax the floor', or 'wax the table', or 'wax the moustache', or alternatively that it can be combined with a very restricted range of adjectives: 'Jane waxed lyrical' 'Jane waxed poetic' (about X). etc. Clearly these different restricted contexts disambiguate two distinct senses of the verb. 'Collocations of a given word are statements of the habitual or customary places of that word.'

It is interesting to see what Manning and Schütze (1999) say about the Firthian notion of distribution, which they clearly see as incompatible with compositionality (phrases prefixed with ?? are judged as odd-sounding in some way):

Collocations include noun phrases like *strong tea* and *weapons of mass destruction*, phrasal verbs like *to make up*, and other stock phrases like *the rich and powerful*. Particularly interesting are the subtle and not-easily-explainable patterns of word usage that native speakers all know: why we say *a stiff breeze* but not *??a stiff wind* (while either *a strong breeze* or *a strong wind* is okay), or why we speak of *broad daylight* (but not *??bright daylight* or *??narrow darkness*). Collocations are characterized by limited compositionality…

Some linguists, in particular Maurice Gross (1975), but also perhaps some later Firth-influenced members of the British corpus linguistics tradition such as Sinclair (2004), have held a position that seems to imply that if the Harrisian notion of distribution is pursued to its extreme, then it collapses into something more like Firth's notion of collocation. More dramatically, the distinction between lexis and grammar is challenged: if careful attention is paid to the facts, then there are no accurate abstractions of the kind encoded in grammatical rules. Gross held that, for example, different types of verb could only be classed as transitive or intransitive etc. by ignoring some of the differences in the context in which they occur. If we were to really individuate classes of verbs in terms of commonality of contexts, we would eventually assign each verb to its own class, and finish up with contexts as specific as those for the two uses of 'wax' above. The Harrisian program involves ignoring idiosyncratic properties of contexts, abstracting away from some semantic differences. The legitimacy of this approach depends on a level of independence between syntactic and semantic properties.

So what does this tension between these two notions of distribution tell us about the prospect of distributional compositional semantics? On the one hand, we need a notion of distribution that abstracts away from semantic properties in order to identify the units that will combine to form phrases and sentences. On the other, we need a fine-grained notion of distribution to identify important semantic properties of individual words, with no guarantee that the contexts that identify an individual sense of the word will form a natural syntactic class. This tension is evident in the following quote from Harris about the relation between meaning and distribution:

. . . if we consider words or morphemes A and B to be more different in meaning that A and C, then we will often find that the distributions of A and B are more different than the distributions of A and C. In other words, difference of meaning correlates with difference of distribution' (p43)

Note that Harris is restricting this to words and morphemes, although even here there are obvious counterexamples; for example, most antonymic pairs of words have very similar, sometimes identical, distributions. It is not at all clear that the correlation will apply to phrases or larger units.

12.5 Distributional semantics

Distributional semantics largely interprets 'distributional' in the Firthian sense, being concerned with word-sense disambiguation. Since the earliest days of computational semantics, this notion has been explored in various guises. The earliest pieces of work of which I am aware that explored the connection between the Firthian notion of distribution and the notion of word-sense disambiguation were the papers by Margaret Masterman and Karen Spärck Jones in the 1950s and 1960s (Masterman, 1957; Spärck Jones, 1986). Many of these ideas could not be adequately tested at the time because of limitations of computing power and the lack of suitable corpora, although arguably later work has showed that many of them were essentially correct (for some discussion, see Pulman, 2010).

An early, and influential piece of work was by Schütze (1998), who was among the first to use distributional techniques to 'learn' word senses on a large scale. Like every approach to distributional methods, Schütze's vector-space theory of word meaning presupposes the availability of very large corpus of words. We describe the various steps of his technique. The very first preliminary step is to decide which words will be 'target' words, and which words will form 'features' or 'context' words. Typically the target words will be a subset of the context words. Target and context words are usually chosen on some frequency-related basis.

The next step is to construct 'word space', by going through the corpus and counting the occurrence of our target words within a particular window of occurrence of the context words. The size of the window is variable but can be as wide as 50 words. We represent the results of these counts as a matrix where the rows are the target words, the columns are the context words, and cells represent the number of times each target word occurred within the window containing the context word. At this point we can perform various tests to make sure that the co-occurrences between the target words and the context words are not accidental. Schütze uses the chi-squared test, but it is also possible to use a measure like the log-likelihood ratio, or some measure based on inverse document frequency. We can describe feature words as:

- 'local': those that are observed to occur in the context of the target word
- 'global': those that are more frequent in the corpus overall.

To illustrate with a small artificial example, given a corpus of sentences including examples like these:

- *Bank* erosion and *stream* widening may occur with strong *water* flow.
- One way of *raising* this *finance* is to go to a *bank*.
- etc.

we might construct a matrix like:

	Context words				
Target	river	stream	money	raise	finance
bank	10	15	25	20	13
water	28	25	2	15	0
cheque	0	0	30	20	25
etc.					

The resulting rows of numbers can be thought of as vectors in a multidimensional space, and can therefore be compared for length and direction, although we are not usually interested in length. Words with similar meanings should have vectors pointing in a similar direction (but as observed earlier, antonyms, since they frequently occur in identical contexts, often do too). Again there are various measures of semantic similarity, but a frequently used one is cosine distance, which ranges between one and zero, where a distance of one means the two vectors are parallel (i.e. identical in their direction) and zero means they are unrelated or orthogonal.

In more detail:

- 'dot product': $\mathbf{a} \cdot \mathbf{b} = a_1 b_1 + a_2 b_2 + a_3 b_3$
- length = $\|\mathbf{a}\| = \sqrt{a_1^2 + a_2^2 + a_3^2} = \sqrt{\mathbf{a} \cdot \mathbf{a}}$.
- similarity = $\cos(\theta) = \frac{A \cdot B}{\|A\| \|B\|} = \frac{\sum_{i=1}^{n} A_i \times B_i}{\sqrt{\sum_{i=1}^{n} (A_i)^2} \times \sqrt{\sum_{i=1}^{n} (B_i)^2}}$

Notice that the word 'bank' is ambiguous between (at least) a 'financial' sense and a 'water'-related sense. However the vectors we have gathered from our tiny corpus do not differentiate these different senses of 'bank' or of any other word. By inspection we might observe that counts for some columns of the vector are systematically different from others, but there is as yet no formal way of distinguishing senses.

The third stage of Schütze's process aims to do this, by constructing 'context space'. We go back to the corpus, where we now have word-space vectors for each of the target and context words (since the former are a subset of the latter). Now, for each occurrence of a target word, we construct the centroid (average) of the vectors for each of the context words occurring within the relevant window. The centroid 'averages' the direction of the set of vectors associated with the context words. So for example in the context of the first sentence in our tiny corpus, we would sum the vectors for 'stream' and 'water' and then average them, whereas in the second sentence we would average the vectors for 'raise' and 'finance'. At the end of this process we have for each occurrence of each target word a vector representing the context of that occurrence.

The fourth step of Schütze's process is to construct 'senses' by clustering the context vectors, using a form of 'agglomerated clustering', where each vector initially forms its own cluster and clusters are repeatedly merged based on some distance criterion until the target number of clusters is arrived at. He experimented with different target numbers between two and ten. Each cluster should correspond to a distinct sense of our ambiguous word, and the sense can be represented by the centroid of the vectors in that cluster.

We can now use the resulting learned senses for ambiguous words to disambiguate new occurrences of the word in a particular context. To do this, we first construct the vector for the words in the context of the target ambiguous word, just as we did in the third step of Schütze's process. For our ambiguous word we will have learned several different sense vectors resulting from the clustering process. If everything works out right then the sense whose vector is closest to the context vector constructed for this particular occurrence of the target word should be the one most appropriate for that context. This is a direct expression of the intuition that the context of an ambiguous word will determine its most likely sense.

In order to evaluate this technique we would need a large corpus in which ambiguous words were tagged with the senses appropriate for each occurrence. Such corpora do not exist on a large scale, although there are some for which this annotation process has been carried out for a small number of ambiguous words. Using such a corpus and also using the technique of pseudo-words, Schütze was able to evaluate his sense disambiguation algorithm. The pseudo-word technique works as follows. We choose two distinct words from the corpus, for example, 'computer' and 'banana' and replace all occurrences of each of them by the pseudo-word 'banana-computer'. We can now evaluate how well the algorithm performs on disambiguating 'banana-computer' by looking at the original form of the corpus. Each occurrence of the pseudo-word will correspond to a single occurrence of one of its components.

In fact Schütze found that in practice it is easier to construct pseudo-words from short compounds rather than individual words, for example 'wide range' + 'consulting firm', 'league baseball' + 'square feet', because individual words are frequently ambiguous in isolation, making the interpretation of the results less clear.

Schütze obtained results for naturally ambiguous words: *interest, space, plant, ...*, as well as pseudo-words, comparing different numbers of clusters and using either local or global context words. A summary of average accuracy is shown in Table 12.1.

These results seem very impressive. However there are a number of issues that need careful consideration. The first of these is that the differing numbers of clusters (i.e. word senses) seems to be entirely unmotivated. It would have been much more impressive to have arrived at a natural number of distinct clusters (there are alternative clustering algorithms that can achieve this by stopping the clustering process at a point that minimizes within-cluster semantic distance and maximizes between-cluster distance). Even better would be to discover that the number of clusters empirically derived conformed to the number of senses as provided by some

TABLE 12.1 Summary of average accuracy

	2 clusters	10 clusters
Natural		
Local	76.0%	84.4%
Global	80.8%	83.1%
Artificial		
Local	89.9%	92.2%
Global	98.6%	98.0%

authoritative dictionary. However, this may be an unrealistic hope: dictionaries are constructed by lexicographers, largely on the basis of their own intuitions, and may not be the best guide to the number of senses that word has in actual use. Comparing entries for the same word in different dictionaries will often yield different numbers of senses. Indeed many researchers, such as Adam Kilgarriff (1997), have doubted that there is any such notion as a distinct sense for a word.

The second thing to notice about these scores is that the results on the artificial pseudo-words are better than those for naturally occurring words. We do not know on what basis he chose the components of the pseudo-words, but it is reasonable to surmise that that they were relatively distinct in their senses and thus unlikely to share very many contexts of occurrence. This would make the task much easier, hence the almost perfect scores.

12.6 Compositional distributional semantics

Despite some caveats over Schütze's work, it has become abundantly clear that distributional models of word meaning can provide a notion of word sense that does many of the things we want it to do. Curran's 2003 demonstration (Curran and Moens), that it was possible to outperform hand-built thesauri like WordNet on a range of different tasks using automatically learned distributional measures of semantic similarity, or indeed to build a useful thesaurus automatically, showed some of the promise of such models. Since then there has been a slew of recent work using these approaches to carry out various semantic tasks: Turney and Pantel (2010) provide a recent overview of many of these.

It is natural then to ask whether such models can move to a level of meaning over and above that of individual words, in particular to ask whether there is a way to combine individual word meanings into meanings for phrases and whole sentences. This would be a very exciting prospect: formal semantics in the Montague tradition supplies a compelling account of the compositional 'glue' that builds phrase and sentence denotations from the denotations of individual words, but, as remarked earlier, has nothing of interest to say about the meanings of most individual words, which are just translated into a non-logical constant whose interpretation is supplied by some unspecified function of the right semantic type.

The attempt to combine compositional semantics of the Montague type with distributional semantics of the Schütze type is what is meant by 'compositional distributional semantics'. However, there are many possible ways of interpreting the notion of combination here. We saw earlier that Harris regarded the correspondence between meaning and distribution as something that applied primarily to words and morphemes. It is not clear that he believed the correspondence would extend fully to phrases and sentences: indeed, at one point Harris (1954, p. 47) remarks that there is relatively little to say about the distributional properties of entire sentences other than some platitudes such as that questions are frequently followed by answers: 'the regularities in a discourse are far weaker and less inter-related than those within a sentence'. So there are many options for how the combination might work, among them:

1. Only words have distributionally derived meanings. Any function that combines the meanings of words into phrases need not preserve any relation between the distributions of the component words and the phrases. This seems on the face of it unlikely, at least for 'small' phrases, given the distributional tests for 'headedness' described earlier: 'fox' is going to occur anywhere that 'big fox' can occur, although 'big' alone will not.
2. There is a connection between the distributions of words and phrases, but it is limited to 'small' (i.e. non-recursive) cases like those just described, and does not extend more generally, or to complete sentences.
3. The connection between composition and distribution is complete and, in principle given the composition function and the distributions of the components (words or phrases), one could predict the distribution of the composed units…
4. …and vice versa: Frege's compositionality principle and his context principle are duals.

12.7 A philosophical digression

Before moving on to consider some attempts to experiment with these different views of composition and distribution, it is worth pointing out that computational linguists are not the first to wonder whether it is possible to combine compositional and distributional approaches to semantics. The philosopher Jerry Fodor, about 15 or 20 years ago, raised the question of whether it was possible to make something like a distributional model of semantics display the properties of compositionality in anything like the usual sense, and in fact the notion of distribution connects with philosophical issues at several different points.

For example, there is an intimate connection between the notion of distribution in the Harris sense and an old philosophical opposition of empiricism versus nativism. Empiricism is, very roughly, the view that humans learn by experience with minimal a priori knowledge, whereas nativism is the view that humans have a fairly rich type of a priori knowledge with correspondingly much less empirical experience needed for learning. Of course, there are weak and strong versions of both positions, and it is pretty clear that extreme versions of either will be implausible. But it is interesting that it is possible to see Harris's distributional hypothesis as a version of an empiricist 'learning procedure' for language if it could be fully automated, since it seems to rely only on judgements that could be obtained by observations of sentences in a corpus, with no recourse to semantic intuitions or other things difficult to observe empirically. Chomsky, who was of course Harris's student, in fact tried at some stage to interpret Harris's distributional hypothesis in exactly this way, but quickly came to the conclusion that there were some important facts about language for which there was no direct evidence in the surface form of sentences, and therefore for which Harris's procedure would be unable to arrive at a satisfactory account of how we learn those aspects of the language.

In more detail, Chomsky's argument (Chomsky, 1964) was as follows: the Harrisian notion of distribution can only lead to what he calls a 'taxonomic' model of grammar, in which any abstract grammatical categories (like NP or VP) in the context of the description of a sentence will correspond to a substring of that sentence. The formal model of grammar consistent with this approach will be context-free phrase-structure grammars, as we pointed out in our earlier discussion of Harrisian distribution. But such grammars, while they can describe one aspect of the structure of sentences like the famous pair:

(a) John is easy to please
(b) John is eager to please

they cannot give the full story about them. At the part-of-speech and phrase-structure level the two sentences are identical, but this conceals many differences and distinctions that cannot be captured in a phrase-structure tree. For example, that 'John' is interpreted as the object of 'please' in (a), but the subject of 'please' in (b). Likewise, such 'taxonomic' descriptions cannot provide, Chomsky says, an account of the fact that there are systematic relations between sentence structures: (a) can be paraphrased as 'It is easy to please John', whereas (b) cannot be paraphrased analogously. Simple techniques of segmentation and comparison are not sufficient to allow the induction of the correct theory (grammar) of such sentences from raw data.

This led Chomsky, as is well known, to embrace a form of nativism, with the associated claim that humans have a rich set of a priori, genetically based assumptions about the forms that human languages can take, meaning that language learning needs only a relatively few clues to trigger learning, rather than simpler inductive methods over large amounts of data.

Although in compositional distributional semantics our concerns may have more the flavour of engineering than of philosophy, these arguments of Chomsky's nevertheless present a challenge to the hypothesis. Surely one of the basic aspects of the meanings of adjectives like these, and even more centrally of verbs, is their argument structure: if you know what a verb like 'hit' means, you know that it stereotypically involves at least two participants. The 'easy/eager to please' examples suggest that it may not be at all straightforward to learn such information by distributional means alone.

The second connection is with an old philosophical distinction between analytic and synthetic truths. Analytic truths are supposed to be exhibited by sentences that are true simply in virtue of their meaning, whereas synthetic truths are those sentences that are true because of the way the world is; because of the facts. The question of whether there are any analytic truths at all (other than stipulative definitions) is one that has been debated for several thousand years and we will not enter on that debate here (my own take on this can be found in Pulman, 2005).

Suffice it to say that in the event that there are any analytic truths, it is by no means clear that they could be learned by anything like the distributional hypothesis, which is never going to result in more than 'constant conjunction' of meanings rather than any necessary connection. Again an alternative nativist explanation has been advanced for the putative existence of analytic truths, with differing degrees of plausibility (see Katz and Fodor, 1963; note that Fodor has changed his mind on these issues several times). But if the distributional story about the acquisition of word meanings is correct, and there are no analytic truths or analytic connections between word meanings, i.e. if *all* such connections are learned via distributional means, then this seems to open the door to an extreme form of semantic relativism (which arguably leads to all types of relativisim, moral and otherwise): how do I know, now, that you understand words in the same way that I do, since it is vanishingly implausible that our empirical experience or exposure to the data for our distributional learning mechanisms to work on will coincide sufficiently for us to arrive at exactly the same hypotheses.

Fodor and Lepore (1999) discussed many of these issues in the course of a debate with connectionist psychologists (e.g. Smolensky, 1987) over the nature of distributed-meaning representations. They argued that any theory of semantic content (word meaning) should support at least the following properties:

1. assignment of truth or satisfaction conditions to sentences—a description of what the world must be like if the sentence is to be true
2. compositionality
3. translation: 'pre-theoretic intuition has it that meaning is what good translations preserve; a semantic theory should provide a notion of meaning according to which this turns out true'
4. intentional explanation: 'a semantic theory should reconstruct a notion of content that is adequate to the purposes of intentional (e.g. belief/desire) explanation'.

They go on to draw a distinction between three contrasting positions concerning word meanings. They describe these as atomism, molecularism, and holism. These positions are fairly coarsely drawn here but atomism can be characterized as the position that no words are connected to any other words by meaning-constitutive relations, only by beliefs. Fodor appears to have held this view from about 1998 onwards. It is implicitly the position held by most working formal semanticists, in the absence of any other extra theoretical apparatus.

Molecularism is the view that at least some meanings are related to or even reducible to others by meaning-constitutive relations, for example if 'I persuade X that P' then 'X believes that P', and if 'X assassinates Y', then 'X kills Y', and furthermore 'Y is dead'. Almost all linguists subscribe to some form of this theory, although in formal semantics these meaning relations are only accounted for by stipulation in the form of meaning postulates.

The third position is holism, the view that all meanings are related and thus defined by relations to all or at least some other meanings or beliefs: an interconnected web of belief. There is no difference between meaning-constitutive and belief-constitutive relations. There are no analytic truths. This is the view held by philosophers such as Quine and Davidson, and I maintain that it is this view that vector-space models of the type we have been discussing can be seen as an implementation of, if taken literally as the embodiment of a philosophical theory of meaning.

12.8 Experiments in compositional interpretation

In this section I provide a (necessarily selective) survey of recent work that claims to be demonstrating compositional effects in combining word meanings. We know from Schütze and Curran that simple vector-space models (i.e. where the vector counts are just based on raw text) can be used for disambiguation and for rankings of semantic similarity between words. There are a variety of things that can be done to make the vector construction process more sophisticated, too: Curran discusses some of them, and several other researchers have also experimented with using syntactically analysed text to form vectors from, the intuition being that words that are in a syntactic relation with a target word rather than merely being nearby in a text are likely to be better indicators of the meanings of those words.

For example, Padó and Lapata (2007) discuss a particular approach to the inclusion of syntactic information: the bases for their vector-construction algorithm consist of paths of lexicalized dependency relations (e.g. '(lorry-noun)-subject-(carry-verb)-object-(apples-noun)'. Clearly such detailed paths will lead to a rather sparse space, and so they define a mapping function to collapse together paths by ignoring some of their properties. Thus the eventual vector space for a word will represent its meaning in terms of the words it is characteristically grammatically combined with in a dependency relation.

They used the well-known Rubenstein and Goodenough (1965) dataset for tuning various parameters of their dependency-based vector-space model. This is a small dataset consisting of just of 65 pairs of nouns, some nearly synonymous ('gem', 'jewel'), others not ('noon', 'string'), but annotated with human judgements of similarity on a five-point scale, and thus a high-quality gold standard test of meaning similarity. Their best performing set of parameters yielded a Pearson coefficient of around 0.6 between the vector-space model and human judgements, suggesting high agreement. In more complex tasks, the syntax-based model was shown to consistently outperform a syntax-free model. For example, the 'teaching of English as a foreign language' (TOEFL) tests rely on the ability to spot synonyms in a multiple choice setting: a candidate would be asked for a suitable alternative to 'intersection' from the choices presented in (a) to (d):

You will find the office at the main **intersection**.

(a) place (b) crossroads (c) roundabout (d) building

Padó and Lapata's dependency-based vector-space model gets the right answers to such questions right about 73% of the time, a substantially better accuracy level than a model trained on raw words. (It is amusing that non-native speakers usually score around 64% on the same task.)

But can we show that vector-based models of meaning can support compositionality? Can we combine the vectors for individual words to get vectors for phrases or sentences? Of course, many people have experimented with vector representations for sentences or even whole documents which are just the addition of vectors for words. But such a simple composition method completely ignores syntactic structure and treats sentences as 'bags of words'; any permutation of the same words would yield the same overall vector. The real test is whether we can get vectors for sentences such that, for example, 'dogs chase cats' differs from 'cats chase dogs'?

The earliest work that I am familiar with to have attempted to display some genuine compositionality is that of Widdows (2003). He experimented with a word-space model using a method similar to that of Schütze but where the dimensions of the vector space were reduced using 'singular value decomposition'. This technique is intended to filter out uninteresting dimensions of the space, resulting in models that are more compact but which do not lose any important information.

Widdows experimented, within an information retrieval application, with logical operations on word vectors, in particular conjunction, disjunction, and negation. Conjunction '$a \wedge b$' can be approximated via the addition of the vectors for a and b, and disjunction '$a \vee b$' by taking the plane spanned by the a and b vectors. Negation can be represented via the notion of 'orthogonality': two vectors are orthogonal if their cosine distance is 0. The orthogonal subspace of a vector a is the set of vectors orthogonal to it. We can capture the effect of (conjunction and) 'negation' '$a \wedge \neg b$' therefore, by projecting vector a onto the orthogonal subspace of b (for normalized vectors, $a \wedge \neg b = a - (a \cdot b) \times b$).

Widdows used this technique to disambiguate different senses of words: note that he was using word vectors, not context vectors. To illustrate the effectiveness of this method, he compared the words most similar to the vectors for 'suit' (ambiguous between the legal and the clothing senses) and the combination 'suit AND NOT lawsuit', which should filter out the legal sense of the word 'suit': the resulting similarity scores are very encouraging, as now the nearest neighbours for the combination are all unambiguously (with the possible exception of 'silk', at least in British English) clothing-related (see Table 12.2).

While Widdows' suggestion covers one type of compositionality, essentially at the sentence level, Clark and Pulman (2007) attempted to find a way of combining words into phrases in such a way that a finer grain of grammatical structure was taken into account. As remarked earlier, obvious vector composition operations such as addition or multiplication are not fully compositional, since they are completely symmetric and thus under any of these operations a sentence like 'a man bites a dog' will mean the same as 'a dog bites a man'. Grammatical structure is completely ignored.

TABLE 12.2 Widdows' disambiguation of word sense

Suit		Suit NOT	lawsuit
suit	1.000000	pants	0.810573
lawsuit	0.868791	shirt	0.807780
suits	0.807798	jacket	0.795674
plaintiff	0.717156	silk	0.781623
sued	0.706158	dress	0.778841
plaintiffs	0.697506	trousers	0.771312
suing	0.674661	sweater	0.765677
lawsuits	0.664649	wearing	0.764283
etc			

Clark and Pulman tried to overcome this problem, following a suggestion by Smolensky (1990), by treating composition as a tensor product operation ($\otimes$) combining vectors both from a word space and an abstract grammatical relations space. The tensor product operation for vectors (it is usually used with whole vector spaces) amounts to a kind of multiplicative cartesian product:

$$[a, b, c] \otimes [d, e] = [ad, ae, bd, be, cd, ce]$$

Assuming a grammatical relations space can be built somehow, their suggestion was to capture the effect of grammatical structure by producing a composite vector that combined word meaning and grammatical relations obtained via a parse of the sentence:

John drinks strong beer =

John $\otimes$ subj $\otimes$ (drinks $\otimes$ obj $\otimes$ (beer $\otimes$ adj $\otimes$ strong))

This approach solves the problem of avoiding a 'bag-of-words' treatment of sentence structure, since the following vectors will not be identical:

man $\otimes$ subj $\otimes$ bites $\otimes$ dog $\otimes$ obj

dog $\otimes$ subj $\otimes$ bites $\otimes$ man $\otimes$ obj

Unfortunately there is another problem with this approach (for which the solution they originally envisaged proves not to be satisfactory), namely that some extra mechanism has to be found to ensure that dimensions of compositional vectors are comparable. The tensor product operation on vectors has the property that dimension(A⊗B) = dimension(A)×dimension(B), so we can only compare vectors for different sentences if they have isomorphic structures, because otherwise the tensor products representing their structures would have different lengths. So we would be unable to get a measure of semantic similarity between a pair of sentences like, say, 'Hamish is teetotal' and 'Hamish does not drink alcohol', although intuitively they are nearly synonymous even though having different syntactic structures.

In an influential paper Mitchell and Lapata (2008) present a general framework for composition of vectors, which they evaluate empirically on a sentence-similarity task. Their general statement of vector composition is:

$$\mathbf{p} = f(\mathbf{u}, \mathbf{v}, R, K)$$

Here, **p** represents the composition of vectors **u** and **v** (not necessarily in the same space as them), and composition can be made dependent on background knowledge K and/or syntactic properties R. Ignoring R and K for simplicity still gives a range of possibilities for $f(\mathbf{u}, \mathbf{v})$. Assuming all three vectors have the same dimensions we could have 'additive' models:

$$\mathbf{p} = A\mathbf{u} + B\mathbf{v}$$

or 'multiplicative' models (where **C** may be some function of **u** and **v**):

$$\mathbf{p} = C\mathbf{u}\mathbf{v}$$

Many other possibilities also exist, including combining weighting functions with the input vectors.

Mitchell and Lapata's evaluation method is based on a sentence similarity judgement task originally presented in a psycholinguistic context by Kintsch (2001). Kintsch was investigating the phenomenon of disambiguation associated with sentence composition. For example, many verbs like 'run' are ambiguous in isolation ('run' can describe a physical motion of something with legs or wheels, or what happens if you put too much paint on a surface, among other things). But in the context of a particular sentence, the subject of the sentence will 'select' the appropriate sense: in *The horse ran* and *The colour ran* there appears to be no ambiguity. This effect can be tested empirically by asking subjects to give intuitive judgements of meaning similarity: a judgement as to whether one of the two sentences is closer in meaning to a relevant related word or 'landmark' like *gallop* versus *dissolve*. For example, the judgement should be that *The colour ran* is closer in meaning to 'dissolve' than to 'gallop'.

Mitchell and Lapata collected human judgements on (after filtering) 15 verbs by 4 noun combinations with 2 landmarks each, similar to the 'run' example, e.g. 'the shoulders slumped' is judged closer in meaning to the word 'slouch' than to 'decline', whereas 'the shares slumped' is closer to 'decline' than to 'slouch'. An obvious question to ask about this data is whether there really are separate senses for these words, in the way that there clearly are distinct and unrelated senses for a word like 'bank'. It does seem plausible that the two 'senses' of 'run' and 'slump' do in fact share many properties of their meanings ('relatively rapid movement', 'movement downwards', respectively). However, this is an issue which is not easy to resolve and we will put it aside for now.

Mitchell and Lapata's semantic vectors were built on a lemmatized version of the British National Corpus (http://www.natcorp.ox.ac.uk), using a context window of ± 5 words, with 2000 context words (selected by frequency). Rather than raw counts, their vector components were set to the ratio: $\frac{P(context|target)}{P(context)}$. These parameters were tuned, using cosine distance as the similarity measure, on the WordSim353 test set, another small but high quality gold standard semantic similarity dataset consisting of human relatedness judgments (on a scale of 0 to 10) for 353 word pairs (Finkelstein et al., 2002).

Mitchell and Lapata experimented with various ways of combining vectors: addition, weighted addition, multiplication, or a weighted combination of addition and multiplication. They achieved the best results (i.e. closest to human judgements) using *pointwise multiplication*, i.e. $[a, b, c] * [d, e, f] = [ad, be, cf]$, which was ahead of all of the other combinations. However, both the human judgements and the pointwise multiplication scores display a considerable amount of overlap, in that a significant number of cases are judged as both low and high similarity examples (see their Figures 2 and 3), perhaps reflecting the worry expressed earlier about the degree of ambiguity actually exhibited by the verbs chosen. Nevertheless, the finding that multiplicative approaches succeed best at this task is an interesting finding.

Erk and Padó (2008) argue, as we have done, that symmetric methods of combining word vectors will be unable to capture the results of different syntactic combinations of words like 'a horse draws (a cart)' and 'draw a horse'. However, they also argue that a single vector could not capture arbitrarily complex sentences: referring to Mitchell and Lapata's general formula for vector composition they say 'The problem is that the vector **c** can only encode a fixed amount of structural information if its dimensionality is fixed, but there is no upper limit on sentence length, and hence on the amount of structure to be encoded.' It is not clear what the logical force of this argument is, in my view, but they draw the conclusion that a single vector will be *ipso facto* insufficient and go on to offer a method of composition which operates on multiple vectors, and which also results in multiple vectors.

In more detail, Erk and Padó learn a 'structured vector' space consisting of triples, for each content word **w**, of form $\mathbf{w} = \langle v, R, R^{-1} \rangle$ where v is a word-space vector of the usual kind; R is a set of vectors representing possible occupants of grammatical or dependency relations of the type 'subject', 'object', etc.; R^{-1} is a set of vectors representing the inverse of relations in R, such as 'is-subject-of', 'is-modified-by'. R are the relations that the word selects, and R^{-1} are relations that the word is selected by: their intuition is that when we have a combination like 'catch the ball' (compare 'catch a cold', 'attend a ball') that the meaning of 'catch' will be constrained by the presence of 'ball' to be in the 'things that can be done with a ball' category, while the meaning of 'ball' will be constrained to be in the 'things you can catch' category.

Word vectors are formed in exactly the same way as for Mitchell and Lapata. The relations in R and R^{-1} are built from a dependency-parsed version of the British National Corpus and are represented as the weighted centroid of the individual words observed to stand in those relations to **w**. So, for a word like 'catch' we might have vector tuples like:

⟨catch, {has-subj, has-obj, ...},{is-comp-of, is-modified-by, ...}⟩

where 'has-subj' would be a vector formed from observed subjects of 'catch' and 'is-comp-of' would be a vector formed from verbs observed to take 'catch' as a complement. Thus R captures the selectional preferences for each grammatical or dependency role that the word governs, and R^{-1} the inverse: the words that this word is among the selectional preferences of.

Composition of word meanings proceeds as follows. Let the representation of word **a** be $⟨v_a, R_a, R_a^{-1}⟩$, and word **b** be $⟨v_b, R_b, R_b^{-1}⟩$. Assume that the relation $r \in R$ is the grammatical or dependency relation holding between **a** and **b**. Now the result of the combination **R(a, b)** is defined as a pair ⟨a', b'⟩, where a' = $⟨v_a \otimes R_b^{-1}(r), R_a-r, R_a^{-1}⟩$, and b' = $⟨v_b \otimes R_a(r), R_b, R_b^{-1}-r⟩$.

So **a** is combined (by whatever operation on vectors is represented by ⊗) with **b**'s inverse of **r**, and **b** is combined with **a**'s preferences for **r**. The **r** relation is removed from the output tuples as it has now been 'used'.

To illustrate, when the following words are combined in the object relation, 'catch the ball', the input structured vectors will be:

⟨catch, {catch-has-obj, catch-has-subj, ... }, {catch-is-comp-of, ... }⟩

⟨ball, {ball-has-mod, ... }, {ball-is-obj-of, ball-is-subj-of, ... }⟩

The result of the composition operation is to replace the original meanings by a new pair of structured vectors:

'catch in the context of ball':

⟨catch ⊗ ball-is-obj-of, {catch-has-subj, ... }{catch-is-comp-of, ... }⟩

'ball in the context of catch':

⟨ball ⊗ catch-has-obj, {ball-has-mod, ... }, {ball-is-subj-of, ... }⟩

In principle, they say, this process could continue recursively across a whole dependency or parse tree, so that, for example, the subject of 'catch' would indirectly be influenced by the properties of 'ball'. However, this process would not yield a vector for a whole sentence; rather, the result is a 'context-adapted vector' for each word in the sentence.

Erk and Pado replicate Mitchell and Lapata's experiment and demonstrate that their syntactically influenced vectors give slightly—but not in fact statistically significantly—better results. They also consider a lexical substitution task: rank the appropriate sense of 'work' in:

- By asking people who **work** there, I have since determined that he didn't. (be employed vs labour…)
- Remember how hard your ancestors worked. (toil vs labour….)

As with Mitchell and Lapata, they tried different vector combination operations, but found that component-wise multiplication gave the best results.

12.9 An empirical test of distributional compositionality

In their paper 'Nouns are vectors, adjectives are matrices' Baroni and Zamparelli (2010) carry out a series of careful experiments, which can be thought of as an empirical test of the hypothesis that distributional models can be compositional, at least up to the 'small phrase' level we discussed earlier. Their idea is that with a large enough corpus, you can learn a distributional model for frequently occurring Adj+Noun combinations, as well as for Adj and Noun separately. We can then test empirically whether the vectors for the Adj+Noun combination are any kind of compositional function of the Adj and Noun vectors.

Baroni and Zamparelli used a 2.83 billion word corpus, POS-tagged and lemmatized. From this they selected the 8000 most frequent nouns, and the 4000 most frequent adjectives (the 'core vocabulary'). They selected their adjective (A), noun (N), and adjective+noun (AN) test sets as follows. First 36 A of different types—descriptive, modal, etc.—were selected by hand. Then 1420 N occurring more than 300 times in a sample corpus consisting of the first 100 million tokens of the original corpus were combined with these to give 26,440 AN combinations, all attested in the sample corpus. The core vocabulary was extended with the 16 A and 43 N that figure in the AN combinations but which are not in the core vocabulary, and with 2500 more ANs randomly chosen from the sample corpus, meeting the condition that the N component is from the same list used for the test set ANs and with an A that occurred at least 5000 times in the sample corpus. This is referred to as the 'extended vocabulary'.

The dimensions of their vectors are the 10,000 N, V, and A that cooccur with the largest number of items in the core vocabulary. They collect counts from the larger corpus of all sentence internal co-occurrences of the dimension words with the items in the extended vocabulary. The raw counts are transformed into 'local mutual information' scores.

To reduce the dimensionality of the resulting large matrix, they use SVD on the 12,000 core vocabulary $\times$ 10,000 dimension matrix, reducing it to 12,000 $\times$ 300. The other row vectors of the larger matrix were then projected onto this reduced space, resulting in a 40,999×300 matrix representing the 8000 N, 4000 A, and 29,000 ANs. The reduced matrix performed slightly better than the full matrix on some standard similarity-judgement datasets.

To see whether the compositionality hypothesis holds for this small construction, Baroni and Zamparelli treat each adjective as a function from N meanings to AN meanings, much as would be done in a formal semantic treatment. Note that each adjective is a *different* function, in distinction to the similar work carried out by Guevara (2010), who learned a single function that applied to adjectives and nouns to produce this syntactic combination of them. This latter is equivalent, in a formal semantic framework, to identifying a semantic combinator with the syntax rule combining an A and N, whereas Baroni and Zamparelli's approach allows each A to potentially have a different effect when combining with an N.

Baroni and Zamparelli tested several different models:

- *Addition*: A + N = A vector + N vector;
- *Multiplication*: A+N = pointwise multiplication of A and N vectors
- *Adjective linear map*, where the A is a matrix, not a vector, and A+N = A*N.

The best—and the most interesting—results involved the adjective linear map matrix version. Recall that when we multiply a column vector by a matrix we take the dot product of the vector with each row of the matrix:

$$\begin{bmatrix} A & B & C \\ D & E & F \\ G & H & I \end{bmatrix} \times \begin{bmatrix} P \\ Q \\ R \end{bmatrix} = \begin{bmatrix} AP + BQ + CR \\ DP + EQ + FR \\ GP + HQ + IR \end{bmatrix}$$

Baroni and Zamparelli have the vector for the N and the vector for the A+N, and they solve (by partial least squares regression), for each A, the problem:

$$\begin{bmatrix} ? & ? & ? \\ ? & ? & ? \\ ? & ? & ? \end{bmatrix} \times \begin{bmatrix} N1 \\ N2 \\ N3 \end{bmatrix} = \begin{bmatrix} AN1 \\ AN2 \\ AN3 \end{bmatrix}$$

The dimensions of the N and AN vectors determine those of the matrix: if they are of dimension x, the matrix must be $x \times x$. The effect of each A matrix is to operate on different aspects of the N vector: more specifically, the j weights in the i-th row of the matrix predict the values of the i-th dimension of the A+N vector, as a linear combination of the j dimensions of the component noun.

The test AN vectors are generated in a 'leave-one-out' manner; that is, they are produced from an A whose matrix was learned from AN' and N' vectors (determined by counts) where N' is any noun other than the N in question in the training data.

Baroni and Zamparelli test each of the 26,440 generated AN combinations by measuring cosine distance between it and all 40,999 items in the extended

TABLE 12.3 Baroni and Zamparelli's results

	25%	50%	75%
Matrix	17	170	1000+
Addition	27	252	1000+
Multiplication	279	1000+	1000+

vocabulary. The rank of the 'correct' (i.e. observed) AN combination is calculated (see Table 12.3). The ranks (i.e. the distance between the generated and the observed combination) are lowest for the matrix multiplication, with the fourth (i.e. top) quartile having an average rank of 17, the median 170, and the first quartile all below 1000 (ranks below 1000 were not calculated in detail).

As Baroni and Zamparelli point out, while perhaps encouraging, these results are far from perfect 'for 27% alm-predicted ANs, the observed vector is not even in the top 1k neighbour set!'. This is rather a disappointing result, in fact: as an empirical test of whether compositionality can be made to work for a reasonably straightforward example where we have the data for a combination as well as the components (we would not be able to do this for more complex examples without an even more enormous corpus), Baroni and Zamparelli argue that data sparseness might in fact already be affecting their results, as there is a inverse correlation between the frequency of an observed AN and the distance between the observed and the generated versions of the AN. They suggest that since the generated AN is based on an A matrix learned from many different examples, it might actually be 'better' than the AN vector based on corpus counts.

This may well be so. However, this claim raises the question of exactly how big a corpus one would need to have to be sure that such effects could be controlled for. It is worth pointing out that a corpus size of 2.83 billion is already thousands of times as big as the number of words it is estimated a 10-year-old person would have been exposed to (Moore, 2003), and many hundreds of times larger than any person will hear or read in a lifetime.

12.10 Compositionality vs disambiguation

There are many other recent attempts to display some notion of compositionality with distributional models; too many to summarize here. Our own group's work is described in the accompanying chapter by Stephen Clark. Clarke (2012) and Rudolph and Giesbrecht (2010) describe sophisticated compositional methods, but without empirical assessment of their accuracy. Thater, Fürstenau and Pinkal (2010, 2011) test an extended version of the approach taken by Erk and Pado, including grammatical-dependency contexts, and get better results on various lexical substitution and disambiguation tasks.

What all of the empirically tested attempts have in common—including some of our own—is that (a) all the word vectors involved are 'first-order' in that all the different senses of a word are in the same vector (unlike Schütze's 'second-order' vectors), and (b) they all get the best results using some form of multiplication. (An exception to (a) is the approach of Reisinger and Mooney (2010), who model word meanings with multiple vectors, one for each sense, and get good results on semantic similarity tasks.)

It is not difficult to see why pointwise multiplication is the vector composition method yielding the best results. The tasks usually amount to a three- (or more) way similarity judgement, which depends on a disambiguation. And the effect of pointwise multiplication is precisely to disambiguate, as we can illustrate with the following artificial example. Assume that:

$$similarity(horse, gallop) > similarity(horse, dissolve)$$

$$similarity(colour, dissolve) > similarity(colour, gallop)$$

The first-order vector for 'run' will have components for collocation with 'horse'-related words and 'colour'-related words. The vectors for 'horse' will have lower values for the context words associated with 'colour' and vice-versa. So when we multiply 'horse' by 'run' the effect will be to reduce the values of the 'colour' components of the vector (see Table 12.4; in real vectors there will be many zeroes, which will totally eliminate the component).

When we combine words into phrases, at least two things happen:

- irrelevant meanings of ambiguous words are largely filtered out
- we get a meaning representing the combination.

These are separate phenomena that much of the current literature fails to distinguish. Most current vector-space models only achieve the first, I maintain. They are doing *disambiguation, not composition*. While it is clear that this kind of diambiguation can be a side-effect of the compositional process, it is questionable whether compositionality is either necessary or sufficient for it. I would argue that this kind

TABLE 12.4 Effects of pointwise multiplication

Context	pasture	feed	gallop	ride	spectrum	bright	durable
colour	1	2	1	1	5	6	8
horse	6	8	7	8	1	3	1
run	5	5	5	5	5	5	5
colour×run	5	10	5	5	25	30	40
horse×run	30	40	35	40	5	15	5

of disambiguation has really little to do with compositionality at all, since such contextual disambiguation effects occur even when the relevant components are in separate sentences:

- John sat on the bank. He was fishing.

And as with other kinds of disambiguation, contextual manipulation can usually recover the 'missing' sense:

- Mary painted a nice watercolour of a horse near a tree. But she spilled water on it and the horse began to run into the tree.

Or even:

- The town was completely flooded. I saw John row past the church and stand on the bank. He was fishing.

12.11 Conclusion: a new program for compositional distributional semantics

In order to make real progress in compositional distributional semantics, I maintain that we need to develop different measures that will separate disambiguation from true composition. In fact, I suggest that we should factor out disambiguation altogether, and use Schütze-type second-order vectors to represent individual word senses. This will involve the use of clustering methods able to produce a 'natural' number of clusters of context vectors corresponding to senses, rather than fixing on some predetermined number.

Once we have constructed vectors that are better approximations of individual word senses, we need some evaluation tasks that are a genuine test of compositionality and which do not confound this with disambiguation. Some possible tasks might include definitions, syntactic variation, and paraphrases.

12.11.1 Definitions

Dictionary definitions produce a phrase or a sentence to illustrate the meaning of a single word. Some examples might be 'a carnivore is something that eats meat', or 'a pudding is a cooked sweet dish served after the main course of a meal', and so on. Given such definitions it would be easy to construct semantic similarity tests like:

- Is 'John is a carnivore' closer to 'John eats meat' than 'John eats vegetables'?
- Is 'a coincidence' closer to 'two events that happen by chance at the same time' than to 'something that happens unexpectedly'? ('i.e. 'a surprise').

This is of course a very challenging task, but a true test of compositionality, because here we may be comparing single words with phrases, or with whole sentences.

12.11.2 Syntactic variation

As have seen earlier, syntax is intimately tied to compositionality. The same words but put together in different syntactic combinations result in sentences with different meanings. We could construct test sets based on syntactic variants, and turn them into semantic similarity tests again:

- Is 'the cat chased the mouse' closer to 'the mouse was chased by the cat' than to 'the mouse chased the cat'?

—and so on for other similar constructions. It would need some care in specifying the details of this task, because you could probably recognize pairs of syntactic variants that preserve meaning using a good parser with no semantics at all. But if the problem could be posed in such a way as to preclude purely syntactic solutions it would be a way of showing that our compositional methods followed the syntactic structure of a sentence in the way that formal semantics compositionality does.

12.11.3 Paraphrases

There exist various corpora of human-generated paraphrases such as those described in Cohn, Callison-Burch and Lapata (2008). Here the task might be to compare the composed vector representation of a sentence, and compare it to those of all others in the corpus; its paraphrases should be significantly closer than non-paraphrases. If the paraphrases are of good quality this would test the ability of a compositional model to compare differing syntactic structures as well as different lexical choices, simultaneously as well as individually.

Until we can demonstrate good performance on a task such as this, I think the jury is still out on whether it is even possible to build truly compositional distributional models of word and phrase meaning.

Acknowledgements

Some of the material in this chapter was first presented at a meeting in honour of Karen Spärck Jones at Downing College Cambridge, 2005. I am grateful to the participants on that occasion, particularly Keith van Rijsbergen, for their comments. The work has also benefited from discussions with Stephen Clark, Edward Grefenstette, Mehrnoosh Sadrzadeh, Dominic Widdows, Manfred Pinkal, Stefan Thater, Marco Baroni, Roberto Zamparelli, and Raffaella Bernardi.

CHAPTER 13

Type-Driven Syntax and Semantics for Composing Meaning Vectors

STEPHEN CLARK
(University of Cambridge)

13.1 Introduction

This chapter describes a recent development in computational linguistics, in which distributional, vector-based models of meaning—which have been successfully applied to word meanings—have been given a compositional treatment allowing the creation of vectors for sentence meanings. More specifically, this chapter presents the theoretical framework of Coecke, Sadrzadeh and Clark (2010), which has been implemented in Grefenstette *et al.* (2011) and Grefenstette and Sadrzadeh (2011), in a form designed to be accessible to computational linguists not familiar with the mathematics of category theory on which the framework is based.

    The previous chapter has described how distributional approaches to lexical semantics can be used to build vectors that represent the meanings of words, and how those vectors can be used to calculate semantic similarity between word meanings. It also describes the problem of creating a *compositional* model within the vector-based framework, i.e. developing a procedure for taking the vectors for two words (or phrases) and combining them to form a vector for the larger phrase made up of those words. This chapter offers an accessible presentation of a recent solution to the compositionality problem.

    Another way to consider the problem is that we would like a procedure that, given a sentence, and a vector for each word in the sentence, produces a vector for the whole sentence. Why might such a procedure be desirable? The first reason is that considering the problem of compositionality in natural language from a geometric viewpoint may provide an interesting new perspective on the problem. Traditionally, compositional methods in natural language semantics, building on the foundational work of Montague (Dowty, Wall and Peters, 1981), have assumed the meanings of words to be given, and effectively atomic, without any internal structure. Once we assume that the meanings of words are vectors, with significant internal structure, then the problem of how to compose them is seen in a new light.

A second, more practical reason is that applications in natural language processing (NLP) would benefit from a framework in which the meanings of whole sentences can be easily compared. For example, suppose that a sophisticated search engine is issued the following query:

- *Find all car showrooms with sales on for Ford cars.*

Suppose further that a web page has the heading:

- *Cheap Fords available at the car salesroom.*

Knowing that the above two sentences are similar in meaning would be of huge benefit to the search engine. Moreover, if the two sentence meanings could be represented in the same vector space, then comparing meanings for similarity is easy: simply use the cosine measure between the sentence vectors, as is standard practice for word vectors.

One counter-argument to the above example might be that compositionality is not required in this case in order to determine sentence similarity, only similarity at the word level. For this example that may be true, but it is uncontroversial that sentence meaning is mediated by syntactic structure. To take another search engine example, the query *A man killed his dog*, entered into Google on January 5, 2012, from the University of Cambridge Computer Laboratory, returned a top-ranked page with the snippet *Dog shoots man* (as opposed to *Man shoots dog*), and the third-ranked page had the snippet *The Man who Killed His Friend for Eating his Dog After it was Killed . . .*[1] Of course the order of words matters when it comes to sentence meaning.

Figure 13.1 shows the intuition behind comparing sentence vectors. The previous chapter explained how vectors for the words *cat* and *dog* could be created that are relatively close in 'noun space' (the vector space at the top in the figure).[2] The framework described in this chapter will provide a mechanism for creating vectors for sentences, based on the vectors for the words, so that *man killed dog* and *man murdered cat* will be relatively close in the 'sentence space' (at the bottom of Figure 13.1), but crucially *man killed by dog* will be located in another part of the space (since in the latter case it is the animal killing the man, rather than vice versa). Note that, in the sentence space in the figure, no commitment has been made

---

[1] Thanks to Mehrnoosh Sadrzadeh for this example.

[2] In practice the noun space would be many orders of magnitude larger than the three-dimensional vector space in the figure, which has only three context words.

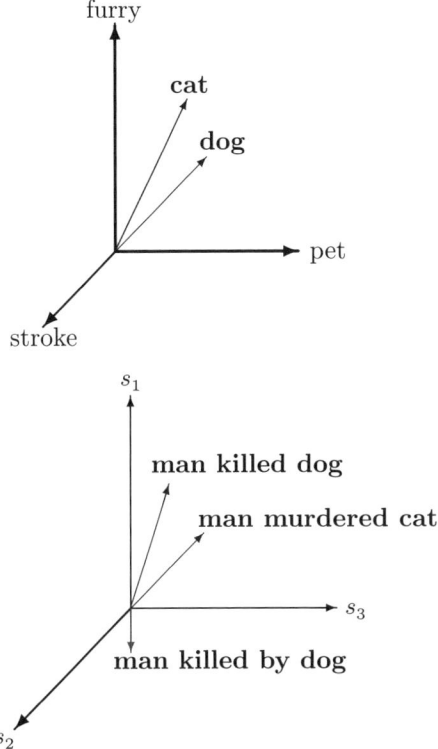

Fig 13.1 Example vector spaces for noun and sentence meanings.

regarding the basis vectors of the sentence space (s_1, s_2, and s_3 are not sentences, but unspecified basis vectors). In fact, the question of what the basis vectors of the sentence space should be is not answered by the compositional framework, but is left to the model developer to answer. The mathematical framework simply provides a compositional device for combining vectors, assuming the sentence space is given. Sections 13.4 and 13.5 give some examples of possible sentence spaces.

A key idea underlying the vector-based compositional framework is that *syntax drives the compositional process*, in much the same way that it does in Montague semantics (see Dowty, Wall and Peters, 1981, and previous chapter). Another key idea borrowed from formal semantics is that the syntactic and semantic descriptions will be *type-driven*, reflecting the fact that many word types in natural language, such as verbs and adjectives, have a relation, or functional, role. In fact, the syntactic formalism assumed here will be a variant of categorial grammar, which is the grammatical framework also used by Montague.

The next section describes pregroup grammars, which provide the syntactic formalism used in Coecke, Sadrzadeh and Clark (2010). However, it should be noted that the use of pregroup grammars is essentially a mathematical expedient (in a way briefly explained in the next section), and it is likely that other type-driven formalisms, for example combinatory categorial grammar (Steedman, 2000), can be accommodated in the compositional framework.

Section 13.3 shows how the use of syntactic functional types leads naturally to the use of tensor products for the meanings of words such as verbs and adjectives. Section 13.4 then provides an example sentence space, and provides some intuition for how to compose a tensor product with one of its 'arguments' (effectively providing the analogue of function application in the semantic vector space). Finally, Section 13.5 describes a sentence space that has been implemented in practice by Grefenstette and Sadrzadeh (2011).

13.2 Syntactic types and pregroup grammars

The key idea in any form of categorial grammar is that all grammatical constituents correspond to a syntactic type, which identifies a constituent as either a function, from one type to another, or as an argument (Steedman and Baldridge, 2011). Combinatory categorial grammar (CCG; Steedman, 2000), following the original work of Lambek (1958), uses *slash operators* to indicate the directionality of arguments. For example, the syntactic type (or *category*) for a transitive verb such as *likes* is as follows:

$$likes := (S \backslash NP)/NP$$

The way to read this category is that *likes* is the sort of verb that first requires an *NP* argument to its right (note the outermost slash operator pointing to the right), resulting in a category that requires an *NP* argument to its left (note the innermost slash operator pointing to the left), finally resulting in a sentence (*S*). Categories with slashes are known as *complex* categories; those without slashes, such as *S* and *NP*, are known as *basic*, or *atomic*, categories.

A categorial grammar lexicon is a mapping from words onto sets of possible syntactic categories for each word. In addition to the lexicon, there is a small number of rules that combine the categories together. In classical categorial grammar, there are only two rules, forward (>) and backward (<) application:

$$X/Y \ Y \Rightarrow X \quad (>) \tag{13.1}$$

$$Y \ X\backslash Y \Rightarrow X \quad (<) \tag{13.2}$$

These rules are technically rule schemata, in which the X and Y variables can be replaced with any category. Figure 13.2, taken from Clark and Curran (2007), gives a derivation using these rules for an example newspaper sentence.

Classical categorial grammar is context-free in terms of its generative power. CCG adds a number of additional rules, such as function composition and

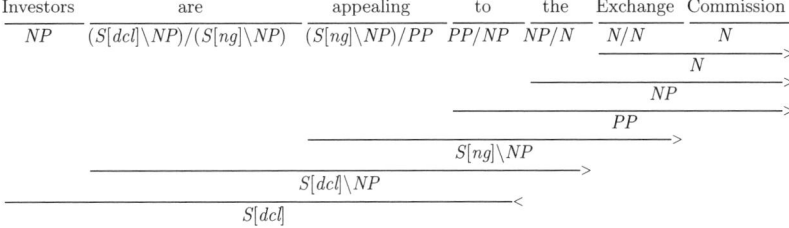

Fig 13.2 Example derivation using forward and backward application; grammatical features on the S indicate the type of the sentence, such as declarative [dcl].

type-raising, which can increase the power of the grammar to so-called mildly context-sensitive (Weir, 1988). This allows the grammar to deal with examples of crossing dependencies seen in Dutch and Swiss German (Shieber, 1985; Steedman, 2000) but whilst still retaining some computationally attractive properties such as polynomial time parsing (Vijay-Shanker and Weir, 1993).

The mathematical move in pregroup grammars, a recent incarnation of categorial grammar due to Lambek (2008), is to replace the slash operators with different kinds of categories (*adjoint* categories), and to adopt a more algebraic, rather than logical, perspective compared with the original work (Lambek, 1958). The category for a transitive verb now looks as follows:

$$likes := NP^r \cdot S \cdot NP^l$$

The first difference to notice is notational, in that the order of the categories is different: in CCG the arguments are ordered from the right in the order in which they are cancelled; pregroups use the type-logical ordering (Moortgat, 1997) in which left arguments appear to the left of the result, and right arguments appear to the right.

The key difference is that a left argument is now represented as a *right adjoint*: NP^r, and a right argument is represented as a *left adjoint*: NP^l; so the adjoint categories have effectively replaced the slash operators in traditional categorial grammar. One potential source of confusion is that arguments to the left are right adjoints, and arguments to the right are left adjoints. The reason is that the 'cancellation rules' of pregroups state that:

$$X \cdot X^r \to 1 \qquad (13.3)$$

$$X^l \cdot X \to 1 \qquad (13.4)$$

That is, any category X cancels with its right adjoint to the right, and cancels with its left adjoint to the left. Figure 13.3 gives the pregroup derivation for the earlier

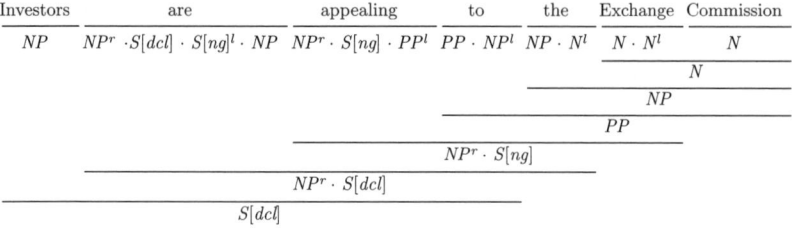

Fig 13.3 Example pregroup derivation.

example newspaper sentence, using the CCG lexical categories translated into pregroup types.[3]

Mathematically we can be more precise about the pregroup cancellation rules:

Definition 13.1 *A pregroup is a partially ordered monoid[4] in which each object of the monoid has a left and right adjoint subject to the cancellation rules above, where $\rightarrow$ is the partial order, $\cdot$ is the monoid operation, and 1 is the unit object of the monoid.*

In the linguistic setting, the objects of the monoid are the syntactic types; the associative monoid operation ($\cdot$) is string concatenation; the identity element is the empty string; and the partial order ($\rightarrow$) encodes the derivation relation. Lambek (2008) has many examples of linguistic derivations, including a demonstration of how iterated adjoints, e.g. NP^{ll}, can deal with interesting syntactic phenomena such as object extraction.

It is an open question whether pregroups can provide an adequate description of natural languages. Pregroup grammars are context-free (Buszkowski and Moroz, 2006) and are therefore generally thought to be too weak to provide a full description of the structural properties of natural language. However, many of the combinatory rules in CCG are sound in pregroups, including forward and backward application, forward and backward composition, and forward and backward type-raising. In fact, the author has recently translated CCGbank (Hockenmaier and Steedman, 2007), a large corpus of English newspaper sentences each with a CCG derivation, into pregroup derivations, with the one caveat that the backward-crossed composition rule, which is frequently used in CCGbank, is unsound in pregroups, meaning that a workaround is required for that rule.[5]

There are two key ideas from this section that will carry over to the distributional, vector-based semantics. First, linguistic constituents are represented using

---

[3] Pregroup derivations are usually represented as 'reduction diagrams' in which cancelling types are connected with a link, similar to dependency representations in computational linguistics. Here we show a categorial grammar-style derivation.

[4] A monoid is a set, together with an associative binary operation, where one of the members of the set is an identity element with respect to the operation.

[5] English is a relatively simple case because of the lack of crossing dependencies; other languages may not be so amenable to a wide-coverage pregroup treatment.

syntactic types, many of which are functional, or relational, in nature. Second, there is a mechanism in the pregroup grammar for combining a functional type with an argument, using the adjoint operators and the partial order (effectively encoding cancellation rules). Hence, there are two key questions for the semantic analysis:

1. For each syntactic type, what is the corresponding semantic type in the world of vector spaces?
2. Once we have the semantic types represented as vectors, how can the vectors be combined to encode a 'cancellation rule' in the semantics?

Before moving to the vector-based semantics, a comment on category theory is in order. Category theory (Lawvere and Schanuel, 1997) is an abstract branch of mathematics that is heavily used in Coecke, Sadrzadeh and Clark (2010). Briefly, a pregroup grammar can be seen as an instance of a so-called compact closed category, in which the objects of the category are the syntactic types, the arrows of the category are provided by the partial order, and the tensor of the compact closed category is string concatenation (the monoidal operator). Why is this useful? It is because vector spaces can also be seen as an instance of a compact closed category, with an analogous structure to pregroup grammars: the objects of the category are the vector spaces, the arrows of the category are provided by linear maps, and the tensor of the compact closed category is the tensor product. Crucially the compact closed structure provides a mechanism for combining objects together in a compositional fashion. We have already seen an instance of this mechanism in the pregroup cancellation rules; the same mechanism (at an abstract level) will provide the recipe for combining meaning vectors. Section 13.4 will motivate the recipe from an intuitive perspective; readers are referred to Coecke, Sadrzadeh and Clark (2010) for the mathematical details.

13.3 Semantic types and tensor products

The question we will answer in this section is: what are the semantic types corresponding to the syntactic types of the pregroup grammar? We will use the type of a transitive verb in English as an example, although in principle the same mechanism can be applied to all syntactic types.

The syntactic type for a transitive verb in English, e.g. *likes*, is as follows:

$$likes := NP^r \cdot S \cdot NP^l$$

Let us assume that noun phrases live in the vector space **N** and that sentences live in the vector space **S**. We have methods available for building the noun space, **N**, detailed in the previous chapter; how to represent **S** is a key question for this whole chapter; for now we simply assume that there is such a space.

Following the form of the syntactic type above, the semantic type of a transitive verb—i.e. the vector space containing the vectors for transitive verbs such as *likes*—is as follows:

$$\overrightarrow{likes} \in \mathbf{N} \cdot \mathbf{S} \cdot \mathbf{N}$$

Now the question becomes what should the monoidal operator (·) be in the vector-space case? As briefly described at the end of the previous section, Coecke, Sadrzadeh and Clark (2010) use category theory to motivate the use of the tensor product as the monoidal operator that binds the individual vector spaces together:

$$\overrightarrow{likes} \in \mathbf{N} \otimes \mathbf{S} \otimes \mathbf{N}$$

Figure 13.4 shows two vector spaces being combined together using the tensor product. The vector space on the far left has two basis vectors, *a* and *b*, and is combined with a vector space also with two basis vectors, *c* and *d*. The resulting tensor product space has $2 \times 2 = 4$ basis vectors, given by the cartesian product of the basis vectors of the individual spaces $\{a, b\} \times \{c, d\}$. Hence basis vectors in the tensor product space are pairs of basis vectors from the individual spaces; if three vector spaces are combined, the resulting tensor product space has basis vectors consisting of triples; and so on.

One feature of the tensor product space is that the number of dimensions grows exponentially with the number of spaces being combined. For example, suppose that we have a noun space with 10,000 dimensions and a sentence space also with 10,000 dimensions; then the tensor product space $\mathbf{N} \otimes \mathbf{S} \otimes \mathbf{N}$ will have $10,000^3$ dimensions. We do not expect this to be a problem in practice, since the size of the largest tensor product space will be determined by the highest arity of a verb, which is unlikely to be higher than, say, four in English.

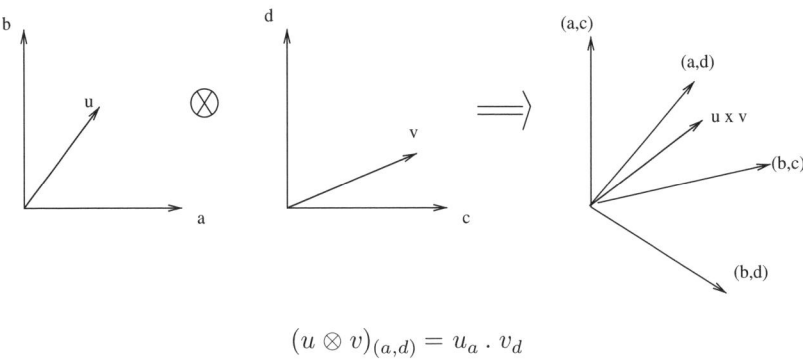

$$(u \otimes v)_{(a,d)} = u_a \cdot v_d$$

Fig 13.4 The tensor product of two vector spaces; $(u \otimes v)_{(a,d)}$ is the coefficient of $(u \otimes v)$ on the (a, d) basis vector.

Figure 13.4 also shows two individual vectors, u and v, being combined $(u \otimes v)$.[6] The expression at the bottom of the figure gives the coefficient for one of the basis vectors in the tensor product space (a, d) for the vector $u \otimes v$. The recipe is simple: take the coefficient of u on the basis vector corresponding to the first element of the pair (coefficient denoted u_a), and multiply it by the coefficient of v on the basis vector corresponding to the second element of the pair (coefficient denoted v_d). Combining vectors in this way results in what is called a *simple* or *pure* vector in the tensor product space.

One of the interesting properties of the tensor product space is that it is much larger than the set of pure vectors; i.e. there are vectors in the tensor product space in Figure 13.4 that cannot be obtained by combining vectors u and v in the manner described above. It is this property of tensor products which allows the representation of *entanglement* in quantum mechanics (Nielsen and Chuang, 2000), and leads to *entangled* vectors in the tensor product space.

An individual vector for a transitive verb can now be written as follows (Coecke, Sadrzadeh and Clark, 2010):

$$\Psi = \sum_{ijk} C_{ijk} (\vec{n_i} \otimes \vec{s_j} \otimes \vec{n_k}) \in \mathbf{N} \otimes \mathbf{S} \otimes \mathbf{N} \qquad (13.5)$$

Here n_i and n_k are basis vectors in the noun space $\mathbf{N}$; s_j is a basis vector in the sentence space $\mathbf{S}$; $\vec{n_i} \otimes \vec{s_j} \otimes \vec{n_k}$ is alternative notation for the basis vector in the tensor product space (i.e. the triple $\langle \vec{n_i}, \vec{s_j}, \vec{n_k} \rangle$), and C_{ijk} is the coefficient for that basis vector.

The intuition we would like to convey is that the vector for the verb is relational, or functional, in nature, as it is in the syntactic case.[7] Informally, the expression in eqn (13.5) for the verb vector can be read as follows:

The vector for a transitive verb can be thought of as a function, which, given a particular basis vector from the subject, n_i, and a particular basis vector from the object, n_k, returns a value C_{ijk} for each basis vector s_j in the sentence space.

The key idea behind the use of the tensor product is that it captures the *interaction* of the subject and object in the case of a transitive verb. The fact that the tensor product effectively retains all the information from the combining spaces—which is why the size of the tensor space grows so quickly—is what allows this interaction to be captured. The final part of Section 13.5 presses this point further by considering a simpler, but conceptually less effective, alternative to the tensor product: the direct sum of vector spaces.

---

[6] The combination of two individual vectors in this way is called the Kronecker product.
[7] A similar intuition lies behind the use of matrices to represent the meanings of adjectives in Baroni and Zamparelli (2010).

We have now answered one of the key questions for this chapter: what is the semantic type corresponding to a particular syntactic type? The next section answers the remaining question: how can the transitive verb in eqn (13.5) be combined with instances of a subject and object to give a vector in the sentence space?

13.4 Composition for an example sentence space

In this section we will use an example sentence space which can be thought of as a 'plausibility space'. Note that this is a fictitious example, in that no such space has been built and no suggestions will be made for how it might be built. However, it is a useful example because the sentence space is small, with only two dimensions, and it is conceptually simple and easy to understand. The next section will describe a more complex sentence space that has been implemented. Figure 13.5 gives two example vectors in the plausibility space, which has basis vectors corresponding to True and False (which can also be thought of as 'highly plausible' and 'not at all plausible'). The sentence *dog chases cat* is considered highly plausible, since it is close to the True basis vector, whereas *apple chases orange* is considered highly implausible, since it is close to the False basis vector.

For the rest of this section we will use the example sentence *dog chases cat*, and show how a vector can be built for this sentence in the plausibility space, assuming vectors for both nouns and the transitive verb already exist. The vectors assumed for the nouns are given in Table 13.1, together with example vectors for the nouns *apple* and *orange*. Note that, again, these are fictitious counts in the table, assumed to have been obtained from analysing a corpus and using some appropriate weighting procedure (Curran, 2004).

The compositional framework is agnostic towards the particular noun vectors, in that it does not matter how those vectors are built (e.g. using a simple window-based method, a dependency-based method, or a dimensionality reduction technique such as latent semantic analysis (LSA; Deerwester *et al.*, 1990). However, for explanatory purposes it will be useful to think of the basis vectors of the noun

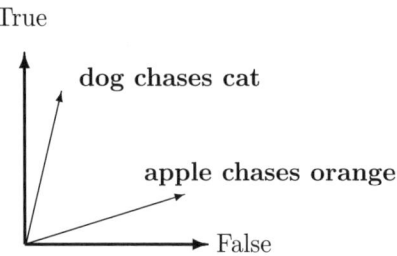

Fig 13.5 An example 'plausibility space' for sentences.

TABLE 13.1 Example noun vectors in **N**

	fluffy	run	fast	aggressive	tasty	buy	juice	fruit
$\overrightarrow{dog}$	0.8	0.8	0.7	0.6	0.1	0.5	0.0	0.0
$\overrightarrow{cat}$	0.9	0.8	0.6	0.3	0.0	0.5	0.0	0.0
$\overrightarrow{apple}$	0.0	0.0	0.0	0.0	0.9	0.9	0.8	1.0
$\overrightarrow{orange}$	0.0	0.0	0.0	0.0	1.0	0.9	1.0	1.0

TABLE 13.2 Example transitive verb vectors in $\mathbf{N} \otimes \mathbf{S} \otimes \mathbf{N}$

	⟨fluffy,T,fluffy⟩	⟨fluffy,F,fluffy⟩	⟨fluffy,T,fast⟩	⟨fluffy,F,fast⟩	⟨fluffy,T,juice⟩	⟨fluffy,F,juice⟩	⟨tasty,T,juice⟩	...
$\overrightarrow{chases}$	0.8	0.2	0.75	0.25	0.2	0.8	0.1	
$\overrightarrow{eats}$	0.7	0.3	0.6	0.4	0.9	0.1	0.1	

space as corresponding to properties of the noun, obtained using the output of a dependency parser (Curran, 2004). For example, the count for *dog* corresponding to the basis vector *fluffy* is assumed to be some weighted, normalized count of the number of times the adjective *fluffy* has modified the noun *dog* in some corpus; intuitively this basis vector has received a high count for *dog* because dogs are generally fluffy. Similarly, the basis vector *buy* corresponds to the object position of the verb *buy*, and *apple* has a high count for this basis vector because apples are the sorts of things that are bought.

Table 13.2 gives example vectors for the verbs *chases* and *eats*. Note that, since transitive verbs live in $\mathbf{N} \otimes \mathbf{S} \otimes \mathbf{N}$, the basis vectors across the top of the table are triples of the form (n_i, s_j, n_k), where n_i and n_k are basis vectors from **N** (properties of the noun) and s_j is a basis vector from **S** (True or False in the plausibility space).

The way to read the fictitious numbers is as follows: *chases* has a high (normalized) count for the ⟨fluffy,T,fluffy⟩ basis vector, because *it is highly plausible that* fluffy things chase fluffy things. Conversely, *chases* has a low count for the ⟨fluffy,F,fluffy⟩ basis vector, because *it is not highly implausible that* fluffy things chase fluffy things.[8] Similarly, *eats* has a low count for the ⟨tasty,T,juice⟩ basis vector, because *it is not highly plausible that* tasty things eat things which can be juice.

Table 13.3 gives the vector for *chases*, together with the corresponding counts for the subject and object in *dog chases cat*. For example, the (*dog*, *cat*) count pair for the basis vectors ⟨fluffy,T,fluffy⟩ and ⟨fluffy,F,fluffy⟩ is (0.8, 0.9), meaning that dogs are fluffy to an extent 0.8, and cats are fluffy to an extent 0.9. The property counts for the subject and object are taken from the noun vectors in Table 13.1.

---

[8] The counts in the table for (n_i, T, n_k) and (n_i, F, n_k), for some n_i and n_k, always sum to 1, but this is not required and no probabilistic interpretation of the counts is being assumed.

TABLE 13.3 The vector for *chases* together with subject *dog* and object *cat*

	⟨fluffy,T,fluffy⟩	⟨fluffy,F,fluffy⟩	⟨fluffy,T,fast⟩	⟨fluffy,F,fast⟩	⟨fluffy,T,juice⟩	⟨fluffy,F,juice⟩	⟨tasty,T,juice⟩	...
$\overrightarrow{chases}$	0.8	0.2	0.75	0.25	0.2	0.8	0.1	
dog,cat	0.8,0.9	0.8,0.9	0.8,0.6	0.8,0.6	0.8,0.0	0.8,0.0	0.1,0.0	

The reason for presenting the vectors in the form in Table 13.3 is that a procedure for combining the vector for *chases* with the vectors for *dog* and *cat* now suggests itself. How plausible is it that a dog chases a cat? Well, we know the extent to which fluffy things chase fluffy things, and the extent to which a dog is fluffy and a cat is fluffy; we know the extent to which things that are bought chase things that can be juice, and we know the extent to which dogs can be bought and cats can be juice; more generally, for each property pair, we know the extent to which the subjects and objects of *chases*, in general, embody those properties, and we know the extent to which the particular subject and object in the sentence embody those properties. So multiplying the corresponding numbers together brings information from both the verb and the particular subject and object.

The calculation for the True basis vector of the sentence space for *dog chases cat* is as follows:

$$\overrightarrow{dog\ chases\ cat}_{True} = 0.8 \cdot 0.8 \cdot 0.9 + 0.75 \cdot 0.8 \cdot 0.6 + 0.2 \cdot 0.8 \cdot 0.0$$
$$+\ 0.1 \cdot 0.1 \cdot 0.0 + \ldots$$

The calculation for the False basis vector is similar:

$$\overrightarrow{dog\ chases\ cat}_{False} = 0.2 \cdot 0.8 \cdot 0.9 + 0.25 \cdot 0.8 \cdot 0.6 + 0.8 \cdot 0.8 \cdot 0.0 + \ldots$$

We would expect the coefficient for the True basis vector to be much higher than that for the False basis vector, since *dog* and *cat* embody exactly those elements of the property pairs that score highly on the True basis vector for *chases*, and do not embody those elements of the property pairs that score highly on the False basis vector.

Through a particular example of the sentence space, we have now derived the expression in Coecke, Sadrzadeh and Clark (2010) for combining a transitive verb, $\overrightarrow{\Psi}$, with its subject, $\overrightarrow{\pi}$, and object, $\overrightarrow{o}$:

$$f(\overrightarrow{\pi} \otimes \overrightarrow{\Psi} \otimes \overrightarrow{o}) = \sum_{ijk} C_{ijk} \langle \overrightarrow{\pi} | \overrightarrow{\pi_i} \rangle \overrightarrow{s_j} \langle \overrightarrow{o} | \overrightarrow{o_k} \rangle \quad (13.6)$$

$$= \sum_j \left(\sum_{ik} C_{ijk} \langle \overrightarrow{\pi} | \overrightarrow{\pi_i} \rangle \langle \overrightarrow{o} | \overrightarrow{o_k} \rangle \right) \overrightarrow{s_j} \quad (13.7)$$

$$\frac{\frac{man}{\frac{NP}{N}} \quad \frac{bites}{\frac{NP^r \cdot S \cdot NP^l}{N \otimes S \otimes N}} \quad \frac{dog}{\frac{NP}{N}}}{\frac{\frac{NP^r \cdot S}{N \otimes S}}{\frac{S}{S}}}$$

Fig 13.6 Example pregroup derivation with semantic types.

The expression $\langle \vec{\pi} | \vec{\pi_i} \rangle$ is the Dirac notation for the inner product between $\vec{\pi}$ and $\vec{\pi_i}$, and in this case the inner product is between a vector and one of its basis vectors, so it simply returns the coefficient of $\vec{\pi}$ for the $\vec{\pi_i}$ basis vector. From the linguistic perspective, these inner products are simply picking out particular properties of the subject, $\vec{\pi_i}$, and object, $\vec{o_k}$, and combining the corresponding property coefficients with the corresponding coefficient for the verb, C_{ijk}, for a particular basis vector in the sentence space, s_j.

Figure 13.6 shows a pregroup derivation for a simple transitive verb sentence, together with the corresponding semantic types. The point of the example is to demonstrate how the semantic types 'become smaller' as the derivation progresses, in much the same way that the syntactic types do. The reduction, or cancellation, rule for the semantic component is given in eqn (13.7), which can be thought of as the semantic vector-based analogue of the syntactic reduction rules in eqns (13.3) and (13.4). Similar to a model-theoretic semantics, the semantic vector for the verb can be thought of as encoding all the ways in which the verb could interact with a subject and object, in order to produce a sentence, and the introduction of a particular subject and object reduces those possibilities to a single vector in the sentence space.

In summary, the meaning of a sentence $w_1 \cdots w_n$ with the grammatical (pregroup) structure $p_1 \cdots p_n \to^\alpha S$, where p_i is the grammatical type of w_i, and $\to^\alpha$ is the pregroup reduction to a sentence, can be represented as follows:

$$\overrightarrow{w_1 \cdots w_n} = F(\alpha)(\vec{w_1} \otimes \cdots \otimes \vec{w_n})$$

Here we have generalized the previous discussion of transitive verbs and extended the idea to syntactic reductions or derivations containing any types. The point is that the semantic reduction mechanism described in this section can be generalized to any syntactic reduction, α, and there is a function, F, which, given α, produces a linear map to take the word vectors $\vec{w_1} \otimes \cdots \otimes \vec{w_n}$ to a sentence vector $\overrightarrow{w_1 \cdots w_n}$. F can be thought of as Montague's 'homomorphic passage' from syntax to semantics.[9]

---

[9] Note that the input to the 'meaning map' is a vector in the tensor product space obtained by combining the semantic types for all the words in the sentence. However, this vector will never be built in practice, only the vectors for the individual words.

Coecke, Sadrzadeh and Clark (2010) contains a detailed description of this more general case.

13.5 A real sentence space

The sentence space in this section is taken from Grefenstette *et al.* (2011) and Grefenstette and Sadrzadeh (2011). The sentence space is designed to work with transitive verbs, and exploits a similar intuition to that used to define the verb vector in the previous section: a vector for a transitive verb *sentence* consists of pairs of properties reflecting both the properties of the subject and object of the verb, and the properties of the subjects and objects that the verb has in general. Pairs of properties are encoded in the $N \otimes N$ space:

$$\overrightarrow{dog\ chases\ cat} \in N \otimes N$$

Given that $S = N \otimes N$, the semantic type of a transitive verb is as follows:

$$\overrightarrow{chases} \in N \otimes N \otimes N \otimes N$$

However, in this section, following Grefenstette *et al.* (2011) and Grefenstette and Sadrzadeh (2011), the semantic type of the verb will be the same as the sentence space:

$$\overrightarrow{chases} \in N \otimes N$$

The reason for restricting the sentence space in this way is that the interpretation of a particular basis vector (n_i, n_k) from $N \otimes N$, when defining a transitive verb, is clear: the coefficient for (n_i, n_k) should reflect the extent to which subjects of the verb embody the n_i noun property, and the extent to which objects of the verb embody the n_k property. One way to consider the verb space is that we are effectively ignoring those basis vectors in $N \otimes N \otimes N \otimes N$ in which the properties of the subject and object (corresponding to the outermost N vectors in the tensor product space) do not match those of the property pairs from the sentence (the innermost N vectors).

Table 13.4 shows the verb vectors from the plausibility space example, but this time with the basis vectors as pairs of noun properties, rather than triples consisting of pairs of noun properties together with a True or False basis vector from the plausibility space. The intuition for the fictitious counts is the same as before: $\overrightarrow{chases}$ has a high coefficient for the ⟨fluffy,fluffy⟩ basis vector because, in general, fluffy things chase fluffy things.

Another reason for defining the verb space as $N \otimes N$ is that there is a clear experimental procedure for obtaining verb vectors from corpora: simply count the number of times that particular property pairs from the subject and object appear with a particular verb in the corpus. Suppose that we have a corpus consisting of

TABLE 13.4 Example transitive verb vectors in N ⊗ N

	⟨fluffy,fluffy⟩	⟨fluffy,fast⟩	⟨fluffy,juice⟩	⟨tasty,juice⟩	⟨tasty,buy⟩	⟨buy,fruit⟩	⟨fruit,fruit⟩	...
$\overrightarrow{chases}$	0.8	0.75	0.2	0.1	0.2	0.2	0.0	
$\overrightarrow{eats}$	0.7	0.6	0.9	0.1	0.1	0.7	0.1	

TABLE 13.5 Combining a transitive verb in N ⊗ N with a subject and object to give a sentence in N ⊗ N

	⟨fluffy,fluffy⟩	⟨fluffy,fast⟩	⟨fluffy,juice⟩	⟨tasty,juice⟩	⟨tasty,buy⟩	⟨buy,fruit⟩	⟨fruit,fruit⟩	...
$\overrightarrow{chases}$	0.8	0.75	0.2	0.1	0.2	0.2	0.0	
dog,cat	0.8,0.9	0.8,0.6	0.8,0.0	0.1,0.0	0.1,0.5	0.5,0.0	0.0,0.0	
$\overrightarrow{dog\ chases\ cat}$	0.58	0.36	0.0	0.0	0.01	0.0	0.0	

only two sentences: *dog chases cat* and *cat chases mouse*. To obtain the vector for *chases*, we first increment counts for all property pairs corresponding to (*dog, cat*), and then increment the counts for all property pairs corresponding to (*cat, mouse*) (where the counts come from the noun vectors, which we assume have already been built). Hence in this example we would obtain evidence for fluffy things chasing fluffy things from the first sentence (since dogs are fluffy and cats are fluffy); for aggressive things chasing fluffy things (since dogs can be aggressive and cats are fluffy); some evidence for things that can be bought chasing things that are brown from the second sentence (since cats can be bought and some mice are brown); and so on for all property pairs and all examples of transitive verbs in the corpus.[10]

Given this particular verb space (N ⊗ N), and continuing with the small corpus example, there is a neat expression for the meaning of a transitive verb (Grefenstette and Sadrzadeh, 2011):

$$\overrightarrow{chases} = \overrightarrow{dog} \otimes \overrightarrow{cat} + \overrightarrow{cat} \otimes \overrightarrow{mouse}$$

where $\overrightarrow{dog} \otimes \overrightarrow{cat}$ is the Kronecker product of $\overrightarrow{dog}$ and $\overrightarrow{cat}$ defined in Section 13.3. This expression generalizes in the obvious way to a corpus containing many instances of *chases*.

Table 13.5 shows how a transitive verb in N ⊗ N combines with a subject and object, using the same compositional procedure described in the previous section. Again the intuition is similar, but now we end up with a sentence vector in a much larger space, N ⊗ N, compared with the two-dimensional plausibility space used

---

[10] For examples where the subject and object are themselves compositional, we take the counts from the head noun of the noun phrase.

earlier. The resulting sentence vector can be thought of informally as answering the following set of questions:

- To what extent, in the sentence, are fluffy things interacting with fluffy things? (In this case to a large extent, since both subject and object are fluffy, and the verb in the sentence, *chase*, is such that fluffy things do chase fluffy things.)
- To what extent, in the sentence, are things that can be bought interacting with things that can be juice? (In this case to a small extent, since although the subject can be bought, the object is not something that can be juice, and the verb *chase* does not generally relate things that can be bought with things that can be juice.)
- And so on for all the noun property pairs in $N \otimes N$.

Again, given this particular sentence space $N \otimes N$, we end up with a neat expression for how to combine a transitive verb with its subject and object (Grefenstette and Sadrzadeh, 2011):

$$\overrightarrow{dog\ chases\ cat} = \overrightarrow{chases} \odot \left(\overrightarrow{dog} \otimes \overrightarrow{cat}\right) \tag{13.8}$$

where $\odot$ is pointwise multiplication. Expressing the combination of verb and arguments in this way makes it clear how both information from the arguments and the verb itself is used to produce the sentence vector. For each property pair in the sentence space, $\overrightarrow{dog} \otimes \overrightarrow{cat}$ captures the extent to which the arguments of the verb satisfy those properties, and the coefficient for each property pair is multiplied by the corresponding coefficient for the verb, which captures the extent to which arguments of the verb in general satisfy those properties. But note that this neat expression only arises because of the choice of $N \otimes N$ as the verb and sentence space; the expression above is not true in the general case.

The question of how to evaluate the models described in this chapter and compositional distributional models more generally is an important one, but one that will be discussed only briefly here. The previous chapter discussed the issue of evaluation and described a recent method of using compositional distributional models to disambiguate verbs in context (Mitchell and Lapata, 2008). The task is to use a compositional distributional model to assign similarity scores to pairs such as (*the face glowed, the face beamed*) and (*the face glowed, the face burned*), and compare these scores to human judgements. In this case the first pair would be expected to obtain the higher score, but if the subject were *fire* rather than *face*, then *burned* rather than *beamed* would score higher. Hence in this case the subject of the intransitive verb is effectively being used to disambiguate the verb.

Grefenstette and Sadrzadeh (2011) extend this evaluation to transitive verbs, so that there is now an object as well as subject, with the idea that having an additional argument makes this a stronger test of compositionality. Here the pairs are cases such as (*the people tried the door, the people tested the door*) and (*the people tried the door, the people judged the door*). In this case the first pair would be expected to get the higher similarity score, whereas if the subject and object were *tribunal* and *crime*, then *judged* would be expected to score higher than *tested*. Grefenstette and Sadrzadeh (2011) find that the model described in this section performs at least as well as the best-performing method from Mitchell and Lapata (2008), which involves building vectors for verbs and arguments using the context method (not treating verbs as relational) and then using pointwise multiplication to combine the verb with its arguments.

The previous chapter discusses this method of evaluation and suggests that it is essentially a method of disambiguation, and so it is perhaps not surprising that multiplicative methods perform so well, since the contextual elements that the verb and arguments have in common will be emphasized in the multiplicative combination. Note that, given the choice of $\mathbf{N} \otimes \mathbf{N}$ as the verb and sentence space, the compositional procedure for the model in this section reduces to a multiplicative combination (eqn 13.8).

In the final part of this section we return to the question of why the tensor product is used to bind the vector spaces together in relational types, by comparing it with an alternative: the direct sum. The direct sum is also an operator on vector spaces, but rather than create basis vectors by taking the cartesian product of the basis vectors of the spaces being combined, it effectively retains the basis vectors of the combining spaces as independent vectors. So the number of dimensions for the direct sum of $\mathbf{V}$ and $\mathbf{W}$, $\mathbf{V} \oplus \mathbf{W}$, is $|\mathbf{V}| + |\mathbf{W}|$ where $|\mathbf{V}|$ is the number of dimensions in $\mathbf{V}$, rather than $|\mathbf{V}|.|\mathbf{W}|$ as in the tensor product case.

If the direct sum were being used for the semantic type of a transitive verb, then the vector space in which transitive verbs live would be $\mathbf{N} \oplus \mathbf{S} \oplus \mathbf{N}$ and the general expression for a transitive verb vector Ψ would be as follows:

$$\Psi = \sum_{ijk} C_i \vec{n_i} + C_j \vec{s_j} + C_k \vec{n_k} \in \mathbf{N} \oplus \mathbf{S} \oplus \mathbf{N} \qquad (13.9)$$

The obvious way to adapt the method for building verb vectors detailed in the previous section to a direct sum representation is as follows. Rather than have the verb and sentence live in the $\mathbf{N} \otimes \mathbf{N}$ space, as before, suppose now that verbs and sentences live in $\mathbf{N} \oplus \mathbf{N}$. Again suppose that we have a corpus consisting of only two sentences: *dog chases cat* and *cat chases mouse*. To obtain the vector for *chases*, we follow a similar procedure to before: first increment counts for all property pairs corresponding to (*dog, cat*), and then increment the counts for all property pairs corresponding to (*cat, mouse*) (where again the counts come from the noun vectors, which we assume have already been built). But now there is a crucial difference: when we increment the counts for the (*fluffy, buy*) pair, for example, there is no *interaction* between the subject and object. Whereas in the tensor product case we were able to represent the fact that fluffy things chase things that can be bought, in the direct sum case we can only represent the fact that fluffy things chase, and that things that can be bought get chased, but not the combination of the two. So more generally the direct sum can represent the properties of subjects of a verb, and

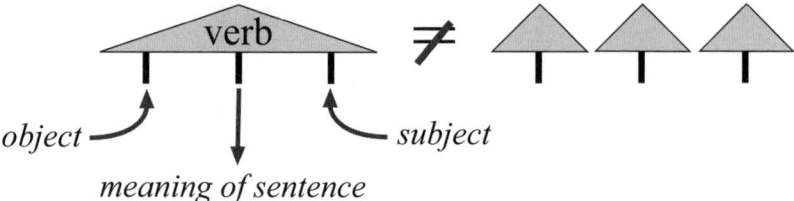

Fig 13.7 The verb needs to interact with its subject and object, which it does with the tensor product on the left, but not with the direct sum on the right.

the properties of objects, but not the pairs of properties which are seen together with the verb.

Figure 13.7 is taken from Coecke, Sadrzadeh and Clark (2010) and expresses the interactive nature of the tensor product on the left, as opposed to the direct sum on the right. The figure is intended to represent the fact that, with the direct sum, the subject, object, and resulting sentence are entirely independent of each other.

13.6 Conclusion and further work

There are some obvious ways in which the existing work could be extended. First, the present chapter has only presented a procedure for building vectors for sentences with a simple transitive verb structure. The mathematical framework in Coecke, Sadrzadeh and Clark (2010) is general and in principle applies to any syntactic reduction from any sequence of syntactic types, but how to build relational vectors for all complex types is an open question. Second, it needs to be demonstrated that the compositional distributional representations presented in this chapter can be useful for language processing tasks and applications. Third, there is a large part of natural language semantics, much of which is the focus of traditional formal semantics, such as logical operators, quantification, inference, and so on, which has been ignored in this chapter. We see distributional models of semantics as essentially providing a semantics of *similarity*, but whether the more traditional questions of semantics can be accommodated in distributional semantics is an interesting and open question.[11] Fourth, there is the question of whether the vectors for relational types can be driven more from a machine-learning perspective, rather than have their form determined by linguistic intuition (as was the case for the verb vectors living in $\mathbf{N} \otimes \mathbf{N}$). Addressing this question would bring together the work in distributional semantics with the recent work in so-called deep learning in the

---

[11] There is some preliminary work in this direction, e.g. Preller and Sadrzadeh (2011*b*); Clarke (2007); Widdows (2004).

neural networks research community.[12] And finally, there is the question of how sentences should be represented as vectors. Two possibilities have been suggested in this chapter, but there are presumably many more; it may be that the form of the sentence space should be determined by the particular language processing task in hand, e.g. sentiment analysis (Socher, Manning and Ng, 2010).

In summary, this chapter is a presentation of Coecke, Sadrzadeh and Clark (2010) designed to be accessible to computational linguists. The key innovations in this work are the use of tensor product spaces for relational types such as verbs, through the use of the tensor product, and a general method for composing a relational type (represented as a vector in a tensor product space) with its vector arguments.

Acknowledgements

Almost all of the ideas in this chapter have arisen from discussions with Mehrnoosh Sadrzadeh, Ed Grefenstette, Bob Coecke and Stephen Pulman. Thanks to Tim Van de Cruys, Marco Baroni, and the editors for useful feedback.

---

[12] See the proceedings for the NIPS workshops in 2010 and 2011 on Deep Learning and Unsupervised Feature Learning.

REFERENCES

Abe, E. (1980). *Hopf Algebras*. Cambridge University Press (transl. of the 1977 Japanese edition).

Abramov, S.A., Le, H.Q., and Li, Z. (2005). Univariate Ore polynomial rings in computer algebra. *J. Math. Sciences*, **131**, 5885–5903.

Abramovich, Y.A. and Aliprantis, C.D. (2002). *An Invitation to Operator Theory*. American Mathematical Society.

Abramsky, S. (2005). Abstract scalars, loops, and free traced and strongly compact closed categories. In *Algebra and Coalgebra in Computer Science* (ed. J. Fiadeiro, N. Harman, M. Roggenbach, and J. Rutten). Springer.

Abramsky, S. and Coecke, B. (2004). A categorical semantics of quantum protocols. In *LICS Proceedings* (ed. H. Ganzinger), pp. 415–425. IEEE Computer Society. Extended version: arXiv:quant-ph/0402130.

Abramsky, S. and Coecke, B. (2009). A categorical semantics of quantum protocols. In *Handbook of Quantum Logic and Quantum Structures* (ed. K. Engesser, D. M. Gabbai, and D. Lehmann). Elsevier: Computer Science Press.

Abramsky, S., Haghverdi, E., and Scott, P. (2002). Geometry of interaction and linear combinatory algebras. *Math. Structures Comp. Sci.*, **12 (5)**.

Abramsky, S. and Heunen, C. (2012). H^*-algebras and nonunital Frobenius algebras: first steps in infinite-dimensional categorical quantum mechanics. In *Mathematical Foundations of Information Flow*. American Mathematical Society and Cambridge University Press. arXiv:1011.6123.

Abramsky, S. and Jung, A. (1994). Domain theory. In *Handbook of Logic in Computer Science* (ed. S. Abramsky, D. M. Gabbay, and T. S. E. Maibaum), pp. 1–168. Clarendon Press.

Abramsky, S. and Tzevelekos, N. (2011). Introduction to categories and categorical logic. In *New Structures for Physics* (ed. B. Coecke). Springer.

Adámek, J. and Velebil, J. (2008). Analytic functors and weak pullbacks. *Theory Applic. Categ.*, **21(11)**, 191–209. Available from: www.tac.mta.ca/tac/volumes/21/11/.

Aguiar, M. and Mahajan, S. (2006). *Coxeter Groups and Hopf Algebras*. American Mathematical Society.

Ajdukiewicz, K. (1937). Die syntaktische konnexität. *Studia Philosophica*, **1**, 1–27.

Albuquerque, H. and Majid, S. (1999). Quasialgebra structure of the octonions. *J. Algebra*, **220**, 188–224.

Aliprantis, C.D. and Burkinshaw, O. (1985). *Positive Operators*. Academic Press.

Alshawi, H. and van Eijck, J. (1989). Logical forms in the core language engine. In *Proc. 27th Ann. Meet. Assoc. Comput. Linguist.*, Vancouver, British Columbia, Canada, pp. 25–32. Association for Computational Linguistics.

Altschuler, D. and Coste, A. (1992). Invariants of 3-manifolds from finite groups. In *Proc. XXth Int. Conf. Diff. Geom. Meth. in Theor. Phys., New York, 1991*, pp. 219–233. World Scientific.

Amelino-Camelia, G. and Majid, S. (2000). Waves on noncommutative spacetime and gamma-ray bursts. *Int. J. Mod. Phys*, **A15**, 4301–4323.

Andreoli, J.-M. (2001). Focussing and proof construction. *Ann. Pure Appl. Logic*, **107**(1–3), 131–163.

Andreoli, J.-M. and Maieli, R. (1999). Focusing and proof-nets in linear and non-commutative logic. In *International Conference on Logic for Programming and Automated Reasoning (LPAR)*. Springer.

Atiyah, M. (1989). Topological quantum field theories. *Publ. Math. IHES*, **68**, 175–186.

Awodey, S. (2006). *Category Theory*. Oxford University Press.

Baez, J.C. and Dolan, J. (1998). Categorification. In *Higher Category Theory* (ed. E. Getzler and M. Kapranov). American Mathematical Society.

Baez, J.C. and May, J.P. (2010). *Towards Higher Categories*. Springer.

Baez, J.C. and Stay, M. (2011). Physics, topology, logic and computation: a Rosetta stone. In *New Structures for Physics* (ed. B. Coecke). Springer.

Balakov, B. and Kirilov Jr., A. (2001). *Lectures on Tensor Categories and Modular Functors*. American Mathematical Society.

Balsam, B. (2010). Turaev–Viro invariants as an extended TQFT II and III. arXiv:1010.1222 & 1012.0560.

Balsam, B. and Kirillov Jr., A. (2010). Turaev–Viro invariants as an extended TQFT. arXiv:1004.1533.

Baltag, A., Coecke, B., and Sadrzadeh, M. (2006). Epistemic actions as resources. *J. Logic Comput.*. arXiv:math.LO/0608166.

Baltag, A., Moss, L.S., and Solecki, S. (1998). The logic of public announcements, common knowledge, and private suspicions. In *Proceedings of the 7th Conference on Theoretical Aspects of Rationality and Knowledge*. Morgan Kaufmann Publishers Inc.

Bantay, P. (2003). The kernel of the modular representation and the Galois action in RCFT. *Comm. Math. Phys.*, **233**, 423–438.

Bar-Hillel, Y. (1953). A quasiarithmetical notation for syntactic description. *Language*, **29**, 47–58.

Barenco, A., Bennett, C.H., Cleve, R., DiVincenzo, D.P., Margolus, N., Shor, P., Sleator, T., Smolin, J.A., and Weinfurter, H. (1995). Elementary gates for quantum computation. *Phys. Rev. A*, **52**, 3457–3467.

Barmeier, T., Fuchs, J., Runkel, I., and Schweigert, C. (2010). On the Rosenberg–Zelinsky sequence in abelian monoidal categories. *Journal f. reine u. angew. Mathematik*, **642**, 1–36.

Baroni, M. and Zamparelli, R. (2010). Nouns are vectors, adjectives are matrices: representing adjective-noun constructions in semantic space. In *Proceedings of the 2010 Conference on Empirical Methods in Natural Language Processing*, Cambridge, MA. Association for Computational Linguistics.

Barr, M. (1979). **-autonomous categories*. Springer-Verlag.

Barr, M. (1995). Non-symmetric *-autonomous categories. *Theor. Comput. Sci.*, **139**, 115–130.

Barrett, J.W. and Westbury, B.W. (1996). Invariants of piecewise-linear 3-manifolds. *Trans. Amer. Math. Soc.*, **348**, 3997–4022.

Barrett, J.W. and Westbury, B.W. (1999). Spherical categories. *Adv. Math.*, **143**, 357–375.

Barwise, J. and Cooper, R. (2002). Generalized quantifiers and natural language. In *Formal Semantics: The Essential Readings* (ed. P. Portner and B. Partee). Wiley-Blackwell.

Bastenhof, A. (2011). Polarized Montagovian semantics for the Lambek–Grishin calculus. In *Proceedings of the 15th Conference on Formal Grammar (Copenhagen, 2010)*. Springer. arXiv:1101.5757.

Baumgärtel, H. and Lledó, F. (2004). Duality of compact groups and Hilbert C*-systems for C*-algebras with a nontrivial center. *Int. J. Math.*, **15**, 759–812.

Beidar, K.I., Fong, Y., and Stolin, A. (1997). On Frobenius algebras and the quantum Yang–Baxter equation. *Trans. Am. Math. Soc.*, **349**, 3823–3836.

Beliakova, A. and Blanchet, C. (2001*a*). Modular categories of types B, C and D. *Comment. Math. Helv.*, **76**, 467–500.

Beliakova, A. and Blanchet, C. (2001*b*). Skein construction of idempotents in Birman–Murakami–Wenzl algebras. *Math. Ann.*, **321**, 347–373.

Benabou, J. (1963). Categories avec multiplication. *Comptes Rendus des Séances de l'Académie des Sciences. Paris*, **256**, 1887–1890.

Bennett, C.H., Brassard, G., Crepeau, C., Jozsa, R., Peres, A., and Wootters, W.K. (1993). Teleporting an unknown quantum state via dual classical and Einstein–Podolsky–Rosen channels. *Phys. Rev. Lett.*, **70**(13), 1895–1899.

Bernardi, R. and Moortgat, M. (2007). Continuation semantics for symmetric categorial grammar. In *Proceedings 14th Workshop on Logic, Language, Information and Computation (WoLLIC'07)* (ed. D. Leivant and R. de Queiros). Springer.

Bernardi, R. and Moortgat, M. (2010). Continuation semantics for the Lambek–Grishin calculus. *Information and Computation*, **208**(5), 397–416.

Bespalov, Y. (1997). Crossed modules and quantum groups in braided categories. *Appl. Cat. Struct*, **5**, 155–204.

Bespalov, Y., Lyubashenko, V., Kerler, T., and Turaev, V. (2000). Integrals for braided Hopf algebras. *J. Pure Appl. Alg.*, **148**, 113–164.

Bezrukavnikov, R. (2004). On tensor categories attached to cells in affine Weyl groups. In *Representation Theory of Algebraic Groups and Quantum Groups* (ed. T. Shoji, M. Kashiwara, N. Kawanaka, G. Lusztig, and K. Shinoda). Math. Soc. Japan.

Birkhoff, G. and von Neumann, J. (1936). The logic of quantum mechanics. *Ann. Math.*, **37**, 823–843.

Blanchet, C. (2000). Hecke algebras, modular categories and 3-manifolds quantum invariants. *Topology*, **39**, 193–223.

Blohmann, C., Tang, X., and Weinstein, A. (2008). Hopfish structure and modules over irrational rotation algebras. In *Non-commutative Geometry in Mathematics and Physics*. American Mathematical Society.

Blute, R.F., Cockett, J.R.B., Seely, R.A.G., and Trimble, T.H. (1991). Natural deduction and coherence for weakly distributive categories. *J. Pure Appl. Algebra*, **113**, 229–296.

Böckenhauer, J., Evans, D.E., and Kawahigashi, Y. (1999). On α-induction, chiral generators and modular invariants for subfactors. *Comm. Math. Phys.*, **208**, 429–487.

Böckenhauer, J., Evans, D.E., and Kawahigashi, Y. (2000). Chiral structure of modular invariants for subfactors. *Comm. Math. Phys.*, **210**, 733–784.

Böhm, G. (2009). Hopf algebroids. In *Handbook of Algebra*, Volume 6. Elsevier.

Böhm, G. (2010). The weak theory of monads. *Adv. Math.*, **225**(1), 1–32.

Böhm, G., Caenepeel, S., and Janssen, K. (2011*a*). Weak bialgebras and monoidal categories. *Comm. Algebra*, **39**, 4584–4607.

Böhm, G., Lack, S., and Street, R. (2011*b*). Weak bimonads and weak Hopf monads. *J. Algebra*, **328**, 1–30.

Böhm, G., Nill, F., and Szlachányi, K. (1999). Weak Hopf algebras. I. Integral theory and C^*-structure. *J. Algebra*, **221**(2), 385–438.

Böhm, G. and Szlachányi, K. (2004). Hopf algebroids with bijective antipodes: axioms, integrals, and duals. *J. Algebra*, **274**(2), 708–750.

Bombelli, L., Lee, J., Meyer, D., and Sorkin, R.D. (1987). Space-time as a causal set. *Phys. Rev. Letters*, **59**(5), 521.

Bonneau, P. and Sternheimer, D. (2005). Topological Hopf algebras, quantum groups and deformation quantization. In *Hopf Algebras in Noncommutative Geometry and Physics*. Dekker.

Booker, T. and Street, R. (2011). Tannaka duality and convolution for duodial categories. arXiv:1111.5695.

Borceux, F. (1994). *Handbook of Categorical Algebra*, Volume 50, 51 and 52 of *Encyclopedia of Mathematics*. Cambridge University Press.

Borceux, F. and Stubbe, I. (2000). Short introduction to enriched categories. In *Current Research in Operational Quantum Logic: Algebras, Categories and Languages* (ed. B. Coecke, D. Moore, and A. Wilce). Springer-Verlag.

Brauer, R. (1937). On algebras which are connected with the semisimple continous groups. *Ann. of Math.*, **38**(4), 857–872.

Brauer, R. and Nesbitt, C. (1937). On the regular representations of algebras. *Proc. Nat. Acad. Sci. USA*, **23**, 236–240.

Briegel, H.J., Browne, D.E., Dür, W., Raussendorf, R., and van den Nest, M. (2009). Measurement-based quantum computation. *Nature Physics*, **5**(1), 19–26.

Brouder, C., Fauser, B., Frabetti, A., and Oeckl, R. (2004). Quantum field theory and Hopf algebra cohomology [formerly: Let's twist again]. *J. Phys. A.*, **37**, 5895–5927. arXiv: hep-th/0311253.

Browne, D.E., Kashefi, E., Mhalla, M., and Perdrix, S. (2007). Generalized flow and determinism in measurement-based quantum computation. *New J. Phys.*, **9**(250).

Bruguières, A. (2000). Catégories prémodulaires, modularisations et invariants de variétés de dimension 3. *Math. Ann.*, **316**, 215–236.

Bruguières, A., Lack, S., and Virelizier, A. (2011). Hopf monads on monoidal categories. *Adv. Math.*, **227**, 745–800.

Bruguières, A. and Virelizier, A. (2007). Hopf monads. *Adv. Math.*, **215**, 679–733.

Bruguières, A. and Virelizier, A. (2008). Categorical centers and Reshetikhin–Turaev invariants. *Acta Math. Vietnam*, **215**, 255–277.

Bruguières, A. and Virelizier, A. (2012). Quantum double of Hopf monads and categorical centers. *Trans. Amer. Math. Soc.*, **364**, 1225–1279.

Brzezinski, T. and Wisbauer, R. (2003). *Corings and comodules*. Cambridge University Press.

Bueso, J.L., Gómes-Torrecillas, J., and Verschoren, A. (2003). *Algorithmic Methods in Non-Commutative Algebra*. Kluwer.

Bulacu, D. and Caenepeel, S. (2012). A monoidal structure on the category of relative Hopf modules. *J. Algebra*, **11**(2).

Bulacu, D., Caenepeel, S., and Torrecillas, B. (2011). The braided monoidal structures on the category of vector spaces graded by the Klein group. *Proc. Edinb. Math. Soc. (2)*, **54**(2), 613–641.

Buszkowski, W. and Moroz, K. (2006). Pregroup grammars and context-free grammars. In *Computational Algebraic Approaches to Natural Language* (ed. C. Casadio and J. Lambek). Polimetrica.

Caenepeel, S. (1998). *Brauer Groups, Hopf Algebras and Galois Theory*. Kluwer.

Caenepeel, S. and De Lombaerde, M. (2006). A categorical approach to Turaev's Hopf group-coalgebras. *Comm. Algebra*, **34**(7), 2631–2657.

Caenepeel, S. and Goyvaerts, I. (2011). Monoidal Hom-Hopf algebras. *Comm. Algebra*, **39**(6), 2216–2240.

Caenepeel, S., Militaru, G., and Zhu, S. (2002). *Frobenius and Separable Functors for Generalized Module Categories and Non-linear Equations*. Springer.

Cameron, P.J. and Majid, S. (2003). Braided line and counting fixed points of $gl(d, f_q)$. *Commun. Algebra*, **31**, 2003–2013.

Carnap, R. (1988). *Meaning and Necessity: a Study in Semantics and Modal Logic*. University of Chicago Press.

Carroll, R.W. (2005). *Fluctuations, Information, Gravity and the Quantum Potential*. Kluwer.

Cartier, P. (2007). A primer of Hopf algebras. In *Frontiers in Number Theory, Physics, and Geometry II* (ed. P. Cartier, P. Moussa, B. Julia, and P. Vanhove). Springer Verlag.

Casadio, C. and Lambek, J. (2002). A tale of four grammars. *Studia Logica*, **71**(3), 315–329.

Chari, V. and Pressley, A. (1995). *A Guide to Quantum Groups*. Cambridge University Press.

Chikhladze, D., Lack, S., and Street, R. (2010). Hopf monoidal comonads. *Theory Appl. Categ.*, **24**, 554–563.

Chomsky, N. (1956). Tree models for the description of language. *IRE. T. Inform. Theor.*, **IT-2**, 113–124.

Chomsky, N. (1964). Current issues in linguistic theory. In *The Structure of Language: Readings in the Philosophy of Language* (ed. J. Fodor and J. Katz). Prentice Hall, NJ.

Chomsky, N. (1965). *Aspects of the Theory of Syntax*. MIT Press.

Clark, S., Coecke, B., and Sadrzadeh, M. (2008). A compositional distributional model of meaning. In *Proceedings of the Second Quantum Interaction Symposium (QI-2008)* (ed. W. L. P. Bruza and J. van Rijsbergen), pp. 133–140.

Clark, S. and Curran, J. (2007). Wide-coverage efficient statistical parsing with CCG and log-linear models. *Comput. Linguist.*, **33**, 493–552.

Clark, S. and Pulman, S. (2007). Combining symbolic and distributional models of meaning. In *Proceedings of the AAAI Spring Symposium on Quantum Interaction*, Stanford, CA.

Clarke, D. (2007). *Context-theoretic Semantics for Natural Language: an Algebraic Framework*. Ph. D. thesis, Department of Informatics, University of Sussex.

Clarke, D. (2012). A context-theoretic framework for compositionality in distributional semantics. *Comput. Linguist.*, **38**(1), 41–71.

Clarke, D., Weir, D., and Lutz, R. (2011). Algebraic approaches to compositional distributional semantics. In *Proceedings of the Ninth International Conference on Computational Semantics (IWCS-2011)*. Association for Computational Linguistics.

Cockett, J.R.B., Koslowski, J., and Seely, R.A.G. (2000). Introduction to linear bicategories. *Math. Structures Comp. Sci.*, **10**, 165–203.

Cockett, J.R.B. and Seely, R.A.G. (1996). Proof theory for full intuitionistic linear logic, bilinear logic, and mix categories. Volume 3, pp. 85–131.

Coecke, B. (2007). Automated quantum reasoning: Non logic – semi-logic – hyper-logic. In *AAAI Spring Symposium: Quantum Interaction*. AAAI.

Coecke, B. (2010). Quantum picturalism. *Contemp. Phys.*, **51**(1), 59–83.

Coecke, B. (2012). The logic of quantum mechanics – take II. arXiv:1204.3458.

Coecke, B. and Duncan, R. (2008). Interacting quantum observables. In *Automata, Languages and Programming, 35th International Colloquium, ICALP 2008, Reykjavik, Iceland, July 7–11, 2008, Proceedings, Part II* (ed. L. Aceto, I. Damgård, L. Goldberg, M. Halldórsson, A. Ingólfsdóttir, and I. Walukiewic). Springer.

Coecke, B. and Duncan, R. (2011). Interacting quantum observables: categorical algebra and diagrammatics. *New J. Phys.*, **13**, 043016.

Coecke, B. and Edwards, B. (2011). Toy quantum categories. *Electronic Notes in Theor. Comput. Sci.*, **270**(1), 29–40. arXiv:0808.1037.

Coecke, B., Edwards, W., and Spekkens, R.W. (2011). Phase groups and the origin of non-locality for qubits. *Electronic Notes in Theor. Comput. Sci.*, **271**(2), 15–36.

Coecke, B. and Heunen, C. (2011). Pictures of complete positivity in arbitrary dimension. arXiv:1110.3055.

Coecke, B. and Martin, K. (2011). A partial order on classical and quantum states. In *New Structures for Physics* (ed. B. Coecke). Springer.

Coecke, B., Moore, D.J., and Stubbe, I. (2001). Quantaloids describing causation and propagation of physical properties. *Foundations of Physics Letters*, **14**, 133–146. arXiv:quant-ph/0009100.

Coecke, B. and Paquette, E.O. (2006). POVMs and Naimark's theorem without sums. In *Proceedings of the 4th International Workshop on Quantum Programming Languages*, pp. 131–152. Elsevier.

Coecke, B. and Paquette, E.O. (2011). Categories for the practicing physicist. In *New Structures for Physics* (ed. B. Coecke). Springer. arXiv:0905.3010.

Coecke, B., Paquette, E.O., and Pavlovic, D. (2010). Classical and quantum structuralism. In *Semantic Techniques in Quantum Computation* (ed. S. Gay and I. Mackie). Cambridge University Press. arXiv:0904.1997.

Coecke, B., Paquette, E.O., and Perdrix, S. (2008). Bases in diagrammatic quantum protocols. *Electron. Notes Theor. Comput. Sci.*, **218**, 131–152.

Coecke, B. and Pavlovic, D. (2007). Quantum measurements without sums. In *Mathematics of Quantum Computing and Technology* (ed. G. Chen, L. Kauffman, and S. Lamonaco). Taylor and Francis. arXiv:quant-ph/0608035.

Coecke, B., Pavlovic, D., and Vicary, J. (2011). A new description of orthogonal bases. *Math. Structures Comp. Sci., to appear*. arXiv:quant-ph/0810.1037.

Coecke, B., Sadrzadeh, M., and Clark, S. (2010). Mathematical foundations for distributed compositional model of meaning. *Linguist. Anal.*, **36**, 345–384.

Coecke, B. and Spekkens, R.W. (2011). Picturing classical and quantum Bayesian inference. *Synthese*, 1–46. arXiv:1102.2368.

Cohn, T., Callison-Burch, C., and Lapata, M. (2008). Constructing corpora for development and evaluation of paraphrase systems. *Comput. Linguist*, 34(4), 597–614.

Cuntz, J. (1977). Simple C*-algebras generated by isometries. *Communications in Mathematical Physics*, 57, 173–185.

Curien, P. and Herbelin, H. (2000). Duality of computation. In *International Conference on Functional Programming (ICFP'00)*. Association for Computing Machinery. [2005: corrected version].

Curien, P.-L. (2012). Operads, clones, and distributive laws. In *Operads and Universal Algebra: Proceedings of China–France Summer Conference* (ed. C. Bai, L. Guo, and J.-L. Loday). World Scientific.

Curran, J.R. (2004). *From Distributional to Semantic Similarity*. Ph. D. thesis, University of Edinburgh.

Curran, J.R. and Moens, M. Scaling context space. In *Proceedings of the 40th Annual Meeting of the Association for Computational Linguistics, July 6–12, 2002, Philadelphia, PA, USA*. Association for Computational Linguistics. Available at: http://aclweb.org/anthology-new/.

Danos, V. and Kashefi, E. (2005). Determinism in the one-way model. In *ERATO conference on Quantum Information Science 2005*. American Physical Society.

Danos, V. and Kashefi, E. (2006). Determinism in the one-way model. *Phys. Rev. A*, 74(052310).

Danos, V., Kashefi, E., and Panangaden, P. (2007). The measurement calculus. *JACM*, 54(2).

Danos, V. and Regnier, L. (1993). Local and asynchronous beta reduction. In *Proceedings of the Eighth Annual IEEE Symp. on Logic in Computer Science*. ACM/IEEE.

Davey, B.A. and Priestley, H. (1990). *Introduction to Lattices and Order*. Cambridge University Press.

Davidson, D. (1984). Theories of meaning and learnable languages. In *Inquiries into Truth and Interpretation*. Oxford University Press. (Reprint of 1964 original.)

Davidson, D. (1984 (original 1967)). Truth and meaning. In *Inquiries into Truth and Interpretation*. Oxford University Press.

Davydov, A. (1998). Monoidal categories. *J. Math. Sci. (New York)*, 88, 457–519.

Davydov, A., Müger, M., Nikshych, D., and Ostrik, V. (2012). The Witt group of non-degenerate braided fusion categories. To appear in *J. Reine Angew. Math.* arXiv:1009.2117.

Davydov, A., Nikshych, D., and Ostrik, V. (2011). On the structure of the Witt group of braided fusion categories. arXiv:1109.5558.

Day, B.J. (1977). Note on compact closed categories. *J. Austral. Math. Soc.*, 24, 309–311.

Day, B.J. (1996). Enriched Tannaka reconstruction. *J. Pure Appl. Algebra*, 108(1), 17–22.

de Groote, P. and Retoré, C. (1996). Semantic readings of proof nets. In *Formal Grammar* (ed. G.-J. Kruijff, G. Morrill, and D. Oehrle). FoLLI.

Deerwester, S., Dumais, S.T., Furnas, G.W., Landauer, T.K., and Harshman, R. (1990). Indexing by latent semantic analysis. *J. Am. Soc. Inform. Sci.*, 41(6), 391–407.

Deligne, P. (1990). Catégories tannakiennes. In *The Grothendieck Festschrift*, Volume 2. Birkhauser.

Dijkstra, E.W. (1975). Guarded commands, nondeterminacy and formal derivation of programs. *Comm. ACM*, 18(8), 453–457.

Dixon, L. and Kissinger, A. (2010). Open graphs and monoidal theories. *Math. Structures in Comp. Sci.*.

Doplicher, S. and Roberts, J.E. (1989). A new duality theory for compact groups. *Inventiones Mathematicae*, 98, 157–218.

Douglas, C., Schommer-Pries, C., and Snyder, N. (2011). Work in progress. See the slides at http://ncatlab.org/nlab/files/DSSFusionSlides.pdf.

Dowty, D.R., Wall, R.E., and Peters, S. (1981). *Introduction to Montague Semantics*. Springer.

Draisma, J. and Kuttler, J. (2009). On the ideals of equivariant tree models. *Math. Ann.*, 344, 619–644.

Drinfeld, V.G. (1987). Quantum groups. *Proc. ICM Berkeley*, 1(2), 798–820.

Drinfeld, V.G. (1989). Quasi-Hopf algebras. *Algebra i Analiz*, 1(6), 114–148.

Drinfeld, V.G., Gelaki, S., Nikshych, D., and Ostrik, V. (2007). Group-theoretical properties of nilpotent modular categories. arXiv:0704.0195.

Drinfeld, V. G., Gelaki, S., Nikshych, D., and Ostrik, V. (2010). On braided fusion categories. I. *Selecta Math. (N.S.)*, 16, 1–119.

Duncan, R. (2006). *Types for Quantum Computing*. Ph. D. thesis, Oxford University.

Duncan, R. and Perdrix, S. (2009). Graph states and the necessity of Euler decomposition. In *Computability in Europe: Mathematical Theory and Computational Practice (CiE'09)* (ed. K. Ambos-Spies, B. Löwe, and W. Merkle). Springer.

Duncan, R. and Perdrix, S. (2010). Rewriting measurement-based quantum computations with generalised flow. In *Automata, Languages and Programming, 37th International Colloquium, ICALP 2010, Proceedings Part II* (ed. S. Abramsky, C. Gavoille, C. Kirchner, F. Meyer auf der Heide, and P. Spirakis). Springer.

Dylan, B. (2004). *Lyrics: 1962–2001*. Barnes & Noble.

Edwards, W. (2009). *Non-locality in Categorical Quantum Mechanics*. Ph. D. thesis, Oxford University.

Eilenberg, S. and Mac Lane, S. (1945). General theory of natural equivalences. *Trans. Amer. Math. Soc.*, 58(2), 231.

Einstein, A., Podolsky, B., and Rosen, N. (1935). Can quantum-mechanical description of physical reality be considered complete? *Phys. Rev.*, 47, 777–780.

Erk, K. and Padó, S. (2008, October). A structured vector space model for word meaning in context. In *Proceedings of the 2008 Conference on Empirical Methods in Natural Language Processing*, Honolulu, Hawaii. Association for Computational Linguistics. Available at: http://www.aclweb.org/anthology/D08-1094.

Erné, M., Koslowski, J., Melton, A., and Strecker, G.E. (1993). A primer on Galois connections. In *Papers on General Topology and Applications* (ed. A. R. Todd). New York Acad. Sci.

Etingof, P. and Gelaki, S. (1998). Some properties of finite-dimensional semisimple Hopf algebras. *Math. Res. Lett.*, **5**, 191–197.

Etingof, P., Nikshych, D., and Ostrik, V. (2005). On fusion categories. *Ann. Math.*, **162**, 581–642.

Etingof, P., Nikshych, D., and Ostrik, V. (2011). Weakly group-theoretical and solvable fusion categories. *Adv. Math.*, **226**, 176–205.

Etingof, P. and Ostrik, V. (2004). Finite tensor categories. *Mosc. Math. J.*, **4**, 627–654, 782–783.

Fauser, B. (2001). On the Hopf-algebraic origin of Wick normal-ordering. *J. Phys. A*, **34**, 105–115. arXiv:hep-th/0007032.

Fauser, B. (2002). A Treatise on Quantum Clifford Algebras. Habilitationsschrift, arXiv:math.QA/0202059.

Fauser, B. (2006). Products, coproducts, and singular value decomposition. *Int. J. Theor. Phys.*, **49**(9), 1718–1743. arXiv:math-ph/0403001.

Fauser, B. (2008). Renormalization: a number theoretical model. *Commun. Math. Phys.*, **277**, 627–641. in press, math-ph/0601053.

Fauser, B. and Jarvis, P.D. (2004). A Hopf laboratory for symmetric functions. *J. Phys. A*, **37**(5), 1633–1663. arXiv:math-ph/0308043.

Fauser, B. and Jarvis, P.D. (2006). The Dirichlet Hopf algebra of arithmetics. *J. Knot Theor. Ramif.*, **16**(4), 1–60. arXiv:math-ph/0511079.

Fauser, B., Jarvis, P.D., and King, R.C. (2010). Plethysms as a source of replicated Schur functions and series, with applications to vertex operators. *J. Phys. A.*, **43**, 405202.

Fauser, B., Jarvis, Peter D., King, R.C., and Wybourne, B.G. (2006). New branching rules induced by plethysm. *J. Phys A*, **39**, 2611–2655. arXiv:math-ph/0505037.

Fenn, R. and Rourke, C. (1992). Racks and links in codimension 2. *J. Knot Th. Ramif.*, **1**, 343–406.

Finkelstein, L., Gabrilovich, E., Matias, Y., Rivlin, E., Solan, Z., Wolfman, G., and Ruppin, E. (2002). Placing search in context: the concept revisited. *ACM Trans. Inf. Syst.*, **20**(1), 116–131.

Fiore, M., Plotkin, G., and Turi, D. (1999). Abstract syntax and variable binding. In *Logic in Computer Science* (ed. M. A. et al.). IEEE: Computer Science Press.

Firth, J.R. (1957). *Papers in Linguistics 1934–1951*. Oxford University Press.

Firth, J.R. (1968). A synopsis of linguistic theory, 1930-1955. In *Selected papers of JR Firth, 1952-59* (ed. J. Firth and F. Palmer). Indiana University Press.

Fodor, J. and Lepore, E. (1999). All at sea in semantic space: Churchland on meaning similarity. *J. Phil.*, **96**, 381-403.

Fredenhagen, K., Rehren, K.-H., and Schroer, B. (1989). Superselection sectors with braid group statistics and exchange algebras I. General theory. *Comm. Math. Phys.*, **125**, 201-226.

Fredenhagen, K., Rehren, K.-H., and Schroer, B. (1992). Superselection sectors with braid group statistics and exchange algebras II. Geometric aspects and conformal covariance. *Rev. Math. Phys.*, **4**, 113-157.

Freedman, M.H., Kitaev, A., Larsen, M.J., and Wang, Z. (2003). Topological quantum computation. Mathematical challenges of the 21st century. *Bull. Amer. Math. Soc.*, **40**, 31-38.

Frege, G. (1980 (original 1884)). *The Foundations of Arithmetic*. Northwestern University Press.

Freyd, P. and Yetter, D. (1989). Braided compact closed categories with applications to low-dimensional topology. *Adv. Math.*, **77**, 156-182.

Freyd, P. and Yetter, D. (1992). Coherence theorems via knot theory. *J. Pure Appl. Alg.*, **78**, 49-76.

Freyd, P., Yetter, D., Hoste, J., Lickorish, W.B.R., Millet, K., and Ocneanu, A. (1985). A new polynomial invariant of knots and links. *Bull. Amer. Math. Soc.*, **12**, 239-246.

Frobenius, F.G. (1903). Theorie der hyperkomplexen Größen. *Sitz. Ber. Akad. Wiss. Berlin*, 504-537.

Fröhlich, J., Fuchs, J., Runkel, I., and Schweigert, C. (2006). Correspondences of ribbon categories. *Adv. Math.*, **199**, 192-329.

Fuchs, J. (2006). The graphical calculus for ribbon categories: algebras, modules, Nakayama automorphism. *J. Nonlinear Math. Phys.*, **13**, 44-54.

Fuchs, J., Runkel, I., and Schweigert, C. (2002). TFT construction of RCFT correlators I: Partition functions. *Nucl. Phys. B*, **646**, 353-497.

Fuchs, J., Runkel, I., and Schweigert, C. (2004a). TFT construction of RCFT correlative II. *Nuclear Phys.*, **B678**, 511-637.

Fuchs, J., Runkel, I., and Schweigert, C. (2004b). TFT construction of RCFT correlative III. *Nuclear Phys.*, **B694**, 277-353.

Fuchs, J., Runkel, I., and Schweigert, C. (2006). TFT construction of RCFT correlative IV. *Theory Appl. Categ.*, **16**, 342-433.

Fuchs, J., Runkel, I., and Schweigert, C. (2007). Ribbon categories and (unoriented) CFT: Frobenius algebras, automorphisms, reversions. *Contemp. Math.*, **431**, 203-224. arXiv:math.CT/0511590.

Fuchs, J. and Stigner, C. (2008). On Frobenius algebras in rigid monoidal categories. *Arab. J. Sci. Eng.*, **33**, 175–191.

Gelaki, S., Naidu, D., and Nikshych, D. (2009). Centers of graded fusion categories. *Alg. Num. Th.*, **3**, 959–990.

Gelaki, S. and Nikshych, D. (2008). Nilpotent fusion categories. *Adv. Math.*, **217**, 1053–1071.

Girard, J.-Y. (1971). Une extension de l'interprétation de Gödel à l'analyse, et son application à l'élimination des coupures dans l'analyse et la théorie des types. In *Proceedings of the Second Scandinavian Logic Symposium*. North-Holland.

Girard, J.-Y. (1987). Linear logic. *Theor. Comput. Sci.*, **50**(1), 1–101.

Girard, J.-Y. (1988). Geometry of interaction 1. In *Proceedings Logic Colloquium Ã•88*, pp. 221–260. North-Holland.

Girard, J.-Y. (1991). A new constructive logic: classical logic. *Math. Structures Comp. Sci.*, **1**(3), 255–296.

Girard, J.-Y. (1996). Proof-nets: the parallel syntax for proof theory. In *Logic and Algebra*. Marcel Dekker.

Golan, J.S. (1999). *Semirings and their Applications*. Kluwer.

Gomez, X. and Majid, S. (2003). Braided Lie algebras and bicovariant differential calculi over coquasitriangular Hopf algebras. *J. Algebra*, **261**, 334–388.

Goncharov, S., Schröder, L., and Mossakowski, T. (2009). Kleene monads: Handling iteration in a framework of generic effects. In *Conference on Algebra and Coalgebra in Computer Science (CALCO 2009)* (ed. A. Kurz and A. Tarlecki). Springer.

Goré, R. (1997). Substructural logics on display. *Logic J. IGPL*, **6**(3), 451–504.

Gottesman, D. and Chuang, I.L. (1999). Quantum teleportation is a universal computational primitive. *Nature*, **402**, 390–393.

Grefenstette, E. and Sadrzadeh, M. (2011). Experimental support for a categorical compositional distributional model of meaning. In *Proceedings of the 2011 Conference on Empirical Methods in Natural Language Processing*, Edinburgh, Scotland, UK. Association for Computational Linguistics.

Grefenstette, E., Sadrzadeh, M., Clark, S., Coecke, B., and Pulman, S. (2011). Concrete sentence spaces for compositional distributional models of meaning. In *Proceedings of the 9th International Conference on Computational Semantics (IWCS-11)*, Oxford, UK. Available at: http://www.aclweb.org/anthology/W/W11/W11-0114.pdf.

Grishin, V.N. (1983). On a generalization of the Ajdukiewicz–Lambek system. In *Studies in Nonclassical Logics and Formal Systems*. Nauka.

Gross, M. (1975). *Méthodes en Syntaxe*. Hermann.

Guevara, E. (2010, July). A regression model of adjective-noun compositionality in distributional semantics. In *Proceedings of the 2010 Workshop on GEometrical Models of Natural Language Semantics*, Uppsala, Sweden. Association for Computational Linguistics.

Haag, R. (1996). *Local Quantum Physics*. Springer.

Haghverdi, E. (2000). *A Categorical Approach to Linear Logic, Geometry of Proofs and Full Completeness*. Ph. D. thesis, University of Ottawa.

Hahn, A.J. (1994). *Quadratic Algebras, Clifford algebras, and Arithmetic Witt Groups*. Springer.

Halmos, P. (1974). *Lectures on Boolean Algebras*. Springer.

Halstead, T.J. (2008). District of Columbia v. Heller: The Supreme Court and the second amendment. *CRS report for congress RL34446*.

Harel, D., Kozen, D., and Tiuryn, J. (2000). *Dynamic Logic*. MIT Press.

Harris, Z.S. (1954). Distributional structure. *Word*, **10**, 146–162.

Harris, Z.S. (1957). Co-occurrence and transformation in linguistic structure. *Language*, **33**(3), 283–340.

Harris, Z.S. (1968). *Mathematical Structures of Language*. Wiley.

Hasegawa, R. (2002). Two applications of analytic functors. *Theor. Comp. Sci.*, **272**(1-2), 113–175.

Heunen, C. (2009). An embedding theorem for Hilbert categories. *Theory Applic. Categ.*, **22**(13), 321–344.

Hewitt, E. and Ross, K.A. (1970). *Abstract Harmonic Analysis. Vol. II: Structure and Analysis for Compact Groups. Analysis on Locally Compact Abelian groups*. Springer.

Heyneman, R.G. and Sweedler, M.E. (1969). Affine Hopf algebras I. *J. Algebra*, **13**, 192–241.

Heyneman, R.G. and Sweedler, M.E. (1970). Affine Hopf alegbras II. *J. Algebra*, **16**, 271–297.

Higginbotham, J. (2003). Conditionals and compositionality. *Phil. Perspect.*, **17**(1), 181–194.

Hines, P. (1998). *The Algebra of Self-similarity and its Applications*. Ph. D. thesis, University of Wales, Bangor.

Hines, P. (1999). The categorical theory of self-similarity. *Theory Applic. Categ.*, **6**, 33–46.

Hines, P. (2002). A short note on coherence and self-similarity. *J. Pure Appl. Algebra*, **175**(1-3), 135–139.

Hines, P. (2003). A categorical framework for finite state machines. *Math. Structures Comp. Sci.*, **13**, 451–480.

Hinman, P. (2005). *Fundamentals of Mathematical Logic*. A.K. Peters.

Hitchin, N. (1997). Frobenius manifolds. In *Gauge Theory and Symplectic Geometry* (ed. J. Hurtubise and F. Lalonde). Kluwer.

Hoare, C.A.R., Hayes, I.J., Jifeng, H., Morgan, C.C., Roscoe, A.W., Sanders, J.W., Sorensen, I.H., Spivey, J.M., and Sufrin, B.A. (1987). Laws of programming. *Communications of the ACM*, **30**(8), 672–686.

Hockenmaier, J. and Steedman, M. (2007). CCGbank: a corpus of CCG derivations and dependency structures extracted from the Penn Treebank. *Comput. Linguist.*, **33**(3), 355–396.

Hopf, H. (1941). Über die Topologie der Gruppen-Mannigfaltigkeiten und ihre Verallgemeinerungen. *Ann. Math.*, **42**(1), 22–52.

Horsman, C. (2011). Quantum picturalism for topological cluster-state computing. *New J. Phys.*, **13**, 095011.

Huang, Y.-Z. (2005). Vertex operator algebras, the Verlinde conjecture, and modular tensor categories. *Proc. Natl. Acad. Sci. USA*, **102**, 5352–5356.

Hyland, M. and Power, J. (2007). The category theoretic understanding of universal algebra: Lawvere theories and monads. In *Computation, Meaning, and Logic: Articles Dedicated to Gordon Plotkin* (ed. L. Cardelli, M. Fiore, and G. Winskel). Elsevier.

Hyland, M. and Schalk, A. (2003). Glueing and orthogonality for models of linear logic. *Theor. Comput. Sci.*, **294**, 183–231.

Jacobs, B. (2010). From coalgebraic to monoidal traces. In *Coalgebraic Methods in Computer Science*. Elsevier.

Jacobs, B. (2011). Probabilities, distribution monads, and convex categories. *Theor. Comp. Sci.*, **412**, 3323–3336.

Janssen, K. and Vercruysse, J. (2010). Multiplier bi- and Hopf algebras. *J. Algebra Appl.*, **9**(2), 275–303.

Janssen, K. and Vercruysse, J. (2012). Kleisli Hopf algebras. Forthcoming.

Janssen, T.M.V. (2001). Frege, contextuality and compositionality. *J. Logic Lang. Inform.*, **10**(1), 115–136.

Jarvis, P.D., Bashford, J.D., and Sumner, J.G (2005). Pathintegral formulation and Feynman rules for phylogenetic branching processes. *J. Phys. A.*, **38**, 9621–9647.

Jarvis, P.D. and Sumner, J.G. (2012). Markov invariants for phylogenetic rate matrices derived from embedded submodels. *J. Math. Biology*, **9**(3), 828–836.

Jones, V.F.R. (1985). A polynomial invariant for knots via von Neumann algebras. *Bull. Amer. Math. Soc.*, **12**, 103–111.

Joyal, A. (1986). Foncteurs analytiques et espèces de structures. In *Combinatoire Enumerative* (ed. G. Labelle and P. Leroux). Springer.

Joyal, A. and Street, R. (1988). Planar diagrams and tensor algebras. Unpublished paper, available from http://www.math.mq.edu.au/~street/PlanarDiags.pdf (accessed 22 August 2012).

Joyal, A. and Street, R. (1991a). Geometry of tensor calculus, I. *Adv. Math.*, **88**, 55–112.

Joyal, A. and Street, R. (1991b). Geometry of tensor calculus II. Unpublished paper, available from http://www.math.mq.edu.au/˜street/GTCII.pdf (accessed 22 August 2012).

Joyal, A. and Street, R. (1991c). An introduction to Tannaka duality and quantum groups. In *Category Theory (Como, 1990)*. Springer.

Joyal, A. and Street, R. (1991d). Tortile Yang–Baxter operators in tensor categories. *J. Pure Appl. Alg.*, **71**, 42–51.

Joyal, A. and Street, R. (1993). Braided tensor categories. *Adv. Math.*, **102**, 20–78.

Kadison, L. (1999). *New Examples of Frobenius Extensions*. American Mathematical Society.

Kallmeyer, L. (2010). *Parsing Beyond Context-Free Grammars*. Springer.

Karaali, G. (2008). On Hopf algebras and their generalizations. *Comm. Algebra*, **36**(12), 4341–4367.

Kasch, F. (1982). *Modules and Rings*. Academic Press. trans. D.A.R. Wallace.

Kasprzak, W., Lysik, B., and Rybaczuk, M. (1990). *Dimensional Analysis in the Identification of Mathematical Models*. World Scientific.

Kassel, C. (1995). *Quantum Groups*. Springer.

Kassel, C. and Turaev, V.G. (1998). Chord diagram invariants of tangles and graphs. *Duke Math. J.*, **92**, 497–552.

Katz, J.J. and Fodor, J.A. (1963). The structure of a semantic theory. *Language*, **39**, 170–210.

Kauffman, L.H. (1991). *Knots and Physics*. World Scientific.

Kauffman, L.H. (1999). Virtual knot theory. *European J. Combin.*, **20**, 663–690.

Kawahigashi, Y., Longo, R., and Müger, M. (2001). Multi-interval subfactors and modularity of representations in conformal field theory. *Comm. Math. Phys.*, **219**, 631–669.

Kaye, P., Laflamme, R., and Mosca, M. (2007). *An Introduction to Quantum Computing*. Oxford University Press.

Kazhdan, D. and Lusztig, G. (1993–1994). Tensor structures arising from affine Lie algebras I–IV. *J. Amer. Math. Soc.*, **6–7**, 905–947, 949–1011, 335–381, 383–453.

Kellendonk, J. and Lawson, M.V. (2000). Tiling semigroups. *J. Algebra*, **224**(1), 140–150.

Kelly, G.M. (1972a). An abstract approach to coherence. In *Coherence in Categories*. Springer.

Kelly, G.M. (1972b). Many-variable functorial calculus I. In *Coherence in Categories*. Springer.

Kelly, G.M. (2005). Basic concepts of enriched category theory. *Repr. Theory Appl. Categ.* (10), vi+137. Reprint of the 1982 original [Cambridge University Press, Cambridge; MR0651714].

Kelly, M. and Laplaza, M. (1980). Coherence for compact closed categories. *J. Pure Appl. Algebra*, **19**, 193–213.

Kempf, A. and Majid, S. (1994). Algebraic q-integration and Fourier theory on quantum and braided spaces. *J. Math. Phys.*, **35**, 6802–6837.

Khovanov, M. (1997). *Graphical Calculus, Canonical Bases and Khazhdan–Lusztig Theory*. Ph. D. thesis, Yale University. i–iv, 1–103.

Khovanov, M. (2006). Link homology and Frobenius extensions. *Fundam. Math.*, **190**, 179–190. arXiv:math/0411447v2.

Khovanov, M. (2010). Categorifications from planar diagrammatics. *Jap. J. Math.*, **5**, 153–181.

Kilgarriff, A. (1997). I don't believe in word senses. *Comput. Humanities*, **31**(2), 91–113.

Kim, S. and Hovy, E. (2006). Identifying and analyzing judgment opinions. In *Proceedings of the Human Language Technology / North American Association of Computational Linguistics conference*.

Kintsch, W. (2001). Predication. *Cognitive Sci.*, **25**(2), 173–202.

Kirillov Jr., A. (2001). Modular categories and orbifold models II. arXiv:math.QA/0110221.

Kirillov Jr., A. (2002). Modular categories and orbifold models I. *Comm. Math. Phys.*, **229**, 309–335.

Kirillov Jr., A. (2004). On g-equivariant modular categories. arXiv:math.WA/0401119.

Kirillov Jr., A. and Ostrik, V. (2002). On q-analog of McKay correspondence and ADE classification of $sl(2)$ conformal field theories. *Adv. Math.*, **171**, 183–227.

Kitaev, A. (2003). Fault-tolerant quantum computation by anyons. *Ann. Physics*, **30**, 2–30.

Kleene, S.C. (1967). *Mathematical Logic*. Dover Publications.

Kock, A. (1971). Closed categories generated by commutative monads. *J. Austr. Math. Soc.*, **XII**, 405–424.

Kock, A. (2011). Monads and extensive quantities. arXiv:1103.6009.

Kock, J. (2003). *Frobenius Algebras and 2D Topological Quantum Field Theories*. Cambridge University Press.

Kong, L. and Runkel, I. (2009). Cardy algebras and sewing constraints. I. *Comm. Math. Phys.*, **292**, 871–912.

Kripke, S. (1963). Semantical considerations on modal logic. *Acta Philosophica Fennica*, **16**(1963), 83–94.

Kuperberg, G. (1991). Involutory Hopf algebras and 3-manifold invariants. *Int. J. Math.*, **2**(1), 41–66.

Kustermans, J. and Vaes, S. (2000). Locally compact quantum groups. *Ann. Sci. École Norm. Sup. (4)*, **33**(6), 837–934.

Lam, T.Y. and Leroy, A. (1988). Vandermonde and Wronskian matrices over division rings. *J. Algebra*, **119**, 308–336.

Lambek, J. (1958). The mathematics of sentence structure. *Am. Math. Mon.*, **65**, 154–170.

Lambek, J. (1961). On the calculus of syntactic types. In *Structure of Language and its Mathematical Aspects* (ed. R. Jacobson). American Mathematical Society.

Lambek, J. (1988). Categorial and categorical grammars. In *Categorial Grammars and Natural Language Structures* (ed. R. Oehrle, E. Bach, and D. Wheeler). Reidel.

Lambek, J. (1993). From categorial to bilinear logic. In *Substructural Logics* (ed. K. Došen and P. Schröder-Heister), pp. 207–237. Oxford University Press.

Lambek, J. (1999). Type grammar revisited. In *Logical Aspects of Computational Linguistics* (ed. A. Lecomte, F. Lamarche, and G. Perrier). Springer.

Lambek, J. (2008). *From Word to Sentence*. Polimetrica, Milan.

Lambek, J. (2010). Compact monoidal categories from linguistics to physics. In *New Structures for Physics* (ed. B. Coecke). Springer.

Lambek, J. and Scott, P. (1986). *Introduction to Higher Order Categorical Logic*. Cambridge University Press.

Lamport, L. (1978). Time, clocks, and the ordering of events in a distributed system. *Commun. ACM*, **21**(7), 558–565.

Laplaza, M. (1977). Coherence in non-monoidal closed categories. *Trans. Am. Math. Soc.*, **230**, 293–311.

Larson, R.G. (1998). Topological Hopf algebras and braided monoidal categories. *Appl. Categ. Structures*, **6**(2), 139–150.

Larson, R.G. and Sweedler, M.E. (1969). An associative orthogonal bilinear form for Hopf algebras. *Amer. J. Math.*, **91**(1), 75–94.

Lawrence, R.E. (1996). An introduction to topological field theory. *Proc. Symp. Appl. Math.*, **51**, 89–128.

Lawson, M.V. (1998). *Inverse Semigroups: the Theory of Partial Symmetries*. World Scientific.

Lawson, M.V. (2007). Orthogonal completions of the polycyclic monoids. *Comm. Algebra*, **35**(5), 1651–1660.

Lawvere, F.W. (1963). *Functorial semantics of algebraic theories and some algebraic problems in the context of functorial semantics of algebraic theories*. Ph. D. thesis, Columbia University. Reprinted in *Theory Applic. Categ.*, **5**, 1–121, 2004.

Lawvere, F.W. (1969). Adjointness in foundations. *Dialectica*, **23**(3–4), 281–296.

Lawvere, F.W. and Schanuel, S.H. (1997). *Conceptual Mathematics: A First Introduction to Categories*. Cambridge University Press.

Leinster, T. (2011). A general theory of self-similarity. *Adv. Mathem.*, **226**, 2935–3017.

Littlewood, D.E. (1940). *The Theory of Group Characters*. Oxford University Press, Oxford. AMS Chelsea Publishing, reprint of the 1976 Dover edn, 2006.

Loday, J.-L. and Pirashvili, T. (1993). Universal enveloping algebras of Leibniz algebras and (co)homology. *Math. Ann.*, **296**, 139–158.

Longo, R. and Roberts, J.E. (1997). A theory of dimension. *K-Theory*, **11**, 103. arXiv:funct-an/9604008.

Lopez Pena, J., Majid, S., and Rietsch, K. (2010). Lie theory of finite simple groups and the Roth property. arxiv:1003.5611.

Lorenz, M. (2011). Some applications of Frobenius algebras to Hopf algebras. In *Groups, Algebras and Applications*. American Mathematical Society.

Lorenz, M. and Fitzgerald Tokoly, L. (2011). Projective modules over Frobenius algebras and Hopf comodule algebras. *Comm. Alg.*, **39**, 4733–4750.

Luauda, A. (2006). Frobenius algebras and ambidextrous adjunctions. *Theory Appl. Categ.*, **16**, 84–122.

Lusztig, G. (1993). *Introduction to Quantum Groups*. Birkhauser.

Lyubashenko, V. (1995). Modular transformations for tensor categories. *J. Pure Appl. Algebra*, **98**, 279–327.

Lyubashenko, V. and Majid, S. (1994). Braided groups and quantum Fourier transform. *J. Algebra*, **166**, 506–528.

Mac Lane, S. (1963). Natural associativity and commutativity. *The Rice University Studies*, **49**(4), 28–46.

Mac Lane, S. (1971). *Categories for the Working Mathematician*. Springer.

MacDonald, I.G. (1979). *Symmetric Functions and Hall Polynomials*. Calderon Press.

Majid, S. (1988). Hopf algebras for physics at the Planck scale. *Classical Quant. Grav.*, **5**, 1587–1607.

Majid, S. (1990). Physics for algebraists: non-commutative and non-cocommutative Hopf algebras by a bicrossproduct construction. *J. Algebra*, **130**, 17–64.

Majid, S. (1991*a*). Braided groups and algebraic quantum field theories. *Lett. Math. Phys.*, **22**, 167–176.

Majid, S. (1991*b*). Doubles of quasitriangular Hopf algebras. *Comm. Algebra*, **19**, 3061–3073.

Majid, S. (1991*c*). Example of braided groups and braided matrices. *J. Math. Phys.*, **32**, 3246–3253.

Majid, S. (1991*d*). Hopf–von Neumann algebra bicrossproduct, Kac algebra bicrosspdoructs and classical Yang–Baxter equations. *J. Functional Analysis*, **95**, 291–319.

Majid, S. (1991*e*). The principle of representation-theoretic self-duality. *Phys. Essays*, **4**(3), 395–405.

Majid, S. (1991*f*). Reconstruction theorems and rational conformal field theories. *Int. J. Mod. Phys. A.*, **6**, 4359–4374.

Majid, S. (1991*g*). Representations, duals and quantum doubles of monoidal categories. *Suppl. Rend. Circ. Mat. Palermo., Series II*, **26**, 197–206.

Majid, S. (1992a). Braided groups and duals of monoidal categories. *Can. Math. Soc. Conf. Proc.*, **13**, 329–343.

Majid, S. (1992b). Rank of quantum groups and braided groups in dual form. In *Quantum Groups*. Springer.

Majid, S. (1993a). Braided groups. *J. Pure Appl. Algebra*, **86**, 187–221.

Majid, S. (1993b). Braided matrix structure of the Sklyanin algebra and of the quantum Lorentz group. *Comm. Math. Phys.*, **156**, 607–638.

Majid, S. (1993c). Free braided differential calculus, braided binomial theorem and the braided exponential map. *J. Math. Phys.*, **34**, 4843–4856.

Majid, S. (1993d). Transmutation theory and rank for quantum braided groups. *Math. Proc. Camb. Phil.*, **113**, 45–70.

Majid, S. (1994a). Algebras and Hopf algebras in braided categories. In *Advances in Hopf Algebras*. Marcel Dekker.

Majid, S. (1994b). Cross products by braided groups and bosonization. *J. Algebra*, **163**, 165–190.

Majid, S. (1994c). Quantum and braided Lie-algebras. *J. Geom. Phys.*, **13**, 307–356.

Majid, S. (1995a). *Foundations of Quantum Group Theory*. Cambridge University Press.

Majid, S. (1995b). Solutions of the Yang–Baxter equations from braided-Lie algebras and braided groups. *J. Knot Th. Ramif.*, **4**, 673–697.

Majid, S. (1998a). Classification of bicovariant differential calculi. *J. Geom. Phys.*, **25**, 119–140.

Majid, S. (1998b). Quantum double of quasi-Hopf algebras. *Lett. Math. Phys.*, **45**, 1–9.

Majid, S. (1999a). Diagrammatics of braided group gauge theory. *J. Knot Th. Ramif.*, **8**, 731–771.

Majid, S. (1999b). Double bosonisation of braided groups and the construction of $u_q(g)$. *Math. Proc. Camb. Phil Soc.*, **125**, 151–192.

Majid, S. (1999c). Quantum and braided group Riemannian geometry. *J. Geom. Phys.*, **30**, 113–146.

Majid, S. (2002). *A Quantum Groups Primer*. Cambridge University Press.

Majid, S. (2004). Noncommutative differentials and Yang–Mills on permutation groups s_n. *Lect. Notes Pure Appl. Maths*, **239**, 189–214.

Majid, S. (2005). Gauge theory on nonassociative spaces. *J. Math. Phys.*, **46**, 3519–3542.

Majid, S. (2007). Algebraic approach to quantum gravity I: relative realism. philsci-archive:00003345.

Makhlouf, A. and Silvestrov, S. (2009). Hom-Lie admissible Hom-coalgebras and Hom-Hopf algebras. In *Generalized Lie Theory in Mathematics, Physics and Beyond*. Springer.

Manning, C. and Schütze, H. (1999). *Foundations of Statistical Natural Language Processing*. MIT Press.

Marshall, A.W., Olkin, I., and Arnold, B. (2010). *Inequalities: Theory of Majorization and its Applications*. Springer.

Martin, K. (2011). Domain theory and measurement. In *New Structures for Physics* (ed. B. Coecke). Springer.

Martin, K. and Panangaden, P. (2006). A domain of spacetime intervals in general relativity. *Comm. Math. Phys.*, **267**(3), 563–586.

Masterman, M. (1957). The thesaurus in syntax and semantics. *Mech. Transl.*, **4**(1/2), 35–43.

Matthews, P. (2001). *A Short History of Structural Linguistics*. Cambridge University Press.

McCrudden, P. (2002). Tannaka duality for Maschkean categories. *J. Pure Appl. Algebra*, **168**(2–3), 265–307.

McCurdy, M.B. (2012). Graphical methods for Tannaka duality of weak bialgebras and weak Hopf algebras in arbitrary braided monoidal categories. *Theory Applic. Categor.*, **26**(9), 233–280.

Mermin, N.D. (2007). *Quantum Computer Science*. Cambridge University Press.

Mesablishvili, B. and Wisbauer, R. (2011). Bimonads and Hopf monads on categories. *J. K-Theory*, **7**(2), 349–388.

Meyer, J.J.C. and van der Hoek, W. (2004). *Epistemic Logic for AI and Computer Science*. Cambridge University Press.

Mhalla, M. and Perdrix, S. (2008). Finding optimal flows efficiently. In *Automata, Languages and Programming, Proceedings of ICALP 2008* (ed. L. Aceto, I. Damgård, L. Goldberg, M. Halldórsson, A. Ingólfsdóttir, and I. Walukiewicz). Springer.

Mitchell, J. and Lapata, M. (2008). Vector-based models of semantic composition. In *Proceedings of ACL-08: HLT*, Columbus, Ohio. Association for Computational Linguistics.

Moerdijk, I. (2002). Monads on tensor categories. *J. Pure Appl. Algebra*, **168**(2–3), 189–208.

Montague, R. (1973). The proper treatment of quantification in ordinary english. In *Approaches to Natural Language*, Volume 49. D. Reidel.

Moore, D.J. (1999). On state spaces and property lattices. *Stud. Hist. Philos. Mod. Phys.*, **30**(1), 61–83.

Moore, G. and Seiberg, N. (1989*a*). Classical and quantum conformal field theory. *Comm. Math. Phys.*, **123**, 177–254.

Moore, G. and Seiberg, N. (1989*b*). Taming the conformal zoo. *Phys. Lett.*, **B220**, 422–430.

Moore, R.K. (2003). A comparison of the data requirements of automatic speech recognition systems and human listeners. In *INTERSPEECH: 8th European Conference on Speech Communication and Technology, Geneva, Switzerland, September 1–4, 2003*. ISCA.

Moortgat, M. (1988). *Categorial Investigations: Logical and Linguistic Aspects of the Lambek Calculus*. Foris.

Moortgat, M. (1996). Multimodal linguistic inference. *J. Logic Lang. Inform.*, **5**(3,4), 349–385.

Moortgat, M. (1997). Categorial type logics. In *Handbook of Logic and Language* (ed. J. van Benthem and A. ter Meulen). Elsevier and MIT Press.

Moortgat, M. (2007). Symmetries in natural language syntax and semantics: the Lambek–Grishin calculus. In *Proceedings of WoLLIC 2007*. Springer.

Moortgat, M. (2009). Symmetric categorial grammar. *J. Philos. Logic*, **38**(6), 681–710.

Moot, R. (2007). Proof nets for display logic. Technical report, CNRS and INRIA Futurs.

Moot, R. and Puite, Q. (2002). Proof nets for the multimodal Lambek calculus. *Studia Logica*, **71**(3), 415–442.

Morrill, G. (1994). *Type Logical Grammar*. Kluwer.

Morrill, G., Fadda, M., and Valentin, O. (2007). Nondeterministic discontinuous Lambek calculus. In *Proceedings of the Seventh International Workshop on Computational Semantics (IWCS7)*, Tilburg.

Müger, M. (2000). Galois theory for braided tensor categories and the modular closure. *Adv. Math.*, **150**, 151–201.

Müger, M. (2003a). From subfactors to categories and topology I. Frobenius algebras in and Morita equivalence of tensor categories. *J. Pure Appl. Alg.*, **180**, 81–157.

Müger, M. (2003b). From subfactors to categories and topology II. The quantum double of tensor categories and subfactors. *J. Pure Appl. Alg.*, **180**, 159–219.

Müger, M. (2003c). On the structure of modular categories. *Proc. Lond. Math. Soc.*, **87**, 291–308.

Müger, M. (2004a). Galois extensions of braided tensor categories and braided crossed g-categories. *J. Algebra*, **277**, 256–281.

Müger, M. (2004b). On the center of a compact group. *Intern. Math. Res. Not.*, **51**, 2751–2756.

Müger, M. (2010a). On superselection theory of quantum fields in low dimensions. In *Proceedings of the XVIth International Congress on Mathematical Physics* (ed. P. Exner). World Scientific. arXiv:0909.2537.

Müger, M. (2010b). On the structure of braided crossed g-categories. In *Homotopy Quantum Field Theory* (ed. V. G. Turaev). European Mathematical Society.

Müger, M. (2010c). Tensor categories: a selective guided tour. *Rev. Un. Mat. Argent.*, **51**, 95–163.

Muirhead, R.F. (1903). Some methods applicable to identities and inequalities of symmetric algebraic functions of *n* letters. *Proc. Edinburgh Math. Soc.*, **21**, 144–157.

Murray, W. (2005). Bilinear forms on Frobenius algebras. *J. Algebra*, **293**, 89–101.

Naaijkens, P. (2011). Localized endomorphisms in Kitaev's toric code on the plane. *Rev. Math. Phys.*, **23**, 347–373.

Naidu, D., Nikshych, D., and Witherspoon, S. (2009). Fusion subcategories of representation categories fo twisted quantum doubles of finite groups. *Int. Math. Res. Not.*, **22**, 4183–4219.

Nayak, C., Simon, S.H., Stern, A., Freedman, M., and Das Sarma, S. (2008). Non-Abelian anyons and topological quantum computation. *Rev. Mod. Phys.*, **80**, 1083–1159.

Nesbitt, C. (1938). On the regular representations of algebras. *Ann. Math.*, **39**(3), 634–658.

Ng, S.-H. and Schauenburg, P. (2010). Congruence subgroups and generalized Frobenius–Schur indicators. *Comm. Math. Phys.*, **300**, 1–46.

Nicholson, W.K. and Yousif, M.F (2003). *Quasi-Frobenius Rings*. Cambridge University Press.

Nida, E. (1946). *Morphology*. University of Michigan Press.

Nielsen, M.A. (1999). Conditions for a class of entanglement transformations. *Phys. Rev. Letters*, **83**(2), 436–439.

Nielsen, M.A. and Chuang, I.L. (2000). *Quantum Computation and Quantum Information*. Cambridge University Press.

Nivat, M. and Perrot, J. (1970). Une généralisation du mono ide bicyclique. *CR Acad. Sci. Paris*, **27**, 824–827.

Oberst, U. and Schneider, H.-J. (1973). Über Untergruppen endlicher algebraischer Gruppen. *Manuscripta Math.*, **8**, 217–241.

Ohtsuki, T. (2002). *Quantum Invariants*. World Scientific.

Ostrik, V. (2003). Module categories, weak Hopf algebras and modular invariants. *Transf. Groups*, **8**, 177–206.

Oziewicz, Z. and Wene, G.P. (2011). S_3-permuted Frobenius algebras. arXiv:1103.5113v1.

Padó, S. and Lapata, M. (2007). Dependency-based construction of semantic space models. *Comput. Linguist.*, **33**(2), 161–199.

Pareigis, B. (1971). When Hopf algebras are Frobenius algebras. *J. Algebra*, **18**, 588–596.

Pareigis, B. (1995). On braiding and dyslexia. *J. Algebra*, **171**, 413–425.

Pati, A.K. and Braunstein, S.L. (2000). Impossibility of deleting an unknown quantum state. *Nature*, **404**, 164–165.

Pavlovic, D. (2009, October). Geometry of abstraction in quantum computation. Technical Report RR-09-13, OUCL.

Pelletier, F.J. (2001). Did Frege believe Frege's principle? *J. Logic Lang. Inform.*, **10**, 87–114.

Penrose, R. (1971). Applications to negative dimensional tensors. In *Combinatorial Mathematics and its Applications* (ed. D. Welsh). Academic Press.

Piron, C. (1976). *Foundations of Quantum Physics*. W.A. Benjamin.

Preller, A. (2005). Category theoretical semantics for pregroup grammars. In *Logical Aspects of Computational Linguistics* (ed. P. Blache and E. Stabler). Springer.

Preller, A. (2007). Toward discourse representation via pregroup grammars. *J. Logic Lang. Inform.*, **16**, 173–194.

Preller, A. and Lambek, J. (2007). Free compact 2-categories. *Math. Struct. Comp. Sci.*, **17**(1), 1–32.

Preller, A and Sadrzadeh, M. (2011*a*). Bell states and negative sentences in the distributed model of meaning. *Electron. Notes Theor. Comp. Sci.*, **270**(2), 141–153.

Preller, A. and Sadrzadeh, M. (2011*b*). Bell states and negative sentences in the distributed model of meaning. In *Proceedings of the 6th International Workshop on Quantum Physics and Logic* (ed. B. Coecke, P. Panangaden, and P. Selinger). Elsevier.

Preller, A. and Sadrzadeh, M. (2011*c*). Semantic vector models and functional models for pregroup grammars. *J. Logic Lang. Inform.*, **20**(4), 419–423.

Presley, A. and Segal, G. (1986). *Loop Groups*. Oxford University Press.

Pulman, S.G. (2005). Lexical decomposition: For and against. In *Charting a New Course: Natural Language Processing And Information Retrieval: Essays in Honour of Karen Spärck Jones* (ed. J. I. Tait). Kluwer/Springer.

Pulman, S.G. (2010). Karen Ida Boalth Spärck Jones 1935–2007. In *Biographical Memoirs of Fellows IX*. Oxford University Press, for the British Academy.

Radford, D. (1985). The structure of Hopf algebras with a projection. *J. Algebra*, **92**, 322–347.

Raussendorf, R. and Briegel, H.J. (2001). A one-way quantum computer. *Phys. Rev. Lett.*, **86**, 5188–5191.

Raussendorf, R. and Briegel, H.J. (2003). Computational model for the one-way quantum computer: Concepts and summary. In *Quantum Information Processing* (ed. G. Leuchs and T. Beth). Wiley.

Raussendorf, R., Browne, D.E., and Briegel, H.J. (2003). Measurement-based quantum computation with cluster states. *Phys. Rev. A*, **68**(022312).

Rehren, K.-H. (1990). Braid group statistics and their superselection rules. In *The Algebraic Theory of Superselection Sectors. Introduction and Recent Examples*. World Scientific.

Reisinger, J. and Mooney, R.J. (2010). Multi-prototype vector-space models of word meaning. In *Human Language Technologies: Conference of the North American Chapter of the Association of Computational Linguistics, Proceedings, June 2–4, 2010, Los Angeles, California, USA*. Available at: http://aclweb.org/anthology-new/.

Reshetikhin, N. Yu. and Turaev, V.G. (1991). Invariants of 3-manifolds via link polynomials and quantum groups. *Invent. Math.*, **103**, 547–598.

Reynolds, J. (1983). Types, abstraction and parametric polymorphism. In *Information Processing* (ed. R. Mason). Elsevier.

Rijsbergen, C.J. van (2004). *The Geometry of Information Retrieval*. Cambridge University Press.

Rota, G.-C. and Stein, J.A. (1994). Plethystic Hopf algebras. *Proc. Natl. Acad. Sci. USA*, **91**, 13057–13061.

Rowell, E. (2005). On a family of non-unitarizable ribbon categories. *Math. Z.*, **250**, 745–774.

Rubenstein, H. and Goodenough, J.B. (1965). Contextual correlates of synonymy. *Commun. ACM*, **8**(10), 627–633.

Rudolph, S. and Giesbrecht, E. (2010). Compositional matrix-space models of language. In *ACL 2010, Proceedings of the 48th Annual Meeting of the Association for Computational Linguistics, July 11–16, 2010, Uppsala, Sweden*. Available at: http://www.aclweb.org/anthology/P10-1093.

Sadrzadeh, M. (2006). Pregroup analysis of Persian sentences. In *Computational Algebraic Approaches to Natural Language* (ed. C. Casadio and J. Lambek). Polimetrica.

Sasaki, U. (1954). Orthocomplemented lattices satisfying the exchange axiom. *J. Sci. Hiroshima Uni A*, **17**, 293–302.

Schäppi, D. (2011a). The formal theory of Tannaka duality. arXiv:1112.5213.

Schäppi, D. (2011b). Tannaka duality for comonoids in cosmoi. arXiv:0911.0977.

Schauenburg, P. (1992). *Tannaka Duality for Arbitrary Hopf Algebras*. Reinhard Fischer.

Schauenburg, P. (1998). Bialgebras over noncommutative rings and a structure theorem for Hopf bimodules. *Appl. Categ. Structures*, **6**(2), 193–222.

Schauenburg, P. (2000). Duals and doubles of quantum groupoids ($\times_R$-Hopf algebras). In *New Trends in Hopf Algebra Theory (La Falda, 1999)*. American Mathematical Society.

Schauenburg, P. (2001). The monoidal center construction and bimodules. *J. Pure Appl. Alg.*, **158**, 325–346.

Schütze, H. (1998, March). Automatic word sense discrimination. *Comput. Linguist.*, **24**(1), 97–123.

Scott, D.S. (1970). *Outline of a Mathematical Theory of Computation*. Oxford University Computing Laboratory, Programming Research Group.

Seely, R.A.G. (1989). Linear logic, ∗-autonomous categories and cofree algebras. *Contemp. Math.*, **92**, 371–382.

Selinger, P. (2007). Dagger compact closed categories and completely positive maps (extended abstract). In *Proceedings of the 3rd International Workshop on Quantum Programming Languages* (ed. P. Selinger), pp. 139–163.

Selinger, P. (2011). A survey of graphical languages for monoidal categories. In *New Structures for Physics* (ed. B. Coecke). Springer. arXiv:0908.3347.

Shieber, Stuart M. (1985). Evidence against the context-freeness of natural language. *Linguist. Philos.*, **8**, 333–343.

Sinclair, J. McH. (2004). *Trust the Text. Language, Corpus and Discourse*. Routledge.

Smolensky, P. (1987). The constituent structure of connectionist mental states: A reply to Fodor and Pylyshyn. *Southern J. Philos. (Supplement)*, **26**, 137–163.

Smolensky, P. (1988). Connectionism, constituency and language of thought. In *Minds, Brains, and Computers* (ed. R. Cummins and D. D. Cummins). Wiley-Blackwell.

Smolensky, P. (1990). Tensor product variable binding and the representation of symbolic structures in connectionist systems. *Artif. Intell.*, **46**(1–2), 159–216.

Socher, R., Manning, C.D., and Ng, A.Y. (2010). Learning continuous phrase representations and syntactic parsing with recursive neural networks. In *Proceedings of the NIPS-2010 Deep Learning and Unsupervised Feature Learning Workshop*. Available at: http://www.proceedings.com/7455.html.

Sommerhäuser, Y. (2002). *Yetter–Drinfel'd Hopf Algebras over Groups of Prime Order*. Springer.

Sommerhäuser, Y. and Zhu, Y. (2007). Hopf algebras and congruence subgroups. arXiv:0710.0705.

Sorkin, R.D. (2003). Causal sets: discrete gravity (notes for the Valdivia Summer School). arXiv:gr-qc/0309009.

Spärck Jones, K. (1986). *Synonymy and Semantic Classification*. Edinburgh University Press. (Original thesis 1964.)

Steedman, M. (2000). *The Syntactic Process*. The MIT Press, Cambridge, MA.

Steedman, M. and Baldridge, J. (2011). Combinatory categorial grammar. In *Non-Transformational Syntax: Formal and Explicit Models of Grammar* (ed. R. Borsley and K. Borjars). Wiley-Blackwell.

Stirling, C. (1991). *Modal and Temporal Logics*. University of Edinburgh, Department of Computer Science, Laboratory for Foundations of Computer Science.

Street, R. (2007). *Quantum Groups: A Path to Current Algebra*. Cambridge University Press.

Sumner, J.G., Holland, B.H., and Jarvis, P.D. (2012). The algebra of the general Markov model on phylogenetic trees and networks. *Bull. Math. Biol.*, **74**(4), 858–880.

Sweedler, M.E. (1968). Cohomology of algebras over Hopf algebras. *Trans. Am. Math. Soc.*, **133**, 205–239.

Sweedler, M.E. (1969). *Hopf Algebras*. W.A. Benjamin.

Szabó, Z.G. (2008). Compositionality. In *The Stanford Encyclopedia of Philosophy* (Winter 2008 edn) (ed. E. N. Zalta). Available at: http://plato.stanford.edu/entries/compositionality/.

Szlachányi, K. (2003). The monoidal Eilenberg–Moore construction and bialgebroids. *J. Pure Appl. Algebra*, **182**(2–3), 287–315.

Szláchanyi, K. (2009). Fiber functors, monoidal sites and Tannaka duality for bialgebroids. arXiv:0907.1578.

Takeuchi, M. (2000). Survey of braided Hopf algebras. In *New Trends in Hopf Algebra Theory (La Falda, 1999)*. American Mathematical Society.

Tang, X., Weinstein, A., and Zhu, C. (2007). Hopfish algebras. *Pacific J. Math.*, **231**(1), 193–216.

Thater, S., Fürstenau, H., and Pinkal, M. (2010). Contextualizing semantic representations using syntactically enriched vector models. In *Proceedings of the 48th Annual Meeting of the Association for Computational Linguistics*, Uppsala, Sweden. Association for Computational Linguistics.

Thater, S., Fürstenau, H., and Pinkal, M. (2011). Word meaning in context: A simple and effective vector model. In *Proceedings of 5th International Joint Conference on Natural Language Processing*, Chiang Mai, Thailand. Asian Federation of Natural Language Processing.

Turaev, V.G. (1989). Operator invariants of tangles, and r-matrices. *Izv. Akad. Nauk SSSR*, **53**, 1073–1107. English translation: Math. USSR-Izv. 35:411–444, 1990.

Turaev, V.G. (1992). Modular categories and 3-manifold invariants. *Int. J. Mod. Phys.*, **B6**, 1807–1824.

Turaev, V.G. (1994). *Quantum Invariants of Knots and 3-manifolds*. De Gruyter.

Turaev, V.G. (2000). Homotopy field theory in dimension 3 and crossed group-categories. Preprint arXiv:math. GT/0005291.

Turaev, V.G. (2010). *Homotopy Quantum Field Theory*. European Mathematical Society. Appendix 5 by Michael Müger and Appendices 6 and 7 by Alexis Virelizier.

Turaev, V.G. and Virelizier, A. (2010). On two approaches to 3-dimensional TQFTs. arXiv:1006.3501.

Turaev, V.G. and Viro, O. (1992). State sum invariants of 3-manifolds and quantum 6j-symbols. *Topology*, **31**, 865–902.

Turaev, V.G. and Wenzl, H. (1997). Semisimple and modular categories from link invariants. *Math. Ann.*, **309**, 411–461.

Turney, P.D. and Pantel, P. (2010). From frequency to meaning: vector space models of semantics. *J. Artif. Intell. Res.*, **37**, 141–188.

van Benthem, J. and Doets, K. (1983). Higher-order logic. In *Handbook of Philosophical Logic*. Reidel Publishing Company.

van Daele, A. (1994). Multiplier Hopf algebras. *Trans. Amer. Math. Soc.*, **342**(2), 917–932.

Vicary, J. (2011). Completeness of †-categories and the complex numbers. *J. Math. Phys.*, **52**(8), 271–293.

Vijay-Shanker, K. and Weir, D. (1993). Parsing some constrained grammar formalisms. *Comput. Linguist.*, **19**, 591–636.

Virelizier, A. (2002). Hopf group-coalgebras. *J. Pure Appl. Algebra*, **171**(1), 75–122.

Vogel, C. (2009). Law matters, syntax matters and semantics matters. In *Formal Linguistics and Law* (ed. G. Grewendorf and M. Rathert). De Gruyter.

Wakimoto, M. (2001). *Infinite Dimensional Lie Algebras*. American Mathematical Society.

Wang, Z. (2010). *Topological Quantum Computation*. American Mathematical Society.

Wassermann, A. (1998). Operator algebras and conformal field theory III. Fusion of positive energy representations of $su(n)$ using bounded operators. *Invent. Math.*, **133**, 467–538.

Weeds, J., Weir, D., and McCarthy, D. (2004). Characterising measures of lexical distributional similarity. In *Proceedings of Coling 2004*. Available at http://aclweb.org/anthology-new/.

Weir, D. (1988). *Characterizing Mildly Context-Sensitive Grammar Formalisms*. Ph. D. thesis, University of Pennsylvania.

Widdows, D. (2003, July). Orthogonal negation in vector spaces for modelling word-meanings and document retrieval. In *Proceedings of the 41st Annual Meeting of the Association for Computational Linguistics*, Sapporo, Japan, pp. 136–143. Association for Computational Linguistics. Available at http://www.aclweb.org/anthology/P03-1018.

Widdows, D. (2004). *Geometry and Meaning*. CSLI Publications.

Winter, Y. (2001). *Flexibility Principles in Boolean Semantics: the Interpretation of Coordination, Plurality, and Scope in Natural Language*. MIT Press.

Witten, E. (1988). Topological quantum field theory. *Comm. Math. Phys.*, **117**, 353–386.

Witten, E. (1989). Quantum field theory and the Jones polynomial. *Comm. Math. Phys.*, **121**, 351–399.

Wolff, H. (1973). Monads and monoids on symmetric monoidal closed categories. *Archiv der Mathematik*, **XXIV**, 113–120.

Wootters, W. and Zurek, W. (1982). A single quantum cannot be cloned. *Nature*, **299**, 802–803.

Woronowicz, S.L. (1989). Differential calculus on compact matrix pseudogroups (quantum groups). *Comm. Math. Phys.*, **122**, 125–170.

Xu, F. (1998). New braided endomorphisms from conformal inclusions. *Comm. Math. Phys.*, **192**, 349–403.

Yamagata, K. (1996). Frobenius algebras. In *Handbook of Algebra* (ed. M. Hazewinkel). Elsevier.

Yetter, D.N. (1992). Framed tangles and a theorem of Deligne on braided deformations of Tannakian categories. *Contemp. Math.*, **134**, 325–350.

Yetter, D.N. (1998). Braided deformations of monoidal categories and Vassiliev invariants. *Contemp. Math.*, **230**, 117–134.

Yetter, D.N. (2001). *Functorial Knot Theory*. World Scientific.

Zadrozny, W. (1992). On compositional semantics. In *Proceedings of the 14th Conference on Computational Linguistics - Volume 1*, Nantes, France. Association for Computational Linguistics.

INDEX

abstract Lebesgue space, 332
aLG, 287, 288
algebra, 96, 119, 152, *see also* monoid
    associative, 323
    Boolean, 328, 329
    context, 322, 325, 330
    enveloping, 33
    real, 324
    representation, 31
    separable, 34, 153, *see also* Frobenius
    unital, 323
analytic truth, 346
antistrophic matrix, 23
arrow of time
    pessimistic, 25
asynchronous formula, 302
atomism, 347
axiom link, 315, 316
axiomatic formula, 295, 301

bases
    orthogonal, 46
bialgebra, 122
bialgebroid, 136
bilinear form
    orthogonality of an, 45
    regular, associative, 39
    transposition of endomap, 40
biproduct, 201
Boolean algebra, *see* algebra, Boolean
braid, 28
braided group, 100

Casimir operator, 47
category, 92
    additive, 147
    autonomous, 118
    braided monoidal, 94, 118
    closed monoidal, 118
    fusion, 150
    monoidal, 92, 117
    pivotal, 149
    ribbon, 164
    semisimple, 147
    spherical, 150
    symmetric monoidal, 15, 118
centre, 95, 154
    braided, 159
    Drinfeld, 159
    symmetric, 156
coalgebra, 100, 119
collocation, 339
comodule, 120
comonoid, 119
composition graph, 311, 315, 318
compositionality, 335, 336, 339, 340, 347, 349, 356–358
    hypothesis, 355
    principle of, *see* compositionality principle
compositionality principle, 335, 345
context algebra, *see* algebra, context
context principle, 335, 345
cosine distance, 342

cosine measure, *see* cosine distance
cotensor link, 291, 292, 295, 298, 301, 312, 316
cut formula, 295, 301

data sparseness, 356
derivation, 33
    inner, 33
diagrammatic algebra, 99
disambiguation, 337, 340, 348, 351, 356
display postulates, 289, 290
distributional generality, 326, 331
distributional semantic models, *see* distributional semantics
distributional semantics, 334, 335, 344
    compositional, 344, 346, 358
distributivity postulates, 287, 289, 290
distributivity principles, 285
dual object, 148
duality
    left, 28
    right, 28

empiricism, 345
entanglement, 47
enveloping algebra, 33

fLG, 301–303, 307, 309
focused display calculus, *see* fLG

formal semantics, 334, 335, 337, 344, 347
compositional, 336
Frobenius
   functors, 35
   homomorphism, 39, 43
      from an Hopf integral, 44
   isomorphism, 41
   manifolds, 48
   structure, 35
Frobenius algebra
   characterizations of, 36
   disconnected, 37
   examples, 23
   graphical definition, 36
   non-associative, 49
   normal form theorem, 37
   special, 37
Frobenius' problem, 23
Frobenius, August Ferdinand (1849–1917), 23
functor
   Frobenius monoidal, 135
   monoidal, 118

general language, *see* language
generated subspace, 324
graphical calculus, 24
   black-white, 46
   coupon, 25
   Morse decomposition, 25
   move, 25
      Reidemeister, 28
      topological, 28
   oriented, 26
   red-green, 46
   tangle, 25
greatest lower bound, 6
Grishin connective, 292, 296
Grishin interaction, 297
Grishin pomonoid, 13
group like element, 43
group matrix, 23

Heyneman–Sweedler index notation, 24
Higman trace, 47
Hochschild cohomology, 34
holism, 347
homothety
   of bilinear forms, 39

Hopf algebra, 91, 116, 122
   finite, 42
   finite, definition of, 24
   group like element, 43
   left/right integral, 43
   primitive element, 43
   quasi-, 133
   quasitriangular, 134
Hopf monad, 125
Hopf, Heinz (1894–1971), 23

ideal completion, 328
information flow, 45
interaction principles, 286

join, 6
Jones polynomial, 47

Kuperberg
   ladder, 38

Lambek–Grishin calculus, *see* LG
language, 324
Lawvere theory, 188
least upper bound, 6
LG, 284, 285, 320
LP, 302, 309

matrix
   antistrophic, 23
   group, 23
   parastrophic, 23
meet, 6
module, 119, 152
moleculralism, 347
monad
   additive, 199
   strong, 187
monoid, 9, 119, 152
   free, 322
   Ordered, 9
monoidal category, 15
   rigid, 28
Montague grammar, *see* formal semantics
Montague semantics, *see* formal semantics

Nakayama automorphism, 40
nativism, 345

NL, 284, 285
norm
   induced by Nakayama automorphism, 41

order, 4
   partial, 4
   strict, 5
   total, 4

parastrophic matrix, 23
partial entailment, 332
picturalism, 46
pomonoid, 9
   Ajdukiewicz–Bar-Hillel, 13
   Grishin, 13
   residuated, 13
pregroup, 14
preorder, 4
primitive element, 43
proof net, 286, 291, 293, 295, 298
proof structure, 291, 292, 295
   abstract, 291, 295–297
protogroup, 12
pseudo-word, 343

quantum double, 160
quantum group, 100

reciprocal basis, 34
Reidemeister moves, 28
representation, 119
   regular, 23
Riesz space, *see* vector lattice

scalar, 184
semantic generality, 326, 331
simple object, 147
singular value decomposition, *see* SVD
sLG, 287, 288, 298
smash product, 27
spider theorem, 37
split sequence, 33
splitting idempotent, 33
string diagrams, 24
structured vector space, 352
SVD, 349, 354
synchronous formula, 302

syntactic calculus, 283, 284, 286
syntactic head, 338
synthetic truth, 346

tangle, 25
  2-categorical, 31
  arity of, 28
  composition of, 26
  critical point, 28
  Frobenius isomorphism, 32
  oriented, 26
  ribbon, 31
  thickened, 30
Tannaka duality, 116
tensor link, 291, 298, 301, 312, 316
tentacle, 291, 312
topological quantum field theory (TQFT)
  disk, trinion, cylinder, 31
trace, 149
trace form, 39
transposition
  non-involutive, 40
tree, 296, 298
  reduction, 312
  rooted, 313
  tensor, 301, 311
Truth
  synthetic, *see* synthetic truth
truth
  analytic, *see* analytic truth

twist morphism, 29

Umkehrabbildung, 23

vector lattice, 326
vector space
  partially ordered, 326
  structured, *see* Structured vector space
vertical composition, 26

Witt group, 181

yanking, 28, 45

zero object, 147, 196